with appreciation and

thanks to all my friends

in East Asian Legal Studies.

Julian Fraser

August, 30, 1983.

Environmental Law in Japan

Environmental Law in Japan

Julian Gresser,

Koichiro Fujikura, and

Akio Morishima

The MIT Press
Cambridge, Massachusetts
London, England

This book was set in VIP Trump Medieval by Achorn
Graphic Services, Inc. and printed and bound by The Alpine
Press, Inc. in the United States of America

Library of Congress Cataloging in Publication Data

Gresser, Julian.
  Environmental law in Japan.

  (Studies in East Asian law;
  Bibliography: p.
  Includes index.
  1. Environmental law—Japan—History.
I. Fujikura, Kōichirō, 1934-   joint author.
II. Morishima, Akio, 1934-   joint author.
III. Title. IV. Series: Harvard studies in
East Asian law;
Law   344.52'046   80-22477
ISBN 0-262-07076-6

Contents

I

The Traditional Setting and Its Transformation

1

Historical Perspective   3

Contents

2

The Transformation of Values and the Legal Process: Moral Bases for Postwar Environmental Policy

II

The Development of the Judicial Role

3

The Legacy of the Four Major Pollution Trials

vi

Contents

4

The Judicial Development of New Rights and Remedies   133

Contents

Contents

III

Environmental Protection Legislation and Its Administration

5

Formulation and Implementation of Environmental Policy: A Comparative Assessment of the Regulatory Process

Contents

6

The 1973 Law for the Compensation of Pollution-Related Health Injury: Theory and Practice   285

Contents

7

The Uses of Conciliation, Mediation, and Arbitration in the Settlement of Environmental Disputes   325

IV

Japan's Environmental Law and Policy in International Perspective

8

Japan's Contribution to International Environmental Protection   353

Contents

Foreword

In *Environmental Law in Japan* the authors present the first comprehensive assessment in any language of Japan's environmental law and policy. Their work provides a unique perspective on the operation of Japanese legal institutions and facilitates cross-cultural comparisons by establishing an analytic framework grounded in law, economics, and public administration. Thus a large international readership of lawyers, legislators, government officials, policy analysts, economists, planners, physicians, and natural and social scientists will find this book useful.

The book begins with an historical summary of environmental problems in Japan. The authors show how conflicts over mining pollution in the nineteenth century led ultimately to the formulation of an integrated environmental strategy in the late 1960s and early 1970s. The discussion will be instructive for both students of environmental protection in the industrialized world and policy makers in rapidly industrialized countries now facing problems similar to those Japan encountered during the postwar period.

Four judicial decisions shaped Japan's approach to environmental protection. These decisions, all involving damage awards to victims of air and water pollution, made innovative use of statistics to demonstrate causation of pollution-induced diseases and developed new doctrine for apportioning legal responsibility among multiple sources of pollution. The materials provided by the authors—which include the only English translations of the four decisions—should encourage practitioners and judges in the United States and other countries to develop similar doctrine.

One of the major lessons to emerge from this study is that Japan has apparently succeeded in meeting some of the world's most stringent environmental requirements without adversely affecting employment, economic growth, and energy supply. This will challenge those who argue that nations must choose between environmental protection and industrial development.

The authors pay special attention to Japan's system for compensating victims of pollution through a fund gathered by levying charges upon industrial polluters. As this is written the U.S. Congress has before it several proposals to compensate victims of hazardous wastes. The major objection to such proposals in the United States is that, although deterrence is promoted by compensation, this approach appears to license injury as the necessary price for economic progress. The authors endorse a maximum effort to reduce pollution but they argue that this effort should not neglect the obligation of all societies to mitigate the suffering of pollution's victims. In their view, compensation is essential to fulfill this obligation.

Although mediation, arbitration, and conciliation are familiar techniques to family and labor lawyers in the United States, nonjudicial dispute settlement has not been widely employed in the environmental field. Yet here again Japan has developed a novel approach that should be of interest to other nations. The authors' analysis of the Law for the Resolution of Pollution-Related Disputes will interest legislative policy analysts and scholars, as well as environmental lawyers and citizens' groups seeking quicker, more flexible relief.

In the final chapter the discussion shifts to Japan's record in the field of international environmental protection and the authors' conclusion gives their presentation a sense of poignancy. Despite her innovative domestic performance, Japan has remained callous to the environmental consequences of her actions abroad.

The authors explore this contradiction and conclude with an appeal for Japan to play a more generous role in environmental protection throughout the world.

This work is a volume in the Harvard University series, *Studies in East Asian Law*.

Jerome A. Cohen
Director
Harvard East Asian Legal Studies Program
Cambridge, Massachusetts

Frederick R. Anderson
President
Environmental Law Institute
Washington, D.C.

Preface

In modern times Japan has looked to the West for inspiration in designing her legal institutions. The Meiji Constitution of 1889, the Penal Code and Code of Criminal Instruction (1880), the Civil Codes of 1890 and 1898, and the old Commercial Code of Civil Procedure (1890) were all of either French or German origin. The prewar Administrative Court was essentially a German transplant. In the post–World War II period, the influence of the United States has dominated. The 1946 Constitution, the antitrust and securities exchange laws, and many other statutes carry the mark of American jurisprudence. The new field of environmental law is particularly interesting because it is, despite some foreign borrowing, a distinctly indigenous Japanese institution. And perhaps for the first time, Western lawyers, judges, legal scholars, and many others can turn to Japan for instruction.

World War II left Japan economically, socially, and politically destitute. The transcendent concern of the Japanese in the early postwar years was economic recovery, from which the nation's rebirth, they hoped, would follow. Petrochemical, paper and pulp, textile, and other manufacturing plants were thrown up in densely populated areas, without zoning, open space, or other land use regulations. Industry and the government paid scant attention to housing, public sanitation, or to the pollution hazards of industrial development.

Japan's economic recovery rested on certain assumptions. It was assumed that industrial growth was an immutable "good" to which all else might be subordinated. Indeed, those who lived in poverty took pride in the smoke rising from the factories' chimneys, for the smoky factories were an emblem of the nation's rebirth. Even in the 1960s, many Japanese continued to view pollution as an unavoidable cohort of economic progress.

The second assumption was that nature eventually would recover from every harm. Before World War II the environment had been seriously degraded, yet nature had always recovered. Confidence in nature's self-healing powers licensed her violation.

The third conviction was that technological invention and scientific discovery would solve all problems, a belief with historical antecedents from before the Meiji era. By the early and mid-1950s the nation's faith in science and technology seemed well justified since only the benefits of industrial developments were evident, and the costs remained disguised.

These proved false assumptions. By the mid-1950s different parts of the country began reporting strange diseases virtually unencountered in the past. Scientists later attributed these illnesses to pollutants released from nearby factories. The maladies were mercury poisoning (Minamata disease), cadmium poisoning (itai-itai, "it hurts—it hurts," disease), and various pulmonary disorders induced by air pollution. Yet in the 1950s Japan's leaders, indeed most people, were indifferent to the victims' suffering.

During the 1960s, however, a great transition began. The victims formed local, then national, protest organizations. Isolated for over a decade in remote fishing and farming villages, some victims now began to meet, share experiences, and plot new strategies. And as the victims' movement grew more radical, the mass media dramatized each new development. The government, and even industry, began to recognize the need for a comprehensive pollution control policy.

In 1967 the Japanese Diet passed the Basic Law for Environmental Pollution Control that erected a framework for later environmental legislation and an extraordinary Diet session in 1970 passed fourteen more specific pollution control laws. In 1971 the Environment Agency was established.

Despite such legislative attention, industry's interests continued to dominate. The new laws left the issue of polluter responsibility unclear, and the laws' deference to scientific uncertainty undermined establishing stringent standards and effectively implementing victim relief. Following the pre–World War II pattern, the government's principal objective was to mollify the public, not to look at underlying causes. Economic growth remained the nation's preeminent concern.

The situation around 1970, however, differed from the earlier pattern in a critical respect. In 1967 the victims of Minamata disease in Niigata City had taken their grievances to court, demanding reparations. Soon victims in Toyama, Kumamoto, and Yokkaichi followed suit. At the time of the courts' decisions in the early 1970s, the victims' movement already had acquired a powerful momentum. There was virtually a national consensus that the companies had been criminally reckless and now must accept full legal responsibility for the victims' injuries.

The courts delivered the society's verdict on issues debated for over a decade. Human life could no longer be compromised for economic growth. Firms would be socially and legally obligated to monitor all risks, adopt the best control technology, and always proceed with extreme caution. When epidemiological, clinical, and experimental evidence suggested a strong correlation between pollution and disease, the courts would presume legal causation despite scientific uncertainty. And polluters henceforth would shoulder legal, as well as moral, responsibility for their victims' rehabilitation.

This indictment renewed the hopes of many other victims' and citizens' groups, who increasingly began to turn to the courts. The pollution trials also fostered more decisive regulatory action. They encouraged those within government who had called for stricter environmental standards; provided an incentive to establish a national compensation system; and stimulated new, needed legislation for factory siting, land use, and nature protection. The judgments compelled a fundamental institutional readjustment that continues today.

We trace these developments with two principal objectives. First, Japan's response to environmental problems, both domestic and international, offers a valuable insight into the role of law and legal institutions in Japan. Second, Japan's successes and failures can help Western industrialized nations design more effective pollution controls. The history of the Japanese approach will also aid developing countries facing seemingly irreconcilable choices between industrial development and environmental protection. How Japan succeeded in implementing environmental policies during the late 1960s and early 1970s without retarding economic growth should be of special interest.

Part I presents the history of Japan's legal response to environmental problems and describes how many Japanese came to view pollution in moral terms. Indeed by the late 1960s and early 1970s many Japanese considered pollution that endangered health as a fundamentally immoral act. Against mindless economic growth the average citizen now advanced new concerns: the sanctity of human life, individual dignity, and integrity of local communities. This transformation in values influenced all subsequent judicial and administrative policies. Part II examines how the courts helped articulate and legitimize these values by recognizing new rights and remedies.

Part III studies the administrative process. In ch. 5 we describe the positive and negative incentives used to induce industry's compliance with environmental policies, the innovative role of local governments, and the Japanese bureaucracy's reliance on "administrative guidance." This general discussion of the administrative process provides the background for three case studies. The most interesting example is the Japanese automobile industry's success during the mid-1970s in meeting the world's strictest auto emission standards. The administration's flexible approach to regulation, the subtle array of incentives, industrial structure, and public attitudes toward pollution were responsible for Japan's success.

Chapters 6 and 7 analyze Japan's compensation system for toxic-substance pollution victims and the present administrative scheme for settling environmental disputes. In light of the increasing evidence of toxic-substance injury throughout the world today, Japan's compensation system demands special attention. Its establishment suggests that serious injury to health may be a necessary incident of industrialization, that the misfortune of the victim must be addressed along with the control of the pollutant.

Even at the height of success, Japan's substantial contribution is marred by a poor international record. Part IV examines how Japan's domestic achievements have been attained at a cost to other countries and to the international environment.

By the mid-1970s, the energy crisis and Japan's efforts to maintain a competitive position in international trade seriously began to influence the implementation of domestic environmental policies. There is now substantial evidence that the country is in the midst of retrenchment and reexamination. For example, as of 1979 Japan has still failed to enact a comprehensive environmental impact assessment law; the Environment Agency, despite public protest, has cut the national ambient standards for NO_x; the compensation system is under assault from all sides, and is being modified; the nation continues to plunge headlong into the development of nuclear fast breeder technology.

Japan is at a crossroads. She can continue to weaken domestic environmental policies. Then there will no longer be contradictions with the international record. Or, after reflection, she can refine her many achievements and begin to defend the environment of Asia and the Pacific.

Acknowledgments

The authors are deeply indebted to Professor Jerome A. Cohen, Director of the Harvard Law School's East Asian Law Program, for his faith and friendship, and to Harvard Law School for sponsoring a two-semester course in 1976–1977 on Japanese environmental law. We also acknowledge with gratitude the support of the Environmental Law Institute under the directorship of Frederick R. Anderson; the Institute's special project on toxic substances, financed by a grant from the Ford Foundation, permitted us to study Japan's compensation system during many field trips in Japan, at a seminar in Hawaii in 1975, and through two conferences in Cambridge and Washington. The collaboration between the Harvard East Asian Law Program and the Environmental Law Institute made *Environmental Law in Japan* possible.

Other organizations also aided, encouraged, and advanced our work: the Japan and Ford Foundations and the Earl Warren Institute of the University of California supported the American author's exploratory visit to Japan in 1972 and the participation of two of the authors at the 1973 Bonn conference on comparative public interest law; the University of Hawaii School of Law and the University of Hawaii's Sea Grant program generously released the American author from some teaching responsibility and provided time to compose and complete this book.

Many individuals offered helpful suggestions on style and substance that we note with pleasure: in Japan, Dr. Michio Hashimoto (Tsukuba University), Katsuhiko Iguchi (Japan Environment Agency), Dr. Jun Ui (Tokyo University), Professor Yoshihiro Nomura (Tokyo Metropolitan University), and Dr. Katsuko Tsurumi (Sophia University). In the United States, Professors Richard B. Stewart (Harvard), Robert E. Keeton (Harvard), Louis B. Sohn (Harvard), Jerrold Guben (Hawaii), Frank Upham (Ohio), Richard Buxbaum (Berkeley), John Craven (Hawaii), John Van Dyke (Hawaii), David Callies (Hawaii), Richard Miller (Hawaii), Sanford Gaines, and William Irwin.

We also remember with gratitude the many others who labored with us—our students at Harvard, Hawaii, Kyoto, and Nagoya; Mrs. Bertha Ezell of the Harvard East Asian Law Program; and especially the wet nurses, Mrs. Helen Shikina and Irene Takahashi, who, conquering all desperation, trudged on through this child's uncertain birth and complicated adolescence.

I The Traditional Setting and Its Transformation

1   Historical Perspective

The history of Japan's response to environmental problems can be divided roughly into four stages. The first stage encompasses the entire pre–World War II period; the second stage ends with the enactment in 1967 of the Basic Law for Environmental Pollution Control[1] that set out a comprehensive, systematic approach; the third stage includes the early implementation of the Basic Law (1967–1973) and a period of intense administrative activity following the courts' decisions in the four major pollution trials (1973–1975). The fourth stage is the present, a period of rationalization and retrenchment.

Section 1 of this chapter deals with Japan's initial experience with environmental problems and begins with an account of pollution at the Ashio mine in the late nineteenth century. The chronicle of the Ashio tragedy is particularly appropriate because in many ways Ashio is a prototype of events to come. Meiji society's perception of environmental problems—the reaction of injured farmers, the government, and the Furukawa enterprise—set a pattern of response that continues to extend its influence today. But Ashio was not the only major environmental dispute before World War II. As the cases in sec. 1.2 describe, Japan's industrialization in the late nineteenth and twentieth centuries was accompanied throughout by bitter conflicts over environmental destruction.

Section 2 discusses the period of the late 1950s and early 1960s when the Japanese government first perceived a need to take remedial action. What we wish to document, and to a lesser extent to analyze, is the process of debate—within the ministries and Diet and among the ministries, industry, and other private interests—that shaped the ultimate direction and character of Japan's postwar legislative and administrative response.

The debate centered on three sets of related issues. The first, and perhaps most critical, touched the society's basic values,[2] for the horrible suffering of Minamata and other victims of pollution-induced diseases necessarily moved the nation to question the shibboleth of economic growth. Many persons by the mid-1960s felt that human health must be given priority at all costs. Others, however, insisted that policies mandating the protection of human health should still be "harmonized" with long-range plans for industrial development. This debate in turn was related to a corollary inquiry into the appropriate allocation of social and legal responsibilities for pollution control.

The second cluster of issues involved the appropriate form of governmental organization, or reorganization. Here the central questions were the allocation of ministerial jurisdiction, the related problem of creating a new, independent environmental protection agency, and finally, the distribution of powers and functions between the central authorities and local governments.

A third set of issues concerned the design and implementation of countermeasures. Should initial regulations focus on particularly injurious polluting industries, or should a comprehensive environment management scheme, emphasizing planning and nature protection, be adopted? How should environmental policies be coordinated with other regulatory measures? What kinds of standards should be set and for what substances or activities? How could pollution controls be enforced effectively and equitably against gigantic industrial combines and economically fragile small and medium enterprises? This introductory chapter explores the forces that helped shape Japan's idiosyncratic solutions to these many problems.

Section 1   The Prewar Period

1.1   The Ashio Case: A Prototype[3]

Copper was first mined at Ashio along the banks of the Watarase River in the beginning of the seventeenth century. During the Tokugawa period, Ashio copper was an important export in Japan's foreign trade with Holland and China, while at home it graced the Tokugawa's Edo Castle and the mausoleum honoring Tokugawa Ieyasu at Nikkō. Although Ashio was originally one of Japan's most productive mines, toward the end of the Tokugawa period and continuing into the Meiji period, the mine was seriously depleted due to excessive exploitation, and it steadily lost money.

In 1877 the mine fell into the control of Furukawa Ichibe, one of the industrial geniuses of the age, and the mine's fortunes immediately began to improve. Furukawa set about unifying the existing piecemeal production process, then under the supervision of various subcontractors.[4] He tunneled new mine shafts; he attracted new investors like the famous entrepreneur, Shibusawa Eiichi; and he substantially expanded the mine's labor force. In 1882, the workers' discovery of an important new vein known as the "Falcon's Nest" rewarded Furukawa's initial investment and in 1884 the miners hit the mine's mother lode. By 1885 Ashio was producing 6,886,000 kin[4a] of copper—six times the production of 1883 and 90 times that of the 1877 output.

As the mine's prospects brightened, Furukawa began to pour large sums into its radical modernization. Central to his scheme was hydroelectric power. In 1888, in consultation with the Siemens Company, Furukawa introduced Japan's first hydroelectric generation station at Matoo in the Ashio complex, and other developments quickly followed. Compressed air rock drills, centrifugal fans for better ventilation, and electric lights greatly facilitated work in the shafts. By 1891 an electric railroad and a system of steel-cabled aerial tramways solved the remote mine's pressing transportation problems.

Modernization sealed Furukawa's success. Ashio was now the most highly industrialized copper mine in Japan, producing twice as much copper at its nearest competitor, the Sumitomo's Besshi mine, with only 66 percent of Besshi's labor force. Statistics for 1891 show that Furukawa produced 47.8 percent of the nation's copper and at that time copper was Japan's third most important export commodity.[5] The harmony between Furukawa's private fortunes and the nation's economic development was now demonstrably complete.

1.1.1   The Beginning of Pollution at Ashio

As early as 1880 farmers and fishermen living along the banks of the Watarase and Tone rivers began to see signs that all was not well. The waters turned a "bluish white," and many people noticed dead fish floating down the rivers. After those who ate these fish became ill, the local authorities banned the sale of fish. By the late 1880s, the livelihood of the Watarase and Tone fishing communities was in ruins.

But the destruction of marine life was only the start of problems. Furukawa's mine expansion program insatiably demanded timber. Lumber was needed to shore up mine shafts, lay ties for the train system, service the mine's steam engines, and most important, provide the material for the tons of charcoal needed in the smelting and refining processes. Because of high transportation costs to this

4

remote area, much of the timber needed by the mine was stripped indiscriminately from the surrounding hills. As a result, much of the watershed at the head of the Watarase was destroyed, and flooding became a serious problem. Records show that as early as 1888 and again in 1890 major floods inundated the Watarase valley and rice fields of Gumma and Tochigi prefectures.

To the superstitious peasants, it seemed the vengeance of an angry god. In former times floods had also occurred, but these floods had contributed to the harvest by covering the agricultural fields with a rich new layer of silt. This was not so now because when the flood waters receded, the chemicals concentrated in the water—sulfuric acid, ammonia, aluminum oxide, magnesia, iron, copper, chlorine, arsenic, nitric and phosphoric acids—shriveled all they touched. No longer could seeds or seedlings grow, and men and women working the affected fields soon developed sores on their feet and hands. Rumors spread that Tochigi Prefecture had been divided into a heaven and hell. Blessed was the Nikkō side of the mountains among the beautiful temples of Tokugawa Ieyasu's mausoleum; accursed was the western land of the Watarase, that the peasants now called the River of Death.

1.1.2   Early Protests and Mollifications

The mounting evidence on the origins of Ashio's pollution spurred the Watarase's fishermen and farmers to action. In the summer of 1891 they drafted a formal petition to the Minister of Agriculture and Commerce calling for the removal of pollution and the temporary closure of the mine. In the same year Tanaka Shōzō, the Tochigi Prefecture representative called on the government, in an impassioned speech before the House of Representatives, to protect the people's constitutional rights in property[6] and to revoke Furukawa's concession.[7] In closing Tanaka put three questions to the ruling Matsukata Cabinet: Why had the government failed to take action? By what means did the government propose to alleviate the victims' suffering? How could a recurrence of the tragedy be prevented?

About six months later the government issued its first public response to the Ashio problem. Although aware of the difficulties in Gumma and Tochigi prefectures, the government intoned, the situation did not warrant revoking the Furukawa concession because it was not "injurious to the public interest" within the meaning of the Mining Law. Replying to Tanaka's questions, the government argued that the central administrators did not have authority to impose the requested remedial measures in the prefectures. The government assured Tanaka that Furukawa had acquired equipment to extract ore dust and was already building sedimentation pools to prevent further pollution. If the fishermen would just be more patient, all would be well.

While experimenting with some technological solutions, Furukawa also undertook to mollify the victims by other means. For example he urged Orita Heinai, Governor of Tochigi Prefecture, to establish an Arbitration Association to indemnify the affected villages.[8] The contracts ultimately concluded contained three basic clauses:

1. By June 1893 Furukawa would install equipment to extract ore dust.
2. Furukawa would pay each village annually a specified sum to indemnify its losses.
3. While the ore-dust extraction equipment was being introduced, the villages would not petition or otherwise complain about their circumstances.

The total cost of the four-year contract for the Furukawa firm was ¥226,288. Considering that the annual profit of the mine in 1885 was ¥240,000 and that by the 1890s profits were six to eight times this amount, the purchase of the villagers' silence was but a paltry sum. In the years after the Sino-Japanese War (1894–1895), Furukawa agents often would visit the fishermen and farmers seeking to induce them to sign "permanent agreements," designed to extend the earlier arrangements indefinitely.

1.1.3   Ashio Becomes a National Issue

Conditions in Gumma and Tochigi prefectures deteriorated even as Furukawa's henchmen negotiated, and new, yet stranger things came to pass. Earthworms, spiders, and even ants vanished; birds disappeared. Infant mortality rose and mothers' milk failed. Even the bamboo and willow groves by the Watarase withered.

But again the region was alive with protest. One cause was the return of villagers who had served in the Sino-Japanese War. For these belligerents the Furukawa Company contracts were a hoax and they denounced them to their more trusting kinsmen. The villagers organized an opposition force centering two offices at Unryōji, a temple in the Watarase Village of Tochigi Prefecture. While one office called for the closing of the mine, the other petitioned the central authorities for relief.

In the spring of 1896 a great flood rolled out of the Watarase Valley inundating Gumma, Tochigi, and parts of Saitama and Ibaragi prefectures. More than 88 villages were affected, 13,000 households were submersed, and in Ashio itself more than 300 people perished. From a political perspective, the floods proved the failure of Furukawa's flood control measures and further aroused the farmers. Backed by villages in Ibaragi and Saitama, the Ashio peasants drew up a new three-point petition. Along with the closing of the mine the farmers now demanded the reduction of taxes in the devastated region and the repair of dikes.

As national attention focused on the region, prominent Meiji scholars, writers, and statesmen at last began to show an interest. In Tokyo a vigorous support movement began and the major newspapers also took up the debate. Sympathizers of all types flocked to rallies where the public began to denounce the government's and the Furukawa Company's failure to provide a remedy. Churches, women's organizations, and student groups organized inspection tours and lantern slide shows to portray to the masses the suffering of the Ashio countryside.

1.1.4   The Government Decides to Act

Despite the protests, the government continued to equivocate. In December 1876 Takeaki Enomoto, Minister of Agriculture and Commerce issued a directive to the Furukawa Company to begin "preventative construction." The order, however, failed to specify what construction would be required and its implementation date was left uncertain. During the spring of 1897 the government reiterated its satisfaction with Furukawa's progress in pollution control, pointing out that the villagers already had obtained relief under the indemnity contracts. It was by now a familiar line.

Yet the times had changed. On March 23, 1897, 800 farmers left Unryōji to petition the Imperial Household Ministry, calling for tax relief and the closing of the Ashio mine. As the marchers proceeded toward Tokyo, they were intercepted by police and a violent skirmish broke out. Although many were injured and arrested, eighty farmers managed to reach the capital where they demanded to meet with Enomoto. Under great pressure from the Tokyo press, particularly the Yomiuri and Yoruzu newspapers, Enomoto agreed to listen to 50 representatives of the group. In a meeting charged with emotion, Enomoto confessed that he had been gravely mistaken about Ashio and would now make every effort to make amends. The next day Enomoto, accepting full responsibility, resigned from the Cabinet.

Faced with a peasant uprising in the countryside and growing concern even within its inner circle, the government now had no recourse but action. On May 27, 1897, the new Minister of Agriculture and Commerce issued a revised schedule for "preventative construction." This time the directive was unambiguous. Under clear deadlines, there would be precipitation and filtration ponds, equipment to control sulfur emissions, and beds for slag. The government warned that if the charges were ignored or incompletely carried out Furukawa would have to close down.

With characteristic industry Furukawa immediately inaugurated a crash program to implement the necessary charges. Soon the filtration and precipitation ponds were built, the slag and tailing heaps were constructed, and a system of "scrubbing" acidic smoke was devised; lime was used to neutralize the acidic pit-water before it was discharged into the Watarase River. The company even introduced dust chambers, acid condensers, and a cement process for extracting fine particles of copper from sand, slime, and water. Many of the essential measures that might have been introduced 10 years earlier were at last put into effect.

1.1.5   The Kawamata Affair

Despite Furukawa's expanded efforts, disillusionment and bitterness spread. Early in 1900 many farmers gathered at Unryōji to plan a final march. At dawn on February 13, two thousand people started for Tokyo. At a nearby town, the police attempted to arrest the mob's ringleaders, but the marchers attacked the police headquarters and released the prisoners. By noon the mob had reached the village of Kawamata on the banks of the Tone River. Kawamata had been a frequent site of conflict between marchers and police and here they clashed again. By the end of the battle, six policemen and fifty marchers had been seriously injured, and one hundred farmers were arrested, fifty-one of them on serious charges of sedition and incitement to riot.

The defendants were arraigned and tried before the Maebashi District Court in the autumn of 1900. The trial centered principally on the legal interpretation of sedition and incitement to riot under the new Criminal Code. The court showed little interest in understanding the motivations behind the protest movement and on December 22, 1900, it found twenty-nine defendants guilty of the lesser charge of resisting public officials; the remaining twenty-two defendants were acquitted.

The Maebashi decision was thereafter appealed to the Tokyo High Court. Unlike the District Court trial, the appellate proceeding traversed almost every aspect of the Ashio problem. Between August 20, 1901, and March 15, 1902, the court held thirty-one public sessions to which the press was invited. In October, on the motion of Chief Judge Kojirō Isoya, the entire court, which included five judges, two prosecutors, and twenty-two defense lawyers, moved to the affected region for an on-the-spot investigation. Press coverage of the expedition was extensive. The court also ordered the first thorough scientific study of the region, engaging some of the country's leading agricultural scholars, doctors, and public health specialists. Finally on March 15, 1902, the court declared all but three of the defendants innocent.

For many Japanese, the Tokyo Court of Appeals' decision demonstrated the government's sensitivity to public feelings and its willingness to abide by traditional concepts of justice. The Ashio problem had at last been fully investigated, and at least indirectly the Furukawa Company's actions had been condemned. Yet for the government there still remained the issue of civil disobedience. Could sedition and incitement to riot ever be condoned, even where the offenders suffered the most extreme, intolerable hardships?

The issue was resolved through a somewhat elaborate procedural maneuver. The government appealed the case to the Court of Cassation,[9] which dismissed both the Maebashi and Tokyo Appeals courts' decisions on the basis that both courts had misinterpreted the terms "sedition" and "riot" under the Criminal Code. At the same time, the Court of Cassation remanded the case to the Miyagi Court of Appeals for retrial. On December 25, 1902, all charges were formally dropped against the defendants on the curious technicality that the prosecutor had failed to write his own brief. At least in a formal sense, the Miyagi Court of Appeals had protected the security of the law while tempering it, as the Tokyo Court of Appeals had sought, with justice.

1.1.6   Ashio in Retrospect

With the termination of the Kawamata trials, the memory of Ashio drifted away from public cognizance. The harvests returned to Tochigi and Gumma prefectures and the nation concerned itself with greater affairs. Pollution in Ashio, however, continued, and although it was contained within acceptable limits, the problem persisted even into the postwar era. A final mediated settlement with the Furukawa Company was reached only in 1974.

In retrospect the Ashio case offers insight into later controversies between the Japanese government and pollution victims. To the Meiji leaders the Ashio mine's astounding productivity served to ratify the official policy of harnessing nationalism to private capitalism. And after the humiliation of the Triple Intervention and the resulting military buildup, the government was understandably reluctant to make an issue of Ashio's pollution. Furukawa copper took on an additional strategic significance during Japan's preparation for the Russo-Japanese war.

Given the importance of the industry, the Meiji government's chief concern was to confine the Ashio agitation to the narrowest local perimeter. Only when the conflict promised national repercussions did the authorities vouchsafe an official inquiry. Yet even then, the government's primary objective was to propitiate the parties or excuse its own inaction.[10]

During both the Meiji and postwar periods, government policy toward pollution reflected certain societal assumptions about the appropriate allocation of property rights, especially rights in common resources. Since the development of mining was deemed a national priority, Furukawa was permitted to utilize the air, water, and land resources freely and with full government support.[11] Although the Meiji Constitution and Mining Law recognized in principle that the exercise of property rights should not infringe on the interests of others, the interests of the pollution victims were routinely subordinated to those of the property rights holder. Moreover, because of constant fiscal difficulties, the authorities were reluctant to expend public funds for environmental protection or to impose effective controls upon private enterprise.[12]

As in most of the postwar period, the government relied heavily on the broad sweep of administrative action to control activities at the Furukawa mine. When the Ashio problem was still a local dispute, administrative actions tended to be abstract exhortations with unclear implementation dates; only when the disturbance became acute did governmental authority move unequivocally, compelling immediate action. Throughout both periods the government's control of enterprise was conducted more in the general spirit of law than according to the exact prescriptions of any specific legislative mandate.[13]

The most distinctive aspect of the victim farmers' response to the Ashio mine is their resort to protests and the political process rather than to the courts. Indeed, during the period between Ashio and World War II, in only one major instance, the Osaka Alkali case,[14] did victim groups seek judicial redress. Even today demonstrations and petitions are the principal means of expressing grievances over domestic and even transfrontier environmental issues.

It is especially important for the Western reader not to misconstrue the Ashio farmers' militancy as early evidence in Japan of the assertion of individual rights, at least in the Western sense of this concept.[15] It is improbable that many viewed their protests to be based upon a legally recognized, individual entitlement to official redress. The demonstrations of the farmers over their injuries more probably expressed a group appeal for official relief similar to the peasant protests of an earlier era.[16] It is interesting that even with recent judicial protection of environmental interests, and with the judicial recognition of new rights and remedies, the allegiance of most victim groups to collective as opposed to individual objectives (that is, the assertion of individual claims) continues.[17]

One of Japan's most insightful writers on environmental problems, Jun Ui, commenting on the Ashio case has identified five "common principles" that, he contends, characterize most pollution conflicts from Ashio to the present.[18] His first principle is that pollution goes through four stages: outbreak, identification, refutation (or denial), and "solution." Ui suggests that after an initial attempt to identify the causes of the outbreak(s) of pollution-induced injury, the responsible parties and their cohorts (so-called neutral experts, such as university professors and administrators) attempt to obfuscate the issue. Usually a "solution(s)" is reached that is favorable to the responsible parties.

The second principle is that third parties do not exist, for there are only two possible sides to these controversies—the side of those who suffer and the side of those who cause their suffering. Ui continues:

It is easy to identify from which side utterances about pollution problems are made. The victims always want to clarify the cause and effect of pollution and they act in an effort to fulfill this desire. The victimizers always act so as to cloud and confuse the issues at stake. This fact is clearly evident in those passages dealing with the definition of pollution in all theses so far published on the subject.

The third principle is that whenever the victims demand the removal of the causes of pollution, the polluter always tries to achieve settlement on a purely monetary basis. And because of superior bargaining power, the polluter usually compels the victim to accept a pittance.[19]

Ui's fourth principle is that "victory or defeat in any dispute is almost always decided at the very outset by the spirit and the content of the movement launched by the victims." If the victims docilely exhaust their endless administrative appeals, they will almost surely fail; but if, emphasizing the injustices done, they directly attack the polluter and the various echelons of government, their chances of success are substantially increased.

9

The final principle is that "all organizations based on concentration of power at the center" are generally useless in solving pollution problems and will eventually prove a stumbling block. All successful antipollution movements, he argues, are decentralized, based on the full, voluntary, and spontaneous cooperation of their members.

1.2   Pollution Disputes Between Ashio and World War II[20]

1.2.1   The Besshi Copper Mine Case

A copper mine at Besshi in Shikoku was first discovered in 1690 and thereafter managed by the Sumitomo family in Osaka. The operation was greatly expanded when a Western style furnace was constructed in Niihama Village on the sea coast in 1884, and a railway was built between the smelting plant and the mine in 1892. Soon after these events, local farmers began to notice crop damages in areas exposed to smoke discharged from the plant. In 1894 negotiations between groups of farmers and the company began on the issues of compensation for damages and pollution control. Throughout the negotiations, Sumitomo denied any connection between the smoke from its plant and the farmers' damaged crops, and as a result the negotiations were delayed. Although the farmers thereafter petitioned the prefectural government, the Osaka Supervisory Office of the Bureau of Mines, and the Minister of Agriculture and Commerce, none of these efforts produced positive results.

In 1895 crops were again extensively damaged in the vicinity of the Sumitomo plant, but this time the aroused farmers marched to Niihama and demanded compensation. Along the way there was a skirmish with local police squadrons and a number of farmers were arrested.

In 1898 the Osaka Supervisory Office of the Bureau of Mines issued a ten-point order based on art. 59 of the Mining Law of 1890 directing the company to relocate to Shisaka Island off the Inland Sea coast, terminate open smelting of copper ore, and improve the condition of its smokestacks. The order to relocate was not altogether unwelcome. For some time Sumitomo had considered relocation, and in 1895 the company had actually petitioned the bureau to approve a plan for resettlement. In 1904 the Sumitomo plant moved to Shisaka Island.

After operations began, farmers in this new area reported damages extending to an even larger area. In fact, one severely affected community, Ōchi County, organized an investigatory committee in 1906 consisting of twelve mayors and village chiefs to urge the company and responsible governmental agencies to take all necessary measures to reduce pollution from the mine.

In 1910 the company finally agreed to compensate the farmers only after the Minister of Agriculture and Commerce intervened and arbitrated the dispute.[21] The agreement stipulated that the company pay an indemnity of ¥ 77,000 a year. The amount represented seven-tenths of what the farmers had demanded and ten times the amount originally offered by the company. In addition, the company promised to curtail its smelting operations in order to reduce the amount of smoke discharged. A yearly ceiling was set at 220,000 tons, and there was a stipulation that furnaces had to be shut down for ten days during the critical period of the initial flowering of the rice crops. The farmers pooled the damage payments received and used the proceeds to improve farmlands and equipment.

In 1915 Sumitomo installed six new stacks with a new gas flue. But because the stacks were only half the height of the ones previously used, high levels of sulfur dioxide were soon observed throughout Shisaka Island. Air pollution thus continued along the Shikoku coast until Sumitomo's introduction of a neutralization plant in 1939. Although the farmers requested the company to continue their indemnity payments indefinitely, the company was able to persuade them to accept a final agreement of termination wherein Sumitomo paid ¥1 million to the farmers and ¥65,000 to the prefecture. Sumitomo's total liability amounted to about ¥1,300,000.[22]

1.2.2   The Asano Cement Company Dust Case

In 1883 the Japanese government transferred a cement plant that it owned and operated in Tokyo to the Asano Cement Company, a private concern. As operations began, and especially after 1903 when a new revolving kiln was introduced, dust from the plant began to cover the surrounding neighborhood. Infuriated local residents soon demanded removal of the plant. For eight years, however, no official action was taken to remedy the problem.

In 1911 a number of Diet members brought the issue of Asano Cement company's pollution to the attention of the House of Representatives. They cited statistics correlating disease with various levels of dust in the district and noted medical analyses of the adverse effects of dust on human health, housing materials, and food. Various Diet members began to demand that the pollution problem be solved.

At length the Asano Company accepted a proposal suggested by some Diet members that the plant be removed by the end of 1916. Asano thereafter began construction of the new facility in Kawasaki. Construction, however, was delayed by the storage of construction materials resulting from the outbreak of World War I. Seizing this opportunity, Asano further postponed the date for relocation, promising to remove the plant by the end of 1917.

After the war, the Asano Cement case was solved by an unexpected event, the introduction of an electric dust collector that turned the dust into potash fertilizer.[23] Convinced that this new technology might finally solve the problem, the residents unexpectedly dropped their demand that the plant be relocated. An association was later formed in Japan to manage the patent, and other companies subsequently adopted this technology.

1.2.3   The Hitachi Mining Company Smoke Pollution Case

Kuhara Fusanosuke, a rich industrialist, bought an old mine in Ibaraki Prefecture in 1905 and started Kuhara (later Hitachi) Mining and Smelting Company. After a few years, the company began to prosper, but at the same time heavy smoke from the smelting works began to damage forests and farmlands near the plants. By 1912 the damage extended to over twenty surrounding towns and villages.

In 1907 farmers alarmed by the injuries to their crops organized a negotiating team, which was later led by Umenosuke Seki, a young intellectual and village leader. In negotiating with Hitachi, Seki frequently pointed to the precedent of the Ashio case. Hitachi reacted immediately by offering to make compensation. As the plant's operations expanded, damages to the surrounding countryside increased. Although by 1913 smoke had invaded thirty villages and seriously damaged agricultural crops and timber, the government did not take action.

By 1913, however, the press began to become interested in the issue of damages from Hitachi Mining Works, and various articles were written on the plight of the area's farming communities. As public concern increased, the government finally notified the mine that it had to provide facilities to control sulfur oxide emissions. On the advice of the government's Mine Pollution Investigation Council, Hitachi agreed to install a low, thick smokestack. Unfortunately the stack only aggravated the problem. Villagers and others concerned about Hitachi pollution called it the "foolish stack" (ahō entotsu), and even today it is disparagingly referred to by this name.

The case was finally resolved in 1914 when meteorological research by mine officials resulted in the discovery of high altitude air currents flowing toward the sea. The company thereafter constructed a stack 156 meters high through which the fumes were released to the prevailing winds.

Modern commentators consider several aspects of this case particularly significant. First, the scientific methodology and technology used were as advanced as any in the early post–World War II period. Second, cooperation between industry and local residents was, for a rare moment, achieved. Indeed throughout the dispute, Seki was able to hold the mutual trust of both company officials and his comrades. Finally, this cooperation produced several new approaches to compensation. Hitachi agreed not only to compensate farmers for damages to crops and trees but also to supplement its payments with an additional sum of 10% of that amount to account for the mental suffering of those affected. Hitachi also distributed without charge 5 million cedar plants and opened the company's health clinic to local residents.[24]

1.2.4   The Suzuki Case

In 1908 the Suzuki Company built a plant in Zushi, in Kanagawa Prefecture to produce ajinomoto (monosodium glutamate), a food additive. After its establishment, the plant began discharging chlorine gas into the air and dextrine wastes into a nearby stream, causing damage to agricultural crops and fish. During 1916–1921 environmental damage became more severe as company's manufacturing operations expanded.

Local farmers thereafter began to protest against the plant's pollution. Because some response was necessary, Suzuki urged the prefectural head to mediate the dispute. A settlement was at last reached wherein the company agreed to compensate the farmers for crops damaged by the chlorine gas and to adopt appropriate abatement measures. Thereafter, pollution gradually began to decline.

In 1928 a flood dispersed the waste materials at the plant and thereby severely damaged the surrounding farmland. At the urging of a farm improvement union and the local police chief, the company negotiated a new compensation agreement with the injured farmers. Such disputes continued sporadically until 1935.

1.2.5   The Osaka Alkali Company Case[25]

By 1900 Osaka had already become an important center of industrial activity. The Osaka Alkali Company operated a copper refining plant that discharged sulfurous fumes, which damaged surrounding farmlands. Thirty-six landlords and their tenants, enraged at their losses, sued the Osaka Alkali Company in the Osaka District Court for negligence.

The District Court held Osaka Alkali liable. The Osaka High Court upheld the District Court in a decision that established an important precedent on the issue of negligence. The Osaka High Court noted that it was virtually impossible for the directors of Osaka Alkali not to have had knowledge or notice of the escape of toxic sulfur oxides from its chemical plant and to have been ignorant of the possibility that such discharges might damage agricultural crops, animals, and people. Even if company officials had not had knowledge of these circumstances, the court held that they should have assessed the impact of their operations on the community. The Osaka Alkali Company was negligent for not having known these facts.

The company contended that since it used the best available pollution control technology, it should not be liable for damages. In rejecting this contention, the court stated that, irrespective of its smoke control measures, the company should be held liable because sulfurous smoke from the company's plant had damaged the plaintiffs' crops.

On appeal to the Great Court of Judicature (Supreme Court),[26] Osaka Alkali again contended that copper refining was a lawful exercise of its rights, and it had adopted the best available technology. The Great Court of Judicature sustained the company's contention and remanded the case to the High Court for further proceedings. In reversing the High Court's ruling, the Great Court of Judicature held that the use of the best available technology would shield the defendant from liability. On December 27, 1919, the Osaka High Court ruled that Osaka Alkali had never used the best available pollution control technology and therefore should be liable for the negligent operations of its factory.

1.2.6   The Arata River Pollution Case

Farmers and fishermen living along the downstream banks of the Arata River during the 1920s suffered damage to crops and fish from discharges from the small and medium textile factories operating in and around Gifu city. The area was frequently assailed with heavy rain, and during these times the rice paddies in the downstream area easily flooded. Stagnant, polluted waters covered farmlands for weeks, spoiling crops and ruining irrigation works.

In 1924 farmers and fishermen of the area filed an administrative lawsuit in the name of the Arata River Irrigation Association against the prefectural government.[27] The association, however, dropped the suit when the government adopted an irrigation canal construction plan.

In 1932 the Irrigation Association established a committee on water quality charged with the duty of investigating and monitoring effluent from each factory in the area. The committee undertook its responsibility seriously. It sent samples of water from the factories to an agricultural testing center, retained scholars to research waste water treatment methods, and later established effluent standards based on their recommendations. Finally, it negotiated with the individual companies and induced them to accept these standards.

For a number of years the Irrigation Association and the local fisherman's union continued to file opinion letters and complaints with the local government, but they were not able to obtain redress. They even petitioned the Diet, and although the Diet later briefly considered the case, it failed to take any curative action.

In 1936, the fishermen began lobbying for special national legislation. This tactic was somewhat unusual for the period and suggests that the fishermen were considerably more influential than Ashio's poor farmers. The proposed statute required factories to bear the costs of waste treatment facilities constructed for them by the fisherman's union to the extent that the factories themselves were unwilling to build their own facilities. Although the law was never passed, many improvements were introduced into the Arata River Basin before World War II as a result of the public and governmental attention given its problems.

1.2.7   Pollution Disputes During World War II

During World War II there were few pollution-related disputes or mass demonstrations in Japan because the Japanese government actively repressed protest; opposition to national policy bordered on disloyalty. Nevertheless, disputes did continue, although at a drastically reduced level. In 1941 smoke pollution at Annaka aroused local farmers to demand compensation from an offending factory. Although the final mediation settlement provided only ¥800 per year, the mediators persuaded the factory to purchase the damaged fields. In another dispute, local farmers requested compensation for damages covering an area of about 24,500 acres and caused by the discharged pulp waste from the Kokusaku Pulp Factory along the Ishikari River. The company paid compensation to the farmers in 1943 and 1949.

1.3   Japanese and American Response to Environmental Deterioration Before World War II

Japanese and American responses to environmental deterioration before World War II are similar in some respects. In the late nineteenth and early twentieth century neither country anticipated the potential seriousness, pervasiveness, or complexity of the problem, and neither appeared greatly concerned with the health effects of pollution or the husbandry of natural resources.[28] Prescriptions for severe environmental disruption favored ad hoc technological "solutions," and there was virtually no attention paid to underlying social, economic, or institutional factors.[29] Both countries also delayed the development of a comprehensive approach to environmental problems until the 1960s. Despite these broad similarities, each country's use of legal institutions to control pollution differed greatly.

After 1880 many people in the United States recognized the need for government action to halt the progressive deterioration of America's environment, and new laws were viewed as the primary means to effect a remedy.[30] Legislatures and courts were the principal legal institutions involved, and law-making was carried out principally at the local level. The common councils of large industrial municipalities, for example, were the first legislative bodies to take action against air pollution.[31] But the early ordinances passed by these bodies were generally primitive in design. Most penalized the discharge of "dense smoke,"[32] and others placed affirmative duties on polluters to remove ashes or cinders from their shops.[33] Some even enjoined the use of soft, high sulfur, bituminous coal.[34] Despite these efforts and the increasing intervention of county air pollution agencies and state legislatures, air quality continued to deteriorate.

After the 1880s the courts also became involved in environmental controversies because polluters often challenged the constitutionality of municipal and state pollution controls. Generally, the courts upheld these regulations.[35] As pollution problems became more severe, citizens,[36] and even states,[37] began to seek judicial redress. In many cases they were awarded substantial damages[38] and equitable relief.[39] Appellate decisions of the period suggest almost unprecedented judicial willingness to direct the course of the nation's industrial growth.[40]

Despite these common law and legislative developments, the legal institutions often proved unsuccessful in arresting environmental destruction.[41] In part the judicial process was at fault. Many judges were woefully unequipped to assess the complex economic, scientific, and technological questions presented by environmental problems, and most legislators failed to understand the complexities of pollution problems. For the most part, legislative solutions were reactive, seeking quick and easy technological remedies but ignoring social and economic considerations. Pollution prevention was often relegated to existing agencies busy with other jobs, lacking specialized skills, subject to conflicting jurisdictional responsibilities, and possessing inadequate authority to enforce their own regulations. Until recently there was virtually no legislative monitoring or assessment of past enactments. Finally, progress in the environmental field was impeded by the exemptions and other dispensations granted special industries.[42]

In contrast, the Diet and courts in Japan only played an insignificant role during the pre–World War II period. Environmental problems for the most part were handled by local authorities or settled privately by the parties themselves. Venerable extrajudicial institutions—mediation, conciliation, and arbitration—helped settle many disputes, while privately negotiated indemnity contracts mollified even serious conflicts. The usual process of dispute settlement proceeded from an initial negotiation to a polluter's confrontation with irate farmers or fishermen, and then to an offer of compensation. This cycle was repeated as long as the particular controversy continued.

The use of compensation payments deserves special mention. From the beginning they served as a time-honored means of addressing the concerns of the injured party, and at the same time permitted a polluting activity to continue. In virtually all recorded cases, the polluters tendered some offer of compensation. There were various terms for compensation—goodwill money; charitable payments; solatium, consolation, and donation money; sympathy money and contributions to the community; and sorry-to-have-troubled-you money. Despite such largesse, polluters consistently refused to recognize legal responsibility to the victims; sympathy money was motivated more by social and cultural reasons than any legal compulsion.

When a pollution dispute became a matter of national concern, protesters would often turn to the Diet. But the Diet served more as a forum to publicize controversy and stimulate an official investigation than as the fountainhead of legislation. Not once did the Diet respond to a pollution controversy by passing remedial legislation.[43]

The principal official means of controlling pollution was the administrative process. Because many enterprises like the Furukawa Ashio Mine were dependent on government support, administrative control through advice, recommendations, or orders could be exercised in accord with changes in official policy. Administrative supervision tended, however, to be directed to specific needs. In pre–World War II Japan few officials thought to develop a comprehensive pollution control policy.

It is interesting to speculate about why the Japanese judiciary failed to play a more active role in environmental protection during this period, especially in light of the Osaka Alkali case.[44] In many respects the Osaka High Court's treatment of sulfur oxide emissions from the Osaka Alkali copper plant is as sophisticated as any of its American counterparts. Yet this decision stands alone. Most Japanese continued to view the judiciary as an unpredictable, alien institution; and until recently a judicial trial was a costly, time-consuming, and ultimately ineffective procedure, especially when compared to the traditional alternative of a privately negotiated settlement.[45] In light of judicial agonizing in the United States over the formulation of sensible rules to judge environmental controversies, Japan's emphasis on extrajudicial dispute settlements does not seem misplaced.

15

For many people in Japan during this period, technology seemed to provide a solution to pollution problems. Although investment in control technology was generally modest,[46] when economic interests and a need for pollution control intersected, a mutually satisfying remedy to the pollution problem was usually found.[47]

Japan's relentless drive toward industrial reconstruction during the early postwar period upset this fragile institutional balance. After the postwar pollution diseases first appeared in the mid-1950s, the government gradually began to recognize that traditional institutions were ill-equipped to respond to the unprecedented number of people severely injured by pollution, the increasing number of polluters, the contribution of governmental bodies themselves to environmental destruction, the scientific and administrative complexities of the problem, and the apparent trade-offs required between environmental protection and other insistent national priorities. Although the chronicle of its awakening is very different, by the mid-1960s Japan like the United States began a common search for new legal techniques to salvage its environment.

Section 2   The Postwar Period

2.1   Local Ordinances and Early Pollution Control Laws: A Prelude to Comprehensive Environmental Legislation

During the early post–World War II period, the prewar pattern of central-level indifference and sporadic local legislative reactions to pollution problems continued. The first of the early postwar, locally based efforts was the Tokyo Factory Pollution Prevention Ordinance of 1949,[48] which required new factories to obtain a permit from the metropolitan government. Because of the government's inexperience with pollution control, however, the ordinance's terms were vague;[49] standards for factory operations were either absent or unclear and there were no penalties for violations. From a regulatory perspective the ordinance was virtually ineffective.

Despite all its defects, similar ordinances were enacted by Kanagawa Prefecture in 1951,[50] by Osaka Prefecture in 1954,[51] and by Fukuoka Prefecture in 1955.[52] These crudely designed enactments constituted the period's official response to the pollution problem.[53]

While local governments sought to deal with pollution by enacting ordinances, the central government remained stolidly passive and uninterested; its overriding objective remained economic revitalization. By 1955, the growth rate of national per capita income exceeded 10%, matching the prewar level. Since national priorities were being met, there was no apparent need for action.

An isolated exception was the 1949 recommendations of the Resource Research Commission (Shigen Chōsa Kai), a body then attached to the Economic Stabilization Board (Keizei Antei Honbu). The commission's report urged the government to enact special water pollution legislation, establish water quality standards for waters in public use and waste water, create a research center for water resources, and provide arbitration for pollution-related disputes.

The Resource Research Commission was about a decade ahead of its time, not only in the practical measures proposed, but also in its appreciation of the use of environmental policies to foster economic growth objectives (that is, water resource conservation). The government's only response to the commission's recommendations was to set up an interministerial body to serve as a liaison between the concerned ministries. Because of its comparative neutrality as a coordinative body, and the government's policy of "clearing" all pollution control measures with economic growth objectives, the Economic Planning Agency (Keizai Kikaku Chō) was entrusted with the primary responsibility of drafting a new water pollution control law. Yet because of jurisdictional and other rivalries, the Economic Planning Agency's progress was slow. In the end the ambitious Resource Research Commission's proposal was discarded.

Gradually, however, public dissatisfaction over the central authorities' indifference to pollution increased. The media began to carry many reports of citizen complaints. Even some government officials urged the central government to take action because pollution, particularly air pollution, was becoming too severe for local authorities to control. Local governments were increasingly placed in an almost impossible position between residents' demands for pollution prevention and local policies designed to encourage industrial development.[54]

In response to such criticisms, in 1953 the Ministry of Health and Welfare (Kōsei Shō) began a two-year investigation of complaints about air, water, noise pollution, and vibrations. After completing its research and close consultations with the Japan Association of Public Health, the ministry began drafting a new law in December 1955. The draft, however, was opposed by Keidanren, the Federation of Economic Organizations, various chambers of commerce, and the other ministries. Indeed, one redoubtable opponent, the Ministry of Finance, actually obstructed the budgeting of the draft law's implementation. For a while the Ministry of Health and Welfare attempted to find a compromise by substantially revising its original draft, but interministerial agreement proved impossible.[55] The draft, like the earlier Resource Research Commission's proposal, was shelved.

About this time, reports of pollution diseases in the provinces began to increase. Minamata disease was discovered in 1955, in 1957 Norboru Ogino publicized his findings of itai-itai disease in Toyama Prefecture, and oily smelling fish were first reported in the Yokkaichi area. The most dramatic episode of the period, however, was the Urayasu incident.

In the mid-1950s the Honshū Paper Mill Company obtained a permit from the Tokyo Metropolitan government to run an integrated pulp factory. Shortly after the company commenced operations, however, fishermen on the Chiba bank of the river began complaining that their fisheries were being damaged. When the company failed to respond to these charges, several hundred fishermen forced their way into the plant demanding reparations. A scuffle ensued, sixty-four people were injured, and over one thousand policemen had to be summoned to suppress the riot.

Erupting on the very borders of Tokyo, the Urayasu incident soon became a major political issue. The Diet began interpellating responsible government officials, and thousands of fishermen demonstrated demanding new water pollution legislation.[56] In response to public demand, the government proposed two "remedial" laws to the Diet, the Water Quality Conservation Law[57] and the Factory Effluent Regulation Law,[58] which passed without significant debate at the close of 1958.

But the new laws soon proved ineffective. Since water quality standards established under the Water Quality Conservation Law applied only to designated areas, undesignated areas soon degraded. Moreover the designation process itself was politically uncertain and invited conflict. Initially the Economic Planning

Agency was supposed to oversee the establishment of water quality standards,[59] but the ministries began bickering and designation was retarded. Three years after the passage of the Water Quality Conservation Law, the Edo River, scene of the 1958 Urayasu incident, was still undesignated.

Enforcement of the new law was checked in other ways. Water quality standards, the basis for factory effluent regulations, were weakened by pressures from other ministries and the manufacturing industry. Because water quality standards were set in terms of density limitations, such as suspended solids (SS) or biological oxygen demand (BOD), and not in terms of the volume of effluent, a polluter could meet the standards by diluting its waste water. Moreover the law contained a clause requiring enforcement to be tempered with the "natural harmonization of industries,"[60] a concept that assured the priority of industrial concerns. Another problem was that enforcement of the Water Quality Conservation Law's[61] standards was relegated to provisions established by the Factory Effluent Regulation Law, the Mining Safety Law, and other laws. Although this separation of standard setting and enforcement permitted MITI to maintain control over manufacturing, it also significantly weakened the management of water quality.[62]

The government's approach to air pollution control was equally feckless. Air pollution in Yokkaichi City is a case in point. In 1955 the central government decided to sell land in Yokkaichi City, the site of a former naval base, to a group of petrochemical and power companies. These enterprises formed a large industrial complex (konbināto) considered to be the nation's most advanced industrial estate. Soon after the factories began full-scale operations, however, Yokkaichi City's residents began suffering from various respiratory diseases. Their maladies, commonly referred to as Yokkaichi asthma, attracted nationwide attention. As industrial production increased, and as Japan switched from using coal to petroleum, other industrial cities also reported air pollution–induced illnesses. In response, the mass media initiated a compaign against pollution, criticizing the government's failure to adopt controls. As before, public concerns over pollution compelled some remedial government action, but this time official notice was granted even more grudgingly. MITI in particular believed that public health considerations should not be allowed to dominate national air pollution policy. The ministry somewhat cavalierly urged greater concern for "healthy" industrial development.[63]

In March 1962 MITI and MHW submitted a joint bill to the Diet for the regulation of smoke and soot. To the outside world the Smoke and Soot Regulation bill seemed a compromise between MITI, its original sponsor, and MHW. Actually, MITI's concern for industry dominated. The bill's purpose clause directed that effort to prevent harm to public health be "harmonized" with measures to promote sound industrial development;[64] although it established emission standards, it neglected standards governing facility design and operations, and controls on noxious gases applied only in emergencies. With all its defects, the bill passed the Diet on May 4, 1962.[65]

In many ways the Smoke and Soot Regulation Law[66] was a template of earlier efforts. As before, polluters in designated areas were required to register with the prefectural governor,[67] monitor the density of pollutants, and comply with emission standards.[68] The Smoke and Soot Regulation Law met with the same criticisms as did the earlier water laws. Its emission standards were attacked as capitulations to economic feasibility with little regard for public health. Even industry was appalled at the law's inequitable enforcement. When standards were set for the Yokkaichi area, large factories were not required to alter their operations, but small pottery manufacturers were required to install control equipment. Moreover MITI preserved its jurisdiction over electrical generating and gas manufacturing facilities in order to ward off efforts by prefectural governors to impose strict controls on these industries.[69] During the 1960s, the Diet eventually passed a few other comparable stopgap measures,[70] but there was no plan or policy, and implementation of the laws remained sporadic.

2.2   Toward the Enactment of a Basic Law for Pollution Control

2.2.1   The Government Awakens to the Need for a Comprehensive Pollution Control Policy

At last in the early 1960s the government began to see the need for a comprehensive pollution control policy.[71] Three factors hastened this realization: widespread public dissatisfaction with piecemeal remedial efforts, the continuing problem of transfrontier pollution; and local government's festering resentment over the central government's preemption of local pollution standards.[72] The scene was set for Mishima-Numazu.

In 1963 the Mishima-Numazu area had been designated a "special industrial development zone" under the National Comprehensive Development Plan of 1962. In the same year Shizuoka Prefecture announced plans for the construction of a petrochemical industrial complex, expected to be the largest of its time. To facilitate the efficient administration of the industrial complex, the governor of Shizuoka sought to merge the cities of Mishima and Numazu with the town of Shimizu.

Local opposition to this idea was intense. Mishima residents, already suffering a shortage of underground water (due to the recent construction of a large paper and pulp plant), organized a campaign against the petrochemical complex, obtained the mayor of Mishima's support, and persuaded local landowners not to sell their land to the advancing industries. Soon after, Numazu's citizens began mass demonstrations against the complex; groups went off to study the air pollution problems of Yokkaichi and other polluted areas; teams of experts began assessing environmental impacts; high schools held teach-ins; and citizens canvassed neighborhoods for signatures on a petition urging opposition to the development. It all overwhelmed Numazu's mayor who finally announced that he too would fight the plan. The battle ended; the central and prefectural governments and the participating companies withdrew their proposal.

The citizen's rejection of the Mishima-Numazu development was profoundly disturbing. The government and industry now saw that without effective measures to control pollution, comparable disturbances could erupt, unexpectedly, in other prefectures. How would industry then find new industrial sites? How could economic growth proceed? The government actively set to work on the formulation of a comprehensive policy.[73]

2.2.2   The Legislative Process: Interactions Between the Diet and the Administration

At this point the Diet entered the fray, for Mishima-Numazu had aroused the opposition. In 1965 the Socialist and the Democratic Socialist parties submitted slightly different versions of a basic pollution control law. The Socialist party's proposal stressed the responsibility of industry for pollution damage and advocated the creation of an administrative commission to consolidate all aspects of pollution control. The Democratic Socialist party's bill underlined the need to "harmonize" industrial development and pollution control. Both bills extended financial aid to prefectural and municipal governments. Although neither bill passed the Diet, the House of Representatives' Special Standing Committee for Industrial Pollution[74] passed a resolution calling for an immediate study of possible legal options. And under great pressure, the Ministry of Health and Welfare decided in 1965 to appoint a commission of experts, known as the Environmental Pollution Commission (Kōgai Shingikai), as an advisory organ to the ministry.[75]

The Environmental Pollution Commission's Interim and Final Reports

The commission's[76] first act was to instruct its general pollution policy section to draft a report, which could serve as a basis for future legislation. The Commission's potential influence was quickly recognized and many groups began to offer suggestions. Local organizations urged legislation that would expand local powers to regulate pollution; major industrial and economic organizations like the Keidanren stressed that pollution should be remedied by technology and that the government, not industry, bore the chief responsibility for countermeasures.

In 1966 at the Fifty-first session of the Diet, the Socialist party and the Democratic Socialist party again submitted bills for pollution control using this opportunity to interpellate various government officials. In responding to the opposition party members' inquiries, Zenkō Suzuki, Minister of Health and Welfare, announced that after the Environmental Pollution Commission finished its study of basic policies, the government intended to draft an organic pollution control law. During the same session, the Special Standing Committee for Industrial Pollution Control in both Houses again passed resolutions calling for the promotion of pollution control.[77]

After obtaining the draft proposals of its general policy subcommittee, the Environmental Pollution Commission submitted its Interim Report to the Minister of Health and Welfare in August 1966. Despite differences of opinion among commission members, the report consistently emphasized the responsibility of industry for pollution control and urged that industry be held strictly liable for pollution damage. The government also was indicted. The report stated clearly that existing pollution regulations were ineffective, and new drastic, comprehensive, pollution measures were necessary. Health was now to be given explicit priority over industrial development.

Almost immediately, the commission's interim report provoked controversy. Although public opinion was generally favorable, some scholars criticized it as too modest. Industry, principally through the Keidanren, called it too radical. The Keidanren's statement, "Opinion on Fundamental Problems in Pollution Control,"[78] stressed the notion of "harmony" between economic development and pollution control.[79] Harmony meant that pollution controls should in no way harm industry's development.

In the fall of 1966 a number of government agencies also submitted their reactions to the interim report. The position of the Ministry of Home Affairs was among the most progressive; in a sudden volte-face, it sought to champion the desperate situation of local governments. The ministry strongly recommended that strict liability be recognized, polluters be compelled to share the costs of public works projects, and local governments be empowered and aided by the central authorities to regulate polluters.[80]

As the final drafting stage approached, the debate within and without the commission intensified. Some commission members, fearing that the implementation of the interim report would have far-reaching adverse economic repercussions, urged significant modification. They argued that because the government itself had delayed investment in pollution prevention, it also should be responsible for control. Hurrying the establishment of ambient standards because of public pressure was ill advised, they warned, for standards not adequately accounting for economic or technological factors could distort production processes.

Other commission members, however, believed that the report was too weak. These members were disturbed that it failed to provide concrete controls for particularly hazardous substances and that it deferred to technological and economic considerations. Although the debate produced a number of compromises, the commission's report ultimately retained its basic philosophy but was rendered less strident. The final report was unanimously approved in a general meeting of the commission and submitted to the Minister of Health and Welfare on October 7, 1966.[81]

In many respects, the commission's report foreshadowed measures that ultimately were undertaken. It stressed that all pollution control programs should be designed with the purpose of achieving ambient standards. Ambient standards themselves were viewed as administrative targets rather than as rigid limitations. The report also recommended the adoption of a strict liability principle and asserted that industry had an obligation to share the costs of public works for pollution control. It recommended the establishment of special pollution-dispute settlement procedures and the establishment of a compensation fund for pollution victims. Finally, it endorsed permitting local governments to regulate sources of pollution more stringently than the central government, and underlined the need for expanded pollution control public works projects, financial assistance to local governments, further development of monitoring systems, and the education of experts and the general public in the tasks of pollution control.

Ministerial Maneuverings

On October 11, 1966, Zenkō Suzuki, Minister of Health and Welfare, explained the final report of the Environmental Pollution Commission to a Cabinet meeting and formally requested the cooperation of other ministries in enacting a basic pollution control law. His appeal was supported by Prime Minister Satō and the other ministers.[82] Because of its experience with public health, the Ministry of Health and Welfare was designated as the principal agency responsible for preparing the basic law.[83]

Pollution control, of course, raised its own peculiar jurisdictional dilemmas. In ordinary circumstances, jurisdiction would have been allocated to the ministry having the most direct relation to the subject of new legislation. In this case, however, a number of agencies claimed important jurisdictional interests. MHW was concerned with the protection of health; MITI possessed regulatory powers over the sources of pollution; the Ministry of Agriculture and Fisheries[83a] asserted an interest in the protection of farming and fisheries; the Ministry of Local Autonomy represented the interests of local government; and so on. Given the disparity of views, the government decided that its own coordinating body, the

Liaison Council for Pollution Control Promotion, should oversee and review the final report.[84]

Having considered various views expressed by other government agencies, the Ministry of Health and Welfare set to work and submitted a Draft Proposal (Shian Yōkō) to the council on November 22, 1966. The draft substantially followed the suggestions made by the Environmental Pollution Commission's report. There were, however, two basic differences. First, the principle of strict liability for pollution that had concerned industry was eliminated. The ministry explained that it was inappropriate to prescribe by statute a rule of civil liability that would create an exception to the general principles of negligence established by the civil code. The ministry also emphasized that it was technically difficult to define the scope of strict liability in the pollution field. Second, the draft proposed creating an independent administrative committee for pollution control. Originally, the Environmental Pollution Commission's report had suggested an administrative organization be responsible for the formulation of basic control policies. The Draft Proposal gave concrete form to this idea.[85]

But the ministry's proposals were short-lived. Relentlessly, the Liaison Council, through its executive subcommittee, began debate on its virtues and its many limitations. Within the council, MITI and the Economic Planning Agency criticized the draft, arguing that its preoccupation with public health and environmental concerns (including resources like fisheries) was short-sighted. Pollution control policy, they urged, had to be compatible with the sound development of the economy.

A compromise was reached. The priority of the protection of human health over economic issues was retained in the purpose clause, but the phrase "harmony with sound economic development" was inserted in the section dealing with the conservation of the living environment.[86] Although the precedence of public health over economic concerns was recognized, the government's expectation that pollution control measures would harmonize with economic growth was made explicit. The policy of harmonization, however, would soon be attacked as dominating the formulation of all pollution control legislation.

The establishment of ambient standards also was controversial. Pro-industry ministries argued that ambient standards should be treated merely as flexible policy goals not as inflexible limits. Because of such pressures, the definition of an ambient standard was changed from "the standard of environmental conditions to be maintained," a phrase employed in the welfare ministry's draft, to "the standard of desirable environmental conditions to be maintained." In addition, the liaison council's final version stipulated that the interests of all industries (including industries causing pollution) should be considered and "harmonized" during the process of standard setting.[87]

The liaison council weakened other parts of the health ministry's draft at the behest of MITI and other agencies. The issues of strict liability and the establishment of an administrative compensation fund were reexamined and shelved, the latter on the pretext that further analysis of scientific and technological considerations was required. The idea of establishing an independent administrative body to control pollution was also temporarily abandoned.[88]

Public dissatisfaction with the liaison council's draft outline was intense, particularly over the harmonization clause and the council's elimination of the principle of strict liability. Nonetheless, most people still favored enactment. In the Special Pollution Control Committees of the Diet, committee members also insisted that the government submit a bill as soon as possible.

While support for a new law increased, industry desperately attempted to detour government policies. For example, the Keidanren asked the government to entrust jurisdiction of the bill to the Economic Planning Agency, hoping that the agency might give priority to economic growth. As in earlier statements, the Keidanren urged the government to share the costs of pollution control.

A Cabinet meeting on February 24, 1967, directed the Ministry of Health and Welfare to incorporate the Liaison Council's proposals into a bill. Although the Liaison Council was in a better position than the MHW to coordinate the disparate ministerial views, it was necessary to designate a responsible ministry or agency to represent the government in the subsequent Diet discussions. Since the Liaison Council by custom was not permitted to perform this function, the MHW was charged with the subtle task of solving the problems of the bill with the other agencies without changing the basic policy.[89] After a few final modifications, the bill was then reviewed by the legislative bureau of the Cabinet[90] before officials assembled from all related ministries; at last on May 16, 1966, the Cabinet endorsed its submission to the Diet.

The Diet Debate

Shortly after the government submitted its bill to the House of Representatives, the Socialist, Democratic Socialist, and Kōmeitō (Clean Government party) parties submitted their own versions.[91] Following Diet practice, the first bill submitted was deliberated first; thus, debate on the opposition parties' bills was deferred until the first bill was defeated. The House of Representatives was the first to address the government's bill. After the Minister of Health and Welfare explained its purposes at a plenary meeting of the House, the bill was sent to the Special Standing Committee on Industrial Pollution Control on June 2. Thereafter, it was presented in a plenary meeting to the House of Councillors, the upper chamber.[92]

As noted previously, the special committees in both Houses had already passed several resolutions for the enactment of a basic law. During the debates in committee, the opposition parties criticized the bill, arguing that the harmonization clause weakened the government's pollution control policies and that a provision for strict liability should be retained. Indeed, the opposition refused to pass the government's bill without this modification. Because of its campaign promise to enact a basic pollution control law, the governing Liberal Democratic party was now willing to compromise in order to secure the bill's passage. The LDP negotiated with the other three parties and agreed to the following major modifications:

1. The purpose clause was divided into two subsections in order to make clear that "harmonization" applied only to the conservation of the living environment and not to the protection of public health.[93]

2. A Council for Environmental Pollution Control would be established at both central and local levels.[94]

3. A subsection prescribing special financial assistance to small industries was added.[95]

On July 17 the Special Standing Committee for Industrial Pollution Control in the House of Representatives passed the modified bill with a supplementary resolution supported by all four political parties;[96] and after the House of Councillors concurred, the Basic Law for Environmental Pollution Control[97] was passed on August 3, 1967. The opposition parties exhorted the government, as soon as possible, to develop concrete measures giving priority to health over industrial development, enact a strict liability law, and establish at some future time an integrated pollution control administration.

Although not legally binding, the Diet resolutions were deeply significant from a political standpoint. They voiced a strong minority's sentiment for decisive executive action and published for society the most salient weaknesses of the government's position. These views were dramatized by the mass media and actively discussed in academic and professional circles. And the resolutions reflected and fueled public criticism of the government's diffident policies. Thus a debate on the amendment of the Basic Law for Pollution Control began on the eve of its enactment.

2.3   The 1967 Basic Law for Environmental Pollution Control and the 1970 Amendments

The Basic Law for Environmental Pollution Control of 1967 was essentially a charter for the control of air, water, noise, and other pollution.[98] The concrete applications of this program were deferred to subsequent legislation and administrative action.[99] But the Basic Law was more than an airy pronouncement of well-meaning intention; it provided an immediate impetus for the government's drafting of new legislation.[100] In the course of subsequent events, the government would often cite many phrases to justify a certain line of action, and citizen groups would appeal to other sections of the law in their struggles to safeguard the environment.

The Basic Law also accurately mirrored the social values and attitudes toward pollution of its time. For example, by 1967 a national consensus had already been reached that the government's pollution control policies had been sorely remiss. The Basic Law was an acknowledgment of this fact and provided a basis for specific remedies.[101] In a number of provisions the Basic Law also reflected the government's gradual recognition of a responsibility to plan. For instance, art. 11 mandated land use controls for environmentally degraded areas, and art. 17-1 stressed the need to include environmental protection provisions in the planning and implementation of regional industrial development. The Basic Law also made explicit the state's responsibility to assist local governments financially and in other ways during the formulation and implementation of local control programs.[102]

The Basic Law also helped to consolidate the administration of pollution control. Articles 25 and 26 established an interministerial Environmental Pollution Control Council (Kōgai Taisaku Kaigi) that replaced the Liaison Council, and several expert advisory bodies were also created.[103] The central advisory council in effect absorbed the functions of existing councils in various ministries and agencies that were abolished at the time of enactment.[104]

Although the state's responsibilities under the Basic Law were fairly clear, industry's obligations were more ambiguous. Enterprises were to be "responsible for taking measures necessary for the prevention of environmental pollution,"[105] and had to shoulder "all or part of the cost" of the government's pollution control works.[106] Who would decide what measures were necessary, or what costs industry should bear was purposely left uncertain.[107] The greatest expression of the government's ambivalence, however, was the "harmony clause."[108] Innocent enough on the face of it,[109] the clause gave industry a means of escaping the Basic Law's sternest intentions.[110]

While the government continued to procrastinate,[111] pollution was becoming more serious. In 1967 the public, already upset by media coverage of Minamata, itai-itai, and Yokkaichi illnesses, was further alarmed by reports that residents around busy Tokyo traffic intersections were being poisoned by lead and Tokyo's photochemical smog was suffocating school children. These domestic events coincided with rising international concern over pollution. President Nixon's statements on the importance of environmental protection, the establishment of an Environmental Protection Agency in the United States in 1970, the United Nations 1968 resolution calling for an international conference on the human environment (the 1972 Stockholm Conference), and the establishment of an Environmental Committee in the Organization for Economic Cooperation and Development (OECD) in 1970, all attracted great attention in Japan.

International concern for environmental protection combined with domestic pressures to catalyze a movement for drastic reform within the Diet and the Cabinet. In the Diet, members of the special standing committees of both houses began to question government officials, and both committees again passed resolutions proclaiming the urgent need for more aggressive government action.[112] In response to the Diet's demands, in July 1970 the Cabinet composed a special Headquarters for Pollution Control (Kōgai Taisaku Hombu)[113] to accelerate the formulation of new policies and measures. Its organization and function overlapped that of the Conference on Environmental Pollution Control, already established under art. 25 and 26 of the Basic Law. The headquarters held a series of intensive meetings to consider the revision of many existing pollution laws, including the Basic Law. In addition to establishing the headquarters, the Cabinet also organized a special Ministers Conference for Pollution Control (Kōgai Taisaku Kankyō Kaigi) to coordinate various agency views more efficiently. Meeting seven times within a three month period, the Ministers Conference issued a policy opinion that was thereafter drafted into a general outline for the Basic Law's revision.

The draft's key points were that
1. the harmony clause should be eliminated;
2. "policy goals"[114] for pollution control should be made explicit; and
3. prefectures should be empowered to set their own specific standards applicable to the social, environmental, and economic conditions of a particular area once national ambient standards were established.[115]
These recommendations were thereafter approved by the Central Council on Pollution Control and referred to the Cabinet. On November 25, 1970, the Cabinet agreed to the proposed amendments and referred the bill to the Diet two days later.

After intense debate,[116] the bill passed the House of Representatives on December 10 and the House of Councillors on December 25 without modification. Once again the opposition succeeded in attaching a resolution criticizing the shortsighted approach of the government's bill and calling for more spirited action. Given the strength of public sentiment supporting this statement, the government reluctantly allowed the resolution to pass.

On balance the revised version of the Basic Law established a fundamental policy for pollution control that was absent in the early and mid-1960s. The harmony clause was finally eliminated, and the authority of local governments (particularly prefectural governments) strengthened.[117]

The infirmities of the Basic Law, even in amended form, exerted an influence on later measures. For example, an important weakness of the 1967 version was its insensitivity to the significance of protecting the natural environment. Although the 1970 amendments evidenced greater concern by including a specific provision on this subject,[118] effective measures were introduced only slowly during the 1970s.[119] The amendments also failed to acknowledge a need for comprehensive environmental planning.[120]

Although concepts like "environmental management," "impact assessment," or "coastal zone planning" belong to a discipline that was still in its infancy at the time of the amendments, the law's failure to anticipate their importance shows the depth of uncertainty and ambivalence toward measures that could slow economic growth. Even the elimination of the harmony clause in the 1970 amendments, hailed by the opposition as a major achievement, did not measurably strengthen the government's commitment to protection of the environment.[121]

2.4   Toward the Establishment of an Integrated Environmental Protection Administration

2.4.1   Genesis of the Environment Agency

During the 1950s and early 1960s, the government was uninterested in establishing an independent, integrated pollution control administration. Thus jurisdiction over a host of environmental problems remained randomly distributed among eleven ministries and at least nine advisory councils.[122] For the most part, the ministries met each new pollution crisis by unilateral, uncoordinated assertions of power without any comprehensive national environmental protection plan or purpose. The Ministry of Health and Welfare, MITI, and the Ministry of Agriculture and Fisheries all claimed primary jurisdiction over the vast areas of public health, industrial operations, pesticide control, and highway construction, and all pursued their own policies virtually without careful consultation. The Economic Planning Agency was given primary jurisdiction over the establishment of water quality standards at the same time that other ministries were entrusted with their enforcement. The Ministry of Health and Welfare and MITI were charged with the joint administration of the Smoke and Soot Regulation Law; they independently established their own industrial pollution sections to implement the law. Confusion finally prompted the government to establish the Liaison Council for Pollution Control Promotion. But this body merely reflected the diverse and conflicting views[123] of its members, and it too failed to develop effective measures for protection of the environment. The council's mutilation of the MHW's proposal for an independent environmental protection agency affords insight into its proclivities.[124]

During the 1967 Diet debates on the Basic Law, the three major opposition parties (the Socialists, Democratic Socialists, and Kōmei) revived the idea of an administrative commission with plenary regulatory powers. Although this was rejected, the opposition again passed a resolution urging further study during the final stages of the enactment of the Basic Law.

Around 1970 the pressure on the government to establish a new, independent agency increased. The Diet's criticism of the government's policies continued. Foreign precedent also played a part. By the end of 1970 Sweden, the United States, and Great Britain had all established independent agencies,[125] and these models were carefully studied in Japan. At last Prime Minister Satō took up the cause and requested the government to prepare the necessary legislation. The

Diet quickly passed the Environment Agency Establishment Law on May 24, 1971.[126]

2.4.2   Development of Other Environment-Related Legislation

The sixty-fourth session of the Diet in 1970 was an extraordinary affair, for in addition to amending the Basic Law it also enacted thirteen other environment-related laws or amendments.[127] Generally these new laws displayed an incrementally greater sensitivity to the protection of the natural environment and they significantly increased the powers of local governments. The sixty-fifth regular Diet session in 1971 and the Diet sessions of 1972 and 1973 substantially completed Japan's present arsenal of environmental protection legislation.[128] The only legislative proposal remaining from this era is the controversial Environmental Impact Assessment Law, whose fate is still uncertain.

2.5   Summary

It is interesting to consider Japan's position in the mid-1960s from the perspective of other countries now experiencing industrialization.[129] Several factors clearly facilitated the Japanese response: the existence of a trained, dedicated bureaucracy; a long tradition of strong, centralized administration; an abundance of skilled economists, medical experts, and others who were able to integrate pollution control policies with economic planning and other national objectives.

Yet Japan was at a great disadvantage in other respects. Because environmental deterioration was only beginning to be viewed as an important world problem, Japan had to conceive and implement many pollution control measures without the benefit of foreign precedent or experience. Moreover her response was dictated by the exigency of circumstance, for public pressure had become so strong by the mid- and late-1960s that many policies and measures were formulated within only a few months. Japan's administrative tradition in a sense impeded the development of an effective response. Throughout, the ministries fiercely opposed the erosion of their traditional jurisdictional prerogatives, and at every turn sought to subvert the development of a new, powerful, and independent environmental protection agency. Japan's formulation of a systematic response to pollution control thus might be studied by many industrializing countries not because it bears unthinking imitation or even emulation but because it presents us with a paradigm of inquiry, a *rite de passage*, that most countries have, or sometime inevitably must, come to face.

2   The Transformation of Values and the Legal Process: Moral Bases for Postwar Environmental Policy

During the 1960s attitudes toward pollution changed radically in Japan. In the early postwar period many Japanese viewed the smoke rising from their newly reconstructed cities with pride because it symbolized the nation's economic rebirth; by 1970 citizens from all walks of life were challenging the morality of environmentally destructive industrial developments. Against economic growth they now weighed new concerns such as the sanctity of human life, individual dignity, and the integrity of local communities.

To what factors can these changes in public attitudes and values be ascribed? This chapter describes how this transformation began with the suffering of the victims of four pollution-induced diseases and traces how these victims became the center of a national uprising against pollution. We are particularly interested in how the Japanese legal system has articulated and adopted the new values generated by the antipollution movement(s), how these values weakened traditional extrajudicial intermediary institutions and motivated the victims to turn to the courts, and how changing values continue to influence legal processes and in turn to be influenced by them.[1]

Section 1   The Four Pollution Cases

The first pollution-induced diseases appeared in Japan in the mid-1950s during a period of growing economic prosperity. The most famous affliction occurred in the vicinity of Minamata Bay, Kyūshū from which this malady took its name. As was officially recognized in 1968, the cause of Minamata disease is methylated mercury. Methylated mercury is a byproduct of the production of acetaldehyde, which was manufactured at the Minamata plant of the Japan Chisso (Nitrogen Manufacturing) Company. For years the Chisso plant had discharged methylated mercury (in its effluent) directly into the bay. Gradually the chemical had become concentrated in the tissues of fish, other marine organisms, and ultimately the inhabitants of the Minamata fishing community. In 1965 a second outbreak of Minamata disease was reported in far-off Niigata City where the predominantly fish-eating community was poisoned by mercury discharged by the Shōwa Denkō plant.

Minamata disease is frightfully painful, causes tremors and paralysis, and is often fatal. Because pregnant women poisoned by mercury transmit the toxin through the placenta to their fetuses, many children born in Minamata during the mid- and late-1950s were afflicted with the disease.

The Minamata communities were poor and isolated, and their suffering and occasional protests went unheeded for years. It took thirteen years for the victims of Minamata City to decide to go to court. But the victims of Niigata Minamata disease were less patient, and on June 12, 1967, they became the first of Japan's postwar pollution victim groups to file suit.[2]

The second pollution-induced disease was discovered in Yokkaichi City, the site of a huge complex of oil refineries and petrochemical and power plants. The first sign of Yokkaichi's environmental problems, as in Minamata, was the sudden death of animals and plants. Thereafter, local fishermen began to notice that fish caught along the shore of this small Pacific coast city emitted a strong, oily smell. When the fish proved unmarketable, the fishermen began protesting to the companies and local authorities.

Around 1961 other, more disturbing effects of Yokkaichi's pollution began to appear. Residents of Isozu Village who lived near the factories reported that asthma, emphysema, bronchitis, and other respiratory ailments were increasing, particularly among the elderly and children. At the same time concentrations of sulfur oxides in Isozu were logged at six times greater than in other urban areas of Japan. In 1963 the Socialist and Communist parties, the City Councilmen's Progressive Group, and the labor unions formed the Yokkaichi Pollution Countermeasures Council in order to seek assistance and organize a campaign against the companies. After months of struggle, the residents with the assistance of the Council finally persuaded the prefectural governor to vouchsafe free medical examinations for the sick and provide a few officially "certified" victims with financial relief. As air pollution in Yokkaichi worsened between 1965–1967, more cases of serious pulmonary disease were reported. Although the victims sought by all manner of maneuvers to induce the companies to reduce emissions, their pleas, as in Minamata, were largely ignored. On September 1, 1967, shortly after one of the victims, Usaburo Kihara of Isozu Village, committed suicide, twelve other victims finally filed suit.[3]

The last of the famous cases of pollution illness involved Toyama itai-itai ("it hurts, it hurts") disease. Itai-itai disease is caused by chronic cadmium poisoning and its symptoms are frightful pain, splintering of bone tissue, disfigurement, and crippling. Death often results. Although cases similar to itai-itai disease were reported in the early twentieth century, there was a sharp increase among farming communities living in the delta of the Jintsū River between 1945 and 1956. These farmers (and their pregnant wives, who proved even more susceptible) had ingested cadmium by drinking the water of the Jintsū River and eating rice taken from paddies irrigated by it. As was later recognized, the Jintsū River had been heavily polluted by the toxic effluent of the Mitsui Mining and Smelting plant that had operated upstream since the 1890s.

Here, too, official recognition of the disease's cause took many years. Although doctors in Toyama insisted that cadmium played a key role, the company contested this assertion by arguing that the disease was due simply to nutritional deficiencies. At last in 1966 the Ministry of Health and Welfare dispatched an investigatory team to Toyama, which concluded that cadmium was merely one of many contributing factors.

Disappointed with the ambiguous findings of the ministry's report, the patients and their families formed the Itai-Itai Disease Countermeasures Council. In 1967 the patients began negotiations with Mitsui on the issues of compensation and effluent control. Their appeals were rejected.

These were tumultuous times and victims' groups throughout Japan were becoming aware of their common plight and the righteousness of their cause. The Toyama victims were not to be so easily put off. On March 9, 1968, several months after the prefectural authorities belatedly established a small relief fund, the victims filed suit against Mitsui.[4] The itai-itai litigation lasted three years and three months, the shortest of the four major pollution trials, during which time twenty-one victims died.

Section 2   The Crucible of Value Transformation

2.1   The Victims

We begin with the suffering. There is now a considerable body of literature, both journalistic and scholarly, and pictorial as well as literary, that depicts the physical agony of the victims of Minamata, itai-itai, and Yokkaichi diseases.[5] Western readers should become familiar with these works, for they convey far better than we can the excruciating physical pain of those who suffered. It is critical to grasp this emotionally because the victims' agony has traumatized Japanese society beyond any Western experience with pollution. Here only a brief description must suffice. This is what Yuki, a young woman afflicted with Minamata disease, said:

When I recovered a little and began to be able to walk again I was walking down the corridor to go to be examined by the doctor, when I found a cigarette end on the floor. I had not smoked since I lost my mental balance, so I was very pleased. "Oh, there's a cigarette end! How lovely! Lovely! Just let me get my hands on it!" With these thoughts in mind, I began to direct my steps carefully toward it. But I could only walk in a zigzag way. I tried to stand still, but the top half of my body kept swaying. Anyhow, I tried to make straight for the butt. "It's about twenty feet away. I must walk straight ahead or I shall miss it."

Without my wanting them to, my legs suddenly began to dash in all directions. I couldn't stop it. I ran right past the cigarette end. "Oh, no! These convulsions! Not again!" In the midst of these thoughts, I began to feel dizzy. I stopped for a moment and looked back. I wanted to go in that direction but my legs wouldn't let me. . . . I was falling down, my dear! My husband took hold of my back. My body was sticking out backwards. When I fall, I fall backward as if I were starting back from something. But before I fell the convulsions came on again and suddenly I started forward again.

Yuki also described her feeling of isolation:

Now I feel as if my body is gradually drifting away from this world. I have no grip. I can't grip anything firmly in my hands. I can't hold my husband's hands in mine; I can't even hold my own dear son in my arms. Well, I might be able to put up with that. But I can't even hold a bowl of rice which is the chief food in my life. I cannot hold my chopsticks. When I walk, I don't feel as if I am walking with both feet on the ground. I feel as if I'm all on my own, a long way from the earth. I feel so alone. . . . My husband is the only one I love. He's the only one I can rely on now. How I wish I could work again, and use my hands and my legs![6]

Similar accounts by Yokkaichi and itai-itai sufferers are no less moving.

For the most part, the victims, rather than nourished and supported in their many troubles, were rebuked; rather than greeted with patience and kindness by their neighbors, they were ostracized.[7] This was especially true in Minamata. At least in the beginning, the heart of the problem was the fact that the diseases were regarded not only as loathsome but also as contagious. Sanitation officials would come to the victims' homes and spray everything, all their belongings, even their children. In the hospitals the victims were relegated to segregated wards where, even there, they were badgered.

Sometimes, to kill time, I would go to the TB patients' ward. We had been in a ward near the TB patients' ward before, but we were avoided even by the TB patients. They said: "Some patients suffering from the queer disease have come from Minamata. Take care not to catch anything from them!" When they had to pass outside our ward, they would dash past covering their mouths with their hands. TB is also an infectious disease isn't it? It made me angry. We didn't ask to be attacked by this disease, did we? There was no reason why we should be shamed in such a pointed manner. They pointed their fingers at us, saying: "Queer disease! Queer disease!"[8]

Because of fear of contagion, storekeepers would often refuse to accept money handled by a member of a victim's household. On crowded buses, victims would be embarrassed by the seats left vacant around them. Victims' families were barred from drawing water from community wells. Taxi drivers were reluctant to transport patients to or from hospitals; the victims' children were ostracized at school; even healthy individuals from victim families encountered difficulties in finding marriage partners; relatives were reluctant to attend a victim's funeral. The appearance of a case of pollution disease might taint an entire village.[9]

The reaction of many victims to such treatment was a mixture of shame and resignation—shame not only because of the stigma that the society imposed but also because the strange disease was regarded as a curse brought upon them for their moral shortcomings.[10]

We came thinking of spending a year or two here. Going to the mountains to hunt or shoot game. But, you see, I was, well, greedy, I guess. Since the time I came to Kumamoto I liked to eat crabs better than anything, those big sea crabs. And when I came here, there were crabs, mullets, all the things I liked; octopus and sea slugs. And I ate and ate. I wouldn't eat a meal without crabs. So, since I ate that much, I guess it was natural that I got the disease.... Looking back over it now, I guess I got this Minamata disease because I was greedy.... Why did I come here, to Minamata, I wonder. Why did I come here...?[11]

While pollution devastated the bodies of the victims, it also corroded the family structure. Victims' families often suffered a sharp decrease in earnings due to the destruction of their fishing grounds or farming areas. To pay the accumulating medical bills, they had to sell their farms, houses, fishing boats, and nets. Many were forced to go on welfare. Because of these stresses some couples were divorced, while other patients fled from their homes, leaving spouses and children behind. Initially the victims turned their anger against themselves; they withdrew from their communities. Politically impotent, they became even more alone.

2.2   Polluters' Attitudes and the Response of the Authorities

The transformation in values can be attributed only partially to the victims' physical suffering, the disintegration of their families, and their estrangement from the community. Had the polluters and the authorities directly addressed these unfortunates' calamities with kindness, had they quickly offered relief and solace, subsequent events might have been different. It was the polluter's callous indifference that first sparked the changes we will shortly describe.

Arrogance dominated all better human qualities. In Minamata, Chisso's arrogance was engrained, historically and institutionally.[12] Minamata had been a company town since 1907; and the city's economic prospects rose and fell on the financial vicissitudes of Chisso, for whom the majority of Minamata's laborers worked. Chisso's founder, Jun Noguchi, expressed an attitude toward his dependents that continued into the postwar era: "Treat the workers as cows and horses," he instructed his company's officials.[13] In their dealings with less skilled workers or members of the Minamata community, Chisso's managerial staff, many of whom were drawn from the elite ranks of Tokyo University, displayed a similarly cultivated arrogance.

Arrogance bred indifference and licensed the polluting companies' resistance to the victims' appeals. When confronted by the itai-itai victims' protests, officials of Mitsui Mining Company would say: "Go ahead and petition the government. We could mobilize forty or fifty Diet members for our defense at any time."[14] During negotiations with the victims, Mitsui officials also maintained a high-handed attitude, declaiming: "There are some forty Diet members seated in the House with the direct help of the world famous Mitsui—over one hundred if you include the indirect help of Mitsui. The Ministry of Health and Welfare will thus not cast even a little doubt on Mitsui."[15]

The posture of Chisso was more belligerent. Company toughs would stand guard before the gates, physically obstructing the victims' efforts to discuss their illnesses with company officials. On one occasion a Kumamoto University research team was physically prevented from taking samples of the factory's effluent.[16] On another occasion, the so-called Goi incident, Eugene Smith, a famous photographer, was badly beaten.

The most frustrating part of the victims' ordeal was the polluters' endless denials of responsibility and their legalistic strategies to frustrate relief. The most famous example of this kind of conduct was Chisso's "settlement" with the Minamata victims where the company extracted a release against all future claims upon payment of a pittance.[17] Outmaneuvered by the companies, the victims often turned to local government for help. But local government remained implacable for years, even at times siding with the polluters.[18]

2.3   Emerging Values

Were Chisso, the other companies, or the government morally obligated to aid those afflicted by pollution? This was the moral question the pollution cases presented. Although Buddhist teachings might have counseled mercy, compassion, and charity, these fragile virtues could not endure against the savage early post-war drive for economic recovery. The polluters surely would not have been deemed subject to any special moral obligation for care or concern for the victims had their actions not created a special obligation.

Yet a large part of Japanese society soon came to regard the polluters' conduct as fundamentally immoral. This was due less to the fact that Chisso and the others had actually caused the victims' injuries[19] than to the polluters' conduct after the diseases' discovery—the companies' arrogance, the opportunity they had of alleviating the victims' suffering and their refusal to do so, their maneuverings, obfuscations, and attempts to frustrate adequate relief and a fair determination of legal responsibility.

The victims expressed their grievances and grief in highly ritualistic and symbolic ways. One example is the "one share movement."[20] Disappointed that the trial proceedings had not permitted them to confront Chisso's executives, a group of Minamata victims decided to buy shares of the company's stock in order to gain access to its shareholders' meeting. Within the first two months of the campaign, 5,000 people had each bought one share. Thereafter, the victims dressed in the traditional white raiments used in times of pilgrimage and marched in a solemn column from Minamata to Osaka where the shareholders' meeting was being held.[21] In Osaka, the victims entered the meeting and, surrounded by supporters, chanted, "We don't want your money. We want you all, one by one, to drink the mercury-filled water." The incident was widely reported in the media and focused national attention on the victims' grievances.

The emerging values had several further dimensions. They demanded respect for human dignity and the sanctity of human life; and they attacked the shibboleth of economic growth and later the rampant materialism that supported it. Because the polluters had for many years manipulated the uncertainties of scientific proof to their own advantage,[22] the victims also came to view all science and technology with suspicion. Like the irreversible injuries they bore, the values forged from their suffering were held to be immutable—never to be compromised, traded off, monetized.[23] Belief in the moral correctness of their cause also helped ward off the bids of politicians and others with alien ideologies or creeds. The victims kept their convictions essentially pure.[24]

2.4   The Victims' Movement and the Organized Bar

The victims' isolation lasted several years. But gradually other residents, women's groups, labor unions, politicians, journalists, grade school teachers, medical and other scientists, legal specialists, and, on occasion, whole townships[25] began to rally to the victims' aid. The campaigns that grew from the interaction of these various groups constituted the victims' movements.[26]

The movements differed in character and organization depending on circumstance. Central, of course, were the victims. In some areas such as Minamata, however, the victims' groups were divided into factions, principally over tactics. Historically, three factions were the most influential: the Mediation faction (Ichinin-ha), consisting of eighty-nine officially certified victims; the pro-lawsuit faction (Soshō-ha), consisting of twenty-nine certified victims; and the Independent Negotiation faction (Jishu Kōshō-ha), which originally was composed of 101 "new" uncertified victims who finally received certification in October 1971. Victims in another three groups were united principally by their recent certification. These factions included the New Mutual Aid Society (Shin Gojokai) with 232 victims, the Peace Society (Heiwakai) with thirty-nine victims, and the Society of Victims (Higaisha no kai) with ninety-four.[27] There were also other factions in Toyama, the site of itai-itai disease, and in Yokkaichi.

Various auxiliary groups clustered around the victims providing financial, technical, political and other kinds of support. Yokkaichi affords a good example.[28] Unlike Minamata City in Kumamoto where the labor unions sided with Chisso against the victims, some unions in Yokkaichi were generally provictim.[29] For example, the Mie Prefectural Chemical Industrial Union Council itself became concerned about its own workers' health after a union survey revealed that as many as two-thirds of its members had been suffering from pollution-related diseases, with 40% displaying physical abnormalities. The union offered valuable assistance to the victims at least during the early stages of preparation of the litigation.

Various community groups also provided various forms of assistance. One important service was ad hoc "citizens' pollution schools." In Isozu more than thirty patients participated in the school. The school's primary purpose was to arouse the residents through group enlightenment sessions to challenge the companies. During the Yokkaichi trial, the Isozu schools passed several resolutions opposing the expansion of the facilities of the Shōwa petroleum company. Another school was established by the Yokkaichi Patients' Association to oppose Mitsubishi Petrochemical's plans to construct a factory in the Kawajiri District of the Kawaharada area. Together with other citizen associations, the coalition lobbied forcefully against the factory, distributed handbills to all households in the district, and attempted to refute company propaganda.

There were often violent confrontations between the victims and their supporters and the companies and the local authorities. One episode was the Yokkaichi Anti-War Youth Association's agitation in the industrial complex. As in the earlier Minamata demonstrations, riot police, plainclothesmen, and factory vigilante committees clashed with the marchers. The marchers, overwhelmingly outnumbered, eventually withdrew.

Although many people attempted to provide assistance, the contribution of the lawyers to the shaping, transmission, and inculcation of the movements' attitudes, positions, and values was perhaps most significant. The victims' lawyers came from diverse professional backgrounds.[30] Some were labor lawyers associated with the Socialist labor unions and many were deeply committed to Socialist political ideology; others maintained a general practice and were politically neutral. Some lawyers viewed vindication of the victims' claims as a human rights issue, whereas others offered assistance simply because they were appalled by the pollution in Yokkaichi and other areas. The different professional backgrounds of the victims' lawyers necessarily influenced their attitudes toward the victims, the movement, and other lawyers. And at times their differences bred conflicts.

The most sensitive area involved the definition of the appropriate professional role. Some lawyers urged detachment, viewing themselves as "specialists," faithful first to their profession, committed to a "scientific" analysis of the facts and to the technical needs of litigation.[31] Teruo Tomishima, one of the victims' lawyers in the Yokkaichi case, provided a glimpse into this sentiment:

There are those who understand the antipollution movement as an ideological struggle, but a pollution case can never be an ideological problem. Rather it should be thought of in terms of dealing with a concrete infringement of a person's life and health, that is, as a strict question of human rights. It is with that assumption, which corresponds to both my personal motivation in becoming a lawyer and the purpose of pollution litigation, that I joined the lawyers' group.[32]

The "independent" professional, however, was decidedly in the minority. In many ways these lawyers' self-image of independence conflicted fundamentally with the victims' moral fervor. After all, the excuse of "objectivity" and scientific detachment had long been used to frustrate the victims' efforts to obtain relief. Such arguments could not be trusted. The lawyers' insistence on neutrality after the abominations the victims had endured appeared to the victims as being of questionable morality.

Most lawyers participating in the pollution cases soon became intensely emotionally involved with the victims and the movement. Tadataka Kondō, founder of the lawyers' team in the Toyama case, moved from Tokyo to Toyama to experience personally the agonies of his clients and to assemble a group of plaintiffs.[33] During this period he devoted himself entirely to the case while his wife supported him. Other lawyers also visited the affected Toyama area.[34]

Many victims' lawyers perceived their professional responsibilities as extending beyond mere representation of their clients' interests. An additional objective was to attack the infirmities of the entire capitalist system. Katsuhiko Bandō, the founder of the lawyers' team in the Niigata case, noted:

Our most fundamental motivation was the clarification of the cause and responsibility [for pollution], that is, to unravel exactly what the intrinsic nature of the repeated outbreaks of pollution actually was. That was the point from which we proceeded. . . . The idea was that the root cause, the essence of pollution was the same as the cause of the problems that the workers were confronting: the labor accidents, the work related diseases, bad working conditions and automation. We came to realize that the core of the effort to eradicate pollution had to be the working class. That realization and the Niigata struggle based on it greatly enhanced the meaning of the final trial session and the judgment itself. [35]

In Yokkaichi the victims' lawyers stressed the needs of the movement and downplayed the importance of ideological considerations. Hiroshi Noro, one of the most active advisers to the Yokkaichi victims, expressed this attitude:

. . . at the time I also felt that a proper understanding of the meaning and significance of initiating litigation was essential in resolving the problem. As Mr. Kitamura said, my decision to try litigating this suit also grew out of a feeling of the suit as part of a movement. That movement, however, was still sort of vague, revolving around the actions of the local governments, the unions, or other group activity. It hadn't yet taken the form of individual victims raising their voices to accuse the responsible companies and government, when the individual patients decided to stand up to the companies one by one and demand that they face up to their responsibilities; this action, well, I think it took the movement one step higher, don't you? [36]

Thus, Noro argued that a lawyer's performance should be judged not only on whether the victims ultimately obtained compensation. Of equal importance was the lawyer's contribution to the political aspects of the movement:

[Explaining why he felt there should have been more contact with the plaintiffs.] After all, this suit wasn't conducted like a pure lawsuit but as one link in the antipollution movement, as the leading edge of the movement. If one thinks in this way, that is, that the litigation itself is part of the people's movement, one has to admit that the work of the lawyers in this suit was a little too isolated. Of course, some would come to Yokkaichi and go to various meetings and get-to-gethers or attend workers' or citizens' study groups, but it was only a limited number of lawyers in the group. . . . Most didn't use the opportunity to become an intimate part of the people's movement. [37]

Noro also urged complete immersion in the movement, for only then might a lawyer effectively represent the clients' interests:

Thus, as I said, we used the method of giving each person a separate aspect of the case to work on, but in retrospect that was a mistake, as we can see from the insufficient contact with and understanding of the plaintiffs and their fellow inhabitants.

In short, this case was an extremely technical trial from the beginning. Because a high level of nonlegal expertise was required and the lawyers' energy was channelled in that direction, the lawyers themselves couldn't express as their own experience the suffering and experience of the plaintiffs' lives. . . . even though it was essential, they couldn't do it adequately. Because of that, sometimes when the plaintiffs themselves were attacked by the other residents in the area as being mercenary and selfish, or they had similar simple problems, the lawyers couldn't get together with them and resolve the problems. Also, although of course they went to the site of the pollution and engaged in various negotiations, the lawyers did not establish an adequate level of mutual understanding with the inhabitants of areas other than Isotzu. My feeling is that this problem is due to the specialized division of tasks.

Therefore, in the next suit like this, whatever happens, the first rule should be to have the lawyers fully experience and appreciate the life of the plaintiffs and other residents. [38]

These attitudes suggest the kinds of conflicts environmental lawyers experienced in working on the antipollution cases—conflicts in allegiance to clients, one's profession, and the broader goal of the society's reform. Although many of these conflicts were subordinated at the time of litigation to the immediate goal of a court victory, such tensions continue in antipollution litigation today and are still largely unresolved. [39]

Section 3   The Victims Turn to the Courts

Two related questions must now be addressed. First, why did traditional avenues of redress—the political process, bureaucracy, extrajudicial dispute settlement—fail to provide the victims with adequate relief? Second, why did the victims turn to the courts? Specifically, what role could the courts perform that these other institutions could not? How did the victims themselves view the judiciary?

3.1   The Breakdown of Traditional Institutions

The time-honored method of settling pollution-related and other disputes had been mediation. Even as the victims' protests intensified in Minamata and elsewhere, the natural tendency of the companies, the authorities, and some victims was to trust that mediation would resolve the conflict. Gradually, however, the victims became dissatisfied with mediation, and in the late 1960s they rejected it and turned to the courts.

A principal objection was that mediation was proving impractical. Often after a lengthy negotiation, the victims would achieve only a nominal settlement. [40] And it was virtually impossible through mediation to persuade the companies to reduce or terminate their discharges.

More disturbing, however, was the polluters' abuse of mediation, for a polluter like Chisso could easily gain control of the negotiations, effectively dictating the terms of settlement. The famous mimaikin (sympathy money) contract that Chisso extracted from the victims is a good illustration. As already noted (n. 17, this chapter), the contract stipulated that the company would be entitled to terminate payments if it were ever determined that Chisso was not the source of Minamata's pollution; and irrespective of any finding of Chisso's ultimate responsibility, the victims would make no further claim for compensation. Unaware of the impropriety of this perfidious contract, for ten years the victims remained in their isolated villages unwilling to challenge the agreement.

Lacking financial and other resources, the victims often did not have an opportunity to influence the selection of the mediator(s). For example, as late as 1967, well after the government had officially recognized mercury as the cause of Minamata disease, relying on the mimaikin agreement, the company still refused to negotiate. Even at this late hour the victims resorted to mediation, petitioning the central government to intercede. The Ministry of Health and Welfare agreed to mediate; but before determining who should serve on the mediation committee, the ministry asked both parties to submit a written promise that they would

37

comply with any result reached by the mediators.[41] Chisso promptly tendered a written pledge. The issue, however, divided the victims. One faction, known as the "leave it up to other people" group, signed the agreement. Another group, preferring direct negotiations with the company, refused to sign. This group thereafter began litigation. The group that had relied on the government's appointment of a mediation committee later obtained a nominal amount of compensation.[42] Not surprisingly, the victims came to view mediation with an abiding, bitter distrust.

As they gained political strength, the victims began to view the very philosophy underlying mediation as antipathetical to the vitality of their cause. Where mediation strove for harmony and mutual concession, the victims sought a final confrontation; where mediation sought to preserve community integrity, for most victims all sense of community had already been destroyed. Traditionally mediation avoided an adjudication on the issues of responsibility or causation of injury, for this determination would support the cause of one party to the disadvantage of another, a result that would permit discord to continue. This is precisely what the victims' movement demanded, and what traditional mediation would not give. This is why mediation served the polluters' cause so well.

Other possible avenues of relief, the political process and the bureaucracy, appeared less hopeful than mediation. Most victims viewed the gambits of the political parties with profound distrust:

[*After describing how Maekawa, a Socialist city councilman, had organized a supporting association, contacted a group of Socialist labor lawyers, and visited the local hospital to organize the victim patients into a possible group of plaintiffs.*] *Then, all twenty-four of us in the Shiohama Hospital got together with our families, about fifty people total, and talked about our disease symptoms and other circumstances. But around that time the Communist and Socialist parties were creating a lot of propaganda without really doing anything for us. Some people said they were just trying to arouse public opinion and then would discard us when they didn't need us any longer. Because of that, a lot of people hesitated and started to leave the group in ones and twos, saying their son works at the konbināto [industrial estate] or their father, things like that.*

[*In response to the question of why he persevered when the others were getting out.*] *When the nine of us made the jump, we felt that whatever we did, it would be bad. There had been all sorts of groups, but they were always run by people in it for their own selfish gains. We had even blocked the discharge pipe with our own hands and still nothing! We were really afraid of the [Japan Communist party] and [Japan Socialist party] organizations. To rely on them was . . . , well, our families were really opposed. But we felt that if they could use us, we could use them; if they could throw us aside, well, we could throw them aside too. . . . If we lose, we're just back where we started; if we win, maybe we'll get something out of it. That's the way we felt.*[43]

The victims considered the assistance of politicians to be of limited value for other reasons. First, few politicians could marshall immediate effective relief. Since few members of the incumbent Liberal Democratic party supported the victims, the only major source of aid was the opposition parties. But the opposition lacked power to press needed remedial legislation through the Diet. As late as 1969 the only major legislation directly benefiting the victims was the Law for the Relief of Pollution-Induced Injury, which proved to be of only limited value.[44] Second, the political process could not easily compel a final adjudication on the issues of responsibility and causation that the victims believed was necessary to establish indelibly the righteousness of their cause.

The victims' suspicion of the bureaucracy stemmed from other factors. The opposition parties' assistance had been essentially ineffective. To the victims, however, the central bureaucracy (particularly officials in MHW and MITI) was basically unconcerned, inaccessible, and in league with industry. The historical record confirms that MITI officials militated against official recognition of the victims' diseases, assignation of the polluters' responsibility, and even an official determination of the diseases' medical cause.[45] The government's ponderous efforts in the mid-1960s to formulate comprehensive legislation for pollution control for most victims came approximately ten years too late.

3.2   The Victims' Perceptions of the Judiciary

Harsh experience caused many victims to mistrust mediation, politicians, and the bureaucracy. However, litigation represented a new, unfamiliar, and even forbidding course. Yet on balance the courts were viewed to be still uncompromised, offering at least a chance of immediate, unequivocal relief. Indeed for most victims, litigation was all that was left.

From the outset, the victims considered litigation to be a group effort. Once the group decided to sue, it was felt that all members should join, for the injuries involved were suffered not only by individuals; they were the collective afflictions of entire communities.

One Yokkaichi patient expressed the feeling this way:

In Isotzu, as for the residents' way of thinking, everyone wants to do only what the others do. In whatever meeting, it's everyone together, following the group. So in the case of pollution damage, too, everyone must act together. If an antipollution suit is started, the whole group has to do it. This consciousness is very strong. Group unity forms very quickly, but it's a different story when it's a question of one of them stepping forward to take some positive action himself.[46]

At times the group would even become fiercely hostile to the individual who disregarded its decision not to sue.[47]

Because I knew this [that the SO_x level would cause great hardship], I told Mr. Fujita that, for the sake of Isotzu, he had to let them say whatever they wanted. At the beginning, they called him a traitor and wanted to ostracize him. Even so, I encouraged Mr. Fujita and told him that I would stick with him to the end. . . .

As for who was suffering . . . everyone in Isotzu was, but their way of thinking was, "Why should I support some trial so they can make money off it?" The Patients' Association was different, however. They would face up to that kind of talk: "What are you saying? Who do you think you have to thank for the higher smokestacks? . . . and the soot collectors? . . . for all the equipment the companies have installed? It wasn't until those nine people [there were nine plaintiffs] came along that the companies woke up, that the stacks were raised. Now, was it? . . .

Back in 1963 you would wash something, hang it out to dry and it would be black immediately. You would wash it again, and a third time . . . and yet we still shut up and took it, didn't we? "For the sake of the country," "for the sake of industrial expansion." . . . There's nothing we can do. No one said anything. Even now you hear, "Those nine people know there's something in it for themselves"—that kind of thinking still exists.[48]

Those who opposed the group believed their action to be a kind of ritual sacrifice for the benefit of their communities.

In my village people felt, "Why should we do any work for you?" In general the first problem we had in relation to filing suit was the feeling that "this is our burden but our lives are already past—we should just endure it." But when it looked like the precious land left by our ancestors might be encroached upon and our grandchildren's generation affected, we could no longer endure. "Now, we must sacrifice ourselves!" became our cry. I think this trial was motivated by that attitude.

Well, anyway, even today that kind of mistaken consciousness [that the trial was selfishly motivated] exists. People often ask me why my wife never comes to court, but I'm afraid that if it becomes a "family trial," then that consciousness of it just being a single family against the konbināto will continue. And I think that's wrong. [49]

Virtually all the victims considered a final personal confrontation with the polluters to be the overriding purpose of litigation. They wanted the courts to make a public spectacle of the wrongs they had so long endured, and proclaim to all the world the moral and legal responsibility of their oppressors.[50] For almost all the victims, winning monetary compensation was only of secondary importance. When Mitsui delivered ¥66 million to the victims at midnight on the day of the court's decision, the victims and their lawyers reportedly received the money in dead silence, with a sense of futility.[51]

The impersonality of the court proceedings mortified the victims as did the companies' presidents' manipulation of the rules of evidence to avoid appearing in court. Eiji Ono, one of the journalists who joined the associations supporting the Yokkaichi litigation, described the reaction of the plaintiffs as they sat through the first trial session:

The oral pleading began. The trial had several surprises in store for those of us who hadn't seen one before. First was the procedure called the adversary system. It started with the physical arrangements of the courtroom. The two groups of lawyers sat opposite each other on either side of the bench. The plaintiffs themselves sat in the gallery in back. As for the defendants' side, no one even showed up—there was just their lawyers lined up looking professional.

The significance the patients attached to this suit—it wasn't just money, just the compensation—it was to make the presidents of the companies that had inflicted this illness on them say just one word, "I'm sorry." The adversary system pretty completely shattered this hope. At the very instant that the plaintiffs' lawyers were denouncing the defendants' crimes, at the very instant that the patients were making their embittered appeal, "Mr. President" was sitting in his nice, deep office sofa, not in court! Is money an excuse for even this?

There was another problem for the plaintiffs' side. Of course no plaintiff can talk as well as a professional advocate. But it's also a fact that, no matter how talented the lawyer is, he can't relate 100 percent the feeling of the patients themselves. There's the opportunity to testify directly, but even then, the person himself is just asked questions, remains a "guest." Sometimes complicated expressions flickered across the patients' faces as they sat in the gallery listening to the give and take of the courtroom, Why must this court thing be so far removed from the common people?

Of course, the suit had to be won, nonetheless, and the defendant companies defeated even if within the court rules, but as long as it was done in this phony way, there would be no real victory. As long as the trial remained in this framework, "Mr. President" would be able to hire his lawyers to make his excuses for him; even if he loses, it won't be anything more than money. The guilty conscience, the pain he should feel as a human being, the recrimination, he will escape it all. [52]

As the trial proceeded the anxieties of the victims mounted over the endless, seemingly mindless ritual.

. . . we were made extremely anxious by the "formalism." To take a living phenomenon like pollution, turn it into a series of documents, then discuss those documents while people are actually suffering and dying seems somehow arrogant or disrespectful. The plaintiffs' side would say, "It's terrible." The defendants' side would say, "It's not terrible." The judge would strike a balance and write his opinion. This is nothing more than bargaining with the plaintiffs' appeal. . . . [53]

The feeling that it was really slow was very strong. Such a lot of boring formality and verbosity—a trial for something that was so clear, so obvious from the beginning! If you went to the courtroom and watched, well, there'd be scholars lining up and spouting a bunch of meaningless jargon, really stupid stuff! You wanted to say, "Stop! Stop! I can't stand it!" [54]

The victims' pleas were only partially heeded. The courts were not about to countenance a personal confrontation between victims and polluters, modify judicial protocol or procedures, or abandon traditional doctrine. On the other hand, the judges were deeply and favorably impressed by the merits of the victims' cases, a conviction to which they gave full expression in four sweeping decisions between June 1971 and March 1973. [55]

Section 4   The Judicial Contribution

The courts have contributed to the development of the new values essentially in three ways. First, their decisions transformed the victims' moral outrage into specific legal doctrine. Second, because of the media's comprehensive coverage of the trials, the movements' attitudes, goals, and motivating values were quickly and effectively transmitted to the entire nation. Finally, the courts' decisions profoundly influenced the subsequent conduct of the victims, industry, bureaucracy, and the Diet.

The doctrinal contribution of the courts' decisions in the four pollution trials can be briefly summarized. In finding the defendants negligent, the courts came as close as was possible to imposing strict liability without explicitly deciding this issue. The decisions noted that because their operations had exposed the public to grave health risks the defendant had the responsibility of: using the best available analytical techniques to detect and measure toxic or hazardous substances in their industrial processes; conducting a program of continuous research on the possible hazardous effects of substances whose risks were unknown; assessing the health risks of hazardous substances to exposed populations before releasing the substances into the environment; conducting appropriate environmental assessments prior to industrial plant site selection; monitoring continuously the effects of such substances on the environment by using the most advanced technology; using the best available technology to control these substances; and interrupting, and, if necessary, terminating these operations to avoid jeopardizing human health.

While describing the responsibilities of enterprises as broadly as was doctrinally permissible, the courts also fashioned an extremely flexible standard for proof of causation. The judges' opinions relied on the plaintiffs' presentation of statistical

41

(epidemiological) evidence and experimental data, especially when proof of the diseases' etiology based on clinical findings in individual patients was difficult. The courts shifted the burden of presenting evidence to the defendants when plaintiffs successfully established a prima facie case, and when medical proof of causation was held to be extremely difficult.

The courts' opinions contributed to the society's acceptance of the movement's values in several respects. First, the decisions set forth an official position on the key concepts of responsibility and causation. Second, the opinions refined these notions, for in the context of the movement they had served more as political slogans than as analytical, operational concepts. Third, the courts' decisions gave legitimacy to the values propounded by the movement by enshrining them as law. Finally, because the trials quickly became a national cause célèbre, they drew attention to the critical issues in the cases. The subsequent scholarly and journalistic debate over the courts' opinions disseminated the values of the movement and ultimately facilitated their wider acceptance.

The victims reacted to the courts' decisions differently. The fact that the sovereign (okami)[56] had at last moved to address the injustices done was for most victims of greatest significance. This was deemed more important than the awarded damages which many victims believed to be grossly inadequate. Virtually all the victims, however, considered the decisions to be no more than a transitional episode. The decisions merely demarcated the end of one era of the victims' interminable struggle with the polluters and the beginning of another.

After each of the court decisions the victims demanded that the presidents of the offending companies make a public admission of guilt and tender an apology. The victims sought the ultimate expression of penitence—that the president kneel before them and bow his head to the floor. The victims' demands were not idle vindictiveness. The act of supplication in a traditional sense symbolized ultimate acceptance of what was done.[57] This was necessary for the beginning of any forgiveness. The courts' judgments that the polluters were legally responsible served as weapons for the victims to compel the companies to accept their social responsibility. In this way the victims translated the courts' decisions into traditional terms.[58]

The court victories also yielded more mundane dividends. Immediately following the announcement of each verdict, the victims would initiate direct negotiations with the companies, using the courts' decisions as a basis for extracting further concessions. The agreements resulting from these negotiations often included a clause expressing the companies' full acceptance of responsibility for the victims' suffering, various stipulations to introduce pollution controls, and the companies' promise to extend compensation to victims not participating in the litigation.

The aftermath of the Niigata mercury poisoning case offers a good example. In Niigata, the court awarded ¥10 million in compensatory damages to the deceased and severely injured, but it reduced awarded damages to the less severely afflicted victims. In total the court granted only ¥270 million of the ¥502 million claimed by the plaintiffs. In March 1972, six months after the decisions, the Niigata Minamata Disease Action Coordination Council, composed of the Society of Victims and the team of plaintiffs' lawyers, began a series of negotiations with the Shōwa Denkō Company. The council demanded annuity payments for all patients and a ¥15 million lump sum payment for newly certified patients. These negotiations dragged on for fifteen months. In July 1972, however, the

Yokkaichi court awarded ¥14 million to the most severely injured victims, and in March 1973 the Kumamoto Minamata court granted ¥18 million to a similar group of victims of Minamata disease. The public outcry aroused by these decisions had a great influence on Shōwa Denkō officials who quickly accepted virtually all demands made by the council.[59] In addition, Shōwa Denkō promised to divulge all information needed to determine further relief measures for the victims, pay ¥15 million to the deceased and the severely injured, and tender ¥10 million to all other certified patients.[60]

Victims of itai-itai disease and the Yokkaichi illnesses were also able to negotiate similar concessions. The itai-itai victims' group prevailed on Mitsui to compensate those victims who had not participated in the litigation at the same level as the courts' award and to promise to restore the contaminated soil. Similarly, through negotiation the Yokkaichi victims persuaded the six defendant companies to compensate all certified victims in the Isozu District and permit their unrestricted access to and inspection of the facilities. The same group negotiated a plan of action for the control of industrial pollution in the area with the governor of Mie Prefecture.

While the victims successfully employed litigation for traditional ends, the polluters were unable to escape tradition's grip. After the announcement of the courts' decisions, the polluters soon realized that mere compliance with the courts' order to pay compensation would not satisfy the social sense of these judgments. Unlike the usual trials, these decisions constituted a moral verdict that required more. The companies' clearest, most demonstrable, unavoidable response was to waive their right of appeal.[61]

Their decision can be construed in various ways. Practical considerations, of course, militated against an appeal since there was little chance of reversal. Yet social and cultural factors were also important. After the courts' announcements of their decisions, most of the society would have deemed as immoral any polluter's effort to postpone assistance to the victims. By this time company officials were also becoming extremely sensitive to media criticism and public ridicule.

It is possible to give the companies' decision a more abstract interpretation. The conflict between the victims and the polluters in a sense reflected a tension between industrialization and the resistance of a traditional way of life to industrialization. The reaction of the companies to the victims' pleas was "modern," technologically advanced, westernized, objective, and impersonal. The victims' response was essentially traditional, based on personal trust, unsophisticated, and intuitive. From this perspective, the polluters' reliance on legal maneuvers—the unconscionable mimaikin clauses and their insistence on an "official" adjudication of the issues of responsibility and causation—is easily understood for this was the objective, external, impersonal code that had sustained, facilitated, and legitimated all industrial development since the Meiji era. In this light the polluters' waiver of the legal right to appeal symbolized a concession to the continuing force of tradition. It was a compromise between an essentially Western legalistic orientation and the traditional imperative. Chisso's president's ritual supplication before the victims may also be viewed as metaphor of this process of adjustment.

It is extremely difficult to assess the extent to which the courts' decisions have actually influenced the long-term pollution control efforts of industry. In Yokkaichi, Shōseki, one of the defendants, canceled long-standing plans to expand its plant.[62] Also as a result of the Minamata decisions, MITI initiated administrative

43

guidance[63] urging Japan's soda manufacturers to convert their plants from a mercury-based process to the diaphragm (non-mercury) cell method. The new process was more expensive, and the industry alleged that it placed Japanese manufacturers in a weak competitive position internationally. Less clear is the extent to which the marked increase in pollution control investments by petrochemical and other industries since 1973 can be attributed to the courts' decisions. The judgments in the pollution trials are perhaps best viewed as having established a baseline for the legal and moral conduct of industry. It is now less likely that a Japanese company would repeat Chisso's flagrant disregard for human health, at least within Japan.[64] And it is certain that the decisions in the four pollution trials, along with changing popular attitudes toward the environment, have motivated some firms to become (or at least to be concerned to be viewed as) good environmental citizens.[65] Beyond this it is difficult to trace the influence of the decisions. As in the United States, it is uncertain whether whole industries will collectively continue to clean up their operations simply because of "social responsibility," especially in these times of economic recession and uncertain energy supply.

The pollution trials in combination with other factors[66] influenced bureaucratic decision making. This effect is most clearly manifested in the area of legislation. During the period of greatest national ferment over the trials, the government promoted the enactment of the strict liability amendments to the air and water pollution control laws.[67] The Pollution-Related Health Damage Compensation Law[68] and the Factory Siting Law[69] were passed as a legislative response to the Yokkaichi decision.

The courts' decisions were influential in other, more subtle ways. In ch. 3 through 6 we trace the change in the administration's attitude toward the handling of scientific data. The administration began to develop more sophisticated tools for risk analysis and became more willing to initiate incisive action on the basis of an incomplete, inconclusive, or otherwise uncertain scientific record. This change in administrative practice is most apparent in the practice of setting standards. In some areas the government has established stringent standards basing its decision on a variety of factors independent of scientific certainty or economic consequence.[70]

The courts' substantial contribution in the four pollution trials poses the question of why the courts acted at the time they did. What forces can be attributed to this dramatic judicial entry into the field of environmental policy-making? The simplest explanation is the judges' personal motives. For the most part the judges were young, progressive, and clearly sympathetic to the plight of the victims. Further, they were subjected to intense public pressure. It is easy to conclude that the courts simply capitulated. Yet this interpretation does not seem warranted because of the judges' repeated assertions that they did not let such factors dominate their fair consideration of the issues, and also because of the existence of other, historic factors.[71]

From an institutional perspective, there was a judicial willingness to depart from rigid conceptualistic reasoning and to serve instrumentalist ends even in the prewar period.[72] Certainly the postwar constitutional imperative for judicial review has forced the courts to consider many broad problems of public policy. Although less dramatic, many judicial decisions predating the four pollution trials in the products liability, medical malpractice, automobile injury, consumer protection, and civil rights fields suggest a postwar trend toward a more active, affirmative judicial role.[73]

Finally, the common law lawyer should be reminded that the courts in the pollution cases were only interpreting long-standing, exhaustively explicated, statutory provisions. There was no need to seek new remedies from the whole body of judicial precedent. Indeed, the judges in these cases always insisted that their decisions respected the well-trodden doctrinal principles of the prewar era.

Section 5   The Citizens' Movement and Its Impact on the Legal System

We have described the development of the victims' movement. Now let us turn to the separate but related[74] protests by local residents against industrial expansion in the early 1960s.

Like the victims' movement, the citizens' campaigns grew from a series of ad hoc public reactions to environmentally disruptive developments. And like the victims' movement, the citizens' movement has continued to remain decentralized, responding to a given crisis, then disbanding and regrouping at a later time.[75] The number of these groups far exceeds the victims' associations, and their membership is more complex. For example, most of those victimized by pollution diseases were poor farmers and fishermen; citizens' groups now include a broad cross section of urban communities, heterogeneous in occupation, political affiliation, education, financial means, and social class.

Perhaps the greatest distinction between the two movements lies in the values that motivate them. The victims acted primarily from moral outrage, demanding compensation in the monetary and social sense, whereas the antipollution residents' associations seek to protect the integrity of the local community. Their efforts have enshrined the virtues of "smallness," rural simplicity, naturalism,[76] and antimaterialism that the earlier victims' protests may have shared but failed to articulate. Also, behind virtually all the citizens' campaigns lies a deep distrust of the central government, particularly projects sponsored by the government in the name of the national or "public" interest.[77] Indeed most local communities have come to view the concept of "public interest" as simply a pretext by which the giant corporations and urban elites extract sacrifices from local communities for their own benefit.[78]

Most scholars trace the origin of the citizens' movement to the successful campaign waged during 1963 and 1964 against the proposed construction of a petrochemical industrial complex in the townships of Mishima and Numazu.[79] The controversy began when the prefectural government, a strong proponent of the project, suggested a merger of the two cities in order to expedite the construction and administration of the facility. By this time, however, the Ministry of International Trade and Industry had already approved blueprints for the project and a public investment of ¥235.5 billion. On hearing news of the proposed merger, local citizens of Mishima formed a "council" to coordinate neighborhood groups, the labor unions, and the political parties. The council presented the prefectural governor with a protest in the form of a series of questions, and demanded an immediate reply:

1. What would be some of the adverse environmental and other consequences of the development on the area?
2. How large would the projected increase in revenues from the development be, and how would these revenues be used for the welfare of the residents?

3. How would cities like Mishima, located downwind of the proposed site, be affected by the activities in the konbināto?

4. How much water was to be used for industrial purposes, and what effects would use of the water have on agricultural irrigation?

5. What provision had been made for farmers who would be forced to abandon their farmland?

6. Was there any possibility that pollution produced by the development would adversely affect work at the National Institute for Genetic Research in Mishima?

The governor's response was evasive:

1. The total amount of local tax revenues would remain unchanged.

2. Pollution would be kept to a minimum. (The governor insensitively noted that even a small public bath house could produce enough SO_x to pollute the air.)

3. A study of the appropriate allocation of water resources would be undertaken in the future.

4. A development on the side of Mt. Hakone would be contemplated for relocating farmers.

Most residents deemed the governor's response to be patently inadequate. The council acted quickly, organizing several bus tours of pollution sites for prominent groups, among whom were members of the chamber of commerce, landowners, the local staff of the Liberal Democratic party branch office, an association of inns and hotels, and several civic and neighborhood associations. These tours had a profound public impact. Farmers owning land within the proposed site refused to sell their farms, doctors and high school teachers organized and conducted several hundred teach-ins, and a series of mass rallies and demonstrations were inaugurated. Mishima's mayor, at last succumbing to this pressure, issued a public statement expressing his opposition to the project.

Thereafter the movement spread to Numazu. In Numazu the council arranged a public debate between a team of meteorological and other specialists working for the residents' group, and some government investigators dispatched to the area by the Ministry of Health and Welfare and MITI. Armed with data collected by the "movement," the residents' specialist team presented a convincing picture of the adverse environmental and social effects of the proposed development. As a result, thirty-six city councilmen in Numazu declared their opposition to the project, an event that prompted another mass rally of over 25,000 people. Under the threat of public recall, the Mayor of Numazu capitulated and published a statement confirming his opposition to the project.

Many subsequent citizens' campaigns have sought to imitate the Mishima-Numazu prototype with varying degrees of success. The central, most recurrent issue has been the government's insularity and insensitivity to local feeling. In many instances local governments have acted in league with the national government or with industry in the promotion of a particular industrial development. To secure the success of these projects, the authorities have often kept all plans secret until the actual point of their implementation. Generally project promoters have not stooped to consult local residents or even notify property holders except on a pro forma basis. By the mid 1970s there were still no clear procedures for public hearings, notice or review, and local residents enjoyed little opportunity to influence the course of a given development. Indeed government officials have often stressed that local communities must accept industrial projects irrespective of their environmental and social consequences.

A good illustration of the frustrations that local residents face is the case of the construction of a giant industrial development at Tomakomai, Hokkaidō. In 1969 when the Hokkaidō Development Agency, the supervising authority, announced the inauguration of the project, 90% of the land for the proposed site had already been acquired and planning for the relocation of 5,000 people from the Yūfutsu residential district, who were caught in the middle of the complex, had already commenced. When the residents challenged the government's failure to hold a single public hearing on the project, the official agency response was that the residents would be informed of the government's intentions only after an environmental impact assessment on the plan had been completed.[80]

Since 1964 projects like Tomakomai have become the casus belli of countless confrontations between local communities and the government.[81] In turn these confrontations have profoundly influenced the attitudes of local governments to industrial growth. Where local governments previously competed aggressively to attract new industries, perceiving these industries as a lucrative additional source of property tax revenues, pollution control has often become the dominant concern.

Local communities' emerging sensitivity to environmental issues has been manifested in a variety of legal, institutional, and political forms. One common response to citizen pressure has been the passage of special ordinances. The Tokyo Metropolitan Government's Pollution Control Ordinance, for example, imposes a variety of procedural and substantive obligations on industries wishing to operate in the city. In addition, it expands the rights of citizens to obtain information held by the government and to participate in the administrative process.

A second institutional development has been the local government's practice of requiring industries to sign antipollution agreements.[82] Occasionally, residents' groups themselves negotiate these contracts directly with the applicant industries, while at other times the "treaties" are tripartite arrangements among local citizens, municipal or prefectural governments, and industry. Many agreements grant procedural rights, such as the right of inspection of polluting factories, to the residents; some establish special trust funds to compensate individuals injured by pollution. Most agreements set additional standards and impose other obligations on the applicant industries beyond those already established by law.

The campaigns of local citizenry have had other more diffuse effects. Many local governments have embarked on ambitious pollution control planning programs, and the presentation of a convincing environmental platform has become a critical component in winning an election in many areas. From industry's perspective, the emergence of a mobilized local citizenry operating in conjunction with these legal and extralegal controls has constituted a formidable obstacle to expansion, especially given the rapidly depleting supply of available and suitable factory sites.

Industry's predicament has been complicated by a further development, the sudden and dramatic increase in citizen-initiated antipollution litigation, principally since the courts' opinions in the four pollution cases.[83]

Litigation's appeal stems from a number of factors. First, the plaintiffs' victories in the pollution trials demonstrated that success was possible, despite a lawsuit's onerous financial burdens. In addition, many citizens' groups have been deeply impressed by the way in which the victims used the courts to extract further concessions from the polluters. Second, citizens' groups have used litigation to delay or prevent an industrial development. Consequently most of the "second generation" citizen suits have sought injunctive or other preventative relief. Third, litigation has served to publicize a dispute, and citizens' groups have employed the trial record in subsequent campaigns. Fourth, litigation has catalyzed new legislation. Many local ordinances have adopted the obligations and remedies estab-

47

lished by the courts. Finally, litigation has helped articulate and legitimate basic values and concerns.[84]

One expression of these concerns is the increasing number of suits alleging infringement of "environmental" and related rights. Thus citizen plaintiffs have argued that they possess a common claim to the protection of the living environment. Based on this theory, the citizens' movement has asked the courts to enjoin public and private developments in over fifty suits.[85]

Such suits have posed problems even more subtle, diverse, and complex than those raised by the four pollution trials. First, unlike the pollution disease cases, citizen groups are not claiming compensation for past injuries; rather, they assert prospective, and at times, merely supposititious harms. Second, citizens have sought comprehensive protection not only of personal but also of possessory, recreational, cultural, and even aesthetic interests. But the latter interests, the courts have long held, are not entitled to the same degree of protection. Third, the environmental rights cases have frequently required the courts to render judgments on the adequacy of local environmental and land use planning, fields in which most judges possess little expertise. Finally, because these cases have failed to inspire the same moral fervor as the victims' injury trials, the courts have not been confident that their decisions will reflect an already established national consensus.

Because of these factors, the courts have moved cautiously, balancing the movement's concerns against commercial, industrial, and other considerations. Although no court has yet unequivocally recognized a "vested" environmental right, some have concluded that a citizen "interest" in environmental protection justifies injunctive relief. And while categorically rejecting environmental rights arguments, many other decisions have developed the cognate doctrine of the "right to sunshine." Although this concept has been construed less expansively than the environmental rights theory, it has provided a basis for an extensive remedial development.

In many cases the courts have also emphasized procedural considerations as a way of reconciling the exigent demands of the environmental rights proponents with the need for balanced industrial development. The citizens' movement, perhaps sensing the need for some compromise, has also gravitated in this direction. While skirting the underlying substantive issues, the courts have recently interrupted projects because of the government's failure to provide local residents with relevant information, plan adequately, consult with local residents, explain a project's environmental and social consequences, hold formal public hearings, prepare a detailed environmental assessment, and accord citizens an opportunity of participating at a critical stage in the decision-making process.

Section 6   The Permissible Limits of Civil Disobedience

During the mid-1970s, two further court decisions addressed what perhaps was always the ultimate jurisprudential question raised by the antipollution protests: When new values begin to supplant the old, may citizens dedicated to these values violate existing law in order to establish new legal principles of conduct?

The first case was a criminal prosecution for "forcible obstruction of business"[86] against two local residents belonging to the Association for the Protection of Urado Bay. At four in the morning of June 9, 1973, the defendants dumped three cubic meters of concrete and twenty-four bags of gravel into two manholes of the drainage pipes of the plant. The defendants' action was both an exasperated pro-

test to the company's twenty-year pollution of the Enoguchi River and a final effort to control the factory's discharges. As a result, about 1000 tons of liquid waste
inundated the national highway running near Kōchi City, and the company's operations were interrupted for fifteen hours.

While admitting the substance of the indictment, the defendants raised the
novel defense that their use of force had virtually been compelled by Kōchi Pulp's
failure to heed their many appeals. Defendants argued that their acts were necessary to protect the river that supported their lives, and in view of their motives
and the insignificance of the harm caused the company, their actions did not
merit punishment.

In March 1976 Chief Judge Akira Itasaka of the Kōchi District Court found the
defendants guilty, qualifying his decision as follows. The court conceded that the
company itself had acted in an almost criminal fashion and that local health
agencies had been negligent in their efforts to control the river's pollution. The
court noted that pollution hazardous to public health and welfare should not be
protected against otherwise illegal acts. The court concluded that in this case, the
injuries suffered by the defendants had not reached crisis proportions, and thus
the defendants' use of force was impermissible under the prevailing legal order.

The second case involved the criminal prosecution of Teruo Kawamoto (age 45),
a Minamata disease victim, and a representative of the Minamata Disease Independent Negotiations faction. In 1971 Kawamoto had led a sit-in before Chisso's
headquarters in Tokyo demanding that the company negotiate more openly with
the victims. On the morning of July 19, 1972, a fight broke out on the stairway
landing of the company's fourth floor. In the shuffle, Kawamoto shoved several of
Chisso's employees and punched one man in the stomach, causing an injury requiring two weeks for recovery. Kawamoto was promptly arrested. On October 25
he was formally indicted and charged with the criminal assault of four of Chisso's
employees.

In January 1975 Chief Judge Mitsuo Funada of the Tokyo District Court found
Kawamoto guilty, providing the following explanation for his decision:

*Even if the defendant had no choice but to employ some force in direct negoti-
ations, he should have halted at a point where the legal order, under common
notions of social order, would not be violated. Acts like striking another person
exceed necessity's limits.*

On June 14, 1977, the Fourth Criminal Division of the Tokyo High Court upheld Kawamoto's appeal and dismissed the indictment. The relevant portion of
Chief Judge Shōji Terao's opinion follows.

*It is said that Minamata disease is the gravest matter in the history of pollu-
tion. Even now many people are tormented and suffering from the disease. No
words can describe their misery. The sole cause of the harm was Chisso; the
residents alone bore the injuries. . . .*

*An important factor in obtaining compensation for injuries suffered from
Minamata disease has been independent negotiations. These have been used to
break [the deadlock] that has resulted in the victims obtaining only a small sum
of compensation by mediation, and [that has allowed] Chisso to evade responsibility. . . .*

*When we consider that the prosecution of this incident has dealt a substantial
blow to the independent negotiations faction, it is undeniable that [the effect] of
prosecution has been to lend support to the opposing party, Chisso. Of course, in
negotiations one party cannot force the other (to its views); still it was difficult
for the victims, who through no fault of their own suffered one-sided injuries, to*

mitigate their feelings toward Chisso. In view of Chisso's attitude, the prosecutors should have exercised a suitable degree of patients. To confront the defendant immediately with punishment was improper. . . .

We cannot ignore the details of the violent acts. Yet, the defendant bore no ill will against the injured parties, individually. Viewing the harmful conduct as an (expression) of the protests of the many victims suffering from Minamata disease, of the victims who are unable to cry out, the defendant could not help his conduct.

In his concluding remarks Judge Terao accepted the defense's arguments that the charges be dismissed based on art. 338.4 of the Code of Criminal Procedure (Abuses of Prosecutorial Authority):[87]

Because the injuries of Minamata disease were caused exclusively by Chisso, had the proper prosecutorial and administrative authorities responded quickly, the damage would have been kept to a minimum. When one thinks about the course of things, the defendant could not help but take action. It was the mute protest of a pollution victim. To confront the defendant with immediate punishment for his acts was an abuse of prosecutorial authority. . . .

Although there are explanations for the actions of both sides, the fact that one side (Chisso) has not been prosecuted at all (suggests that the present prosecution) is biased, unfair, and startlingly opposed to legal justice. . . .

The court vacated the judgment of the Trial Court that had imposed a ¥50,000 fine and dismissed the remaining charges against Kawamoto. It was the first dismissal of an indictment based on an abuse of prosecutorial authority.[88]

The *Kōchi Pulp* and *Kawamoto* decisions suggest that the courts have begun to identify narrow exceptions to the rigid prohibitions against civil disobedience based on the special circumstances of environmental degradation. Without directly endorsing defendants' conduct, both courts conceded that some otherwise illegal actions can be brought within the ambit of legal protection. The Kōchi District Court recognized a citizen's affirmative right of self-help under emergency circumstances where the authorities have failed to act. Although not condoning Kawamoto's acts, the Tokyo High Court seemed to grope for an excuse. Thus the court emphasized the unique circumstances of the case, that is, the long and dismal history of Chisso's operations, the chilling effect of the prosecution on sensitive victim-company negotiations, and the mens rea of the defendant.

In motivating a judicial reconsideration of what constitutes criminal behavior, these two lower court decisions in a sense mark the legal outer limits of the transformation of values this chapter has described. Although the decisions hardly demonstrate a trend, they provide additional insight into the flexibility of the Japanese judiciary system in responding to the radical social, environmental, and other changes now underway.[89]

Section 7   Summary

Japan's postwar environmental movement was founded on these moral propositions: Industry shall never endanger human life for profit; chemical hazards must always be avoided at any cost; and polluters and the government should bear forever the responsibility to care for the afflicted. The tenacity of this perspective brought environmental questions in Japan close to questions of human rights.

The courts played a critical role in this transformation of values. They provided a national forum, clarified the issues, and gave strength to the victims' claims.

Although many important legislative and administrative changes preceded the judgments in the four pollution trials, the courts' decisions exerted a powerful influence. They precipitated new remedial legislation and administrative initiatives, strengthened the hand of the victims in their negotiations with the companies (and later with the government), and motivated political action.

In the early and mid-1970s the environmental movement's concerns, however, grew more complex. The early, fiercely moral perspective was extended, differentiated, refined, and alloyed by other grievances and ills. Over time the spirit tired.

We have suggested how legal institutions helped shape and were shaped by the values of the environmental movements. What then were the limits of the process? It is clear that the victims' resort to the courts advanced their cause in many ways. But it also involved them in a formal, impersonal institution that consumed time and spirit and caused many to abandon their most baleful intentions. The legal system struck a bargain for social stability, but at what price? This will be a continuing theme and inquiry in the following chapters.

II The Development of the Judicial Role

3   The Legacy of the Four Major Pollution Trials

Part II introduces Japan's rich judicial legacy during the past seven years. The key to this development is the four pollution trials, the subject of this chapter. As noted in ch. 2, these were Civil Code tort actions seeking damages brought by the victims of Minamata, itai-itai, and Yokkaichi illnesses. The saga of these cases, not merely the courts' decisions, represents the first generation[1] of postwar environmental trials.

The second generation is the period of creative judicial exploration in the civil and administrative fields culminating with the decision on November 27, 1974, of the Osaka High Court in the Osaka International Airport case. This period is characterized by several important trends that include: increasing judicial emphasis on the duties of the sovereign to protect public health and the evironment; the courts' consideration of new rights like the "right to sunshine" or the "environmental right," and the correlative judicial development of "new" remedies; and the courts' increasing concern with procedural safeguards, particularly in the administrative law field.[2]

The third generation is the present. Although there are too few decisions to reach any meaningful assessment, the courts appear to be proceeding with less daring than before; an impression, that if proved correct, is in accord with the more general pattern of retrenchment of recent years.

In analyzing the cases in ch. 3 and 4, we will be concerned with three sets of issues. The first is doctrinal development. How successfully have traditional legal principles responded to environmental conflicts and what adjustments have been considered necessary? Second, what are the policy implications of these decisions viewed from the perspectives of individual justice, efficient resource use, technological development and other objectives? How have conflicts between these objectives been harmonized? Third, how has the judicial institution functioned, or, more precisely, to what extent have judges in Japan begun to modify the traditional judicial role to suit new environmental needs? Because we believe U.S. courts can learn from judicial developments in Japan during the past few years, part II concludes with a comparative analysis of Japan's contribution.[3]

Section 1   The Itai-Itai Case

1.1   Introductory Note

As early as 1910 a strange ailment of unknown origin was reported in some areas of Toyama City, Fuchu, and several other towns situated at the lower and middle reaches of the Jintsū River on the Japan Sea side of Honshū. The disease was accompanied by such extreme pain that patients often would cry out "itai-itai" ("it hurts, it hurts"). As a result, the inhabitants later came to call this fearful disease "itai-itai." When cases of itai-itai disease began to be reported with increasing frequency in the early 1950s, it was first diagnosed as a form of neuralgia or rheumatism. But by 1955 local scientists began to believe it was a separate and new sickness. Because many sufferers were poor, hard-working farmers, it was initially thought that the disease was the result of undernourishment and overwork.

A local physician, Noboru Hagino doubted this theory, however, and traced the cause of the disease to pollution from mining.

The initial research of Hagino and some other scientists revealed through chemical analysis that the bone tissues of some patients contained high concentrations of cadmium, lead, and zinc. From this data, Hagino inferred that itai-itai disease might be the result of heavy metal pollution in the area. By 1961 Hagino and his colleagues had decided that cadmium was a primary suspect and that the Mitsui Mining and Smelting Company operating upstream at the Kamioka mine was probably the culprit. The researchers presented their findings to medical conferences but the scientific community ignored them.

Gradually, however, an official investigation was launched. Thus between 1961 and 1966 various research teams at the local and national levels were formed, including the Toyama Prefecture Special Local Disease Countermeasures Committee and groups in the Ministry of Health and Welfare and the Ministry of Education. The debate even reached the attention of the United States National Institute of Health, which granted Hagino and his colleagues $30,000 to continue their animal experiments to isolate the disease. In Japan, scientists still continued to disagree on the disease's cause(s).

Finally in 1968 the Ministry of Health and Welfare published the report of its investigations, which concluded that itai-itai disease was primarily caused by cadmium in conjunction with other factors such as nutritional deficiency and loss of calcium related to hormonal imbalance. Thereafter, the ministry designated itai-itai disease as a pollution-related illness.

The ministry's report was accompanied by two important events. In January 1968 Toyama Prefecture finally established a relief system for patients of itai-itai disease that registered those undergoing diagnostic study and supplied patients with medical care at public expense.[4] On March 9, 1968 eight patients and the successors of six deceased itai-itai victims filed a suit against Mitsui.[5]

The itai-itai disease suit was a damage action based on a theory of strict liability under art. 109 of the Mining Law.[6] The critical issue in the case was proof of the cause of the disease. As in all the major pollution cases, skillful use of expert testimony was a critical factor in the plaintiffs' success;[7] and as in other trials, the opinions of experts were conflicting. The defendant urged that an explanation of all scientific factors was legally required, and the plaintiffs attempted to narrow the focus of the court's analysis. The court's decision to adopt the plaintiffs' experts' testimony provoked a bitter attack. Mitsui accused the court of bias[8] in denying its request to demonstrate the scientific cause of bone softening, and in open court requested the entire judicial panel to withdraw from the case. The defendant's request was denied.[9]

On June 30, 1971, after three years' consideration, the Toyama District Court awarded the plaintiffs a total of ¥57 million in damages. Mitsui appealed to the Nagoya High Court, Kanazawa Branch, and plaintiffs also appealed with the hope of increasing the amount of damages. On August 9, 1972, the High Court upheld the District Court's decision. The opinion emphasized the importance of epidemiology in determining legal causation in pollution cases. In its appeal, Mitsui argued that other factors contributed to the victims' injuries (such as loss of calcium during pregnancy) and that these factors should be considered in determining the amount of damages. In support of its argument, Mitsui relied on art. 113 of the Mining Law, which permits the courts to deny liability or mitigate damages where a plaintiff's own actions contributed to the harm asserted. The High Court, however, rejected this argument, ruling that the victims could not legally be responsible for physiological functions of their own bodies.

The court also increased the amount of damages from ¥5 million to ¥10 million to each family of a deceased victim[10] and from ¥4 million to ¥8 million for living patients. Because of difficulties of proof in individual cases, plaintiffs did not seek compensation for pecuniary losses. The court held, however, that it would take into consideration a plaintiff's failure to plead pecuniary losses in assessing damages for pain and suffering. Attorney's fees were established as 20% of awarded damages.[11]

Although no further appeal was taken, the day after the Nagoya High Court's decision the representatives of the plaintiffs in the second to seventh suits extracted a promise from the president of the Mitsui Mining and Smelting Company that the company would compensate all other patients to the same extent as the parties in the suit.[12] Later the company agreed to pay for monthly nursing care and other expenses even in cases where patients were under the care of family members.

1.2   Decision of the Toyama District Court; Judgment for the Plaintiffs[13]

I Parties

[*The parties do not dispute that all plaintiffs have been living along the Jintsū River, that the defendant Mitsui Mining Company has duly acquired mining rights under the Mining Law and has engaged in the mining and smelting of lead and zinc ore at the Kamioka plant in Toyama Prefecture.*]

II Findings of Facts

[*The court makes extensive factual determinations on defendant's operations and disposal of wastewater, and how the processing and smelting at its plant have damaged agriculture and fishing along the Jintsū River (fig. 1).*

The court notes that the Kamioka plant now discharges a relatively small amount of wastewater and heavy metals such as cadmium because it now uses control technology for slag piling and wastewater treatment. The present concentration of cadmium in the water of the Jintsū River is as low as prevailing U.S. standards for drinking water. The slag piling and waste treatment facilities previously used, however, were insufficient in size, technology, or capacity to treat wastewater effectively. For these reasons the court points out it is not difficult to infer that a substantial amount of wastewater had previously been discharged from the Kamioka plant into the Takahara River, especially between 1910 and the 1940s.]

1. *Heavy metals such as cadmium, lead, and zinc are widely dispersed in the soil of rice paddies that are irrigated through the network of ditches by water from the Jintsū River. The concentration of the heavy metals is considerably higher in rice paddies in the delta than in areas that have other water sources. As a result, agricultural crops like rice and soybeans grown in the delta are contaminated by these heavy metals.*

2. *A higher concentration of heavy metals in rice paddies is found at places where water from the river is introduced, and lower concentrations are found in the middle of the river and at the outlet. A higher concentration is observed in*

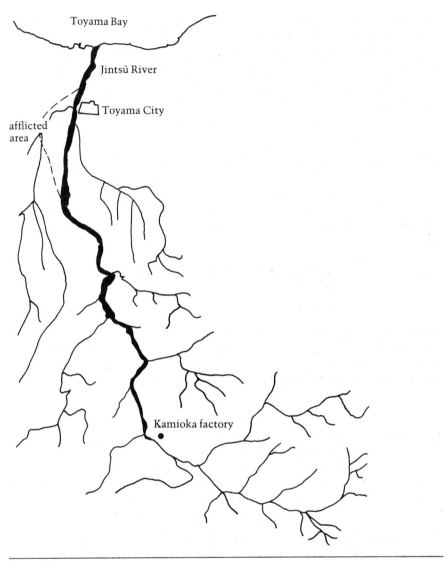

Figure 1
Affected area.

the upper layers of the soil, and a lower concentration in the middle and lower layers. From the pattern of dispersal and concentration of heavy metals, it is clear that these heavy metals have been brought downstream into the rice paddies by the Jintsū River directly and through irrigation channels.
3. The fact that the distribution of heavy metals in the soil of the paddies is inconsistent with the geological formation of soil strata in the Jintsū delta shows that these heavy metals were not accumulated when the strata in the delta was formed.
4. The concentration of heavy metals in the water and mud of portions of the Jintsū River system near the defendant's plant is particularly high.

The above facts are established. It is true that some of the heavy metals such as cadmium, lead, and zinc in the paddy soil and the river water and mud may come from natural sources, but the level of their concentration must not be greatly different from that in comparable areas. The main source of these heavy metals can, therefore, be none other than the defendant's plant. Considering the state of processing and smelting operations by the defendant and its predecessors, the accumulation of heavy metals seems to have been caused by the heavy metal contaminated wastewater discharged from their operations and by water which oozed from the piled slag into the upstream of the Jintsū River for a substantially long period from the 1920s to the 1940s. The court cannot find any evidence contrary to these findings. (45)

III Causation

Epidemiological Observations on Causation

In pollution cases where human life and health are affected by extensive air or water pollution resulting from industrial and other human activities, the issue of natural (factual) causation between these activities and subsequent damage becomes more crucial and heatedly contested than in usual tort actions. This is not only because there is often a big gap in time and space between the act and the occurrence of the harm but also because an indefinite number of persons over an extensive area may become affected. In pollution cases, therefore, in order to judge and clarify the existence of a causal relation between an act and an injury, clinical and pathological examinations are not sufficient and it is often necessary and essential to adopt an epidemiological approach. (46)
[The court thereafter determines that patients were found only in the delta areas irrigated by water from the Jintsū River, that cases of the disease had occurred since the 1920s, and that their number had sharply increased around the end of World War II. In addition, the court notes that the cadmium content in the urine of the patients and residents in the area was higher than that of other areas, that patients suffered kidney malfunction, and that even apparently normal residents suffered some abnormality of the kidneys. Most of the patients were middle-aged or old women.]

Clinical and Pathological Observations on Causation

[The plaintiffs show symptoms of softening bones induced by kidney malfunction. In a typical case, the patient's bones lose their calcium content and break easily.]
At the beginning, the patient feels acute pain in joints in the hips, waist,

shoulders, back, and knees, followed by rheumatic pain in various parts of the body. Because of the excruciating pain, the patient is forced to walk like a duck. These conditions usually continue for several years and in some cases for more than ten years. Thereafter, the patient by some slight bruise or sprain suddenly becomes unable to walk and is bedridden. The condition of the patient thereafter steadily deteriorates. Acute pain is experienced not only when the patient tries to walk or get up but also when he just tries to move about in bed. Because of the incessant pain, patients are unable to sleep, have difficulty breathing and laughing, and lose their appetites. Death frequently occurs from exhaustion with the patient crying "itai-itai." (70)

[In one case, the height of a patient shrunk by 30 cm because of broken back bones. In another, seventy-two broken bones were observed in the patient's body, twenty-eight in her ribs.

The court analyzes clinical and pathological symptoms and introduces in detail the results of several medical studies conducted by individual doctors and by a joint study group organized by the Japan Association of Public Health. It also refers to experiments on cadmium poisoning performed on animals in foreign countries and in Japan.]

Summary and Conclusions on the Issue of Causation

From all the clinical, epidemiological and other evidence before us, this court has traced the natural (factual) causal relation as fully as possible. We restate briefly our findings on the issue of causation-in-fact. A multiple and high incidence of the disease has been observed only in the delta between the Kumano and Ida rivers through which the Jintsū River flows. First, we examined the results of epidemiological studies and research on the cause of the disease. The affected area is essentially agricultural with rice paddies where water from the Jintsū River is supplied through a well-developed network of irrigation ditches such as the Osawano irrigation ditch. . . .

Heavy metals such as cadmium, lead, and zinc were brought in with the water, contaminated the soil, and damaged crops. There has often been agricultural damage in this area since 1900, and there have been many disputes over the contamination of the area from mining activities upstream. Not only have the residents in the delta eaten rice, soybeans, and other crops containing heavy metals, especially cadmium, but until a water system was completed recently, they have had to drink contaminated water either directly or indirectly as it was supplied through wells from the river. (It is difficult to determine from the evidence before us if the major intake of cadmium was from food or drinking water. In any case, it cannot be denied that the residents ingested cadmium through the mouth.) The residents of the affected area, therefore, ingested and accumulated more cadmium in their bodies than residents of comparable areas. There are no particular circumstances other than these noted which are commonly found throughout the delta area affected by the disease but not in other areas. The cadmium content is distinctively high among heavy metals found in crops grown in the area and in the urine of the residents as contrasted to those of other areas. The contrast for cadmium is clearer than for lead or zinc. Therefore, epidemiologically, the only reason that the patients of the disease are concentrated in and limited to the delta between the Kumano and Ida Rivers is to be found in the high concentration of cadmium in their food and water. Factors such as malnutrition, pregnancy, childbirth, nursing, hormones, and aging, however, may also have contributed to the occurrence of the disease. This can be clearly inferred from the following. The number of patients increased drastically around the end of World War II; the patients tend to live on a diet consisting largely of rice and little else; their families usually live on less protein and calories and have lower incomes than the average rural family in Toyama Prefecture; and the patients are mostly old and middle-aged women who have borne more children than the average woman, engage mostly in farming, and have little time to recuperate after each birth.

Next we examined clinical and pathological observations of the disease. The major symptoms of the disease correspond to those of the Fanconi syndrome resulting from a malfunctioning of the capillary tubes in the kidney. Such extensive malfunctioning of the kidney may be caused by genetic factors (such as adult Fanconi syndrome . . .), internal disorders (such as nephritis . . .) as well as by external poisoning (by agents such as heavy metal . . .), and vitamin D deficiency.

In the case before us, all but poisoning by heavy metals are eliminated because of the genetic characteristics (adult Fanconi syndrome . . .), symptoms (Wilson's disease . . .), preconditions (nephritis), or the impossibility of intake or deficiency in this area (tetracycline, . . . vitamin D). Among the suspected heavy metals, zinc does not easily accumulate in the human body although it is necessary for physiological metabolism. The toxicity of zinc is also extremely weak. Although some cases of acute poisoning have been seen, there has never been an observed case of chronic poisoning. There are reports that lead produced softening of the bones, but the lead content detected in the urine of the residents is within the normal range, and there have been no discoveries through physical examinations of the residents of the characteristic symptoms of lead poisoning. Consequently, only cadmium remains a possible suspect as the causal agent for the kidney malfunction, one of the major symptoms of the disease. This corresponds to the results of the epidemiological survey which show that the area has been especially contaminated with cadmium. Similarly, we observe that similar cases of kidney disease caused by ingestion of heavy metals have been reported in some foreign countries. By a process of elimination, cadmium would seem the cause of the kidney malfunction observed in the patients of itai-itai disease. . . .

Another major symptom of the disease is softening of the bones. It may be regarded as a secondary effect of the malfunction of the kidneys. Some Fanconi cases may not result in softening of the bones, but softening of the bones is one of the symptoms of the Fanconi syndrome. Therefore, softening of the bones in itai-itai disease can be a symptom of Fanconi syndrome. Since the cause of the kidney malfunction is considered to be cadmium, the cause of softening of the bones in itai-itai disease is also considered to be cadmium. However, in order for the kidney malfunction to result in softening of the bones, we find that it was necessary for there to have been both a loss of calcium caused by kidney malfunction and the existence of such supplementary factors as pregnancy, childbearing, nursing, and malnutrition. These latter elements increased the demand for calcium. (88)

[The court next examines the results of studies on cadmium poisoning in foreign countries.]

Finally, we note the results of animal experiments. There have not been any published foreign reports of softening of the bones in experiments on animals. However, in several experiments which have been carried out in Japan, cadmium fed to small animals such as rats and mice for a period of a few months to a year has produced symptoms such as decomposition of calcium, kidney malfunction, and softening of the bones. The results of experiments further confirm the conclusion reached through epidemiology and clinical and pathological observations that the kidney malfunction and softening of bones observed in the disease are caused by chronic cadmium poisoning. (89)

[The court, in addition, ruled as follows:
1. Plaintiffs have given sufficient proof that cadmium was ingested and then absorbed in body organs, causing the malfunction of capillary tubes in the kidneys. Kidney malfunction alone could cause the softening of bones.
2. It was thus unnecessary to prove with mathematical accuracy the following points in order to decide the causal relation between cadmium and the disease:
(a) How much ingested cadmium is absorbed in the human body;
(b) What duration of time was needed to produce the malfunction of the kidneys that occurred;
(c) Whether ingested cadmium accumulated in bone tissue.]

We do not deny that there are some points in the pathology of this disease that require future study. It is true that we could ascertain clearly the cause of the disease if we knew all its mechanisms. It is not necessary, however, to establish the entire mechanism of the disease in order to ascertain the cause of the disease. In the case before us, since the causal relation between cadmium and the disease has been confirmed by epidemiological studies, clinical and pathological observations, and the results of animal experiments, we should be satisfied at this stage that an adequate explanation of the mechanism of the disease is possible. (89)

A major source of the cadmium contamination, which is identified as the cause of the disease, is the defendant's plant. It is reasonable to find that cadmium existed in the water (including muddy waste), discharged as a result of its processing and smelting operations and from runoff from its slag piles, and that water containing cadmium kept flowing into the Jintsū River especially for the period from 1900 to 1950. . . . (89)

The soil of the rice paddies in the delta that were irrigated by the river water became contaminated with cadmium and other heavy metals. The residents in the delta ingested cadmium contained in the rice and other crops and water for a long period of time. As a consequence, cadmium was transferred, accumulated in their bodies, and caused them to suffer from the disease. (89)

IV Liability of the Defendant

Article 109(1) of the Mining Law (1950, Law No. 289) provides that when damage to other persons results from drilling, discharge of pit water or wastewater, slag or slag piles, or smoke in connection with mining operations, the holder of the mining right at the time of the occurrence of damages shall be liable for damages. Article 109(3) of the Law also provides that when the mining right is transferred after the occurrence of the damage, the holder at the time of that occurrence and the subsequent holder are jointly liable for the damage. The intent of these provisions is to protect the injured party. The wastewater referred to in the statute consists mainly of water used and discharged in the process of dressing and smelting but also includes rainwater draining from wasted ore or slag piles and water mired with untreated mire from the dressing process. The discharge need not be intentional. We also regard oozings from the piles of slag or waste as discharges under the Mining Law.

It is undisputed that the defendant holds the mining right as specified in the appendix to this opinion [omitted], and has engaged in the mining, dressing and smelting of lead and zinc ore at the Kamioka plant. As a result of its activities found under heading II, the defendant caused the disease as found under heading III, and inflicted damage on the plaintiffs. Therefore, the defendant is liable under art. 109(1) of the Mining Law as the present holder of the mining right. Defendant is also jointly liable under Article 109, paragraph 3, with previous holders for damage which occurred before the date of transfer of the right, May 1, 1950. (90)

1.3   A Written Pledge Concerning Compensation for Itai-Itai Disease

August 10, 1972

Shinpei Omoto
President

Mitsui Metal Mining Company, Inc.
Nihonbashi Muromachi
2-1-1, Chuo-ku
Tokyo

Mr. Toshihisa Komatsu
Chairperson
Council on Itai-Itai Disease

Mr. Kinosuke Shōriki
Chief of Attorney Group

1. Mitsui Metal Mining Company (hereinafter, the Company) admits that itai-itai disease was caused by cadmium and other heavy metals discharged by the Company's plant and, hereinafter, pledges to refrain from disputing that issue by words and actions either in or outside of court.

2. (a) The Company shall pay by the end of this month to the plaintiffs of the litigating groups (second through seventh) the amount as claimed in their amended Complaint, filed with the court, on August 8, 1972.
(b) The Company shall bear all costs of each litigation.

3. In case a plaintiff-patient of the litigating groups (first through seventh) dies after receiving the compensation specified in paragraph 2(a), the Company shall pay to his survivors the difference between the amount of compensation already paid and the total amount of compensation owed to the deceased.

4. The Company shall in good faith provide compensation for patients of itai-itai disease and those who are suspected of having the same disease and are under medical observation, who are not in the litigation groups when the certification of the Governor of Toyama Prefecture is presented to the Company by the Council on itai-itai disease.

5. The Company shall also provide compensation for patients of itai-itai disease and those who are suspected of having the same disease and are under medical observation in the event that they are certified in the future. As for a newly certified patient who has already received compensation as one who is under medical observation, the Company shall pay a difference between the amount already paid and the total amount set for a patient.

6. Upon request, the Company shall pay to patients and those who are under medical observation the total amount of future medical expenses including transportation to and from the hospital, hospitalization, recuperation at hot springs, and other expenses relating to medical treatment and rehabilitation.

7. The parties shall confer and make a separate agreement as to the manner of payment under items 3 through 6 above.
The Company solemnly pledges to carry out the above.

1.4   Agreements Between the Plaintiffs and Defendants Reached as the Result of Direct Negotiations After the Court's Decision

1. The Company, recognizing that discharges by the Company of cadmium and other heavy metals have caused itai-itai disease, covenants not to dispute this fact by word or deed, in a judicial or nonjudicial context, from this day forward.
2.1. The Company shall, by the last day of this month, pay each plaintiff of the second through seventh itai-itai disease lawsuits, the amount claimed in the increased demands of August 8, 1972.
2.2. The Company shall bear the entire litigation costs of the above mentioned lawsuits.
3. If a plaintiff victim of the first through the seventh itai-itai disease lawsuit dies after having received compensation under 2.1 above, the Company shall pay the family of the deceased the difference between the amount already paid and the amount of damages owed to the person who has died from itai-itai disease.
4. The Company shall, in good faith, compensate victims of itai-itai disease who are not plaintiffs in the itai-itai disease lawsuits and persons who need observation, on the basis of certification issued by the Toyama Prefectural Governor and submitted to the company by the Itai-Itai Disease Countermeasures Council.
5. The Company shall compensate persons newly recognized as victims of itai-itai disease or as persons needing observation in the same manner as prescribed in the preceding paragraph. However, where a victim of the disease has already received compensation as a person needing observation, that amount shall be deducted from the compensation to be received by the person as a victim of the disease.
6. The Company shall henceforth pay all costs claimed by itai-itai disease victims and persons needing observation for medical care, hospitalization, hot springs for recuperation, and other expenses incurred in connection with medical treatment.
7. The manner in which payment shall be made pursuant to paragraphs 3 through 6 shall be determined by separate deliberations.

> *We covenant the above*
> *August 10, 1972*
> *1-1, 2-Chōme, Nihonbashi Muromachi, Chūō-ku, Tokyo, Mitsui Metal Mining Co.*

Mitsui Metal Mining Company
1-1, 2-Chōme, Nihonbashi Muromachi
Chūō-ku, Tokyo

By: Omoto Shinpei
*   Representative Director, President*

To: Itai-Itai Disease Countermeasures Council
*   Chairman: Komatsu Yoshihisa, Esq.*

*   Itai-Itai Disease Lawsuits Plaintiffs' Attorneys' Group*
*   Principal Attorney: Shōriki Kinosuke, Esq.*

Section 2   The Niigata and Kumamoto (Mercury Poisoning) Cases

2.1   Introductory Note

Although the outbreak of Minamata disease in Niigata occurred in 1965, nine years after the first reports of this malady in Kumamoto, Kyūshū, the Niigata victims were the first to litigate. In June 1967 seventy-seven victims and their families sued Shōwa Denkō (Electrical Chemical) Corporation and in September 1971, the Niigata District Court awarded them approximately $920,000 in damages. The second Minamata suit was brought in July 1969 by afflicted persons from Kumamoto Bay against the Chisso Corporation; on March 20, 1973, the Kumamoto District Court citing the Niigata and Yokkaichi courts' decision (sec. 3) awarded plaintiffs approximately $3.4 million in damages. Neither decision was appealed.

Both actions were based on a theory of negligence under art. 709 of the Civil Code, which states:

A person who violates intentionally or negligently the rights of another is bound to make compensation for damages arising therefrom.

Although Japanese courts had experience in construing this provision in general tort litigation unrelated to pollution, the Niigata case was unprecedented.[14] The Niigata District Court consequently lacked guidelines for interpreting the critical elements of a negligence action, duty of care, foreseeability, and causation.

Despite its inexperience, the court unequivocally accepted plaintiffs' claims. Sympathizing with their onerous burden of proving causation, the court developed a general rule applicable to other pollution cases on its own. It also held Shōwa Denkō to an extremely high standard of care because the chemical company's operations posed substantial health risks to the public. Finally, we believe that the court broke new ground in its assessment of damages.

By the time of the Kumamoto Minamata trial, causation was no longer seriously in dispute. Rather, the case turned on the issue of foreseeability and the effect of the mimaikin contract, which the plaintiffs claimed was unconscionable.

The Kumamoto District Court's rulings were devastating. As a chemical manufacturer, the defendant Chisso, the court noted, must be on the alert for any, even a remote, risk to human health. The court sternly pointed out that Chisso not only knew of the hazards of its activities, but it also actively suppressed information pertinent to discovery of the cause of Minamata disease. The court quickly voided Chisso's compensation contract as violative of public policy.

Chapter 2 has already suggested the complicated social history of Kumamoto's Minamata victims. Unlike the situation in Niigata, in Minamata there were many factions divided in part over the issue of how Chisso should make amends. One group accepted the assistance of the Minamata Disease Compensation Management Committee established by the Ministry of Health and Welfare; a second group sought conciliation through the Central Committee for Pollution Disputes Settlement; a third turned to litigation; a fourth independently began direct negotiations with Chisso.

After the Kumamoto District Court's decision, one group of Minamata victims negotiated a further agreement with Chisso. By this agreement Chisso acknowledged its responsibility for causing Minamata disease, expressed its deep apologies to the victims, and promised to conclude a pollution abatement agreement (Kōgai

Bōshi Kyōtei) with concerned local government(s). The agreement also provided for the following:
1. Payments of ¥18,000, ¥17,000, and ¥16,000 million to designated patients (and for the family of the deceased in the first case) to be ranked in three grades;
2. Medical care as specified in the Law for the Relief of Pollution Disease (subsequently repealed);
3. Nursing expenses (as specified in the above law with an additional payment of ¥10,000 per month);
4. Adjustment fees of ¥60,000, ¥30,000, and ¥20,000 per month;
5. Funeral expenses of ¥200,000.
In addition, after the court's decision, Chisso established a fund of ¥300 million for the relief of patients.

The courts' decisions on Minamata disease also sparked some important changes in government policy. For example, as a result of the cases, emission controls on mercury were substantially tightened. More dramatically, the Environment Agency and MITI initiated a program of "administrative guidance" to induce the caustic soda manufacturing industry to switch from its mercury-based method of production to the environmentally safer diaphragm-cell method.[15] As discussed in later chapters, the government's actions subsequently caused considerable financial and technical problems for the industry and even produced a serious international imbroglio with an American manufacturer doing business in Japan.

2.2   **Decision of the Niigata District Court**[16]
              Judgment for Seventy-six Plaintiffs. [The claim of one
              plaintiff was denied.]

              I Parties

              (A) The Plaintiffs

[*The plaintiffs are seventy-seven persons who live in the basin of the Agano River that flows through the Niigata Plain to the Sea of Japan. The plaintiffs are the victims of Minamata disease and the relatives of persons who have died therefrom.*]

              (B) The Defendant

[*The defendant Shōwa Denkō is a producer of diversified chemicals who owned and operated the Kanose factory located in Kanose-chō, Niigata Prefecture. The synthesis of acetaldehydes using mercury as a catalytic agent began at the Kanose plant around November 1936 and continued until January 1965, when the operation was moved to a new plant and the Kanose plant was closed. Between 1936 and 1965 wastewater from the production process was continuously discharged into the Agano River.*]

              II Organic Mercury Poisoning Along the Agano River

              (A) The Discovery of the Patients

1. On November 12, 1964, the plaintiff, Kazuo Imai, age 65, was referred to and hospitalized in the cranial neurosurgery section of the Niigata University Medi-

cal School in the Kuwana Hospital in Niigata City. He was suffering from a disease of unknown cause.

Imai's major symptoms included loss of peripheral vision, difficulty with speech and walking, a general loss of feeling and unusual sensations particularly around the mouth and extremities, and impairment of hearing. He was unable to walk, write, or change his clothes and had difficulty in carrying on a conversation.

In January of 1965 Tadao Tsubaki of Niigata University Medical School examined the patient and diagnosed his symptoms as similar to those of a typical case of organic mercury poisoning. He sent the patient's hair to Tokyo University's Department of Pharmacology to be tested and received a report that it contained 320 parts per million (ppm) of mercury.

2. Around March 27, 1965, Kishimatsu Ono (since deceased) was referred to the Medical School (from Kuzuzuka Hospital in Hōei City) as a patient suffering from an unknown disease.

Tsubaki examined the patient and observed the impairment of sensory nerves around the mouth and finger tips, loss of coordination of body movements, loss of hearing, and the stiffening of muscles. An examination of his field of vision was not possible at that time because the patient was suffering from a nervous breakdown. The patient's hair was tested on April 14 and contained 232.6 ppm of mercury.

3. In May of 1965 Niigata University made an investigation of the homes of both patients, taking samples of sewage, well water, mud from the Agano River, and hair from family members. The university team also checked the types of agricultural chemicals used. Through this investigation it was learned that another person in the same village as patient Imai had died with similar symptoms on October 29, 1964.

4. On May 14 the plaintiff Yukimatsu Hoshiyama, age 11, of Eguchi, Niigata City, was brought in and examined by Tsubaki. He showed similar symptoms and his hair contained 570 ppm of mercury.

5. Finding several patients with similar symptoms, all residing along the Agano River, Professor Tsubaki and others came to the conclusion that they were suffering from a disease similar to the organic mercury poisoning previously found in Minamata City in Kumamoto Prefecture. The doctors felt, however, that further deliberation was necessary in order to confirm their diagnosis. In view of the unusual and rare nature of the disease, on May 31 they reported to the Public Health Department of the Niigata Prefectural Government that "there have been sporadic incidents of mercury poisoning of unknown cause among residents along the downstream portion of the Agano River." (113)

(B) Investigation and Research by Niigata University and Prefectural and City Governments

1. Establishment of an Organization. *(a) Upon receiving the report, the public health department immediately went into action to deal with the situation. After learning that Tsubaki and others suspected that factories using agricultural chemicals containing mercury or mercury compounds were possible sources of poisoning, the public health officials on June 1 started investigating factories using mercury and the agricultural use of mercury.*

On June 4 a joint meeting of the city and prefectural governments and the university was held to coordinate efforts to find the source of the poisoning and the remaining patients. The meeting concluded that close cooperation among the three would be necessary and decided that the university would physically examine residents in the affected area and find undetected victims, that the city would facilitate the physical examinations by assembling the residents and would make arrangements for boats necessary for taking samples of water and sludge from the Agano River, and that the prefecture would investigate the use of mercury-based agricultural chemicals, the operational processes of factories, and the industrial use of mercury. These three organizations then started their investigations.

(b) On June 13 newspapers and other media reported the incidents of mercury poisoning in Niigata and attracted wide social attention. By June 16 the number of diagnosed patients along the lower Agano reached fourteen. On the same day, the prefecture established the Niigata Prefecture Mercury Poisoning Research Center (the name was changed on July 31 to the Niigata Prefecture Organic Mercury Poisoning Research Center after the causal substance was practically confirmed as organic mercury) with the lieutenant governor and the dean of the medical school as codirectors. . . .

(c) On June 21 another organization, the Niigata Prefecture Mercury Poisoning Action Center, was set up within the prefectural government structure. It was based on a prefectural plan to establish a center to provide proper care and welfare to the patients and their families. In addition the Niigata Prefecture Mercury Poisoning Liaison Conference was formed to coordinate the several related organizations with the governor, the president of the university, the mayor, and the president of the Prefectural Medical Association, among others.

2. Restrictions on Fishing. On June 16, having observed that all patients regularly ate fish caught in the Agano River, Tsubaki and others issued the statement that "the cause of the poisoning is to be found in Agano River fish."

On the 28th the Action Center issued administrative guidance warning residents and fishermen not to catch fish and shellfish within 14 kilometers of the mouth of the river. The warning was published in the newspapers and channeled through the fishermen's unions in the area.

3. Investigations to Find the Cause of Poisoning. From around June 2, the prefectural government started a village by village investigation of the use of mercury-based agricultural chemicals on both sides of the Yokogumo Bridge. At Nemuro Village [where houses were flooded by a tidal wave (tsunami) caused by the Niigata earthquake (June 16, 1964, magnitude 7.5)], a house by house investigation was made on June 19 of the disposal of soaked and residual agricultural chemicals. An investigation was also made on June 20 of warehouses around Niigata Port to find out what kind and what quantity of agricultural chemicals were stored at the time of the earthquake and subsequent tidal wave and to what extent damage to the contents had occurred.

Next came the industrial surveys. There were a dozen factories such as Nihon Gas Chemicals, . . . which used mercury in the Niigata City area, but as the investigation proceeded, suspicion was narrowed to four factories, including the defendant's Kanose plant. As a preliminary measure, telephone calls were made to these four factories in the name of the governor to "be careful about the treatment of wastewater." Officials also went to each factory to take samples of discharged water from the drainage pipes and sludge from the bottom of treatment pools.

4. Efforts to Find Undetected Victims. (a) The first health examinations of the residents were undertaken by the university in cooperation with the Public

Health Department. In early June initial surveys were made in affected areas of Niigata City. . . . About 2,800 persons in 412 households were interviewed individually. Each person was asked about the presence of subjective symptoms, the extent of river fish in his diet, his drinking water, agricultural chemicals, circumstances of any deaths since the beginning of 1964, and any abnormalities in domestic animals. Physical examinations were given to persons with subjective symptoms.

(b) A second survey was carried out in late August by the public health clinics, the city, and the prefecture. Twenty-five thousand persons along the Agano River were questioned and persons with suspicious symptoms were sent to the medical school for medical examinations.

(c) [As a result of the two examinations, a score of new victims were found.] The hair of about three hundred persons—those with symptoms and their family members whose diet consisted of large quantities of fish—were tested for mercury. Those whose hair contained about 200 ppm of the mercury, called "holders," were hospitalized, given medicine that removed the mercury from their bodies, and kept under observation.

(d) In order to find and prevent fetal (congenital) Minamata disease, 384 infants in the affected area were examined in August by the medical school team. The hair of ten mothers who reported having eaten large quantities of fish and eighty-one pregnant women from the area were tested for mercury content. Women of child-bearing ages (from sixteen to fifty years old) whose hair contained more than 50 ppm of mercury were also visited regularly by public health nurses who gave them instructions concerning pregnancy, childbirth, nursing, and particularly birth control.

(e) There was still the possibility of patients living outside of the affected area. So health officials visited various medical institutions in the prefecture and went through medical records for suspect cases to ascertain the extent of the affected area. (114)

(C) National Government Investigations and Reports

1. **The Investigations by the Ministry of Health and Welfare.** *Upon receiving a report from the Niigata Prefectural Government, the Ministry of Health and Welfare on June 14, 1965, sent a team of officials to investigate the incident in cooperation with the prefectural government and Niigata University. On June 21 and July 2, more officials of the ministry visited the area as part of the investigation.*

2. **The Establishment of a National Investigation Organization.** *From around July 21, the Ministry of Health and Welfare started soliciting the opinions of scholars in various fields. Then on July 30, a liaison conference consisting of related government agencies was held. The conference participants decided to cooperate with the Action Center in the investigations. On September 8, 1965, the government established a research organization drawing officials from related departments and specialists from universities. The Science and Technology Agency budgeted for research projects in the following areas: (1) epidemiological field studies; (2) studies on the pattern and state of mercury compound pollution; (3) studies on the distribution of mercury-contaminated aquatic plants and animals; (4) diagnostic techniques for mercury poisoning; and (5) coordination of this research and investigation.*

Project (3) was administered by the Japan Sea District of the Marine Research Institute of the Ministry of Forestry and Agriculture and project (5) by the Coordi-

nation Bureau of the Science and Technology Agency. The rest were under the jurisdiction of the Ministry of Health and Welfare, which then organized within the ministry a Special Research Group on Niigata Mercury Poisoning (hereinafter Special Research Group). The group was subdivided into three subgroups on (1) epidemiology, (2) experiments and testing, and (3) clinical studies.

These subgroups carried out the following investigations: . . .

3. Investigations and Reports of the Special Research Group. . . .

On March 24, 1966, the Special Research Group issued an interim report in which the epidemiology subgroup summarized its findings as follows:

It is presumed that dimethyl mercury discharged from the acetaldehyde plant at Kanose polluted the water of the Agano River. The change in the course of the river and the disruption of the riverbed caused by the Niigata earthquake contributed to the contamination of river fish. The dimethyl mercury accumulated in the bodies of the fish. Persons who ate a lot of river fish were affected in their central nervous systems and fell victims to this peculiar poisoning.

After the interim report was issued the defendant, through the Ministry of International Trade and Industry (MITI), disputed the findings of the report, especially those of the epidemiology subgroup, and presented material in opposition.

Thereafter, the Special Research Group continued its work by analyzing documents that the defendant had submitted and by refining its previous work through making use of a supplementary investigation by the prefecture. Finally, each subgroup made final reports at the end of March and submitted them to the Ministry of Health and Welfare on about April 7.

On April 18, 1967, at the same time that it presented it to the Science and Technology Agency, the ministry made public the final report of the Special Research Group. The conclusions of the epidemiology subgroup were stated as follows:

The Niigata poisoning incidents resulted from the frequent eating of river fish contaminated by dimethyl mercury compound in the Agano River and should be regarded as a second instance of Minamata disease. The source of pollution is the Kanose plant located upstream on the Agano River. The process of contamination began when dimethyl mercury compound, a by-product arising out of the operation of the acetaldehyde plant, and part of the effluent from the plant, was discharged into the Agano River. As the production of acetaldehyde increased, so also did the amount of mercury compound, which, after being discharged into the water, accumulated in the fish in the Agano River. Some residents who live along the river caught and ate fish consistently, thereby accumulating the mercury compound in their bodies. As a result, they became afflicted by the disease.

4. The Statement of the Technical Opinion of the Government. *The Ministry of Health and Welfare was asked by the Science and Technology Agency to provide a comprehensive statement on the findings of the Special Research Group. On April 20, 1967, the ministry referred the matter to the Council of Food Sanitation, an advisory body to the Minister of Health and Welfare.*

The council established a nine-member committee, which examined the final report and materials submitted by the defendant company and listened to witnesses. These witnesses included technicians of the defendant company and Professor Kitagawa of Yokohama National University, who suggested that fertilizer was the cause of the poisoning. On August 30 the committee submitted a report to the minister. This report was sent to the Science and Technology Agency as the official statement of the Ministry of Health and Welfare. It stated in brief:

(i) Long and extensive pollution of the Agano River by wastewater discharged from the Kanose plant provided the basis for the disease's appearance.
(ii) It is possible that the patients have appeared solely by reason of (i) above. However, many patients were intensively afflicted with the disease during a particular period. It is considered to have been caused, aside from reason (i), by a comparatively rapid and great concentration of mercury compounds including dimethyl mercury in the patients' bodies. At the present stage the cause of this concentration is unknown.

On September 26, 1969, the Science and Technology Agency issued its own technical report after sending the Health and Welfare Ministry report to related ministries and agencies and sounding out and coordinating their views. It contained the following statements:
(1) Dimethyl mercury compound is the agent which caused the poisoning in the instant cases. It has been accumulated in river fish and the disease resulted from eating such fish continuously and in great quantity.
(2) The pattern of pollution suggests that the Agano River was in fact contaminated for a long period. In addition, there might have been a highly concentrated short-term contamination in a single period. In any event, long-term contamination has made a fundamental contribution to the appearance of the disease, although the extent of that contribution is not known.
(3) The source of the long-term contamination is primarily discharge water from the Kanose plant. The effect of fertilizer administered in the Agano River area is considered negligible.
(4) It was argued that fertilizer spilled at the time of the earthquake was the source of the concentrated short-term contamination. There is no evidence to support this theory. It is difficult to judge from the available information and materials whether faulty management of the Kanose plant was the cause of the pollution.

The government adopted the report as the technical statement of its official position. (115)
[The court proceeds to refute the defendant's argument that these reports reflected the bias of some research members who participated in the study and who had also taken part in the investigation of Kumamoto Minamata disease.]

(D) Patient Certification by Administrative Agencies

1. At the beginning of the poisoning incident, Professor Tsubaki had to discover and treat the patients, but on December 8, 1965, the prefectural government organized a team of doctors to diagnose the patients, distinguish between those who needed treatment and those who needed guidance, give the patients medical advice, and carry out studies for these purposes. It was called the Committee to Examine Patients of Niigata Prefecture Organic Mercury Poisoning. (116) [Tsubaki and others were appointed by the governor as members of the committee.]
2. On December 23, 1965, the committee held its first examination. The committee discovered that there were twenty-six patients of the disease (including five deceased) and nine who needed medical observation because of high concentrations of organic mercury in their bodies. . . . The Committee held meetings once every two or three months and diagnosed poisoned patients . . . . (116)
3. In December of 1969 the Law Concerning Special Measures for the Relief of

Pollution Related Patients (1969, Law No. 90) and its implementation order
(Cabinet Order No. 319) were promulgated. Mercury poisoning was called
"Minamata disease" and designated as one of the pollution diseases. A portion of
Niigata City [in addition to other areas] including the districts where the pa-
tients were concentrated was designated under art. 2 of the law. . . . On February
6, 1970, a patient certification review board was established as an advisory
organ to the governor and mayor who were made responsible for final certifica-
tion. [The names of doctors are listed.] . . . (117)
4. By September of 1970, the board had examined and certified a total of forty-
nine patients (including six deceased). (117)

III Relationship Between the Poisoning and the Defendant's Actions

[The following is a summary of sections (A)–(E) of the court's opinion.] The
plaintiffs argued that the methyl mercury compound from the defendant's
Kanose plant caused the same disease as Minamata disease (the so-called "fac-
tory discharge theory"). At the same time the defendant contended that methyl
mercury did not come from its plant but rather was part of the agricultural
chemicals washed from storage houses in Niigata Port by a tidal wave that fol-
lowed the Niigata earthquake of 1964.

The parties agreed that the poisoning in the instant case was caused by methyl
mercury, particularly low-grade alkyl mercury. The court then described at
length the symptoms of Kumamoto Minamata disease that had preceded the
Niigata incident and compared these symptoms to the clinical and epidemio-
logical characteristics of the disease in the present case. It found a similarity be-
tween the two and found the disease in the Niigata incident to be Minamata dis-
ease. It utilized the scientific data collected in Kumamoto to a great extent.

The court refused to call scientists as expert witnesses to carry out new sci-
entific investigations and experiments. It admitted that it was very difficult to
assess accurately the validity of differing scientific explanations of the chemical
processes at issue without calling new expert witnesses. The court was con-
cerned that calling new expert witnesses would entangle the proceeding in an
unresolvable and endless scientific debate. And the court felt it was unnecessary
to engage in further scientific debate to determine legal causation. Although it
declined to decide exactly how the defendant's production process had produced
methyl mercury by-product, it concluded the evidence on this question was
sufficient as long as the theoretical possibility of production existed and as long as
the methyl mercury compound was actually detected at the defendant's factory.

It then made, inter alia, the following factual findings:

1. The Amount of Acetaldehyde Production.

1957	6,251 (tons)
1958	6,630
1959	9,143
1960	11,800
1961	15,552
1962	17,734
1963	19,043
1964[17]	19,476
1965[18]	543

72

2. The Wholesale Destruction of Records by the Defendant Company. *Toward the end of 1965, the defendant company destroyed all of the records of its operations as well as the blueprints and technical drawings of the plant. The defendant did not provide any plausible explanation for their destruction.*

3. The Loss of Mercury at the Plant. *Over the eight years prior to its closing, 34 tons of mercury were lost in the acetaldehyde production process at the Kanose plant. Without considering the increased amount of production in recent years, simple arithmetic shows that on average 11.6 kg of mercury was lost each day for eight years. If we assume that the loss of mercury corresponds to the amount of the acetaldehyde produced, 20 kg of mercury on the average were lost each day during 1963 to 1964.*

4. Accumulation of Dimethyl Mercury Compound in Fish and in the Human Body. *The highest mercury concentration found in a patient's hair was 570 ppm (220 ppm of dimethyl mercury). The patients formerly ate fish caught around the mouth of the Agano River. Fish were found to contain 21.0–23.6 ppm of mercury (fish samples caught through June to October). When these fish were dried, the mercury content became 140 ppm (40 ppm of organic mercury).*

5. Disposition of the Defendant's Argument that the Source of Pollution was Agricultural Chemicals Washed Away by the Tidal Wave Caused by the Niigata Earthquake in 1964. *[The court thereafter refuted the defendant's assertion that, in view of the local currents, there was only an extremely small possibility that chemicals had washed into the Agano River (apart from the small quantity of chemicals that might have entered Niigata Port at the very mouth of the river). The defendant's argument also conflicted with the evidence that some patients had been affected by the disease before the earthquake. The court, however, had some difficulty in explaining why most patients were found in 1964 and 1965 and why they were concentrated in the area along the mouth of the Agano River.]*

(F) The Cause of Poisoning

1. In tort cases the aggrieved party has the burden of proving causation between the occurrence of the injuries and the alleged harmful acts. In what are called pollution cases (we use here the definition provided in art. 2 of the Basic Law for Environmental Pollution Control), it is often extremely difficult for the aggrieved party to explain scientifically each of the causal links between the harmful acts and the injuries suffered. This is especially true in cases involving the chemical industry where the discharge of chemical substances causes disease in many people (hereinafter chemical pollution). To require of the victims in such a case a highly scientific explanation of every issue would effectively eliminate the courts as a means of civil relief. In the final analysis, the issues presented by causation theory in this instance can be considered as threefold:
(1) the characteristic symptoms of the disorder and its causal (etiological) agent;
(2) the pathway by which the pollutant reached the victim (the path of pollution); and
(3) the discharge of the causal agent by the wrongdoer (the mechanism of production and discharge).

Aggrieved parties may demonstrate (1) medically with the cooperation of clinicians, pathologists, epidemiologists, and other medical specialists. Yet as the Kumamoto Minamata example bluntly demonstrates, it is necessary first to have a substantial population of patients and a large number of deaths and autopsies before the characteristic symptoms of the disorder and its agent become known. With regard to (2), apart from some instances where discharged effluents can be distinguished by color or smell, most chemicals cannot be detected by their appearance. It is not simply that it is difficult for persons not directly involved with the particular company to ascertain correctly the types of effluents, their characteristics, and the levels of their discharge; between the discharge of the effluent and its effect on the victim, various natural phenomena or other complex factors may intervene. Ordinarily, victims as well as third parties remain unaware of the pollution mechanism. (One characteristic of this kind of pollution is that because it is an unseen poison, a large number of persons are unknowingly exposed and their health unwittingly impaired.) With regard to (3), the company usually cites the need to protect "trade secrets" as a reason for absolutely refusing to make public what materials it uses or what effluent it discharges. Even governmental regulatory agencies are denied full cooperation and are not allowed to enter the plant to conduct tests or take samples. Ordinary persons without authority or influence cannot conduct plant inspections or take samples. With these constraints, the construction of a scientific explanation is next to impossible. On the other hand, the victims of this kind of pollution are not in a position to engage in the same type of polluting activity. Thus, to acquire the "trade secrets" of the company and to understand its content and implications is an unattainable goal for them. In contrast to the victims, the polluting company in many instances enjoys exclusive possession of the technical knowledge of the formation process and discharge of the effluent. With regard to (3), the company technicians are in the best position to clarify the matter most efficaciously.

Considering the above, it would be inappropriate from the standpoint of equity, the fundamental basis of the tort system, to require an aggrieved party in a chemical pollution case such as this one to provide a scientific explanation of these issues. Therefore with regard to issues (1) and (2), we conclude causation may be proved by an accumulation of circumstantial evidence if that explanation is consistent with the relevant scientific disciplines. When the above level of proof concerning issues (1) and (2) is obtained and the search for the source of the pollution leads, so to speak, to the very doorstep of the factory, unless the defendant company proves with respect to (3) that its factory cannot be the pollution source, it shall be factually presumed that the factory is the source, and legal causation shall be fully established.

2. Accordingly, in applying these findings to the case at bar, we note that the toxicosis has been previously shown by animal experiments as well as by clinical and pathological studies to be a low-grade form of alkyl mercury toxicosis called Minamata disease. The toxic agent has been scientifically demonstrated to be low-grade alkyl mercury, in particular methyl mercury. As a consequence the establishment of (1) is complete. With regard to (2), it has been clearly demonstrated that the patients living near the mouth of the Agano River consumed large amounts of fish contaminated with methyl mercury. The source of the contamination of the fish may not have been fully explained with scientific accuracy, but the following evidence is sufficient to support the plaintiffs' contention that factory discharge was the pollution source. The Kanose plant continuously allowed wastewater from the production of acetaldehydes to drain into the Agano River. Mercury compounds and similar substances were detected both on the plant's acetaldehyde processing equipment and in the bog moss at the mouth of the discharge pipe. Even with an extremely low level of contamination in the water, it is possible for the pollutant to accumulate in the food chain and be-

come highly concentrated in river fish. It has also been demonstrated that the flora and fauna contaminated upstream floated downstream and were deposited as sediment in the tidal estuary at the mouth of the river. The plaintiff's contention can provide a consistent explanation for the pollution of the Agano River through time and space, although there are some phenomena left unexplained by the contention. It is completely free of any contradiction with the scientific disciplines upon which it is based. For these reasons we hold that the plaintiffs have met the standard of proof discussed in (1). As for the defendant's contention that agricultural chemicals were the source of pollution, the only possible pollution pathway is the salt water wedge. However, not only is this explanation itself scientifically questionable, but it is also inconsistent with relevant scientific explanations of the chronological and geographical extent of the pollution. Finally, even if the agricultural chemical theory were established as true in fact, it would not be sufficient to preclude the truth of plaintiff's factory discharge theory.

Thus we turn to (3) discussed above. The defendant is unable to deny that its Kanose plant produced and discharged methyl mercury compounds. On the contrary, the existence of that theoretical possibility has been confirmed. In addition, as previously stated, the source of the methyl mercury compounds detected has been proved to be within the factory or among the bog moss growing around the wastewater discharge pipe. Therefore, we cannot but conclude that the acetaldehyde synthesis process at the Kanose plant produced the methyl mercury compounds which were mixed with other factory wastes and then dumped into the Agano River.

Because of the foregoing, we hold that the defendant's discharge into the Agano River of effluents contaminated with the by-products of acetaldehyde synthesis was the legal and proximate cause of the subsequent occurrence of the poisoning.

Furthermore, we do not think that the approach to the issue of causation which we have adopted places an undue burden on the defendant company. The defendant has burned the production flow charts of acetaldehyde at the Kanose plant, has not preserved any sample of materials from the production process, and has removed the entire plant. The defendant, on the other hand, spent an exorbitant amount of money and mobilized an unprecedented amount of human as well as physical resources in order to disprove causation in the case before us. The defendant knew that the Kanose plant was suspected as the source of poisoning by the time the defendant destroyed relevant materials and facilities. Had the defendant preserved those materials (needless to say, it would have been very easy for the defendant to do so) the defendant could have offered them as evidence to relieve our doubts. In that case, the defendant could have proved item (3); and, if the truth were as the defendant contends, it could easily disprove the causation of the case.

3. In summary, the defendant's Kanose plant continuously discharged wastewater, which contained dimethyl mercury, into the Agano River thereby contaminating fish in the river and causing alkyl mercury poisoning among the residents along the river who consumed a large quantity of contaminated fish. The course and cause of the disease are very much like what was observed in Minamata. It may properly be called a second Minamata disease. (158)

IV. The Defendant's Liability

(A) Intentional Torts

The plaintiffs charge that the defendant intentionally inflicted mercury poisoning (Minamata disease) on the residents along the Agano River thereby causing physical harm and death to the plaintiffs. According to plaintiffs, the defendant knew by November of 1959 at the latest that the discharge into the Agano River of untreated wastewater containing dimethyl mercury from the Kanose plant might result in physical harm and death to the residents along the river who depended on fish and shellfish for food. Knowing the possible consequences, the defendant nevertheless continued the discharges. It should therefore be liable for its intentional torts. We shall deal with this issue first.

1. The Publication of the Research of Kumamoto University. Beginning in 1956, the research team of Kumamoto University made steady progress in determining the mechanism of Minamata disease. On November 4, 1956, the team announced that Minamata disease was caused by heavy metals taken into the human body through fish and shellfish. By July 22, 1959, the organic mercury theory had been reported, and on November 12, 1959, the Council on Food Sanitation of the Ministry of Public Health reported to the Minister that "the primary cause of Minamata disease is an organic mercury compound." These findings were made public in the form of research reports and articles in medical and other specialized journals and were frequently reported by local newspapers in the Kumamoto area. During this period, experiments [whose results were adverse to Chisso] on cats were being carried out by Hosokawa in the Chisso Minamata plant. Without publicizing those results, the company tried to refute the organic mercury theory and made its opinion public in October of 1959. But the Kumamoto University research group continued its clinical, pathological, and chemical experiments and verified the organic theory. Thus the Chisso Minamata plant became identified as the source of pollution.

2. The Minamata Fishery Disputes. Around 1957 Minamata disease patients and injured fishermen became convinced that discharges from the Chisso plant had caused the disease and started demanding that the company terminate these discharges, establish a treatment facility, compensate fishermen for their economic losses, and provide patients with relief. After July of 1959 when Kumamoto University announced that organic mercury was a possible cause, these demands intensified. In mid-August 1959 members of fishermen's unions in the area broke into the plant twice. That November they entered again, destroyed plant facilities and instruments, and battled with dispatched police. The fight resulted in more than one hundred injuries on both sides. The Minamata disease problem had reached a very serious stage. On November 25 of that year, the Mutual Aid Society of Patients of the Minamata Disease and their Families [the Society], an organization of patients of Minamata disease, demanded a total of ¥230 million [$770,000] from Chisso and sat in for more than one month at the entrance of the plant.

Under these circumstances forty-seven fishermen's unions along the Shiranui Coast (excluding the Minamata City fishermen's union which had already settled the compensation problem) accepted a proposal for settlement made by the Minamata Disease Conciliation Committee that (1) Chisso install a purifier within a week after the settlement and that (2) Chisso pay ¥35 million [$120,000] as compensation for fishery damage and ¥65 million [$220,000] as a recovery fund. On December 30, the Society also accepted a settlement proposal that required Chisso to pay a lump sum of ¥300,000 for deceased patients and an annuity to surviving patients. With these settlements the disputes over compensation and water treatment temporarily subsided.

This series of incidents were reported mainly by local newspapers.

3. A Circular Issued by MITI. A circular entitled Request for the Investigation and Report on the Quality of Factory Effluents dated November 10, 1959, and signed by the Chief of the Light Industrial Bureau of MITI was sent secretly to companies producing acetaldehyde and vinyl chloride, including the defendant. It stated:

Since 1953 the Government has made several investigations mainly through the Ministry of Health and Welfare to find the cause of the so-called strange Minamata disease, which appeared in the Minamata area of Kumamoto Prefecture, but has as yet been unable to obtain a definite result. Recently, however, a theory propounded by local Kumamoto University (Departments of Medicine and Science) has reportedly been gaining force. It speculates that a very small amount of mercury flowing from the Minamata plant of the Shin Nippon Chisso Fertilizer Inc. may have caused fish and shellfish in the Minamata Bay to become poisonous. The products, the production of which it is thought might result in the discharge of mercury, are acetaldehyde and vinyl chloride monomer. However, at other factories in various parts of Japan which engage in the production of these products, there have been no problems up to the present concerning the factories' effluents. This poses a major problem for the mercury theory. Therefore, it is thought necessary to investigate at every plant in Japan which engages in the same production as Chisso such matters as the treatment of mercury, the mercury contents in the discharged water, places in the production process where the loss of mercury might occur, and the state of mercury (element or compound) when flowing out of the plant. It is essential that we compare these data in order to determine the cause of Minamata disease as early as possible. We are sorry for taking your time when you are busy, but we would appreciate it very much if you would investigate and supply us with information on the matters appearing on the attached sheet [deleted]. Please send your reply in a confidential letter to the Chief of the First Organic Chemistry Section by November 30.

Because of the present politicization of the strange Minamata disease problem, it has been decided that this investigation should be carried out in secret. Under the circumstances, we hope that you will handle the investigation and the related matters carefully and sensibly inside as well as outside of your company.

After receiving the circular at the defendant's Kanose plant, the Chief of the Production Department called together the technicians dealing with organic chemistry above the rank of section head and discussed the fact that the Chisso plant was suspected of causing Minamata disease. However, no special action was undertaken at the plant. The defendant only took samples of the wastewater and sludge from the channel 50 meters away from the point where the wastewater from the organic and inorganic plants merged. The samples were analyzed and no mercury was detected. [The court proceeds to describe the method of analysis.] (159)

4. Investigation and Research by the Tamiya Commission. (a) (1) The Association of Japan Chemical Industries (hereinafter the Association) is a voluntary organization of producers, makers, and sellers of chemical products. The defendant and Chisso are among its members. In the summer of 1959, Chisso reported to the Association that it was having a difficult time because it was under suspicion of causing Minamata disease and because of disputes over compensation for

fishermen and others. In September, Takeji Oshima, Executive Director of the Association, went to Minamata to campaign for Chisso, distributed pamphlets to interested parties, and criticized the organic mercury theory of Kumamoto University.

(2) In November in view of the trouble at the Minamata plant, the Association established a special subcommittee (which included Chisso and the defendant) within its Special Committee on Industrial Wastewater Policies to study the treatment of industrial vinyl chloride and acetic acid wastewater by other members of the industry. . . .

Hearing from the President of Chisso on the situation in Minamata at its first meeting in mid-November, the subcommittee decided to leave Chisso to deal with the local problem. However, since the subcommittee found the study of Kumamoto University unpersuasive, it decided to organize a group of the nation's leading authorities including scholars from Kumamoto University to study the disease's causes. It also decided to fund the research of this group by contributions from the subcommittee member companies. On December 8 the subcommittee decision was confirmed by the Executive Board of Trustees of the Association. (159)

(3), (4) [The court then discusses how Takeo Tamiya, chairman of the Japan Medical Association, was chosen to head the group (Tamiya Commission) and how he selected its members, largely from Tokyo University and other Tokyo-based universities. Thereafter, four professors from Kumamoto University also joined the commission in 1961. The first meeting was held on April 8, 1960, and Masao Anzai, president of the defendant company, and the president of Chisso attended as representatives of the industry.

At the meetings some members named other chemicals as possible causes of the poisoning and remonstrated with those members who suspected organic mercury as the disease's cause. After the four members from Kumamoto University joined the commission, however, the organic mercury theory began to dominate. The member companies of the Association were informed of the activities of the commission either through Executive Director Oshima or by the journal of the Association. The information they received, however, was not completely accurate, for the committee's increasing acceptance of the organic mercury theory was not reported. Rather the organic theory was represented to be of questionable value.]

(b) As indicated in the above findings, the origin and progression of Minamata disease had gradually been uncovered by the Kumamoto University research team. As a consequence the fishermen's protests and demands against the company intensified and developed into a social problem. The government agencies reacted. MITI in particular notified other factories in the same industry as Chisso of the connection, advanced by the mercury theory, between mercury and Minamata disease and asked for reports on the treatment of mercury. Under these circumstances, it was only natural that the problem of Minamata disease caused concern and debate in the Association which was, after all, an organization of chemical companies engaging in the same production and trade as Chisso. Meanwhile, as for the academic question, the Tamiya Commission was continuing to conduct research on and discussions of the problem. It is reasonable to assume that member companies of the Association [and particularly members of the special subcommittee] were kept accurately informed, if not completely, about the general progress of the commission's research. The defendant was not only a member of the Association and the special subcommittee,

but the defendant's president was also the chairman of the Special Committee on Industrial Wastewater Policies. Moreover, he also attended the first meeting of the Tamiya Commission and discussed the Minamata disease with members from various universities. It is inconceivable that the defendant, the third-ranking company in acetaldehyde production, remained uninterested in the problem of the Minamata disease (from which some deaths had already resulted), a suspected source of which was the wastewater from the Chisso plant, the first-ranking company in the same industry. In fact, technicians at the Kanose plant did discuss the Minamata disease when they received the MITI circular. Therefore we can assume from the above that by the end of the Tamiya Commission's activities and by the end of 1961 at the latest, the defendant was seriously interested in, although it might not have wholly believed, the organic mercury theory. The company directors were aware of the fact that a connection between the occurrence of Minamata disease and wastewater from acetaldehyde was debated among scholars. It is clear that the defendant knew that wastewater from the Chisso Minamata plant was suspected to be the source of Minamata disease.

However, there is not enough evidence to find, as the plaintiffs allege, that the defendant "was indifferent to the fact that its Kanose plant's wastewater contained dimethyl mercury and that it could cause Minamata disease among residents along the Agano River." It is true that the results of research by Kumamoto University and others were published from time to time in professional medical journals or research reports and distributed among concerned persons. It is difficult, however, to say from that that the defendant was aware of the professional opinions in these sources and specifically that it believed in the organic mercury theory. Although the research developments and fishery disputes were reported in newspapers, those reports were primarily centered in the Kumamoto region. The defendant eventually learned of the problem of Minamata disease for certain from the MITI circular and through its activities in the Association. (161)

[The court then notes the testimony of several witnesses and concludes that there was not enough evidence to hold that the defendant acted intentionally and with precise knowledge.]

(B) Negligence

1. We turn now to the question of the defendant's negligence. Generally speaking, enterprises within the chemical industry produce vast amounts of products through the use of daily advancing chemical technology. In the process of chemical reactions, not only the intended product but also many kinds of by-products may be produced. Among these by-products some have the potential for posing grave danger to the lives of animals, plants, and human beings if discharged from the factory in their untreated form (hereinafter hazardous substances). Chemical companies, therefore, have a continuous duty to operate their factories safely so that hazardous substances are not discharged.

Consequently when a chemical company wants to discharge wastewater into a river or a stream, it must use the best analytical techniques to determine whether it contains any hazardous substances, what characteristics those substances possess, as well as the amount or concentration of the substance in the effluent. Based on the results, it must take the strictest safety precautions to prevent even the slightest danger to humans and other living things. Concrete *measures to avoid such harm will correspond to the nature of the hazardous sub-stance and the extent of its discharge.*

If despite the utilization of the most advanced technology, however, danger to human health remains, curtailment or cessation of operations may be necessary. Industrial activity can only be permitted in harmony with the preservation of the general populace's living environment. There is no reason to safeguard the profits of industry at the sacrifice of what must be called the people's most fundamental human right, the right to life and health.

2. We turn now to the case at bar. The Kanose plant of the defendant chemical company used inorganic mercury as a catalyst in the production of acetal-dehydes and disposed of the wastewater from that process by discharging it into the Agano River. In view of this method of disposal, the hazardous substances which arose as by-products of the process should have been continuously monitored for safety and quantity. Based on these results, there was a duty to take whatever measures necessary to prevent even the slightest injury to riparian residents using the river. As is clear from our earlier findings, the defendant was aware, especially by the end of 1961, of the existence of the organic mercury theory in the Kumamoto incident, that is, that wastewater from the Minamata plant of Chisso, the world's leading producer of the same products as the defen-dant, was the cause of Minamata disease. The defendant company, therefore, which was producing the same end products from the same raw materials and which was aware of the imminent danger to human life, had a duty to pay par-ticular attention to its wastewater, utilizing the most advanced analytic tech-niques to detect the presence of hazardous substances, determine their special properties, and measure the amounts present in the wastewater. As long as wastewater was discharged into the Agano River, there was a continuous duty to consider the hazards mentioned above.

Nevertheless, the defendant company paid no heed to the organic mercury theory advanced by Kumamoto University and simply regarded Minamata dis-ease as someone else's problem. It failed to fulfill the duty of care noted above in regard to the sampling and analysis of its wastewater and took no notice of the fact that it was producing methyl mercury, even tiny amounts of which are dangerous to plants, animals, and people. Instead, it continued discharging un-treated mercury together with wastewater into the Agano River where it con-taminated fish and caused Minamata disease among the plaintiffs, who, as ripar-ian residents, had eaten the fish.

It is true that the defendant company analyzed water and sludge in the dis-charge channel at the Kanose plant after the issue of the MITI circular and again in the spring and autumn of 1960, and mercury was not detected. The samples, however, were taken from a spot 50 meters downstream from the point where the wastewater from the organic and inorganic plants merged. By this point, the acetaldehyde wastewater had been diluted between 120 and 150 times by 50,000 tons of factory wastewater. This method of examination and analysis is not satisfactory. If the company had analyzed the wastewater before dilu-tion as it should have, it could have detected methyl mercury compounds in the wastewater even with less developed analytical techniques than were available at that time. The defendant company, however, closed the plant without ever car-rying out such an analysis.

The defendant company argues that its activities were not illegal because in discharging wastewater from the Kanose plant into the Agano River it abided by statutory standards, and the content of the methyl mercury in the wastewater did not exceed existing effluent standards. However, the mere observation of regulations cannot be deemed determinative in a damage suit as long as an activity injures human life and health. The laws referred to by the defendant are administrative regulations and are not decisive in determining civil liability. Furthermore, the several measures taken by the defendant for the recovery of mercury did not work effectively to remove methyl mercury compounds. It also appears from testimony that the cyclator installed by Chisso was equally ineffective. In view of these facts and all the evidence before the court, it appears that it was extremely difficult with existing technology to find a completely effective method of removing methyl mercury compounds. If because of that, danger to human health was foreseeable, the company had the duty of avoiding such a result by stopping its operations if necessary.

3. As a consequence of the defendant chemical company's negligence described above, the plaintiffs ingested methyl mercury and contracted Minamata disease. As described in sec. V these plaintiffs, the victims of the disease and their families, were injured. Under art. 709 of the Civil Code, defendants have a duty to compensate the plaintiffs for the harm inflicted upon them. (162)

V. Damages

(A) The defendant is liable for its wrongful acts and is obliged to compensate the plaintiffs for damages. In the case of tort liability, the amount of damages is determined in order to achieve a fair, reasonable, and rational allocation of loss between the parties involved.

(B) The following special characteristics of pollution cases should be considered in determining the amount of compensation.

1. In pollution cases, unlike automobile accidents and other personal injury cases, the injured have no possibility of becoming the injurers. No reciprocity exists between the parties. Usually the company has at its disposal capital and human and physical resources, while local residents have no equivalent resources and are in no position to obtain them. It is unlike automobile accidents where everyone has a fair chance to be both victim and injurer.

2. Pollution is accompanied by the destruction of the natural environment. The residents who live in the area are unavoidably and adversely affected and have no choice but to endure their injuries. It is unlike industrial or airplane accidents where victims may have some opportunity to avoid an accident. In cases of air and water pollution, it is almost impossible for residents to avoid the adverse effects produced by industrial activities. In most cases there is no contributory negligence on the part of the victims.

3. Pollution affects an indefinite number of residents over an extensive area. Usually the number of victims and the extent of the affected area result in a serious social impact beyond the extent of individual injury. As a result, the offending industry may become liable for a vast sum of compensation payments.

4. Pollution affects the living environment of the residents in a general way and affects and injures them indiscriminately because they all are sharing the same environment. There are cases where all or almost all of the members of a family are affected and their family life is entirely destroyed.

5. Pollution arises from industrial activities from which enterprises derive profit. Residents are not directly benefited from these activities at all. Enterprises may contribute to the society in general through their activities, but they cannot be permitted to destroy the local environment. Their activities are permissible only in so far as they are in harmony with the preservation of the living environment of neighboring residents.

(C) In determining the amount of compensation, these characteristics must be considered in arriving at a fair, reasonable, and rational result. In the case before us, the following facts must be taken into consideration.

1. Minamata disease irreversibly affects the patients' central nervous system. Effective methods of cure and treatment have not been found. Various treatments to remove organic mercury from a patient's body have been attempted. Corticosterone and various vitamins and tranquilizers are used to treat symptoms of the disease. They are somewhat effective in improving symptoms of acute poisoning, but are far from a cure for the disease. Some patients in Kumamoto show some improvement but others show worsening conditions after a dozen years. We find that the conditions of patients are still unstable and drastic improvements cannot be expected.

2. In the case before us the plaintiffs are claiming only pain and suffering (isharyō). It is apparent from the evidence and arguments before the court that they do not intend to seek damages for loss of present or future income. Under these circumstances factors such as the difficulty of cure or amelioration of Minamata disease, the duration of hospitalization and rehabilitation, age, the number of remaining working years, income, and living conditions are taken into consideration in determining the amount of damages for pain and suffering awarded each plaintiff.

(D) We now examine each plaintiff's damages. We find it reasonable to classify the patients of Minamata disease into five categories according to the extent to which their handicap affects their coping with daily life and their capacity to perform physical work. The amount of solatium is computed with reference to the following categories:

(1) A patient who is unable to maintain daily life without the help of others and suffers from psychological distress comparable to one close to death.

(2) A patient who is seriously handicapped in maintaining daily life.

(3) A patient who is able to maintain himself in daily life but is unable to perform physical work except simple and light chores.

(4) A patient whose capacity to perform physical work is considerably limited.

(5) A patient who is suffering from slight symptoms of Minamata disease and is in constant discomfort.

[The court then classified patients and awarded compensation according to the following ranks:

Deceased	¥10,000,000
1	¥10,000,000
2	¥5,000,000
3	¥4,000,000
4	¥2,500,000
5	¥1,000,000.

The relatives of the deceased and rank 1 patients were awarded compensation according to the following categories:

Dependent child	¥1,500,000 (in the case of deceased)
Spouse	¥700,000
Parent of young child	¥700,000
Child compensation allowance	¥500,000 (in the case of parents deceased)
Pregnant woman with possibility of congenital injury to fetus	¥300,000

Some other items such as the loss of opportunity for marriage were thereafter added. The total amount of compensation awarded was ¥245,699,800 [$819,000]. In addition to the compensation payments to the plaintiffs, a sum equal to 10% of the compensation was awarded as attorneys' fees. [19]
Under public pressure, the defendant surrendered its right of appeal.]

2.3 Agreements Between Plaintiffs and Defendants Reached as a Result of Direct Negotiations after the Niigata District Court Decision

Agreement

The Association of Niigata Minamata Disease Victims and the Joint Council to Combat Niigata Minamata Disease agree with the Shōwa Denkō Corporation as to the following in connection with the settlement of the question of compensation for the victims of the disease.

1. The Shōwa Denkō Corporation recognizes that by polluting the natural environment of the Agano River by its negligence concerning sewage treatment at its Kanose factory, even though it knew of the Kumamoto Minamata disease problem, the Corporation caused a second outbreak of Minamata disease. The Corporation has become aware of and has reflected upon the fact that, besides causing great injury to persons whose deaths were directly caused by the disease, to patients and their families, and to persons connected with the fishing industry, it has done regrettable deeds in the course of the lawsuit and the investigation of the cause of the disease and has burdened society by prolonging the settlement; the Corporation offers its heartfelt apologies to the victims and to society.

2. The Shōwa Denkō Corporation promises to perform the requirements of the agreement mentioned below in order to carry out its responsibilities as the perpetrator of harm and to compensate victims for the remainder of their lives and for damages to health inflicted by the Corporation. At the same time, the Corporation properly appreciates that the greatest wish of the victims is recovery of health and a complete cure, and so the Corporation shall energetically seek to discover methods of treatment and shall use the best measures to preserve their health.

3. Having reflected deeply upon the course which led to its having caused the Niigata Minamata disease, the Shōwa Denkō Corporation covenants that it shall endeavor to prevent—by the use of rigorous inspection and management of the industrial wastes (sewage, exhaust, refuse, etc.)—the creation of pollution by its own factories and those of related companies which it controls, including the Kanose Denkō Corporation; when it clearly perceives that one of the above fac-

tories may become a source of danger, it shall suspend operations in the factory and shall not permit the slightest harm to the residents of the area.

4. Since Minamata disease is a destroyer of human life hitherto unexperienced by mankind, there are still many aspects of its pathology and the harm it causes that are not yet elucidated. This agreement is based on the best present understanding of the situation, and although the compensation issue has been settled by the conclusion of this agreement for the present, it is understood that compensation to victims is not to be resolved on the whole solely by the payment of the compensation money established by this agreement. Should, for instance, the harm caused by the Minamata disease spread and intensify in the future, the problem shall be discussed and settled through negotiations with the Association of Niigata Minamata Disease Victims and a Joint Council to Combat Niigata Minamata Disease, in the spirit of the preceding paragraphs.

5. In view of the recently reported third and fourth outbreaks of Minamata disease, both the Shōwa Denkō Corporation and the Joint Council to Combat Niigata Minamata Disease are under a duty further to elucidate the nature of Minamata disease and leave for posterity more data for the prevention of the disease. Moreover, it is imperative as fellow human beings that we extend a helping hand to the victims. In order to accomplish this, the Shōwa Denkō Corporation shall submit all relevant data, endeavor to the extent possible to discover potential patients, and responsibly render aid to all discovered victims in the spirit of this agreement.

6. The Shōwa Denkō Corporation covenants to implement in good faith each of the following provisions of the agreement in the fundamental spirit mentioned above.

In witness whereof, the parties hereto set their signatures and seals, and each party keeps one original of this contract.

June 21, 1973

Association of Niigata Minamata Disease Victims
Agent of the Chairman: Hashimoto Jūichiro

Joint Conference to Combat Niigata Minamata Disease
Chairman: Watanabe Kihachi

Shōwa Denkō Corporation
Director, President: Suzuki Haruno

Provisions of the Agreement

The Shōwa Denkō Corporation shall render compensation in the form of lump-sum compensation payments, continuing compensation payments, special lump-sum compensation payments, and measures for changes in symptoms.

Part I. Lump-Sum and Continuing Compensation Payments

1. **Lump-Sum Compensation Payments.** *Such payments shall be made in the following manner to certified patients, excluding patients who joined the lawsuit as plaintiffs.*

(a) Persons who die of the disease or those who are incapable of living normal lives without others' assistance (hereafter serious cases) shall receive a flat ¥15,000,000.

(b) Other persons (hereafter general certified patients) shall receive a flat ¥10,000,000.

2. Continuing Compensation Payments. *Such payments shall be made to living patients for their entire lives, including patients who joined the lawsuit as plaintiffs.*

(a) The amount is to be a flat ¥500,000 per year. This figure is arrived at by taking the 1972 figure for Niigata given in the "Composite Index for Metropolitan-Class Districts in which Prefectural Offices are Located" in the Annual Consumer Goods Index *prepared by the Bureau of Statistics under the Prime Minister's Office, and revised to take account of significant fluctuations of over 30%.*

(b) This provision takes effect January 1, 1973. Half the amount of the annual payment is to be paid on the 31st of March and half on the 30th of September to patients currently certified under these data.

(c) Persons who are certified as patients prior to the effective date of this provision shall receive appropriate amounts of continuing payments for the period from the date of certification until December 31, 1972, calculated as monthly shares of the annual ¥500,000.

(d) The details of implementation shall be set forth in the appended agreement.

Part II. Provisions Concerning Changes in Symptoms

1. In the event that a patient having joined the lawsuits as a plaintiff, or one of the general certified patients, becomes a serious case or dies (the cause being Niigata Minamata disease, complications or side effects of Niigata Minamata disease, or accidents related to Niigata Minamata disease), the difference shall be paid so that the amount of compensation equals the lump sum stipulated in part I-1a.

2. Where Niigata Prefecture or Niigata City has decided to pay a person a nursing allowance based on law, the confirmation that a person has become a serious case is to be determined thereafter.

Part III. Provisions Concerning Patients Having Joined the Lawsuit as Plaintiffs

1. The families of persons who die during the course of adjudication (or after adjudication, or those who die prior to April 21, 1973, the date on which the confirmation was concluded) and those who are recognized as serious cases in the judgment shall each receive a special lump-sum payment of compensation, the difference between the amount determined by the court and the amount paid to the families of the deceased serious cases.

2. Other patients who joined the lawsuit as plaintiffs shall receive, as special lump-sum compensation payments, the difference between the amount determined by the court and the amount paid to general certified patients.

Part IV. Provisions Concerning Costs of Medical Treatment

Medical treatment costs are not included in the compensation to be paid under this agreement.

Part V. Provisions Concerning Negotiations to be Conducted Hereafter

1. With the exception of matters stipulated in this agreement, the following shall be negotiated and determined by the Association of Niigata Minamata Disease Victims and the Joint Council to Combat Niigata Minamata Disease.
(a) Matters concerning handling of future certified patients.
(b) Matters concerning implementation of this agreement.
(c) Matters concerning prevention of pollution at the Agano River, including removal of pollution.
(d) Matters concerning maintenance of patients' health and their rehabilitation.
(e) Other necessary matters.
2. Whenever either party to this agreement proposes to enter negotiations as to the preceding provisions, the other shall respond promptly.
3. Either the Joint Council to Combat Niigata Minamata Disease or the Shōwa Denkō Corporation may initiate negotiations.
4. Negotiation rules and details as to the content of discussions shall be set forth in the Appended Memorandum [omitted].

Part VI. Provisions Concerning Payment of Compensation

Details as to methods and times for payment of the compensation prescribed by this agreement shall be set forth by separate agreement [omitted].

2.4   Decision of the Kumamoto District Court[20]

I. Parties

[*The plaintiffs are Eizo Watanabe and 137 others. The defendant is the Chisso Corporation.*]

II. Causation

(A) Neither party disputes the following facts: Minamata disease, a poisoning of the central nervous system, has resulted from the consumption of great quantities of fish and shellfish from in and around Minamata Bay and over a long period. All patients of Minamata disease (and, for patients born with congenital disease, their mothers) have engaged on a full- or part-time basis in fishing in Minamata Bay and in the neighboring Shiranui Sea.
The defendant, however, formally disputes the causal relation between the discharge of wastewater from its plant and the occurrence of Minamata disease. We shall thus examine the issue of causation below. (63)
(B) Relevant Facts in Finding Causation.
1. Discovery of Minamata Disease. *At the end of 1953, some incidents of a disease of unknown cause affecting the central nervous system were observed in certain districts on the outskirts of Minamata City. Eight patients in 1954 and about five in 1955 were observed when the disease's cause was still unknown. On April 21, 1956, a 6-year-old girl with symptoms of brain damage was brought by her mother to the pediatric section of the hospital attached to the defendant's factory (hereinafter defendant's hospital). She was promptly hospitalized and given various medical tests. While she was there and still not diagnosed, her 3-*

Table 1
Incidence of Minamata disease.

Year	Number of Incidents	Number of Adult and Child Victims	Number of Congenital Cases[a]	Number of Deaths
1953	1	1	0	0
1954	12	12	0	5
1955	14	9	5	3
1956	51	44	7	10
1957	6	0	6	2
1958	5	3	2	5
1959	18	16	2	7
1960	4	4	0	2
1961	0	0	0	1
1962	0	0	0	2
1963	0	0	0	0
1964	0	0	0	0
1965	0	0	0	4
Total	111	89	22	41

[a]In congenital cases the year of birth is regarded as the year of incidence.

year-old sister was admitted to the hospital on April 29, with identical symptoms. A doctor questioned the mother in detail and learned that there was a patient with the same symptoms in the next house. Sensing the seriousness of the situation, the hospital reported the incidents to the Minamata Health Center on May 1 and asked it to make a survey of the district where the patients had come from. In the meantime a hospital doctor went with health officials to the patients' homes and found that there were a considerable number of other patients (about forty) in the neighborhood. The individuals were being concealed, because they had been quickly labeled as suffering from a communicable disease and the conventional belief persisted that patients of communicable diseases should be hidden from sight. The doctor also learned that the affected district was on the shore of Minamata Bay, that large fish had sometimes floated to its surface, and that cats which ate those fish became afflicted with the same disease.

Thereafter on May 28, 1956, the Action Committee on the Strange Disease at Minamata was formed by the Minamata City Doctors' Association, the health center, the city government, the city hospital, and the defendant's hospital. The committee was instructed to treat the patients and investigate the cause of their disease. Kumamoto University Medical School was also asked to investigate and determine the cause of the disease. On August 24 the medical school established a Minamata Disease Medical Research Team to begin the search for the cause of the disease.

2. The Circumstances of the Outbreak of the Disease. *By the end of December 1965, there were 111 diagnosed patients of Minamata disease. (63) [See the distribution of patients by year [table 1] and by area [table 2].]*

The first case occurred on December 15, 1953, and the last case on October 9, 1960. There were a large number of patients in 1954 and 1955 and a dramatic increase in 1956. The number decreased in 1957 and 1958 but rose again in 1959. In

Table 2
Distribution of Minamata disease patients.

Area	1953	1954	1955	1956	1957	1958	1959	1960	Total
Deguki	1	2	3	13					19
Yudo		1	2	16	1	1		1	22
Tsukiura		2	3	7				2	14
Hyakken		4	1	5					10
Myogin		2	2	2					6
Marushima				3		1			4
Umedo			2	1		1	1		5
Sakaguchi				1					1
Tatara		1							1
Modo				2	3	2	2	1	10
Hirashita				1					1
Yawata							5		5
Tsunagi							5		5
Taura					1				1
Yuura							1		1
Ashikita							1		1
Igumi			1		1		3		5
Total	1	12	14	51	6	5	18	4	111

1957, based on the results of an investigation by Kumamoto University, administrative guidance was issued which strongly urged residents to restrain themselves from eating seafood from Minamata Bay as part of the effort to deal with the situation. At the same time many residents voluntarily stopped eating fish altogether out of fear of the disease. These precautions may help explain the decrease in 1957 and 1958.

As is clear from table 2 from 1953 to 1958 most of the incidents of the disease occurred around Hyakken Port and Minamata Bay. However in 1959 patients were found in villages far north and south of the Minamata Bay area. In the north patients were found at the mouth of the Minamata River, some at Tsunagi, and one at Ashikita. In the south incidents were observed as far as Yonenotsu in Izumi City. From September 1958 to September 1959, the defendant altered the course of discharged wastewater from its acetaldehyde plant from Hyakken Port to the mouth of the Minamata River via Yawata Pool. Some patients afflicted with the disease in 1959 were known to have eaten fish and shellfish caught at the mouth of the river. A singularly prevalent phenomenon associated with this large outbreak of the disease was the death of many cats and other domestic animals with symptoms similar to those of the patients. (64) [One count taken by a house-to-house survey showed that 50 out of 60 cats that lived in 40 patients' households died during the four years from 1953 to 1956; 24 out of 60 in 68 nonpatients' households died during the same period.] A house cat usually died a couple of months before some member of the same household was taken ill. Autopsies of these cats indicated that they died from Minamata disease. (64)

3. Clinical Symptoms of Minamata Disease.

(a) Major Symptoms. *Major common symptoms of the disease [details deleted][21] are identical to those of the organic mercury poisoning called Hunter-Russell*

syndrome. (65) (The disease was reported in Britain in 1940 by doctors Hunter and Russell. Some factory workers in England who handled various mercury compounds became affected with the symptoms by intake of mercury through the skin and by breathing.) However it should be noted that since Minamata disease results from the concentration of mercury through the food chain and its victims range from fetuses to the elderly, it shows a wide variety of symptoms that were not observed in the Hunter-Russell case. . . . (65)

(b) Characteristics of Congenital Minamata Disease. . . .[22]

(c) Mercury Content in Hair. *Various poisonous agents taken into the human body may accumulate in hair. Between December 1959 and January 1960 a Kumamoto University Medical School team examined the mercury content in the hair of twenty-five patients. (67)*

[*The highest figure was 705 ppm, and samples of hair taken from adult persons of normal health who lived outside of the Minamata area showed from 0.14 to 4.42 ppm of mercury.*]

4. Pathological Observations of Minamata Disease. . . .[23]

5. Experiments with Animals. *At the defendant's factory, doctors Hosokawa and Kojima of the factory hospital undertook experiments on cats in cooperation with the engineering section. Beginning in May of 1957, the doctors kept adult cats weighing 2.0 to 4.5 kg in a kennel established on the hospital premises. The doctors fed the cats fish caught in and around Minamata Bay. By August 1959 a total of 292 cats had been experimented on and within one or two months 118 had developed symptoms similar to those occurring naturally in cats in the infected area. The symptoms of Minamata disease in cats can be separated into preliminary and typical symptoms. At the onset the cats lost their appetite and sat still all day. Their hair lost its shine. After a couple of days of these preliminary symptoms, the cats began to show such typical symptoms as spasms, a disjointed walk, loss of coordination, abnormal compulsive movements of the head, running around, bumping into walls, standing upside down, "dancing," uncontrolled salivation, and narrowing vision.*

The following findings were made as a result of close observation of the 118 cats with these symptoms:

(1) The fish and shellfish that induced typical symptoms in cats were all caught in the Minamata Bay or within 2 km of its entrance.

(2) There were many different kinds of poisonous fish and shellfish in the bay. Shellfish transplanted from other areas into the bay became poisonous within a month. Fish such as sardines, which move about, were more poisonous than shellfish rooted in the bay.

(3) Both the internal organs and the meat of the fish were poisonous. Either part fed separately to the cats could induce the disease.

(4) Fish and shellfish heated and dried for preservation were still poisonous. The poisonous section would not dissolve in water or acetone.

(5) Shellfish in the bay were poisonous in all four seasons. No seasonal change of toxicity was detected.

(6) It took twenty days in the longest case before a cat showed typical symptoms after the termination of feeding. During that period no outward change was observed. Once cats were taken ill, they showed no improvement of symptoms and would die within twenty days.

(7) Feeding cats with sludge from the bottom of the bay or acetone extracted from it could not induce symptoms. (69)

6. The Examination of the Manganese Theory and Other Theories Concerning the Suspected Causal Substance of the Disease. *In the search for causes of the disease, various theories pointed to different substances such as manganese (November 1956), manganese-selenium-thallium (July 1957), and thallium (May 1958) as the suspected cause prior to the publication of the organic mercury theory by the Kumamoto University Medical School in July 1959. A bomb-explosion theory was also propounded by Takeji Oshima, acting trustee of the Japan Association of Chemical Industries (September 1959). (69) [This theory suggested that some bombs from the naval base at Minamata that had been dumped into the bay at the end of World War II had leaked due to deterioration of the bomb casings and released their contents into the bay. Between October 1959 and February 1960, the defendant attempted to verify this theory by having scuba divers search for bombs but nothing was found. The court then examined this and several other theories, but rejected them all because the supporting data were unreliable.]*

7. Organic Mercury Compounds Arising Out of the Defendant's Acetaldehyde Plant. *[The court discusses how the defendant used mercury as a catalytic agent in manufacturing acetaldehyde (CH_3CHO), and how methyl mercury compounds (CH_3-Hg-X) were formed in the production process and allowed to drain into the wastewater. The court next points out that since the defendant's plant began producing acetaldehyde in 1932, production increased as follows:*

1946	200 tons per month
1947	200
1948	300
1949	370
1950	370
1951	520
1952	520
1953	700
1954	750
1955	900
1956	1,300
1958	1,600
1959	2,500
1960	3,300

In 1964, defendant Chisso Company completed a petrochemical plant to produce acetaldehyde in Chiba Prefecture and in May 1968 terminated the production of acetaldehyde at the Minamata factory.]

8. Defendant's Factory's Wastewater, Especially Acetaldehye and Vinyl Chloride. *[The court further explains how the defendant discharged wastewater from its acetaldehyde production process through the Hyakken Channel into Minamata Bay from February 1946 to September 1958, and how thereafter wastewater was discharged into the mouth of Minamata River via Yawata Pool.*

The court continues its analysis by noting that according to figures which the defendant published in October 1959, the factory used 11,964 kg of mercury in its acetaldehyde production process and retrieved 10,305 kg of it during the

months of September and October of the same year. A total of 1659 kg of mercury was thus lost. At about the same time, 296 kg of mercury were used in the vinyl chloride production process and 65.5 kg of mercury were lost. The court emphasizes the significance of the fact that the discovery of patients near the mouth of the river began around 1959 when the flow of wastewater from the acetaldehyde production process was directed into the river in great quantities.]
9. The Conclusion of Kumamoto University's Research and the Government's Statement. [The court reports that in 1961 the researchers of the public health section of Kumamoto University Medical School extracted crystals of an organic mercury compound from samples that were taken from a pipe connected to the acetaldehyde reaction tank in August 1959 and October 1960. The researchers analyzed the crystals and identified them as CH_3HgCl. Thereafter they raised short-necked clams in a sea water tank that contained CH_3HgCl and found that the clams produced the same organic mercury compound as had been found in clams caught in Minamata Bay.]

On September 26, 1968, the government (the Ministry of Public Health) issued the following official statement:

Minamata disease results from the poisoning of the central nervous system. It is caused by the consumption of fish and shellfish from Minamata Bay in large quantities over a long period. The causal substance is organic dimethyl mercury. Dimethyl mercury compound is a by-product of the defendant's acetaldehyde plant. It is contained in its wastewater and was discharged into Minamata Bay where it contaminated fish and shellfish. The residents in the area have been afflicted with this disease by eating those fish and shellfish which have accumulated and concentrated dimethyl mercury in their bodies.

Also, Eiichi Nishida, who was the plant manager of the defendant's factory from January of 1957 to May of 1960, offered testimony at the tenth oral proceeding before the court on February 5, 1971, which clearly confirmed the statement made by the government. (73)
(C) We have found no sufficient evidence to contradict the findings stated above in headings 1 through 6 and in 8 and 9. From our findings in 1 through 9, we decide the issue of causation as follows: The substance causing Minamata disease was a methyl mercury compound created within the defendant's acetaldehyde factory. The mercury in the wastewater was discharged into Minamata Bay and the surrounding sea and accumulated in the bodies of different marine organisms. The residents of that area who ate seafood in large quantities over a long period of time were those who contracted Minamata disease. Therefore, the causal relationship between the defendant's discharge of acetaldehyde wastewater from its factory and the outbreak of Minamata disease is sufficiently confirmed. Evidence contradicting this finding is insufficient. (73).

III. The Defendant's Liability (Negligence)

(A) As noted above, the defendant's factory began the production of acetaldehyde for use in the manufacture of an acetic acid compound in 1932. As the demand increased, the defendant gradually increased production. Production particularly increased each postwar year after 1946. The amount of wastewater discharged from the factory consequently also increased markedly after this year. It has been clearly demonstrated that the plaintiffs, as is noted for each individual in part VI, were stricken with Minamata disease from 1953 to 1961 as a result of exposure to discharged wastewater containing dimethyl mercury compounds.

The plaintiffs' contentions, set forth as follows, assume that a discharge of contaminated wastewater is permitted only when its safety is assured. As long as it is foreseeable that discharges of contaminated water could result in injury to another's legally protected interests, there is a duty to ascertain beforehand whether such discharges are poisonous in content and to assure that the wastewater is harmless. Since the defendant was able to foresee the risk and failed to take precautions, it cannot escape liability for negligence. The defendant answers that until the middle of 1962, when it was detected and confirmed at the medical school of Kumamoto University, the fact that methyl mercury chloride, CH_3HgCl, the causal substance of the disease, would be produced in the production process of acetaldehyde was totally unknown by the chemical industry and academicians. The defendant contends that it therefore follows that it could not have known of the existence of a dimethyl mercury compound in its wastewater, the process of its accumulation in fish and shellfish, or that residents who consumed these fish would contract Minamata disease. The defendant argues that there should be no liability for negligence where there was no foreseeability of the consequences.

(B) Since the production process of the chemical industry generally utilizes large quantities of dangerous substances such as raw materials and catalysts, there is an extremely high probability that unpredictably harmful by-products such as unreacted materials, catalytic agents, intermediate products, or the finished product itself will be in the factory's wastewater. When these dangerous materials are discharged into the rivers and seas, harm to plants, animals, or people can be easily anticipated. Therefore when a chemical plant discharges wastewater, it must always use the best knowledge and technology to determine whether harmful substances are present and what effect there might be on plants, animals, and humans. In addition to assuring the safety of its wastewater, if by any chance harm becomes apparent or there arises doubt about its safety, the factory should immediately suspend operations and adopt the necessary maximum preventive measures. Especially with regard to the life and health of area residents, the factory must exercise a high degree of care to prevent harm before it happens. It must bear alone the obligation of guaranteeing the safety of the lives and health of residents since there is no way for residents to know what or how things are produced or what kind of wastewater is being discharged. Certainly the factory did not tell them. After all, no factory of whatever kind should pollute or destroy the environment through its operation; even less should it infringe on the health and lives of the residents or allow them to be sacrificed.

The defendant claims that foreseeability is limited to the foreseeability of the production of the specific causal agent and contends that it did not violate any duty since it could not possibly have foreseen this specific outcome. But if one were to proceed along these lines, the degree of danger could only be proven after the environment was polluted and destroyed and lives and health of people harmed. Until that point, the discharge of dangerous wastewater would have to be tolerated. The inevitable consequence would be that the encroachment on the lives and health of residents could not be stopped. Since this would be tantamount to allowing the residents to be human experiments, it is clearly unjust. (73)

[*The court notes that the amount of production of acetaldehyde and vinyl chloride increased greatly during the 1950s, and that the defendant's factory became one of the leading chemical plants in Japan by its use of the most advanced*

technology in the field. As the defendant's facilities were enlarged and its pro-
duction expanded, so also did the risk of dangerous substances forming in the
production process and mixing in the discharged wastewater increase every
year.]

In order to comply with the high degree of care expected of one of the nation's
leading chemical synthesis plants, the defendant should have conducted con-
tinuous research and investigation. At the same time that it should have fre-
quently analyzed and investigated the quality of its wastewater to insure its
safety, it should have studied the topography, tides, etc., of Minamata Bay into
which the wastewater was being discharged and noted any changes therein. It
cannot be denied that the defendant was expected to assure the complete safety
of its wastewater. (74)

[From the submitted evidence and testimony of witnesses, the court found
that, had the defendant researched available technical literature carefully, it
could have learned even before 1955 that a methyl mercury compound soluble
in water might be formed in the production process of acetaldehyde. The court
found that the defendant should have known from its own experience that unex-
pected by-products were often formed by chemical reactions in the production
process. The court found no evidence that the defendant researched available lit-
erature on the catalytic function of mercury before 1959. Occasionally, the de-
fendant had analyzed wastewater discharged from its factory for the purpose of
improving the efficiency of the production process or of ascertaining whether it
complied with administrative regulations for pH, suspended solids, biochemical
oxygen demand (BOD), chemical oxygen demand (COD), and dissolved oxygen.
The court emphasized that the defendant had failed to undertake the necessary
measures to ensure public health and safety.]

Despite the fact that the defendant's factory had some of the best equipment
and technology in the nation, it had failed to comply with the required standard
of care before heedlessly discharging its wastewater. It can be held liable for
negligence on this point alone. (74)

(C) We will now examine how the defendant reacted to events like the observ-
able changes in the environment, disputes over compensation for the fishing in-
dustry, research into the cause of Minamata disease, treatment of factory
wastewater, and animal (cat) experiments. We examine the defendant's behav-
ior concerning these matters as factors in our decision on the issue of negligence.

1. Unusual Changes in the Environment. *The amount of fish caught drastically*
decreased in and around Minamata Bay between 1953 and 1954. Fish such as
snapper, gray mullet, and scabbard were often observed floating dead in the bay.
From 1954 to 1956 many cats in the villages around the bay [names omitted]
died with symptoms of nerve damage. In those three years more than fifty cats
died, some pigs and dogs displaying similar symptoms died, and in some areas
birds became unable to fly or walk and dropped dead. These unusual and strange
phenomena became almost common occurrences. . . . (75)

2. Disputes over Compensation for Damages to Fisheries (Part IV, 8.2). . . .

3. Efforts to Determine the Cause of the Disease. *As we have already mentioned,*
after 1956 the defendant's plant came under strong suspicion because its dis-
charge water was viewed as connected with the occurrence of the disease. The
defendant, therefore, before anyone else, should have made every effort to inves-
tigate and to identify the cause of the disease. . . . However, the defendant did
not undertake any investigation worth noting, and in no instance did it make the
results of its investigations public. (77) . . .

Even after Kumamoto University developed the organic mercury theory, there is no indication that the defendant, whose factory used mercury in its production process, made any effort to investigate or to analyze its wastewater [that is, the acetaldehyde and vinyl chloride wastewater] for the presence of organic mercury compounds. (77)

Moreover, the defendant prolonged the university's efforts to identify the cause of the disease by its unwillingness to disclose an overall picture of its production layout, the processes of the factory, and other relevant information such as materials used, catalytic agents, by-products in the production processes, and the method of wastewater treatment. It is no exaggeration to say that the defendant contributed largely to the growth of the number of Minamata patients. The defendant failed to cooperate with the university teams and failed utterly to take proper independent action to determine the cause of the disease. (77)

4. **The Treatment of Wastewater and the Amount of Mercury Discharged.** *According to the testimony of the plant manager, Nishida Eiichi, a total of 60 tons of mercury was estimated to have been lost by an account made in July, 1959. The basis of this estimate is unclear, but the daily production records reveal that more mercury may have been lost from the factory production process. [During 1954, 38,058 tons of mercury were used in the production process; of this, 28,069 tons were retrieved, meaning that 9,939 tons were lost in the process. During 1955, 51,716 tons were used, 39,701 tons retrieved, and 12,015 tons lost.] In October 1969, Professor Namba of Kumamoto University Medical School estimated that a total of 600 tons of mercury had been discharged in wastewater from the defendant's factory. The total amount of mercury discharged, if 600 tons is exaggerated, clearly exceeded by far the 60 tons estimated by the defendant. The defendant cannot escape criticism that in order to protect itself it made a too conservative, and even misleading, estimate of mercury used and lost. (79)*

5. **Experiments on Cats, Especially Cat Number 400.** *[As noted above, some factory hospital doctors with the cooperation of the factory's engineering department undertook an experiment of feeding cats with fish caught in the bay. In the middle of 1959 they adopted a direct feeding method where cats were fed with food soaked with water discharged from different production processes such as acetaldehyde and vinyl chloride. The experiment continued until December 1962, and the number of cats experimented on reached 900. The condition of the cats was observed daily and recorded in detail. Every cat was classified and registered in the so-called cat registry.]*

In mid-July 1959 Hosokawa himself took a sample of wastewater from the mouth of the acetaldehyde plant's discharge pipe and starting on the 21st, poured 20 cc's of it daily on the food for cat number 400. By October 6th or 7th the cat had developed a light paralysis in its hind legs and thereafter Minamata disease symptoms such as spasms, salivation, shivering, dancing, and running in circles appeared. Its weight decreased from 3 kg at the beginning of the experiment to 1.8 kg and it became weaker every day. On October 24, the doctor killed the cat, performed an autopsy, and sent the samples to Kyūshū University for pathological analysis. (80) . . .

[The court notes that Hosokawa testified that he had informed the engineering department of the results of the experiment on cat number 400, and that he was subsequently ordered to terminate direct feeding experiments. Witnesses for the defendant, including personnel employed at the time in the engineering department, strongly denied these allegations. The court noted that the journal of the cat experiments corroborated the termination of direct feeding.]

It is reasonable to assume that the officials of the engineering section were aware of the fact that cat number 400 was directly fed food soaked with the discharged water from the acetaldehyde plant and that it had developed symptoms almost identical to those of Minamata disease. . . . (81)

We find that at least those who were in the engineering section of the defendant's factory knew the results of the experiment on cat number 400 by October 1959. On November 30 the chief of the section ordered the termination of that experiment. Since the results of the experiment on cat number 400 were not made public, the researchers of Kumamoto University Medical School were misled as to the direction of their research. The termination of Hosokawa's experiment clearly delayed the identification of the cause of Minamata disease. The defendant's reponsibility on this point is extremely grave. Even if those in the engineering section did not in fact know the result of the experiment on cat number 400 and did not order the experiment terminated, if they regarded the experiments as important and were following their progress, they could have known about cat number 400. In fact they should have directed and promoted such experiments themselves and by doing so made every effort to find and identify the cause. By failing to do so, the defendant cannot escape liability. (82)

(D) We summarize here what we stated under headings B and C. The defendant's factory was a leading chemical plant with the most advanced technology and facilities. As such, the defendant should have diligently researched the relevant literature and should have assured the safety of its wastewater before discharge by analyzing it for the presence of hazardous substances. Also the defendant should have cast a watchful eye on the environmental conditions of the area into which the discharged water flowed and noted any changes therein. Defendants should have made sure that no harm whatsoever came to the residents in the area from the discharged water. Had the defendant not failed to exercise this duty, it would have been possible to foresee the risk from the discharged water to humans and other living things. The defendant could have prevented the occurrence of Minamata disease or at least have kept it at a minimum. We cannot find that the defendant took any of the precautionary measures called for in this situation whatsoever. There were many signs, such as strange environmental phenomena, the fishery disputes over compensation, the investigation into the disease's cause, concern over the waste treatment facilities, and experiments with cats. We cannot find even one measure taken by the defendant that was either adequate or satisfactory. We find absolutely unsupportable the defendant's contentions that factory officials gave full cooperation to the investigation team of Kumamoto University, that management of wastewater was adequate, or that the defendant completely treated these wastes.

Judging from the above, the presumption that the defendant had been negligent from beginning to end in discharging wastewater from its acetaldehyde plant is amply supported. Even if the quality of wastewater was within legal and administrative standards and the facilities and methods of treatment at the defendant's plant were superior to those of other factories in the same industry, it is not enough to overcome this presumption. The discharge of wastewater occurred as a result of the defendant's industrial activity, and the defendant cannot escape liability for negligence in that context.

Therefore the defendant, by the discharge of a dimethyl mercury compound in wastewater from its plant, inflicted Minamata disease on the plaintiffs, thereby imposing on the plaintiffs, victims and their families, the damage described in sec. VI of this opinion. Furthermore, the discharge of wastewater was part of the defendant's industrial activities. This is not a case of vicarious liability of a corporation where a representative agent (art. 44(1)) or employee (art. 715(1)) in the course of performing his duty causes damage to a third person. The defendant company, while engaged in its enterprise, discharged wastewater and thereby became directly liable for damages under art. 709 of the Civil Code. (82)

IV. Mimaikin Contracts

(A) [The defendant company had given the victim-plaintiffs money as a token of sympathy after entering into the seven separate agreements with them.]

Agreement:

First	12/30/59
Second	4/26/60
Third	12/27/60
Fourth	10/12/61
Fifth	12/27/62
Sixth	8/12/64
Seventh	6/16/69

(B) Circumstances Leading to the Signing of the First Contract on December 30, 1959.
1. From around 1953, Minamata disease patients began to appear. Although at that time they were treated as being afflicted with a contagious disease known only as the "strange" disease, in February 1957 the public health department of the Kumamoto Prefectural Government issued a warning, based on the report of the Minamata Disease Research Team of Kumamoto University, that fish and shellfish from Minamata Bay were dangerous. About the same time, due to an epidemiological study undertaken by Kumamoto University Medical School, the industrial wastewater discharged from the Minamata factory came under suspicion as a possible polluting source. However, the exact cause and source of the pollution were not yet identified. Finally on July 22, 1959, Takeuchi Tadao of Kumamoto University Medical School announced that the causal substance in Minamata disease was a "certain kind of organic mercury." The media reported that inorganic mercury contained in the wastewater from the Minamata factory and concentrated in fish and shellfish might be transformed into poisonous organic mercury. The general public as well as the people who lived in the area thus strongly suspected the factory to be the source of pollution. A little more than three months later, on November 12, 1959, the Investigatory Council on Food Sanitation of the Ministry of Health and Welfare reported to the minister that the primary cause of Minamata disease was a certain type of organic mercury compound. (83)
2. The dispute over the compensation for damage to the fishing industry continued. On August 6, 1959, when its members were unable to sell their fish, the Minamata City Fishermen's Union demanded ¥100 million compensation from the defendant company. When their demands were not met, they forcibly entered the factory to protest. On the 30th, the factory settled the dispute with the fishermen by paying them ¥35 million. Meanwhile the presence of Minamata disease patients was confirmed in an area extending north of Minamata City.

Thereafter, no one was willing to buy fish caught along the Shiranui coast. On October 17, 1959, 1,500 fishermen from six different unions in the affected area gathered under the name of the Kumamoto Fishermen's Mass Protest Meeting to make the following demands of the factory:
(a) to stop operations until a treatment facility was completed;
(b) to clean the sludge off the sea bed of Minamata Bay;
(c) to pay compensation for damages suffered by them.
When they handed this statement to Nishida Eiichi, the plant manager, an incident took place in which a few plant guards were injured. The factory replied in a message to the chairman of the Federation of Kumamoto Fishermen's Unions (hereinafter the Federation) that they could not stop operations nor would they clean the sea bed nor pay compensation so long as the cause of the strange illness was unknown. On November 2 the Minamata Disease Investigation team of the Diet visited Minamata City and recommended that the factory make an earnest effort to determine the cause of the illness by cooperating with Kumamoto University Medical School and complete the wastewater treatment facility as soon as possible. On the same day 1,700 fishermen held a demonstration in the city to petition the Diet team. Several hundred demonstrators, who were discontented with the factory's refusal to stop operations, forced their way into the factory, destroying windows and office facilities, burning documents, and assaulting company employees. More than 300 policemen were called in to suppress the riot. More than fifty persons including fishermen, employees, and policemen were injured.

In this way the compensation of fishermen became a large social issue. On November 5, a special committee of the Kumamoto Prefectural Assembly passed a resolution to consider a prefectural ordinance to stop the discharge of wastewater and propose to the fishermen and the factory that they accept the intervention of the governor of Kumamoto Prefecture. That same day a message based on this resolution was sent from the chairman of the Prefectural Assembly to the plant manager and the chairman of the Federation requesting that they accept the governor's mediation. Both parties took the advice and by the middle of the month had asked the governor to intervene. Governor Teramoto Hirosaku felt the best way to settle the dispute was to establish a committee, and on November 24 he appointed Iwao Yutaka, chairman of the Prefectural Assembly; Nakamura Todomu, mayor of Minamata City; Kozu Torao, chairman of the National Association of Town and Village Heads; and Izu Tomito, president of Kumamoto Daily Newspaper: He also named Kawase, Fukuoka Bureau Chief of MITI, and Oka, trustee of the Federation of Fishermen's Unions, as observers. The committee was headed by Governor Hirosaku and was called the Conciliation Committee on Disputes Concerning Fisheries on the Shiranui Coast.

On December 18 after three conciliation meetings both parties signed an agreement in which the factory agreed to complete the wastewater treatment facility (circulator, sediment-floater) within one week of signing the agreement, to pay the Federation ¥35 million for damage compensation, and to provide a loan of ¥65,000,000 to be used as a relief fund. Later in May 1960 the factory also concluded an agreement with fishermen's unions of Kagoshima Prefecture and in October, with the Minamata City Fishermen's Union. Thus disputes relating to the compensation of the fishermen of the Shiranui coast were resolved. (84)
3. On August 15, 1957, victims of Minamata disease and their families (including the survivors of deceased patients) formed the Mutual Aid Society of Patients of the Minamata Disease and their Families (hereinafter the Society) in order to

negotiate with the defendant factory and give each other mutual assistance. Watanabe Eizō, one of the plaintiffs, was elected to be chairman. Thereafter other newly certified patients and their families joined the Society, and at a general meeting later in August, officers were elected. [Names omitted.] Starting in November 1959, they repeatedly petitioned the prefectural government and assemblymen elected from Minamata City to use their best efforts to determine the cause of the disease.

4. The Society held a special general meeting on November 25, 1959, and those present unanimously decided that the cause of the disease was the water discharged from the factory. It was decided that they should demand from the defendant the payment of a total of ¥234 million as compensation for seventy-eight patients (¥3 million per patient). They produced the following written request: "It is common knowledge that Minamata disease, which has afflicted many people since 1953, and has caused the death of some people, is caused by water discharged from your factory. Therefore, we demand compensation for seventy-eight victims in the amount of ¥234,000,000 and expect your response by November 30." It was addressed to Nishida Eiichi, the manager of the factory. The victims selected six members to negotiate their demands, and on the same day four of the six visited the factory to deliver the written request. The request was accepted by the chief of the General Affairs Section, Kawamura Kazu, who orally responded at the time that the factory could not comply with the request because it was not proven that Minamata disease was related to water discharged from the factory. Three days later the factory confirmed the same response in writing signed by the factory manager.

On receiving the reply on November 28, the members of the Society put up a tent and staged a sit-in by the side of the factory's main gate. Their sit-in continued until the 27th of December.

5. On December 1, 1959, a group of people with at least one person from every family of the then certified patients (including those who had died), accompanied by Fuchigami Sueki, chairman of the Minamata City Council, visited the Governor of Kumamoto Prefecture to ask for help with their predicament. They requested that the Conciliation Committee on Disputes Concerning Fisheries on Shiranui Coast (hereinafter Conciliation Committee) take up the problem of compensating the victims and indicated that they were asking ¥3,000,000 for each patient. The mayor of Minamata City made a similar request to the governor at about the same time.

6. Upon receiving these requests, the governor consulted with the Conciliation Committee and on December 7 went to Tokyo to meet with Yoshioka Kiichi, president of Chisso. The governor tried to persuade him to be receptive to the patients' demands by saying that the denial of any relationship between Minamata disease and the defendant was no longer acceptable to the people of Kumamoto Prefecture. Yoshioka finally agreed to compensate the patients but insisted that the payment be termed "mimaikin" [sympathy money]. No specific mention of the amount of payment was made at that meeting.

7. Around December 12, the governor instructed Takano Tatsuo of the Industry and Mining Section of the prefectural government to formulate a draft mediation proposal with reference to compensation cases in fields such as automobile, industrial, and railroad accidents. The scale of social welfare payments was also taken into consideration.

On December 15 when the Conciliation Committee met for the third time to discuss compensation of the fishermen, the governor presented the draft proposal to the other committee members. Yokota, chief of the Industry and Mining Section explained the content of the draft. No particular opinions or comments were offered by the other committee members.

8. On the next day, December 16, Yokota showed the draft proposal to the factory officials and to Watanabe and the other representatives of the Society. Yokota met with the latter group at the city hall and at a local inn with the mayor and the city council chairman as observers. He explained the draft and left a copy with them.

Watanabe and the others went back to the sit-in tent and read the entire draft proposal aloud to the patients and their families. During the discussion with those present, strong dissatisfaction was expressed over the amount of the annuity for patients who were minors at the time they were stricken by the disease. The amount proposed was ¥10,000 [$27.80] annually. There was strong feeling that they should get at least half of the amount provided for an adult patient (¥100,000 annually). Two days later Watanabe and more than ten other representatives again visited the governor and a member of the Industry and Mining Section. At the same time, they complained about the [minor patients'] problem and proposed that a lump-sum payment rather than an annuity be made for all surviving patients and that there be differentiation in the amount of payment according to the severity of symptoms. (85)

9. On the 17th, the governor met President Yoshioka and other company officials in Kumamoto City. The governor finally was able to persuade them to accept the basic features of the draft proposal, although they were reluctant to do so, and they still strongly argued that the cause of Minamata disease had not been proved.

10. Around the 23rd, the governor and other committee members informed Watanabe through the Minamata City office that the amount of payment to patients who were minors when struck by the disease [see 8 above] would be increased from ¥10,000 to ¥30,000. They then inquired whether the Society would accept the draft proposal.

11. On December 25, representatives of all of the patients and their families met and voted on the draft proposal. Many were still dissatisfied with the ¥30,000 proposed for minor patients. The proposal was rejected by a margin of one vote. The families were clearly split, and at one point the negotiation committee members expressed their desire to resign. Thereafter, however, those who had voted against the proposal met again to reconsider the matter and concluded that under the circumstances they had no choice but to accept it. Thereupon, at 8 P.M. on December 27 all members finally agreed to accept the draft proposal. Watanabe so informed the committee.

12. On the 27th, the patients' families ended the sit-in, and on the 28th they removed the tent. The negotiation committee sent thank-you notes to factory officials and others for their help and support during the sit-in and negotiation.

13. The Conciliation Committee, thereupon, drafted the final conciliation proposal. On the 28th at about 4 P.M., Morinaga, chief of the Commerce, Industry, and Fishery Bureau of the prefectural government with Yokota and Takano of the Industry and Mining Section carried the final draft to the factory in Minamata and met Manager Nishida and others at the factory club. They explained that the payments to minor patients would be increased from ¥10,000 to ¥30,000 and that the annuities for all patients were to be revised if the cost of living should rise in the future.

The factory officials wanted to include a provision whereby, if Minamata dis-

ease was determined not to be caused by the factory wastewater, the mimaikin would be terminated as of that date. Morinaga telephoned the Conciliation Committee to relate the factory's demand and, when he added it as art. 5, he then told the factory officials of the inclusion and gave them a list specifying the amounts of the mimaikin payments for each patient.

14. At about 10:30 A.M. on the 29th, both parties [names deleted] met for the final round of negotiations in the mayor's office. Mayor Nakamura represented the Conciliation Committee. The City Council Chairman Fuchigami and two councilmen were present as observers. Morinaga explained the content of the final draft proposal provision by provision. The patients' families expressed the opinion that the amounts of the mimaikin were too low and asked about the termination clause. They were satisfied with Morinaga's explanation. Then each side retired to a separate room and examined the final draft proposal. At 1:50 P.M. they met again and both sides agreed to accept the proposal. Takano and other officials from the prefectural government explained the contract document and the accompanying memorandum drafted according to the accepted draft proposal. By 3:00 P.M. all provisions of the agreement and seven clauses of the memorandum were explained and agreed upon by both parties. When they came to the eighth clause, however, the factory officers strenuously objected. This clause provided that "when the amount of the government pension payment is revised due to changes in the prices of commodities, the annuity shall be revised accordingly." The factory insisted on changing it to read: "Both parties shall consult and revise the amount of annuity only if there is a remarkable change in the prices of commodities." The debate on this point went on past midnight. It was finally settled by adopting the factory's version and by treating this provision as a separate note on a mutually understood matter to be attached to the contract and memorandum.

15. On the 30th at noon, both parties signed the contract, memorandum, and accompanying note. Mayor Nakamura and Morinaga and others from the prefectural government were present. Nishida, representing the factory, and Watanabe and four others (one of the members of the negotiating committee was not present but was represented by the others) representing seventy-nine certified patients of Minamata disease, signed and put their seals to the documents. There were forty-eight living patients and thirty-one deceased victims whose family members, surviving parents, spouses, and children were to receive the mimaikin payments. (85)

(C) Validity of the First Agreement. [*The plaintiffs contended that the agreement was concluded by others without lawful authorization. The court denied plaintiffs' contention, holding that they had known about the whole process of negotiation and the results of the agreement and had freely received mimaikin. The court held, however, that some plaintiffs who were relatives of patients were not bound by the contract because their names were not listed in it.*]

(D) Conclusion of Second to Seventh Contracts and Contract Amendments. [*After the first agreement was concluded, some other patients were certified as patients of Minamata disease. These patients concluded identical agreements. On April 17, 1964, May 21, 1965, and March 6, 1968, payments were amended. The court held that these agreements and their amendments were binding on all parties, with the exception of persons not noted in the contracts.*]

(E) The Nature of Mimaikin and the Mimaikin Contracts.

1. The defendant contends that the mimaikin contracts are settlements based on art. 695 of the Civil Code; the plaintiffs argue that there are no provisions of the Civil Code that cover mimaikin contracts.

Limited to the context, however, it is not necessary to determine this question. These mimaikin contracts, as mentioned above, were signed by the victims and the defendant. If they are considered valid, their content will have effect, and it goes without saying that both parties will be held to their terms. This [result] is not dependent on whether the agreements are Civil Code art. 695 settlements or contracts not provided for in the code.

2. On the other hand, there exists a substantial dispute between the parties as to the nature of the mimaikin paid under the terms of these agreements. The agreements contained the following terms: "If in the future, it is decided that Minamata disease is not caused by the wastewater of the defendant's factory, compensation will be terminated at the end of that month" (sec. 4); and "even if in the future it is decided that Minamata disease is caused by the wastewater of the defendant's factory, there will be no further claim for compensation whatsoever" (sec. 5, Waiver of Rights). Whether or not Minamata disease was attributable to wastewater from the defendant's factory had not yet been decided. Judging from this language, it is clear that these contracts were concluded on the assumption that the causal relationship between the factory's wastewater and Minamata disease was unclear, that is, that there was no recognized duty of the defendant to pay compensation. Thus payments under these contracts, even though in actuality treated as compensation to the patients, must be legally interpreted literally as mimaikin payments rather than compensation payments. Furthermore, the defendant itself admits that it did not sign these agreements as an admission of its liability for harmful acts and that the payment of mimaikin was not the payment of damages based on a duty to compensate.

3. Therefore, the defendant made mimaikin payments to the victims which were not prescribed as compensation for injury. It was agreed that in return for these payments, if it became clear that Minamata disease was caused by the wastewater from the defendant's factory and the defendant's liability was confirmed, the victims would request no compensation other than that stipulated by the terms of the agreement. It goes without saying that even this type of contract is valid if its terms are lawful. (89).

(F) The Determination of the Issue of Infringement of Public Policy and Good Morals by the Mimaikin Agreements.

1. The plaintiffs argue that these contracts are invalid since they violate public policy and good morals. By the time the first agreement was concluded, the defendant already knew from the results of a series of experiments on cats that Minamata disease was attributable to the wastewater of the factory. However, this fact was concealed and these contracts were concluded with the victims on the premise that it was uncertain whether or not Minamata disease was attributable to the factory's wastewater. Furthermore, [the defendant] took advantage of the victims' distress to get them to waive all rights to compensation in return for an extremely low amount of mimaikin. The Civil Code requires that one exercise one's rights and fulfill one's obligations sincerely and in good faith (art. 1(2)). Even when a contract specifies the content of the parties' rights and duties, it will be invalid if it violates society's general concepts of order and morality (public policy and good morals) (Art. 90).

These same principles of course apply to the payment of damages arising out of wrongful acts. They require the wrongdoer to fulfill his obligation to compensate the victim in good faith without betraying the latter's trust. If a wrongdoer, therefore, groundlessly denies his obligation to compensate and, taking advantage of the victim's ignorance and distress, causes the victim to waive his right to just compensation for his injuries, that contract will be null and void for being contrary to public policy and good morals. (89) [The court then makes detailed findings regarding various conditions relating to the defendant and each plaintiff that led to the conclusion of the mimaikin agreements.]

V. The Defense of the Statute of Limitations

(A) and (B). [The defendant contended that the plaintiffs' cause of action was barred by the statute of limitations because three years had elapsed from the time that plaintiffs had first become aware of their injuries (and the fact that defendant had caused them) and the date of commencing suit. The court rejects the defendant's argument.]
(C) . . .
1. In contrast to the prescribed period of ten years for the statute of limitations for general obligations (Civil Code art. 167 (1)), Civil Code art. 724, first column, sets a shorter, three-year period for tort liability. One reason is that as time passes, problems of proof arise which make it difficult to establish the prerequisites of tort liability and to calculate the total amount of damages. The main reason, however, is that if the victim has not exercised his tort cause of action after he has had knowledge of the injury and the wrongdoer's identity for three years, the victim's emotions have been soothed and the wrongdoer forgiven. To complicate the patients' relationship thereafter is considered inappropriate. Even if we fix the tolling of the statute in this case, that is, the time when the patients became aware of both their injuries and the wrongdoer, at November 25, 1959, as the defendant argues or alternatively at the time of their certification as disease victims, the patients except those already deceased have continued to suffer incessantly from their symptoms and various other difficulties, as will be noted in sec. VI of this opinion. Where there is personal injury involved in this way, damage continues to arise even after the injurious act [has been completed]. It would be entirely unreasonable, therefore, to adopt the interpretation that holds that the statute of limitations for the entire injury begins to run from the time the victim first becomes aware of a portion of his injury and the identity of the perpetrator. Under this interpretation, the statute would begin to run in conjunction with injuries, which the victim cannot foresee and for which he could not possibly claim compensation. This would violate the meaning of the special provision in art. 724, which expressly states that the statute shall run from "the time the injury is known." The facts are that symptoms continued for extended periods and aftereffects beyond the imagination of the victims at the time later appeared and increasingly inflamed their emotions. In this type of situation, it would be contrary to the legislative intent of said article to hold that, because three years had passed since the discovery of initial harm and of the wrongdoer's identity, the right to claim compensation for damages is extinguished.

In any event, it is correct to say that in this instance the statute of limitations begins to run only for that portion of the injury which the victim knew about or could have foreseen. As for the remainder of the injury, the statute does not run until each portion is known or becomes foreseeable. (99)

VI. Damages

[The court next distinguishes pollution cases from other kinds of damage actions and notes that five factors are critical in assessing damages:

(i) Unlike ordinary personal injury situations, such as traffic accidents, pollution is caused solely and unilaterally by industries. Moreover, the position of the victims and the wrongdoers is not easily interchangeable.

(ii) Pollution destroys the environment. It is almost impossible for the residents near an industrial factory to avoid injury, and in most instances there is no contributory negligence on the part of the victim.

(iii) The residents who conduct their daily activities in a polluted environment sustain a common damage of different degrees. A great many persons are harmed, which often results in the total destruction of family life.

(iv) Since pollution harms a large number of unspecified residents throughout a considerably wide area, along with having serious social ramifications, it is presumed that the industries must bear an immense burden of injury compensation.

(v) Certain industries stand to profit from the industrial activities that cause pollution. However, for the residents who are the victims, there are no benefits.]

(A) . . .

1. In the present case, plaintiffs sued only for emotional and mental suffering and attorneys' fees. It is apparent that they have no present or future intention of any claim of damages for loss of income. If they had claimed pecuniary damages including loss of income, matters of proof would have been complicated, the trial period lengthened, and the victims' relief delayed. Therefore, in calculating the amount of damages for mental suffering, it is permissible to consider [economic loss] as an additional factor in the calculation of emotional and mental anguish.

When loss of income and other forms of economic damage are included in the calculation of pain and suffering, various circumstances such as social status, income, number of working years, occupation, and age must be considered in addition to the length of hospitalization, different life expectancies, and symptoms, and so forth of each patient. The plaintiffs' request that uniform damages be awarded each victim without consideration of their individual circumstances is contrary to the concept of tort law which aims to provide fair, appropriate, and rational compensation.

Related to this point, the plaintiffs do not distinguish between the harm suffered by Minamata disease patients' families and the symptoms of individual patients. Actually, however, the families' circumstances and the patients' symptoms do differ. (103) . . .

[The court then analyzes each plaintiff's circumstances and makes individual damage awards.]

2.5   The Mimaikin Contract

Shin Nihon Chisso Fertilizer Company (hereinafter referred to as A) and Watanabe Eizo [and five others] (who represent living patients and, in the case of deceased patients, their successors and surviving parents, spouses and children, listed on the attached list of names [not included] of Minamata disease patients; hereinafter referrred to as B) have accepted a conciliation proposal on the matter of compensation for Minamata disease patients offered by the Conciliation Committee on Disputes Concerning Fisheries on the Shiranui Coast, on December 29, 1959, and have reached an amicable settlement on the same day. Hereby, A and B conclude a contract as follows.

Article 1

A shall grant the patients of Minamata (including those who have died; hereinafter referred to simply as patients) the following amount of money as mimaikin, calculated by the method described below.

1. In the case of the deceased:

(1) In the case of a person who was an adult when stricken by disease: ¥100,000 multiplied by the number of years between the start of the person's disease and death, plus ¥300,000 condolence money, plus ¥20,000 funeral expense. This shall be paid in a lump sum.

(2) In the case of a person who was a minor when stricken by disease: ¥30,000 multiplied by the number of years between the start of the person's disease and death, plus ¥300,000 condolence money, plus ¥20,000 funeral expense. This shall be paid in a lump sum.

2. In the case of living patients:

(1) In the case of a person who was an adult when stricken by disease:

(a) ¥100,000 multiplied by the number of years between the start of the person's disease and December 31, 1959; this shall be paid in a lump sum and

(b) ¥100,000 shall be paid annually after 1959.

(2) In the case of a person who was a minor when stricken by disease:

(a) If that person became an adult before December 31, 1959, ¥30,000 multiplied by the number of years between the start of the person's disease and his reaching adulthood, plus ¥50,000 multiplied by the number of years since that person reached adulthood to December 31, 1959. This shall be paid in a lump sum; and

(b) ¥30,000 for a minor and ¥50,000 for an adult shall be paid after December 31, 1959.

(3) In the case of death of a recipient of an annuity:

Payment of condolence money and funeral expenses provided for the deceased shall apply and shall be paid in a lump sum, and the annuity payment shall be terminated as of the month of the person's death.

(4) A lump-sum payment by request:

(1) A patient whose symptoms are judged to have been stabilized or minimized by the Minamata Disease Examination Board (or a guardian in case of a minor patient) may apply for a lump-sum payment instead of an annuity. A shall terminate annuity payments at the month of application and shall pay a lump-sum of ¥200,000 provided: (1) that the period of application shall be limited to the six-month period following the conclusion of this contract; (2) a person who received a lump sum payment under (1) shall be deemed to have waived all claims to subsequent mimaikin.

Article 2

As to the mimaikin payments by A to B provided by the preceding article, the necessary amount shall be delivered to and distributed by the chief of the Minamata Office of the Kumamoto Branch of the Japanese Red Cross.

Article 3

A shall grant mimaikin separately to those patients who might appear (and be approved by the board) after the conclusion of this contract, in accordance with the contents of this contract.

Article 4

If in the future, it is determined that Minamata disease is not caused by water discharged from A's factory, A shall terminate the grant of mimaikin as of that month.

Article 5

Even if it is determined in the future that Minamata disease is caused by water discharged from A's factory, B shall make no further claim for compensation whatsoever.

To prove that both parties agree to the contract, two sets of this contract shall be made one each for A and B to hold. (December 31, 1959   Names of the parties signed and sealed)

Section 3   The Yokkaichi Air Pollution Case

3.1   Introductory Note

This case was a tort action brought by air pollution victims from Isozu District and several family members of the deceased against six petrochemical and power companies. The plaintiffs asserted that their illnesses were caused by sulfur dioxides discharged from the defendants' factories and requested compensation for loss of earnings, solatium, and attorneys' fees, totaling ¥200,586,300.[24] On July 24, 1973, the Yokkaichi District Court delivered its decision holding the defendants jointly and severally liable, and awarding plaintiffs ¥88 million. After the decision, the defendants reportedly negotiated among themselves how the burden of this award should be allocated. The details of this secret accord, however, have not been made public to this day.

The legal bases of the Yokkaichi suit were art. 709 and 719 of the Civil Code. As already noted, art. 709 was the basic tort provision in issue in the Minamata trials. The text of art. 719 is as follows:

If two or more persons have by their joint unlawful act caused damages to another, they are jointly and severally liable to make compensation for such damage; the same shall apply if it is impossible to ascertain which of the joint participants has caused the damage.

Instigators and accomplices are deemed to be joint participants.

The Yokkaichi decision differed from the earlier trials in several important respects. In the Niigata case, for example, the foreseeability of harm was not a central issue because the nation had been alerted to the hazards of mercury since the poisonings in Kumamoto nine years earlier. In the Yokkaichi case, however, the defendants argued that there was no scientific evidence pointing to the possibility of a major health hazard from low concentrations of sulfur dioxides.[25]

Another problem was the absence of precedent on joint tort liability in the pollution cases. In an earlier decision, the Sannogawa case [22 Minshū 964 (Sup. Court April 23, 1968)] a national alcohol plant had been held liable under art. 719 to local farmers for pollution-induced damage to their rice fields. Scholars, however, debated the value of this case as a precedent under art. 719 because plaintiffs probably would have prevailed in any event under art. 709. In the case plaintiffs were able to show that although other sources had polluted the river, the defendants' discharges alone probably would have been sufficient to cause damage. In Yokkaichi, however, the discharges of some of the plants were not sufficient in themselves to cause compensable harm.

As in the other pollution cases, some residents of Isozu who had not been parties to the original suit commenced negotiations with the defendant companies after the court's decision. Under a subsequent agreement, the companies pledged not to expand their factories without prior consent of the residents. The companies also agreed to submit to the residents' inspection of their factories in the future.

On November 30 of the same year, the defendants and another group of Isozu pollution victims (again individuals who were not parties to the original action) concluded a further agreement. This arrangement stipulated that the companies would pay ¥569 million in compensation to 130 victims (¥6.5 million for a hospitalized adult, ¥10 million for the family of a deceased victim, and ¥2 million for a child).

The Yokkaichi decision like the other pollution trials caused a great stir within the government. For example, it spurred MITI to expand its monitoring of major air pollution sources throughout the country and prompted the director general of the Environment Agency to announce that the government would reexamine its land use policies. After the Yokkaichi trial the government also revised the Industrial Location Law (1959, Law No. 224) to read as follows:

Article 6. *When a plant is to be newly built or enlarged in a designated area, a report must be submitted to the Minister of International Trade and Industry on the maximum volume of pollutants to be discharged from the plant and on the pollution control measures that will be taken to ensure that this volume will not be exceeded.*

Articles 9 and 10 provide in part that:

If it is judged that synergistic pollution may occur, recommendations and orders to make changes will be issued by the Minister of International Trade and Industry.

The Yokkaichi decision also accelerated Japan's establishment of a national compensation system.[26]

3.2   Decision of the Tsu District Court, Yokkaichi Branch[27]

[Judgment is for the plaintiffs. Each defendant is ordered to pay damages to the plaintiffs; each defendant is responsible for the total amount of compensation.]

I. Parties

[The nine original plaintiffs are residents of Isozu District in the southeastern part of Yokkaichi City. Two of the plaintiffs died during the trial and four heirs succeeded to their actions.

The defendants are six companies located to the north of Isozu across the Suzuka River [see fig. 2].
Defendant Shōseki is an oil refinery.
Defendant Chūden is a power plant.
Defendant Ishihara is a manufacturer of chemical fertilizer,
*   titanium dioxide, etc.*
Defendant Mitsubishi Yuka is a manufacturer of ethylene.

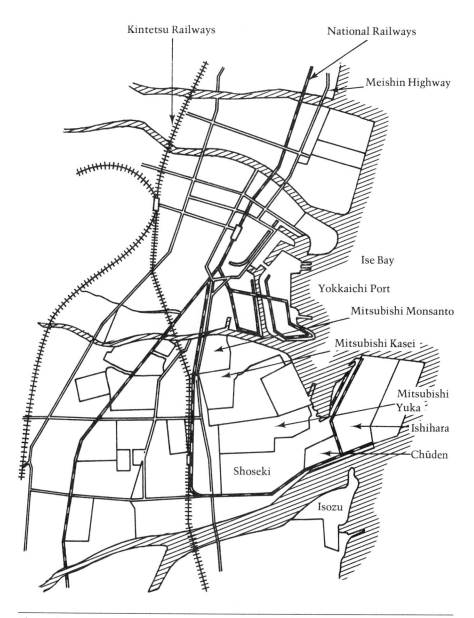

Kintetsu Railways

National Railways

Meishin Highway

Ise Bay

Yokkaichi Port

Mitsubishi Monsanto

Mitsubishi Kasei

Mitsubishi
Yuka

Ishihara

Chūden

Shoseki

Isozu

Figure 2
Yokkaichi konbināto area.

Defendant Mitsubishi Kasei is a manufacturer of diethyl hexane,
carbon black, etc.
Defendant Mitsubishi Monsanto is a manufacturer of vinyl chloride,
etc.]

II. Wrongful Act

(A) The Process of Establishment and Operation of the Defendants' Factories

1. The Process of Establishment and Operation.
(a) The Petrochemical Industry and Combines [Konbināto]
After World War II the development of synthetic high polymer technology and the expansion of markets for its products, such as synthetic resins and fibers, created a shortage of basic chemical raw materials. In 1954 shortage of carbide and tar derivatives was predicted. Consequently the need to develop a petrochemical industry to provide new sources of supplies became urgent.

Meanwhile in 1949 the occupation forces had permitted the reopening of oil refineries along the Pacific Coast, and the petrochemical industry had begun the reconstruction of refineries. By 1954 the initial stage of modernization had been completed and the industry was able to supply some petrochemical materials, particularly naphtha.

Against this background the Japanese petrochemical industry set the foundation for large scale production beginning in 1955. Encouraged by the protective policy of the government, the industry began implementation of its plan in 1956 and actual construction started in 1957. (33)

[*The court next discusses how the industry built modern oil refineries that provided chemical companies with naphtha and other basic materials and describes how, in a typical petrochemical plant, liquid naphtha is transformed into ethylene (C_2H_4) or butylene (C_4H_8), and then is made into secondary products like polyethylene, vinyl chloride, diethylene, or hexanol.*]

Since most materials and products of these processes are in liquid or gaseous form, there are technical as well as economic advantages in concentrating the refinery's manufacturing processes for several derivative products in one area and systematically connecting each process with pipes to supply naphtha and other materials to the other areas. However, for one company to operate all the production processes would be extremely difficult financially, and thus it became standard practice to establish industrial complexes.

If we refer to a group of companies in a line of sequential processing of basic materials, such as that described above, as a vertical combine, the diversification caused by the mutual utilization of by-products among several companies may be termed a horizontal combine. As each type grows and develops, however, the vertical broadens into the horizontal and the horizontal leads to the vertical, and both develop into a comprehensive combine so that not only the basic materials but also by-products produced along the way are utilized most efficiently. In this way, the appearance of industrial combines became necessary to the growth and development of the petrochemical industry. (33)
(b) Industrial Changes in Yokkaichi: The Entry of Defendant Kasei into Yokkaichi and the Establishment of the Monsanto Plant. *Yokkaichi was originally a town of ceramic factories, small paper mills, and similar industries. In 1883 and 1907 respectively, however, a cotton mill and a woolen goods factory located*

there. During the Taishō and Shōwa eras textile factory after textile factory moved in and Yokkaichi became known nationally as a textile manufacturing city.

During the time from 1934 to 1941, in response to the efforts of local financial circles to attract industry, a plate-glass factory, a copper-smelting plant, a sulfuric-acid plant, and an oil refinery were built in Yokkaichi. The Second Naval Fuel Depot, with Japan's largest oil refining capacity, was built between 1939 and 1943. In this way Yokkaichi, which had been a light manufacturing center specializing in textiles and ceramics, began to emerge as a center of the petroleum and chemical industry. In 1945 in an air raid by the U.S. Air Force the naval fuel depot and a large portion of the city were destroyed by fire.

After the war, with the reopening of oil refineries and the sudden rise of the petrochemical industry, the former naval base with its good port facilities and spacious location became a likely target for a new petrochemical plant site. A group of Mitsubishi companies in cooperation with an international oil consortium began to promote the formation of a group of petrochemical factories.

[Next the court describes how the defendant Mitsubishi Kasei expanded its business into the newly developing chemical industry and became interested in Yokkaichi. In 1953 Mitsubishi Kasei merged with a chemical company and gained control over another company in Yokkaichi. The latter became the defendant Mitsubishi Yuka in 1956. In 1952 Kasei established the defendant Mitsubishi Monsanto Kasei in cooperation with Monsanto Chemical, Inc., of the United States. Meanwhile in 1952 Shell Oil Company, which had invested in Shōwa Oil, applied to the government along with Mitsubishi Oil for the purchase of the former navy site. But the proposal failed due to the opposition of the domestic oil companies. In September 1953 the government decided to sell the site to eight domestic oil companies including the Shōwa and Mitsubishi oil companies for the joint operation of an oil refinery. But under pressure from Shell Oil, the defendant, Kasei and the Mitsubishi group, the government later changed its policy and decided informally in April 1955 to sell the site to Shōwa Oil Company for the construction of a petrochemical plant. For this purpose a survey team from Shōwa Oil, including Shell Oil technicians, visited the site several times between May and August 1955.]

On August 26, 1955, the government revealed as a Cabinet Order its decision to sell the site and facilities thereon to Shōwa Oil and to extend loans to the same company as necessary for the establishment of an oil refinery. At the same time the government revealed its policy of encouraging combines by the statement, "In the future when the Mitsubishi group and the Shell group have begun commercial production of petrochemicals, it is expected that there will be close cooperation with the refining facilities of the Shōwa Oil Company to be built on this site." (33)

(c) Establishment of the Defendants Mitsubishi Yuka and Shōseki. *[The court describes how the Mitsubishi group, including the defendant Kasei, was established to produce naphtha. Thereafter, in 1957 Shōwa Oil Company, Shell Oil Company, and the Mitsubishi group cooperated to establish the defendant Shōseki.]*

(d) Development of the First Combine in Yokkaichi. *[Details of expansion and development process omitted.]*

(e) The Defendant Chūden. *In 1955 the defendant Chūden built the Mie Power Plant at its present site. In December it was completed and Chūden began operation of its number one generator with a 66,000 kW/h capacity. Thereafter in Feb-*

ruary 1957 and June 1958 the second and third generators with capacities of 75,000 kW/h were added. Generator number 4 with a 125,000 kW/h capacity was added in October of 1961. (34)

[*The expansion of the plant's generating capacity paralleled and supported the development of the first combine.*]

(f) The Defendant Ishihara. [*Ishihara started copper smelting and production of sulfuric acid in Yokkaichi in 1941. In 1949 it began to produce chemical fertilizer and in 1954 titanium oxides.*]

2. The Functional Relationship Among Defendants' Factories. [*The defendants exchange materials and products through pipelines. The defendant Mitsubishi companies, namely Yuka, Kasei, and Monsanto, have a particularly intimate relationship. They jointly invest capital, exchange employees, and share telephone facilities. The defendant Yuka produces steam not only for its own use but also for the use of Kasei and Monsanto. See fig. 3*]

(B) Air Pollution by Soot and Smoke (Particularly Sulfur Oxides) in Isozu District

1. The Discharge of Soot and Smoke by the Defendants' Factories.

(a) Article 2 of the Air Pollution Control Law defines soot and smoke, automobile exhaust, and particulates as substances which cause air pollution and are subject to regulation. In regard to smoke and soot, art. 2, par. 1, states the following:

[*Par. 1*] *In this law the term "soot and smoke" means substances enumerated in the following items:*

(1) Sulfur oxides generated as a result of the combustion of fuel and the like;

(2) Soot and dust generated as a result of the combustion of fuel and the like;

Paragraph 4 of the same article, defines particulates as:

Any substance discharged or scattered as a result of mechanical treatment, such as the crushing or selection of materials, or substances dispersed from piles of materials.

The reason why item (1) of par. 1 provides for sulfur oxides generated by combustion is that the great bulk of the sulfur oxides that cause air pollution result only from combustion processes. Furthermore, the sources of sulfur oxides are interpreted as being limited to this. (63)

[*The court points out that the amounts of sulfur oxides (mostly SO_2) discharged can be calculated on the basis of the amount of fuel burned and the sulfur content of the fuel. The amount of sulfur oxides discharged by each defendant was as shown in table 3.*

Some of the defendants argued that they were able to eliminate SO_2 in their production processes. The court, however, rejected this argument because of the insufficiency of evidence.]

2. The Increase of the SO_2 Level in Isozu District and Its Distinctive Characteristics.

(a) The Increase of SO_2 Levels.

(1) Around 1960, the monitoring of SO_2 levels in Yokkaichi and the levels in Isozu District made the Yokkaichi City government conscious of poor air quality as a pollution problem in Yokkaichi. In order to get basic data for its pollution control policies in October, the city asked the Public Health Laboratory of Mie Prefectural University Medical School to investigate the state of pollution. In November the laboratory started measuring falling particulates and sulfur

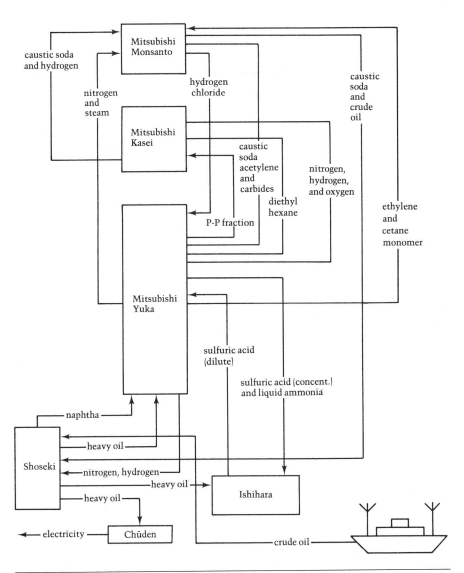

Figure 3
Yokkaichi petrochemical konbināto (July 1968).

Table 3
SO$_x$ discharges (tons).

	Shōseki	Mitsubishi Yūka	Mitsubishi Kasei	Mitsubishi Monsanto	Chūden	Ishihara	Total
1959				1.35	5,292		5,293.35
1960	4,202	2,310		0.86	15,318	1,517	23,347.86
1961	5,286	3,257		0.78	20,321	1,619	30,483.78
1962	5,472	2,510	18.4	0.86	29,141	2,370	39,512.26
1963	6,230	2,473	41.8	0.78	19,430	2,374	30,549.58
1964	7,454	4,031	310.8	14.24	11,823	3,513	27,146.04
1965	11,510	4,690	502.6	107.46	7,847	3,879	28,536.06
1966	11,144	5,472	524.2	110.08	4,172	3,979	25,401.28
1967	12,756	5,746	477.2	126.80	3,012	4,764	27,882.00
Total	64,054	30,489	1,875.0	363.21	116,356	24,015	237,152.21

oxides at eleven monitoring points. (Later the Mie Prefecture Health Department added seven more points bringing the total to eighteen.) (65)

[The court then analyzes in detail the results of the monitoring reports and finds that the average concentration of SO$_2$ in Isozu was much higher than the ambient standards set by the Cabinet in 1969. Isozu's average winter concentration (November-April) was usually twice or three times as much as the government ambient standard (yearly average: 0.05 ppm) (table 4). In Isozu, a north wind prevails in the winter and a south wind in the summer. The concentration of SO$_2$ in Isozu, located to the south of the defendants' plants, is higher in the winter and lower in the summer. When a strong north wind blows in Isozu, a high concentration of SO$_2$ is temporarily observed (peak type pollution). The court finds that pollution of this type may be particularly hazardous to the health of the residents.]

3. The Cause of Air Pollution in Isozu District.

(a) The Location and Distance between Defendant Factories and Isozu District. The defendants' factories are located from the north-northwest to the northeast and adjacent to Isozu District as described in I [parties].

(b) The Chronological Correlation between the Defendants' Operations and Air Pollution in Isozu District.

(1) Yokkaichi City's first combine, which comprises the core of the defendants' factories, went into full operation during the period between 1958 and 1959. As noted above, the city government became aware of air pollution as a problem in 1960 when an abnormally high concentration of SO$_2$ was recorded in Isozu and adjacent areas. The period of the worsening of air pollution coincides with the full operation of the defendants' facilities. (68)

(2) [Until 1963 when the second combine started operations in Umaokoshi District, there had not been a major polluting source in Yokkaichi other than the defendants' plants.]

(c) The Relationship Between Annual Levels of SO$_2$ in Isozu District and the Annual Amounts of SO$_2$ Emitted by the Defendants' Factories. [The yearly change in the total amount of SO$_2$ discharged by the defendants generally coincides with changes in SO$_2$ concentration during the winter in Isozu.]

(d) The·Relationship Between Wind Direction and SO$_2$ Levels. [The court finds

Table 4
Average seasonal concentration and yearly discharge of SO$_2$.

Year	Season[a]	SO$_2$ Reading (mg/day/100 cm^2 PbO$_2$)	Average Seasonal Concentration (ppm)[b]	Yearly Discharge (tons)
1960	11–4	1.48	0.079	30,483.78
	5–10	0.88	0.046	
1961	11–4	2.20	0.116	39,512.26
	5–10	1.36	0.072	
1962	11–4	2.75	0.142	30,549.58
	5–10	1.06	0.056	
1963	11–4	2.38	0.125	27,046.04
	5–10	1.28	0.067	
1964	11–4	2.15	0.113	28,388.06
	5–10	1.05	0.056	
1965	11–4	2.07	0.109	25,344.28
	5–10	1.63	0.086	
1966	11–4	2.06	0.109	26,875.00
	5–10	1.23	0.065	

[a]Winter includes months 11 (November) through 4 (April); summer includes months 5 (May) through 10 (October).
[b]The average seasonal concentration of SO$_2$ is found from the equation 19 × SO$_2$ reading = average seasonal concentration (ppm).

that a 50% concentration of SO$_2$ in Isozu is more observable in winter than in summer because in the winter the prevailing winds blow from the north and west. The district is then leeward of the complex. There is a correlation between the frequency of the north and west winds and the concentration of SO$_2$ in Isozu. At Mihama Primary School, located on the north side of the complex, high concentrations of SO$_2$ have been recorded in summer (80% more than in winter) during the time when the prevailing winds blow from east to southeast. This contrasts with the situation in Isozu that is located to the south of the complex.]

(e) [A contour map of SO$_2$ concentration shows that the closer a site is to the complex, the higher the concentration. The concentration decreases as one moves away from the complex.]

(f) As described earlier, a distinctive feature of Isozu's air pollution is a peak concentration of SO$_2$ when a relatively strong north wind is blowing. With this as a basis, the following explanation is possible. Because of the concentration of large fuel-consuming factories in the combine, large amounts of gas including SO$_2$ are discharged and, without much dispersal, flow in a leeward direction. When the wind speed is high, a low pressure area is formed on the leeward side of the plant buildings, and the gas discharged from the factories is drawn into the space and causes a high concentration of SO$_2$ in the adjacent area.

(g) From findings (a) through (f) and consideration of evidence and testimony, it is clear that air pollution in Isozu district is caused primarily by soot and smoke (particularly SO$_2$) from the defendants' factories. (69)

(h) The Defendants' Counterargument. [The defendants argued that smoke diffusion is a complex phenomenon dependent on the combination of dispersal and flow, and diffusion cannot be determined simply from the relative location of the parties and the direction of the wind. The court finds that the causal relation between air pollution in Isozu and smoke discharged from the defendants' factories *could be determined without finding the exact processes of dispersal and flow patterns in the air. As it demonstrated in (a) through (e), the causal relation between pollution in Isozu and smoke discharged from the complex was established. The court then states that the burden was on each defendant to prove by elaborate theories of dispersion and flow that smoke from its own factory did not reach Isozu.*]

(C) The Relationship Between Air Pollution and the Plaintiffs' Illnesses

1. The Increase in Respiratory Illness in Yokkaichi City (Particularly Isozu District) and Its Causes.

(a) Preliminary Remarks. *It has been said that because of the special characteristics of pollution cases, an epidemiological approach plays a particularly important role in the determination of the cause of disease.*

In the instant case as well, proof of the relationship between the plaintiffs' illnesses and air pollution is central to the claims of the parties. The court will examine the existence of a relationship between the increase in respiratory ailments in Yokkaichi City (particularly Isozu District) and air pollution from an epidemiological perspective. Before that, however, a simple explanation of epidemiological methods is presented.

The term respiratory illness, as used by the American, Goldsmith, describes the general situation in industrial countries whereby the amount of air passing through the windpipe is lessened by the effects of air pollution. This term includes chronic bronchitis, asthma, and emphysema.

The following methods are used in an epidemiological analysis to determine cause: (1) the descriptive epidemiological method, (2) the analytical epidemiological method, and (3) the experimental epidemiological method.

In the descriptive method one observes the way in which the disease spreads in the natural world and records and studies any special characteristic of this process.

In the analytical method theories concerning the cause of the outbreak of a disease are generated from data gathered by the descriptive method and then carefully examined.

The experimental method, while reconfirming the findings of the analytical method through animal experiments, has the particular purpose of determining the precise mechanisms in the causal process.

If, through the use of descriptive and analytical methods, an agent appears to have a high correlation [with the illness], it will be necessary to examine comprehensively observed data to conclude whether there exists a cause and effect relationship between the agent and the illness.

For a cause and effect relationship to be established, the following four conditions are necessary.

a. The agent has been (present) for a certain period before the outbreak of the disease.

b. An increase in exposure to the effects of the agent coincides with an increase in the incidence of the disease.

c. Special characteristics of the spread of the disease observed by the descriptive method can be explained consistently with the vicissitudes and geographic variations in the agent. (If the agent is removed, the incidence of disease drops; if the population has no contact with the agent, the incidence of disease is very low.)

d. The precise mechanism of causation by the agent can be explained without contradiction with the biological sciences.

Of the four, the second is the most critical and usually is expressed as the relationship between dose and effect (that is the stronger the dosage, the higher the incidence of disease). So, if this principle is interpreted in its broadest sense, the third principle is included therein.

For the fourth principle, experiments may be used. It is not necessary, however, to have experimental validation so long as a biological explanation of the process of causation originating in the agent is possible.

The experimental method is the way to verify an epidemiological fact under predetermined conditions.

We can agree with all the above, but what we must examine here is legal causation. In finding legal causation, of course, the existence of natural causation is a precondition, but it goes without saying that the discussion of natural causation should be limited to the extent necessary and sufficient to find legal causation.

Although the purpose of the experimental method is to discover a precise mechanism for causation, this determination is not always necessary for a finding of legal causation.

For a legal determination, it is enough if the relevant agent can be confirmed as the cause. It is not necessary to determine the process of the agent's operation.

Also, experiments that verify theories developed by the analytical method may in principle be required to determine legal causation. In a legal determination, however, the degree to which such experiments are required varies on a case-by-case basis. Several factors such as the extent to which a theory has been substantiated by the descriptive or analytical methods should be considered in determining the necessity for experimentation. (71)

[The following is a summary of pp. 71–84 of the opinion:

(1) A study of national health insurance payment claims was made and showed an increase in respiratory illness in Yokkaichi for seven years after 1961. The study was sponsored by local governments (prefectural and city) and was carried out by the Public Health Laboratory of Mie Prefecture University Medical School. The laboratory picked ten (later thirteen) districts in the city and made monthly tallies of the number of visits by patients with respiratory illness in each district. The results show that the number of visits per 100 persons (the incidence rate) in polluted districts is two or three times that of unpolluted districts, and the incidence rate rises with the SO_2 concentrations. Elderly and very young children are more adversely affected than young adults.

(2) Medical Examination of Residents. In 1964 the laboratory also conducted medical examinations of 9,000 people over forty years old in six districts in Yokkaichi (three polluted districts, three unpolluted). The results were similar to those obtained from (1) above.

(3) Medical Examination of Public School Pupils. The laboratory surveyed the effects of air pollution on the health of children in 1965. It examined a total of 738 pupils in polluted districts and unpolluted districts. Several adverse effects on children's health were found.

(4) A study of mortality and the causes of death in Yokkaichi.

(5) An intensive medical examination of Isozu residents in 1964.

(6) A study of the relationship between the number of attacks of patients and SO₂ concentration.

The laboratory examined the correlation between the number of attacks of seventeen asthma patients in Isozu and the SO₂ concentration in Isozu in 1963. There was a high correlation. The court also examined other epidemiological surveys.

Finally, the effects of low concentrations of SO₂ on mice, rats, and guinea pigs were observed in experiments carried out by doctors at the Mie Prefecture University. Questions had been raised on the effects of low SO₂ concentrations by the results of some foreign experiments.]

(f) Conclusion.

(1) As has been noted above, it is true that the defendants' counterarguments in regard to the epidemiological studies and animal experiments are partially true.

The results of many epidemiological surveys and studies of effects on the human body show beyond a doubt that there was a dramatic increase in respiratory illness in Yokkaichi City, especially in Isozu District, from 1961 on and that the reason for this increase was air pollution, particularly sulfur dioxide. This finding conforms to the four epidemiological principles set forth in the previous section. The finding (together with the results of animal experiments), the state of controls on sulfur dioxide discharges at the time, and the testimony of witnesses establishes the fact that there was a dramatic increase in respiratory illness due to air pollution consisting mainly of sulfur dioxide acting in synergy with particulates and other pollutants.

As noted above, the results of animal experiments have diverged. The evidence of such experiments, however, should be evaluated comprehensively and from the perspective of the extent to which the hypothesis generated by the analytic epidemiological method is corroborated. In the instant case there have been experimental results that confirm the influence of low concentrations of sulfurous acid gas (H_2SO_3) on living things. This experimental proof of the possibility of such effects should be considered significant.

This conclusion is also supported by the indisputable fact of the drastic increase in respiratory illness from 1961. No other hypothesis than the one which attributes the cause of disease to air pollution can better explain the phenomenon. (84)...

3. Cause of the Plaintiffs' Illnesses.

(1) [The court then examines and refutes the defendants' arguments denying the influence of SO₂ on human health.]

(2) There are numerous factors besides air pollution that may have contributed to the outbreak of respiratory diseases. These factors have different effects on these diseases.

The question, however, is the existence of legal causation between air pollution and the plaintiffs' illnesses (including its ongoing effects). In order to establish that relationship, it is necessary and sufficient to establish that each plaintiff's condition would not have existed or have been aggravated but for the pollution. When a causal relationship is established between illness and air pollution, as a rule there is liability for damages even though other factors may exert an influence. An exception occurs when the additional factor is the victim's fault, and the problem becomes one of contributory negligence.

As long as each plaintiff has the characteristics of all residents of Isozu (that is, residence in this area and exposure to air pollution), we may recognize the (causal) influence of air pollution on a person's disease unless it can be shown that another stronger agent has caused or worsened the disease irrespective of the existence of air pollution. (86)

[*The court then examines the symptoms of each plaintiff and finds that no other particular agent had independently caused or worsened the disease.*]

(D) Joint Tort Liability

1. As determined in sections B and C, it is found that the combined soot and smoke from the six defendants' factories were the main sources of pollution in the Isozu area and that this pollution caused and aggravated the plaintiffs' respiratory ailments. In order for a joint tort to lie, however, it is necessary that the actions of each party contain the requisite elements of a tort and that a joint relationship exist among the actors. (91)

2. Causation in Joint Torts. Since each person's act must contain the elements of a tort, intent, negligence, capacity [ability of the person to comprehend the act], and the illegality of the act must be proved.

In the narrow definition of a joint tort found in the first column of art. 719 (1), a causal relationship must exist between the actions of each person and the result.

In regard to causation, however, even if the act alone could not have produced the result, it is sufficient to establish that the injury would not have occurred but for the act. It is not necessary for the act alone to have been able independently to cause the result. The interpretation, which holds that joint tort liability is not a true joint and several obligation, and therefore requires a showing that each defendant's conduct could independently have caused the result, would negate the entire purpose of providing for art. 719 in addition to art. 709.

If the injured party proves that the defendants are associated and that their joint actions culminated in the injury, the law will assume causation between the act of each defendant and the result. Therefore, unless the defendants can prove that there was no causal relation between their acts and the result, they cannot avoid liability. This principle stems from the second sentence of art. 719 (1) that states that even if the chain of causation between the acts of each defendant and the result is unclear, all joint actors must assume joint liability. The relation of the above principle to the burden of proof in litigation is that when a victim asserts a claim of joint liability of tortfeasors (regardless of the obscurity of their identity), it will generally not be necessary for the plaintiff to establish causation for each of the defendants' acts.

The defendants each argue that the soot and smoke from their factories did not extend to the area in question or that the amount that did reach there was so small that no causal link could exist between the discharge and the result. This defense will be examined in sec. E below [omitted]. (91)

3. Joint Association.

(a) Weak Joint Association.

(1) In a joint tort action, it is sufficient to have an objective joint association among the defendants' acts.

The substance of this requirement is that these acts, which have given rise to the harm, can be perceived to be one distinctive act according to the common understanding of society.

As was previously determined, it seems reasonable to find an objective joint association in the instant case because the factories of the six defendant companies are all located together on the site of the former navy fuel depot next to

Isozu, began operations at about the same time, and have continued operations in the same way since.

In the case of members of an industrial complex, this type of objective joint association can usually be established by the fact of membership alone, although the scope of the relationship is not necessarily limited to members.

(2) As noted, a joint tort may also exist when the acts of each defendant are insufficient to cause the result if they produce the result in combination with other acts. In this type of case, however, we hold that since the injury arises under special circumstances, it is necessary to require a showing that the defendant had foreseen or was able to foresee that his actions joined with those of the others was likely to produce the resultant injury.

If we apply this to the present case, foreseeability is established for the following reasons. As was seen before, all of the defendants' factories are located at the same site and operate as a unified industrial complex. Because of this proximity, each defendant was aware of the general form, content, and size of the other factories. Since all factories burned the same type of fuel oil, it was naturally possible for them to know that the others would also discharge soot and smoke containing elements like sulfur dioxide.

In light of the location and distance of the factories in relation to Isozu District, the direction of the prevailing winds and other meteorological conditions in Yokkaichi City, injury to the residents of the district also was foreseeable. The defendants should have known that the discharge from each of their factories would combine and reach the area where the plaintiffs resided. In addition there was some indication of possible adverse effects on the human body from soot and smoke. (92)

(b) Strong Joint Association. *When the relationship among the defendants' factories is closer and more unified than the one described, even if the amount of discharge from a given factory is so small that no causal relationship between it and the damage can be established, liability will still be recognized.*

It has been established that there was a particularly close and dependent relationship between defendants [Mitsubishi] Kasei and [Mitsubishi] Monsanto.

Each of these companies had separate tasks in an integrated production process. For example, defendant Yuka fractionates naphtha to produce ethylene and other substances basic to petrochemical manufacture. Defendants Kasei and Monsanto employ these basic materials to manufacture secondary products such as vinyl chloride and diethyl hexanol. Each of these companies obtain or obtained steam essential to the manufacturing process from defendant Yuka.

Additional exchanges of manufactured and primary products from Monsanto to Yuka and Kasei and from Kasei to Monsanto occur as described above. These products and steam are transferred by pipeline. Thus it is extremely difficult or impossible technically and economically to obtain them from other sources. The relationship of the three companies is so close functionally, technically, and economically that it would be impossible for one to change its operations without considering the others.

In this way the three companies are very closely associated and incorporate each others' production into their operations, and each discharges soot and smoke. Among these three companies there is not only a strong association but the financial and historical relationship described earlier also exists. Because this relationship exists, even if the discharge from one of the three companies is insufficient to establish a causal relationship with the plaintiffs' injuries, its involvement with the discharge of soot and smoke from the other two companies is grounds for imposing full liability for the injuries. (92)

118

4. The Defendants' Counterarguments. [*The defendants denied joint tort liability. One company denied joint commission of a tort, asserting that the association of companies was for the purpose of conducting a lawful activity. The court decided, however, that soot and smoke necessarily would be discharged from their operations. The court next found that the association had operated fortuitously. Another defendant argued that the sale and purchase of materials and products among them did not constitute a joint tort. The court also rejected this argument, noting that the defendants constituted one system of production, and the sale and purchase of products were not mere transactions but an essential part of the system's operations.*

By using a theoretical dispersion equation each defendant tried to prove that the smoke and soot from its own factory either passed over or did not reach Isozu. For that purpose the defendants applied different dispersion coefficients that were favorable to the conclusion advanced by each company. Yet a coefficient that was advantageous to one defendant contradicted the conclusions of other defendants. The court points out that theoretical dispersion equations were still not sufficiently reliable to ascertain the real situation, and the data used in the defendants' equations were not trustworthy. The court also expresses its doubts on the reliability of the results of a wind tunnel test.

The defendants Kasei and Monsanto argued that they should be liable only for damages attributable to their own acts (divisible liability) because they discharged only small amounts of soot and smoke. The court, however, finds them jointly liable for all damages because of their strong association with the defendant Yuka.]

III. Liability of the Defendants

(A) Negligence

1. Foreseeability. [*The court next takes judicial notice of several previous incidents of sulfur oxide pollution, citing the Ashio copper mining incident, the Besshi copper mining incident, and the Hitachi mining incident. It also refers to respiratory ailments induced by sulfur dioxide occurring as occupational diseases among mine workers. The court finds that the harmfulness of high concentrations of sulfur oxides is now generally known. It continues that defendants should have foreseen the injuries in the present case because the harmfulness of even low concentrations of sulfur oxides were under scrutiny as early as 1955 when the Japan Association of Public Health published a report in which the threshold concentration was designated as 0.1 ppm.*]

Although it may be true, as defendants claim, that there is no definitive scientific explanation for the harmfulness of sulfur oxides, it was foreseeable that some kind of hazard existed to human health from low concentrations of sulfur oxides. Foreseeability to this extent is sufficient to hold the defendants liable. (98)

2. Breach of the Duty of Care.

(a) Negligence in Site Selection. On the selection of a site for a new factory or on beginning operations, enterprises like those of the defendants, which use oil as a raw material or as a fuel, have a duty to exercise due care in operating oil refining, petrochemical, chemical fertilizer, and electrical generating plants, which inevitably produce polluting by-products such as sulfur oxides. This is especially true in the case of factories built in a group such as in an industrial complex. With these enterprises there is a danger that pollution will jeopardize the lives and health of residents in neighboring areas. The duty involves locating the op-

119

eration to avoid this danger. The choice should be based on studies of factors such as the character and amount of the discharge, the location and distance of residences from the discharging facility, and atmospheric conditions including the speed and direction of the wind.

However, judging from the plaintiff's evidence and testimony of the witness Miyamoto Kenichi, the defendants established their plants without taking any measures to determine the possible effects of their operations on neighboring residents' health. We find, therefore, that the defendants, with the exception of Ishihara, were negligent in selecting this site.

(b) Negligence in Operation. As long as the defendants engaged in operations that resulted in the production of soot and smoke, they were under a duty to study the possible effects of the discharge on residents of nearby areas and operate their factories so that there was no danger to the lives of these residents. Compliance with this duty was even more important because the defendants have gradually expanded their factories since beginning operations.

According to the evidence, the six defendant companies failed to study the effect on the residents of nearby areas. (Defendant Ishihara, during its prewar copper manufacturing period, conducted a fixed-point observation. We take notice of this precautionary measure. However, no survey has since been made even though the content and form of its operations have changed since the war.) Also, negligence for the continued haphazard operation of the plants is established. (99)

(B) Intentional Torts

The plaintiffs argue that the defendants' torts became intentional by March of 1964 at the latest when the Kurokawa Investigation Group recommended that Yokkaichi pollution be controlled.

As the plaintiffs asserted, that recommendation pointed out the intimate relationship between air pollution in Yokkaichi, particularly sulfur oxides, and the respiratory diseases in Isozu.

In the case before us, however, the soot and smoke discharged from defendants' different factories combined to cause the damage to the plaintiffs. Furthermore, defendants abided by emission standards. Under these circumstances, the court cannot hold the defendants liable for an intentional tort. There is no other evidence that supports the defendants' liability for an intentional tort. (99)

IV. Defendants' Argument of the Lack of Illegality

(A) Argument of Nonexistence of Illegality

As far as tort illegality is concerned, the situation of the victim, including the interests invaded, is comprehensively weighed against the situation of the injurer, including its injurious actions. Illegality is not established unless the harm exceeds the level that the victim is expected to endure according to generally accepted social norms.

The court will examine each defendant's argument in light of the above principle.

1. Defendant Shōseki. Defendant Shōseki argues that the discharge of soot and smoke from its plant was not illegal due to the public nature and the social value of oil refining activity, its compliance with regulations imposed by clean air laws, the small quantities of sulfur oxides that reached Isozu, and the oversensitivity of some of the victims.

(a) **The Small Amounts of SO$_x$.** *The small quantities of SO$_x$ which reached the victims. We denied this argument above.*

(b) **The Public Nature of the Activity.** *The alleged public nature of the activity does not excuse the wrongfulness of the defendant's acts where, as in this case, such acts inflict injuries on fundamental and irreplaceable interests such as human life and health.*

(c) **Compliance with Emission Standards.** *Even if the defendant Shōseki abided by emission standards, this act only immunizes the defendant from administrative sanction. We cannot say that because the standards were followed, the victims must naturally endure [their fate]. In view of the serious injuries in this case, it is completely impossible to intepret them as being within the scope of human tolerance.*

(d) **The Sensitivity of the Victims.** *We do not find that all plaintiffs in this case were particularly sensitive. Even if they were, there are usually some people in a polluted area who are sensitive to pollution. Mere sensitiveness to the pollution should not be considered as a factor which removes the illegality of the defendant's acts. (99) [The court next examines similar arguments developed by other defendants and holds that their activities caused injury beyond a level of human tolerance. For example, the defendant Ishihara argued that it had located in the area before some of the plaintiffs arrived in Isozu, and therefore it was unjust to impose liability on it. The court holds that who came first should not be taken into consideration if damage involves human health.]*

(B) The Defenses of the Impossibility of Avoiding Injury and the Use of the Best Avoidance Measures

1. The defendants claim that since negligence is based on a breach of the duty to avoid foreseeable consequences, they are not liable since the consequences of their actions were impossible to avoid and they took the best possible antipollution measures.

There are divergent theories that may be used in judging the defendants' claim. If a finding of responsibility is dependent solely on a showing of whether or not the best or appropriate preventive measures were used, the result would not properly reflect the purpose of tort law, which is to effect a fair distribution of the burden of damage. Other elements should be incorporated, and a finding should be based on whether there has been damage that has exceeded tolerable limits.

If we apply this test to this case where damage is to invaluable, irrecoverable interests, i.e., the life and health of neighboring residents, it is difficult to find that such damage falls within bearable limits. Even where other factors such as the public value of their industries are added to the defendants' claim that they have used the best available antipollution measures, their claim should fail. (100)

2. Even if we assume for the moment that, as the defendants claim, negligence is interpreted as the duty to avoid foreseeable consequences and there is no liability when the best or appropriate preventive measures have been taken, reference to the revisions of the Basic Law for Environmental Pollution Control indicate that [the standards should be severe]. The revisions have eliminated the section on economic harmony and have emphasized two goals: the protection of life and the preservation of the environment. It is consistent with this section, at least in cases where the discharge involves substances known to endanger life and health, to expect industry to set aside economic considerations and take preventive measures that utilize the best techniques and knowledge available in the world. If these measures are neglected, industry cannot avoid fault.

We now examine the defendants' preventive measures in light of this interpretation. (101)

(a) Site Selection. *[The defendants next claimed that the central as well as local government had urged them to build factories at Yokkaichi. The court recognizes that local governments at that time had given priority to economic growth and had not considered the possibility of serious pollution. The court rules that the government's own fault should not shield defendants from liability because the defendants had actively competed with other companies for the site.]*

(b) The Possibility of Prevention and the Use of the Best Preventive Measures.

(1) Possibility of Prevention. *[The defendants claimed that because of the slow development of desulfurization technology, they were unable to use fuel with a low sulfur content. The court refutes this claim. The defendants failed to take effective preventive measures because of economic reasons. They could have erected high smokestacks and might have adopted other preventive measures.]*

(2) The Possibility of Prevention in Joint Torts. *[The defendants claimed that they could not prevent damage caused by their joint polluting acts because one company could not control the emissions of other companies. The court holds that the defendants were still obliged to control their own emissions in order not to injure residents of neighboring areas. In the case of the Mitsubishi group, the court finds the companies could have cooperated to control pollution.]*

(3) Best Preventive Measures. *[Finally, in describing the defendants' pollution control efforts, the court concludes that they failed to employ the best preventive measures. without consideration to economic feasibility.]*

V. Damages

(A) Compensation

As determined above, the defendants must compensate the plaintiffs for damages resulting from their joint tort. Special characteristics of liability for damages in pollution suits include the following. First, in pollution cases, in contrast to ordinary instances of personal injuries such as traffic accidents, there is no interchangeability between tort-feasors and victims because a defendant can never be a victim. Second, because pollution destroys the environment, the inhabitants of neighboring areas cannot avoid its effects. Third, since damage from pollution occurs over a wide area, its effect on society is profound and the amount of damages the company is required to pay is usually very large. Fourth, pollution is created by a company's production process, from which it plans to profit. On the other hand, there is no direct profit from these activities for residents of adjacent areas, that is, the victims. Fifth, in pollution cases, the inhabitants of a specific area are all injured by a common cause, and in this sense they are all injured equally. These characteristics are basically applicable to the instant case.

In addition, in considering compensation the following distinctive features of the illness in the instant case will be taken into account. All the plaintiffs' illnesses, with the exception of chronic emphysema, are generally reversible.

However, there is as yet no perfect treatment for any of these illnesses. Although allopathic treatment is used, it serves mainly to relieve the patient through confinement in a room with purified air, injections during attacks, and inhaling medication. Furthermore, even if a victim improves because of treatment during hospitalization, as long as the air remains polluted, he will sooner or later have to return to the hospital.

Depending on the seriousness of the illness, one cannot distinguish between a normal person and a victim who is not undergoing an asthma attack. Asthmatics may also do limited amounts of work. On the other hand, it is difficult to predict when an asthmatic attack will occur. There is always an uneasiness caused by the fear of an attack. If an attack does occur, the victim must stop working and return for treatment. (103)

(B) Loss of Earnings

1. The following is a discussion of the plaintiff's common problems in the area of loss of earnings. We consider the special characteristics of pollution cases noted above and the special features of the illnesses suffered.
(a) As was seen above, the victims in the instant case are almost indistinguishable from normal people except when they are experiencing an attack. It is necessary to limit physical labor since there are adverse burdens put on the heart when a certain limit is exceeded. In this way, this case is different from instances such as amputation where there is a visible indication of loss of ability to work.

It is thus unusual to find a rapid decrease in income immediately after contracting an illness. Instead there is a tendency for this decrease to occur over a long period. In these cases, awards for loss of earnings based on the difference in income before and after contracting the disease are difficult to calculate and unfair in their results.

Of course, in cases where a victim is forced to work to maintain self and family even when it is medically inadvisable to do so, it would be unfair to deduct that income from an award.

Basically, a calculation of loss of earnings should be concerned with the amount of earning potential a person has remaining; in other words, the amount of potential that has been lost, instead of the amount actually earned after the disease was contracted. A determination of whether or not the victim's remaining ability is actually used, or the way it is used, is not required.

For the above reasons, in the instant case it is proper to predicate damages on the loss of the ability to work.
(b) In pollution suits, as was mentioned above, residents of the polluted areas suffer extensively from the damages. Victims living in a common environment are equally affected in the sense that they have been injured by the same wrongdoing. In order to make an award to victims as fairly and quickly as possible, it is necessary to standardize the item of damage in calculating the amount of damages.

In this respect, the plaintiffs' assertion that the amount of damage should be calculated on the basis of average wages of all workers classified by sex, age, and classification is reasonable. (103)
[The court relies on the expert testimony of Dr. Sagawa in calculating compensation for the individual plaintiff's disability.][28]

(C) Pain and Suffering

1. In calculating the amount of damage for pain and suffering, we take the following factors into consideration as common elements among the plaintiffs.
(a) The above mentioned characteristics of pollution cases and the characteristics of the particular disease in this case.
(b) Physical pain. *The plaintiffs became ill between 1961 and 1965 and have been suffering from six to as long as ten years. Furthermore, there is no prospect of their recovering from the disease. Asthma attacks result in difficulty in breathing due to the constriction of the windpipe. The pain of such attacks is extremely great.*
(c) Mental Suffering. *Mental anguish resulting from a long struggle against the disease is extremely great, particularly from long hospitalization and accompanying anxiety, never knowing when or even if one will be able to leave. Furthermore, the sudden death of Miyako Seo, the youngest among the plaintiffs, seriously shocked all the plaintiffs and they became fearful of their own sudden and unexpected deaths.*
(d) The Breakdown of Family Life. *Long hospitalization not only deprived plaintiffs of the pleasure of family life, it also caused financial difficulty in cases where a victim was the breadwinner of the family. As a consequence, the plaintiffs' families are exposed to the fear of financial ruin. (108) [The court examines the circumstances of each plaintiff and awards between ¥2 million and ¥5 million in damages for pain and suffering.]*

(D) [omitted]

(E) Attorneys' Fees

[The court orders defendants to pay a sum equal to 10% of the amount awarded to plaintiffs for attorneys' fees.]

Section 4   The Legacy of the Four Pollution Trials

Although preceded by a few less significant decisions, the four major pollution trials essentially established the basic legal principles governing claims for compensation of health injuries from air and water pollution.[29] In effect, the decisions rewrote the code of operations for the chemical industry, thereby establishing standards far more stringent than any administrative regulation at the time.[30] They also hastened the government's implementation of new and needed pollution laws and regulations. This section analyzes the doctrinal contribution of the decisions from the perspective of environmental policy.

4.1   Problems of Polluter Liability

4.1.1   Fault, Strict Liability, and Willfulness

It is critical to recall that three of the pollution decisions (the two Minamata cases and the Yokkaichi decision)[31] were incontrovertibly predicated on fault, not strict liability. This may surprise some Western readers because the standards of care and conduct the courts imposed are so high (perhaps impossibly high) that they encompass and blame industry for virtually any chemically induced injury, even an unavoidable one. Why did not the courts simply hold defendants strictly liable, justifying their actions according to familiar notions of cost spreading, economic efficiency, or incentives for technological development?[32] There are at least two explanations. First, the courts were constrained to some extent because these were negligence actions based on the orthodox Civil Code. Thus, traditional doctrine bound the discretion of the judges. Since art. 709 of the Civil Code explicitly and unmistakenly specifies liability based on negligence, the courts simply could not impose strict liability, at least under this provision.[33] Second, although doctrinal considerations were undoubtedly important, the social and cultural issues presented may have appeared even more pressing. Given the dominant sentiment of the period that the polluters had to be blamed for the victims' injuries, it was necessary for the courts to reinforce and legitimate this consensus by law. In a sense the justice of the nation's laws was on trial.

Why then did not the courts proceed one logical step further and hold the defendants' actions to be willful? Indeed many people at the time branded the companies' conduct criminal, and the judges were under great social pressure to concur. Moreover, some scholars note that the facts of the cases, particularly the Niigata decision, would have permitted a determination of willfulness because of the defendants' awareness of the possible causes of disease. The apparent inconsistency has been explained by some simply as a concession to industry, already enormously burdened by liability for damages,[34] and by others as due to the courts' unfamiliarity with industrial operations.[35]

4.1.2   Standard of Care

Perhaps the most devastating part of the courts' decisions for industry was the standard of care they imposed. Let us study the implications of the Yokkaichi decision that a company undertake pre-siting investigations. Essentially, the Yokkaichi court imposed three correlative duties. The first was a company's pre-siting duty to study not only the character and amounts of possible future emissions but also atmospheric and environmental conditions. The second was the duty to assess future effects, essentially the impact of the proposed factory's operations on the health of local residents. The third obligation was to update assessments, especially when expanding operations. This was a daring prescription, for at the time there was no explicit statutory provision mandating the court's decision. Although the government and some industries under administrative guidance had investigated the environmental effects of large-scale developments for years, these studies were generally rudimentary and superficial.[36] Before the Yokkaichi decision, there was no legal obligation for pre-siting assessments.

Moreover, the reader should not assume that Japanese industry at the time possessed the scientific and technical expertise to perform the required assessments. The art of impact analysis was still underdeveloped even in the United States, where Congress had passed comprehensive legislation three years earlier. And the decision left industry with only a vague idea of specifically what should be assessed, how various options should be quantified or otherwise evaluated, and what constituted an adequate initial review and a satisfactory later assessment.

The decisions' failings perhaps are defensible if we reconsider how the pollution courts viewed themselves at the time. Their critical task was to lead the nation in a new direction; the specific details of implementing their mandate, the courts obviously hoped, would be worked out between the administration and industry.

In this sense, the four pollution decisions may be viewed as much as an exhortation for government action as a command for industrial obedience.

Yet integrating the holdings in the three decisions, the courts imposed a formidable array of specific obligations on the actual conduct of chemical manufacturers. First, they ruled that companies must be aware of the amount and nature of the chemical used, the concentrations in the factories' effluent, the existence of by-products, and the possible toxicities individually and synergistically of all original substances, by-products, and reagents. Second, the courts required chemical manufacturers not only to be aware of the dangers of already studied toxic substances but also to peer into the unknown. To meet these obligations, they demanded that a chemical manufacturer employ the best available techniques to analyze the contents of its effluent, the best monitoring equipment to ascertain the effects of discharges on the health of exposed populations and the environment, and the best control technology available anywhere in the world. They required industry to keep a continuous vigil for adverse health or environmental effects and to be familiar with all current scientific theories and research on the toxicity of chemicals employed or produced. Indeed the courts even advanced two steps further. The Niigata Court instructed industry: (1) to "take the strictest safety precautions to prevent even the slightest danger to humans and other living things"; and (2) "if a danger to human health remains, curtailment or cessation of operations may be necessary."

Even for those unfamiliar with the internal workings of the chemical industry, the courts' decisions raise profoundly difficult questions. From a scientific and technological perspective, industry's assignment, if not an impossibility, was certainly an immensely difficult undertaking. Chemical manufacturers today generally still are not entirely familiar with every reaction taking place in the devil's brew that constitutes a major factory's effluent. Moreover, to ascertain the actual, not to mention putative, risks of these processes is quite clearly beyond the present scientific and technical competence (and proclivity) of industry and government anywhere.

Equally perplexing are the economic implications of the courts' holdings. As noted, the courts insisted that industry employ the "best available control technology." This standard, if precisely and immediately implemented, would clearly have forced the bankruptcy of many concerns.[37] Moreover, the courts' requirement that industry operate essentially at no risk, and terminate operations if risk cannot be avoided, raises serious questions from the perspective of traditional economic analysis.[37a] *All* industrial activities involve some risks, and the cost of preventing all risks in most cases is prohibitive. Finally, the notion that a company must shut down at all costs simply because its operations create risks also makes little "economic" sense. Undoubtedly, the companies' operations created jobs that fed their workers (also a life-giving function) and conferred many other benefits on the nation.

But our analysis is not intended to criticize the courts for their naiveté from the economist's perspective. Rather, we wish to emphasize that they were motivated by an entirely different concern—justice for individuals. The opinions suggest that the objective of justice may at times be antipathetical to notions of economic efficiency. When human health or life is concerned, the courts suggest, justice may simply preclude balancing.

Once we grasp the courts' philosophical position, its rejection of strict liability also becomes more understandable. Strict liability assumes that costs and benefits can be weighed together; that to achieve efficiency we need simply extract by the theory's cold, impersonal calculus the "correct" solution. This was precisely the kind of immoral rationalism the courts were endeavoring to correct.[38]

4.1.3   Joint Liability Under the Yokkaichi Decision

For its time, the Yokkaichi court's decision was a doctrinal tour de force, unexpected by industry and government. The case represented the first judicial analysis of pollution-related health injuries resulting from the combined acts of several tortfeasors where the discharges of some companies were patently insufficient in themselves to cause injury. As described, the court divided the defendants into two groups. The first group consisted of the three Mitsubishi affiliates. This group, the court held, was in a "strong relation," because the three separate companies were functionally one entity measured by their coordinate financing and symbiotic technological relationship. Although the emissions of one member of this group might not have been sufficient to cause plaintiff's injuries, this "innocent" party would be held liable because the actions of other members could be attributed to it. The attribution theory, however, could not be applied fairly to companies less closely associated, hence in a "weak relationship." Thus the court resorted to the concept of foreseeability, ruling that this group of defendants should have foreseen that their factories' emissions reacting together presented a serious health hazard.[39]

For all its inventiveness, the Yokkaichi decision leaves several intriguing questions unanswered. First, should a company be held liable if it does not discharge any pollutants but is affiliated in a strong relationship with other companies that do? The Yokkaichi court's reasoning seems to suggest an affirmative answer because it was not the amount of emissions but rather the company's tainted association that was decisive. From the victims' perspective, there is no difference. From the perspective of policy, the objective is the same in both cases—to induce the "innocent" enterprise to take greater precautions, perhaps by withdrawing from the association or influencing the conduct of its "partners."

Second, would the Yokkaichi rule apply to companies not operating as a konbināto but perhaps in a looser association within the same industrial area? The answer seems to depend on the circumstances, particularly the degree of harm, and the extent to which this harm was foreseeable.

A third problem is why the court permitted the Yokkaichi defendants to continue their operations. Was this life-endangering activity not precisely what the Niigata court's decision had interdicted? Indeed, the Yokkaichi decisions for damages for such grievous harm had the same effect as granting the defendants a permit to purchase human lives. Was this not contrary to the court's own philosophy?

Perhaps the most plausible explanation for this inconsistency is that the victims and their lawyers simply did not seek injunctive relief. We must remember that in 1969, the time of filing suit, the Japanese courts were far less willing to grant injunctive relief than they are today. No doubt the victims' lawyers felt the case presented more than enough complexities and that obtaining injunctive relief would have been all but impossible.[40]

4.1.4   Illegality: The Fault Problem Reconsidered

From the perspective of our analysis of fault, the Yokkaichi and Niigata decisions raise two further complicating issues. First, in both cases the defendants proved that they had complied with all existing emission regulations. Second, in the Yokkaichi case the defendants also urged that their activity was sufficiently "public" to deserve exculpation. The defendants had initially located in Yokkaichi under administrative encouragement; and, given the nation's desperate requirements for oil, their refinery operations did confer substantial public benefits. Both courts rejected the former argument, noting that compliance with an administrative regulation would only shield a tortfeasor from administrative sanction, not civil liability. The Yokkaichi court rejected the latter argument by emphasizing the defendants' culpability where "fundamental and irreplaceable interests such as human life and health" were involved, leaving open the question of how it might have responded if only property interests were at stake.

From the perspective of economic analysis, the courts' treatment of these two defenses seems correct. It may be perfectly appropriate to have two sets of standards—an administrative regulation representing an outer limit on permissible activity, and a liability rule to "charge" polluters for the risks imposed by otherwise legal activities. In theory this "double standard" may contribute to efficiency both at the firm's level and more generally because the firm will have an additional incentive to reduce costs (both its own as well as society's) either by changing its operations, developing new technology, or by other means.

What is surprising, again, is that the court uses fault rather than strict liability to effect these adjustments. For presumably the government had determined the "safe" level before establishing emission standards, or itself conducted "appropriate" pre-siting assessments before encouraging industry to locate in Yokkaichi. In fact, as the Yokkaichi case reveals, the emission standards were patently unsafe and pre-siting procedures demonstratively inadequate. The court's use of fault thus seems motivated by an additional objective to those already offered. Without doubt the Yokkaichi court is also focusing on the government's own responsibility, warning it of the dangers of the free and loose practice of administrative guidance, seeking thereby to stimulate needed reforms.

4.2   Causation

Orthodox tort principles required the plaintiffs to prove causation. This demanded a demonstration of three subsidiary issues: (1) the characteristic symptoms and etiology of the disorder (i.e., the "dose effect"); (2) the pollutants' pathways from their sources to the victims; (3) the defendant's discharge of the toxic substances. This was a formidable task, because the orthodox doctrine required proof with scientific precision, but each case posed issues of substantial scientific uncertainty, which uncertainty had been used to delay, dispirit, and defraud. The courts sought to rectify the inequities inherent in an unthinking subservience to scientific rationalism, striking a new balance between scientific sense and basic fairness.

The principle contribution of the Toyama court was the way in which it made use of epidemiological data to show the dose effect. Adoption of epidemiological evidence was critical to the case, because factors such as nutritional deficiency, climate, overwork, and vitamin D deficiency substantially weakened the victims' proof of causation solely from clinical and pathological data. The court adopted four epidemiological principles as rules of evidence. The plaintiffs were required to demonstrate that: (a) discharges of the polluting agent had preceded the outbreak of the disease; (b) increased exposure to the agent resulted in increased occurrence of the disease; (c) areas of low pollution were associated with low prevalence and incidence; (d) statistical inference of causality was not contradicted by clinical or experimental evidence. After the plaintiffs successfully demonstrated these points, the court held that cadmium was the legal cause of itai-itai disease, and that the victims did not need to explain the dose effect with further mathematical or scientific precision.

A second contribution of the Toyama court was its acceptance of the analysis of how cadmium released from the factory came to be ingested by the victims over an uncertain period. The court's acceptance of the plaintiffs' differential analysis of cadmium concentrations in sludge deposits around the company's outfall and in the rice fields by the river represented a substantial advance in judicial sensitivity to the processes of transmission of toxic substances in the environment.[41]

The Niigata decision refined the Toyama court's analysis by its use of probability theory and legal presumptions. Causation posed a substantial burden in the case because transmission (factor 2) and origin (factor 3) were unclear.[42] Transmission was at first obscure because methylated mercury, unlike cadmium, could not be detected simply from deposits in the river bed. Detection was also impeded because of mercury's concentration in the protein of the river's fish, a process at first unrecognized and poorly understood. By necessity, the plaintiffs sought to show transmission by epidemiological and statistical analysis of mercury concentrations in different types of fish, percentages of contaminated fish eaten, and differential studies of incidence of Minamata disease correlated with the type of fish eaten. The Niigata trial represents the first attempt in Japanese law at showing proof of physical injury by epidemiological analysis of indirect causation via the food chain.

The origin of the toxicant also became an issue in the Niigata case when the defendant was able to show that an earthquake in 1964 had caused an explosion at a nearby chemical plant, causing mercury and other toxicants to be discharged into the river. Further uncertainties were introduced by the plaintiffs' inability to explain why most of the victims had contracted the disease in the short period 1964–1965, and the defendant's failure to explain why some patients had developed the disease prior to the earthquake.

The court disposed of these problems by stressing the importance of statistical probability and its relation to the burden of proof. In effect the court held that if the plaintiffs could demonstrate a high statistical correlation between the defendant's activity and the occurrence of the disease, this would be sufficient. In addition, plaintiffs had to show: (a) the nature of the disease and the material that caused it; and (b) the mechanism by which the agent entered the host. Proof of these elements would create a rebuttable presumption shifting the burden to the defendant of demonstrating that their activities were not the original source of the causative agent.

In many ways the Yokkaichi trial addressed even more complex issues, for unlike the earlier cases several diseases were involved, attributable to multiple factors; at issue also were the emissions of several factories. The Yokkaichi court extended the analyses of the Toyama and Niigata courts. It artfully drew the distinction between the causal effects of the defendant's activities when viewed as a whole, and when analyzed individually. With respect to the plaintiffs' proof of factors 2 and 3, the court accepted evidence of "origin" based on statistics of regional pollution levels (Yokkaichi's was four times that of other areas): SO_x studies by year, seasonal fluctuation of pollutants, emission discharges from individual facilities, and studies of the type of pollution equipment used. "Transmission" was identified by data on yearly and seasonal wind patterns affecting SO_x discharges from polluting factories. These and other elements were corroborated by statistics on differential incidence and prevalence of victims controlled for age, sex, and physical condition; by comparison of hospital admissions and outpatient statistics for pulmonary diseases for polluted and nonpolluted areas; by statistics on the increase of pulmonary diseases over time, control data for smoking and other potential causal factors, overlay map studies of pollution disease; and by clinical data (results of animal experiments with sulfur oxides under polluting conditions). In light of this evidence, the court adopted the probability test of the Niigata Minamata court and concluded that defendants' polluting discharges were the legal cause of the plaintiffs' injuries.

Although their handling of the causation problem was highly innovative, the courts' analyses in the four pollution trials left important questions unanswered. For example, the Niigata court never expounds on its use of probability, so that it is impossible to extract any sense as to what level of probability can be said to ceate a legal presumption; the Toyama and Yokkaichi decisions are similarly imprecise on the specific kinds of epidemiological correlations deemed legally significant. As ch. 6 describes, the unresolved issue of the correlation between pollution levels and pulmonary disease has continued to pose problems in the implementation of Japan's administrative compensation system, established directly after the Yokkaichi decision.

The cases also raise fundamental questions about the courts' role in scientifically complex environmental controversies. Is it appropriate for a judge who ordinarily lacks expertise in engineering, medicine, or the natural sciences to adjudicate what are in many cases predominantly scientific questions?[43] In some cases, of course, considerations of equity are so demanding that liberties with science seem justified. For example, the Niigata court's handling of presumptions makes sense both from the perspective of fairness as well as efficiency.[44] The defendant controlled the technical and scientific information essential to an explanation of the discharge of mercury, and the defendant had destroyed this important evidence. It would have been prohibitively expensive and unjust to require impecunious victims to explain the factory's operations, when even an administrative agency's investigation of pollution at the time depended on a company's full cooperation. In other cases, however, judicial freewheeling with science might be less justified; and the four pollution decisions do not provide clear guidelines as to where judicial restraint must begin.

4.3   **Damages**

The four pollution cases also established important precedents in their treatment of damages, particularly regarding loss of income. The reader will recall that in three of the cases, the two Minamata decisions and the Yokkaichi decision, the plaintiffs were reluctant to seek compensation for disability. There were several reasons. First, because of their need for urgent relief, the victims could not afford to waste time proving individual pecuniary losses. Second, it was difficult to calculate future income and to assess probable life expectancy, even from actuarial tables.[45] Finally, the plaintiffs themselves were disinclined to differentiate their damages, afraid that this might undermine the group's belief that the wrongs were collectively endured.

The Toyama, Kumamoto, and Niigata courts attempted to circumvent these doctrinal and other barriers by awarding lost income under the rubric of damages for mental suffering. In its analysis, the Niigata court emphasized the following factors, present in all four cases: (1) the positions of victim and polluter could not

130

be exchanged; (2) pollution covered a substantial area entailing risk of great damage; (3) pollution injuries affected not only the victim, but entire families; (4) industry extracted profits from the activity, but the activity did not directly benefit the victims; (5) the victims were not contributorily negligent; (6) cultural and economic impediments virtually prevented the victims from escaping injury.

From these essentially equitable guidelines the Niigata court attempted to fashion fair, proper, and reasonable awards; because damages were not easily computed, it classified the victims' injuries into five categories according to the degree of difficulty in carrying out a day's work. The Kumamoto decision rejected as unfair the idea of a uniform computation heedless of each plaintiff's individual circumstances. Rather, it assessed damages for mental suffering by looking to the extent of each victim's affliction, income, the amount of mimaikin received, and other factors.[46]

Unlike the other cases, the Yokkaichi court ventured into the thicket of assessing individual income losses. After testimony on the issue, the court ruled that lost income should be calculated from a patient's lost earning potential according to sex and age; victims did not have to prove that they actually suffered these losses. Finally, three of the pollution trials (Yokkaichi, Kumamoto, and Toyama) also established important precedents for the award of attorneys' fees in environmental litigation,[47] with the Kumamoto and Yokkaichi courts approving an award of 10% of claims sustained, and the Toyama appellate decision, an award of 20%.

Upon reflection, at least two aspects of the courts' treatment of damages are troublesome. First, the courts' analyses to some extent reinforce existing social inequalities. For example, considering earning capacity in the computation of damages for mental distress suggests that human suffering is a function of social status. The Yokkaichi decision has also been attacked particularly by women's groups for its implicit support of the practice of sex discrimination in employment.[48]

A second problem is the subjectivity of the criteria for entitlement. In the Toyama case, the sympathies of the appellate court apparently so deepened that it increased the amount awarded. Although this increase was related to a clear policy objective—to discourage polluters from harassing indigent pollution victims or otherwise delaying relief—the reader may imagine other situations where this practice could be abused. In cases involving damages to the property of parties in a more equal bargaining position, we certainly would not blame an enterprise for seeking to defend legitimate business interests despite damages caused by its pollution.[49]

4.4   The Mimaikin Contract and the Statute of Limitations

Perhaps the most significant aspect of the Kumamoto decision is the court's treatment of the mimaikin contract. As noted, the court struck down the contract as a violation of public order and good morals, deeming it a duplicitous attempt to secure an advantage from the victims' ignorance and predicament.[50] This theory, one commentator suggests, can be applied to other extrajudicial settlements in the traffic accident, medical accident, or food-related injury fields.[51] For here also the unscrupulous take advantage of the victims' distress, ignorance, and inexperience.[52]

But the court's analysis, as in other areas, leaves important issues unattended. For example, are all mimaikin contracts voidable when a victim does not fully

131

understand their content? Or only those settlements containing waivers of future claims? Cannot mimaikin be treated simply as the first installment, or an advance payment, of a damage award, once the waiver provision is eliminated?

The Kumamoto decision also established an important precedent in its interpretation of the statute of limitations. The court held that given the severity of injury and the fact that the damage to health was a continuing process, it would be wholly inappropriate to let the statute run from the point at which victims first became aware of their injuries (or a part of their injuries) and the identity of the polluter. To so hold would require the victims to grasp all subsequently assembled scientific evidence elucidating the disease's cause and the acts of the tortfeasor. Indeed, to adopt this rule would invite political and administrative interference, and would effectively bar relief.[53] The rule adopted by the court is far fairer because under it the statutory time period begins when a victim becomes aware of part of his injury but applies *only* to that discovered part of the injury; the statute is tolled for subsequently discovered aspects of the disease or for new evidence of the defendant's conduct.

The court's analysis, of course, does not completely end the matter. One can easily imagine that allocating a victim's awareness in the way suggested may at times not be possible, and linking the statutory time period to an administrative determination does not completely eliminate the dangers of political interference.

4.5   Summary

Beyond their doctrinal contribution, the legacy of the four pollution trials lies in their influence on judicial, administrative, social, and political attitudes. They motivated administrative action to protect the environment; they stimulated new legislation; they demonstrated to everyone how serious, indiscriminate, and widespread pollution injury could be, how subtle and insidious were the biological processes of contamination. This awareness brought thousands to the courts claiming new environmental rights, seeking alternative, more effective remedies. The public's awakening in the early 1970s reinforced changes in judicial thinking evident even before the last of the four courts' decisions.

4   The Judicial Development of New Rights and Remedies

Unlike the four major pollution trials, the cases in this chapter, with the exception of the Osaka International Airport case, largely involve local, less publicized issues. The plaintiffs' primary concern here is the preservation of a way of life threatened by pollution. The defendants are usually government units or public development corporations (kōdan) and the most frequent remedy sought is the injunction.

The plaintiffs and their lawyers are different than those in the major trials. Whereas the victims in the earlier, celebrated suits were poor, unsophisticated, and socially disadvantaged, these plaintiffs are more often urban and economically well off. Gone also is the ideologically motivated labor lawyer, affiliated with the Communist party, for his place is taken by new champions of the public interest who, although devoted to the environmental movement, maintain a diverse practice and stay away from direct involvement in political parties.

The plaintiffs' confrontations with governmental entities have meant facing or circumventing the intricacies and burdens of Japanese administrative law. In most cases the plaintiffs have chosen circumvention, seeking to accomplish through civil injunctive litigation what they could attain administratively only with hardship. Because the plaintiffs' success in managing the considerable procedural problems in civil and administrative litigation has been one of the hallmarks of these second and third generation environmental cases, the remainder of this chapter gives careful attention to the principal doctrinal problems in issue.

Section 1   Barriers to Administrative and Civil Litigation

Although Japan's separate administrative court system was abolished in the Occupation's reforms,[1] a strong traditional distinction still exists between public and private law. Litigation concerning official action is principally governed by the Administrative Litigation Law,[2] which applies public law principles that are significantly different from those of private law litigation. Yet because many Japanese scholars and jurists have been unwilling to permit substantial liberalization of this law, its usefulness in controlling abuses of official action has been limited. Perhaps the most intransigent impediments are the doctrines of standing and ripeness.

1.1   Standing

In Japan, as in other countries,[3] standing (uttae no rieki) has continued to restrict citizen review of administrative action. Under the orthodox view, plaintiffs must demonstrate possession of a right recognized by statute, in addition to proving a causal relation between the contested administrative action and the alleged adverse effect on legal interests.[4] Japanese doctrine also distinguishes between legal interests and "reflex interests," or those interests shared by the general public that may incidentally be adversely affected by an agency's action.[5] Traditionally the courts have not recognized the standing of a party with only a reflex interest. Thus the effect of this doctrine is to narrow substantially the class of judicially protected interests.[6] Although standing remains the principal procedural barrier, as the cases included in sec. 3 suggest, some courts have very gradually begun to loosen the requirement of a statutory right and to expand the zone of legally cognizable environment-related interests.

1.2   Ripeness

After standing, the principal barrier in administrative litigation is ripeness, or proof of a legally recognizable administrative act. Here the focus of controversy is whether an official action is "in the nature of a disposition" (shobunsei); if not, the case is dismissed without further inquiry.

By American standards the construction of the shobunsei requirement has been extremely narrow. The Supreme Court, for example, has ruled that only official acts that "create rights and duties in citizens [i.e., specific individuals], or confirm their scope" satisfy these requirements. By this definition intra-agency supervisory orders, permissions, and approvals may not be challenged, for they do not directly create rights and duties. Moreover, "general dispositions" like regulations, decrees, or other "actions in the form of law" are also not judicially cognizable, unless a plaintiff can show that before implementation a disposition immediately and concretely affected a specific person's legal right or obligation. As a result of the shobunsei requirement, most goverment plans and policies, no matter how formal or final, have not been deemed judicially cognizable.[7]

1.3   Other Aspects of Administrative Litigation

Having surmounted threshold barriers like standing or ripeness, a plaintiff must next chart a correct course through the various substantive forms of action prescribed by the Administrative Litigation Law. The most important challenges are the direct "appeal suits" (kōkoku soshō), which are divided into four principal categories.[7a] A party may invoke three of the four types of suits when an administrative agency is likely to perform an act that would infringe private rights. In theory these suits are designed to prevent injury in advance by giving the courts jurisdiction to prohibit it or to affirm the nonexistence of the administrative authority to make the disposition.[8]

The first suit seeks revocation of the original or basic administrative decision itself.[9] The second suit seeks to revoke a determination or adjudication of an administrative appeal from the original administrative act.[10] Ogawa notes that these represent the "ordinary" kōkoku remedies. Generally these suits must be filed within three months after the action or adjudication.[11]

The third suit requests a declaratory judgment that an administrative act is void.[12] This third type of claim is usually raised collaterally as a supplementary action to the first two types of appeals. This action is regarded as extraordinary since it is not subject to any limitation period, and a special decree of illegality is required to invoke it.[13]

The final type of action is designed to ensure that a certain disposition by which a party will receive some benefit (e.g., a license) is taken. Usually the suit requests a coercive judgment affirming the duty that the authorities must perform.[14]

Although there has been a long-standing academic debate over whether the latter "duty-imposing" suits, especially those seeking to mandate specific administrative acts should be recognized in Japanese law, Naohiko Harada notes the gradual consolidation of scholarly and judicial opinion in support of an affirmative interpretation:

Today, many court cases and scholarly theories are in agreement, although with differences, in permitting duty-imposing or preventive suits in cases where (a) the duty of the administrative authorities to act or not act is unmistakably clear under the law so that the right of the authorities to make the initial decision is no longer recognized as necessary; (b) the people are threatened with serious damage or imminent danger on account of the action or inaction of the administrative authorities; and (c) no other appropriate recourse for relief is available. It would be safe to say that a decision based on consideration of various interests lies behind this consensus. It follows then that, in view of such precedents and theories, the question of whether a suit in the form of an innominate kōkoku appeal prior to the administrative action is proper and the scope thereof can now be regarded, in terms of procedural law, as having been resolved, at least as to the theoretical issues.[14a]

Although an increasing number of Japanese scholars like Professor Harada are urging liberalization, the growth in this area of the law has been slow and the plaintiffs' victories recorded in sec. 4 have been exceptional. Because of the formidable doctrine barriers discussed, plaintiffs have preferred to face the complexities of civil litigation.

1.4	Procedural Problems in Civil Litigation

Although the standing barrier in civil cases is less formidable than in administrative litigation,[15] technical Civil Code rules concerning injunctive relief engulf many otherwise worthy suits. Under the orthodox interpretation, courts may issue injunctions only for the infringement of property rights (bukkenteki seikyūken). Articles 197–202[16] stipulate that injunctive relief may be used to recover possession, to preserve possession, and for other actions necessary to control or protect private property. For example, under this rule owners or occupiers of land or buildings could enjoin a trespass by a polluting activity that actually infringed rights of ownership or possession. In such cases the courts generally felt it unnecessary to assess the extent of injury before granting injunctive relief. Although useful to the property rights holder, the theory was obviously useless for persons who did not possess property rights. It was also not helpful where a trespass to land took the form of noise, smoke, or vibrations originating from an activity outside one's property. Until World War II few persons were really concerned about the limitations of traditional doctrine, because few activities seemed to pose risks of substantial physical harm. Thus an independent doctrinal basis for protecting persons seemed unnecessary. As the hazards of industrial activities gradually became recognized during the 1960s, however, especially after the four major pollution trials, the legal world grew more concerned with the possibility of using injunctions to protect "personal rights."

Section 2	The Personal Rights (Jinkakuken Riron), Tort (Fuhōkōi Riron), and Environmental Rights (Kankyōken Riron) Theories

Although serious invasions of personal interests were customarily recompensed, the term "personal rights" does not appear anywhere in the Civil Code. The only major rights of this nature explicitly identified in the code are contained in art. 709, 710, and 723,[17] which relate to name, honor, or reputation. Recently, however, some judges and legal scholars have argued that personal rights are a composite of interests associated with physical and mental well-being; and that the law protects individuals from polluting activities injurious to health.[18]

A second doctrine related to personal rights is the tort theory. Like the personal rights doctrine, this theory does not condition injunctive relief upon proof that a statutory right has been infringed. Rather, injunctive relief is deemed appropriate when a legally protected interest is infringed by a wrongful act. "Wrongfulness" is inferred from a violation of a statute or an ordinary duty of care. The tort theory argues that it is only logical to grant injunctive relief in order to safeguard the interests of persons who, when injured, would ordinarily be entitled to monetary compensation under the Civil Code. But the standard of "illegality" required for issuing an injunction under this theory is higher than that required in a suit for money damages.

The most radical of the new doctrines is the environmental rights theory. In its most extreme interpretation this theory holds that where environmental quality is threatened, an injunction should be granted irrespective of whether there exists a direct danger to health. The theory's proponents argue that arts. 13 and 25 of the Japanese Constitution[19] implicitly recognize environmental rights.

2.1   Balancing Interests

Under the property right, personal right, and tort theories, it is necessary for the court to compare the nature of the wrongful act and the nature and degree of the interests harmed. An injunction would be granted ordinarily under the personal right and tort theories only when an injury exceeds the limits of ordinary tolerance. Under the environmental right theory, however, balancing is not necessary. If an environmental right is determined to have been infringed, the court must issue an injunction without balancing the interests.

The standards for measuring each element under the balancing test are extremely difficult and often obscure. Generally, under the personal right and tort theories health risks are heavily weighed. The greater the demonstrated risk, the more likely the parties are to obtain injunctive relief.

But this general proposition only frames the more difficult question of how precisely the courts should balance health against other concerns. This problem has been addressed by the "tolerable limits" or juningendoron theory. Juningendoron is directed primarily at clarifying the central question of what specific interests are protected by arts. 709 and 710 and what rights individuals possess to protect the environment. Basically the juningendoron theory holds that a defendant's conduct is not illegal (ihō sei), hence enjoinable, if the harm the plaintiff suffers exceeds the injury society expects its members to tolerate. The touchstone of the "limits of tolerance" is the basic tort balancing test. But if the harm to plaintiffs exceeds tolerable limits, the defendant's conduct will be found illegal, irrespective of its social value.

Although it was originally formulated to expand liability in pollution cases, juningendoron has been widely criticized for its indifference to the harms to be endured and its exaggeration of the social benefits of defendants' conduct. Its critics allege this has predisposed the courts to favor industrial activities. Some argue that the theory does not even broaden the scope of protected interests beyond those previously recognized under art. 710.

Where a direct and immediate risk to human health is not in issue, the question of balancing has been much less controversial and usually left to the broad discretion of the court. Traditionally the Japanese courts have emphasized the following considerations in assessing whether to issue an injunction: (1) the benefits of the environmental amenity and the probable extent of its impairment; (2) the characteristics of the local environment; (3) whether the victims lived in the polluted area before the polluter's beginning operations; (4) whether the injured parties voluntarily exposed themselves to the risks of pollution and whether they were aware of these risks; (5) how highly society esteems the benefits of the activity; (6) the losses to be suffered by the polluter if an injunction is issued; and (7) whether the polluter has adopted (or is about to adopt) the best practical measures to prevent pollution.

More recently, courts have also begun to emphasize "newer" procedural and regulatory considerations. These include: (1) whether a developer has conducted an environmental assessment; (2) the adequacy of the developer's site selection procedures (for example, did the developer investigate alternative sites); (3) the development of a practical and effective antipollution plan (some courts have carefully scrutinized the effectiveness of the plan); (4) evidence of prior consultations with local residents; (5) evidence of prior consultations with local governments; (6) the developer's execution of an antipollution agreement with the local authorities; (7) the adequacy of existing environmental and other standards; (8) the character of administrative guidance; and (9) supplementary standards and other controls set by local pollution control agreements.

2.2   Types of Injunctive Relief

In order to understand the Japanese innovations in this area, it is important for the reader to grasp the relationship between the courts' analysis of interests and the available remedies under the Code of Civil Procedure. There are two principal classes of injunctive relief in Japan; provisional dispositions (karishobun) and permanent dispositions (sashidome).

2.2.1   Provisional Dispositions

The purpose of a provisional disposition is to provide a means for prompt relief, usually to preserve an existing state of affairs. Often this is necessary when circumstances change. For example, even after a successful judgment, a party may not be able to obtain satisfaction from a bankrupt defendant; or the plaintiff may suffer irreparable harm by the time a final decision is delivered. Provisional relief is perhaps best compared to the temporary injunction in the United States.[20]

The basic code provisions authorizing provisional relief are art. 755 and 760 of the Code of Civil Procedure. Article 755 states:

Provisional disposition relating to the object in issue shall be permitted in cases where the realization of the right of one party will be otherwise impossible, or where there is a risk of considerable difficulty in undertaking it.

Article 760 states:

Provisional disposition may also be made to establish a provisional state of affairs regarding the relation of a right in issue; provided that said disposition shall be made only in such cases as necessary to avert considerable damage to the lasting relation of rights, or to prevent an imminent violation thereof, or for any other reason.

Requirements for Provisional Disposition

A party may ask a court to issue a provisional disposition before or after filing suit. Provisional dispositions, however, usually are conditioned on the plaintiffs litigating the merits of a case. Affairs "frozen" by a provisional disposition may be later approved retroactively once a judgment is rendered.

Types of Provisional Disposition

Article 758 of the Code of Civil Procedure provides a flexible standard for judicially imposed provisional relief, and the remedy may take various forms. Article 758 states:

The court shall, in the exercise of its discretion, decide the disposition necessary to attain the purpose of the motion. Provisional disposition may be effected by appointing a sequestrator, by ordering a party to perform an act, by prohibiting the act, or by ordering a party to render performance.

In a provisional disposition proceeding, a party may either enjoin an activity temporarily or request provisional payment of part of the damages sought. A court can also require a party to establish a fund for the payment of the plaintiffs' putative damages.

The actual content of a provisional disposition depends on the situation. Courts have prohibited discharges of toxic substances above a judicially imposed level; they have directed polluters to adopt specific measures to eliminate harmful substances; they have even closed down some industrial operations. At times a court will create a remedy beyond the relief originally requested. Because of the time and expense involved, many enterprises subjected to a provisional disposition quickly tender concessions. For this reason the provisional disposition has strengthened the bargaining power of local residents.

The Relationship Between Provisional Dispositions and the Development of New Environmental Rights

Traditional doctrine requires a plaintiff to assert a clear statutory right in order to obtain a provisional disposition. Under a recent trend of court decisions, however, a plaintiff's demonstrating merely protected interest has been regarded as adequate for provisional relief. In many cases citizens' groups have availed themselves of the ex parte procedures permitted under art. 757-2 of the Code of Civil Procedure[21] and the lighter evidentiary burdens associated with provisional relief to persuade the courts to recognize less well formulated legal interests. And as these interests have gradually come to be recognized in provisional relief cases, more plaintiffs have argued that the interests should be transmuted into vested rights.

The development of a "right to sunlight" discussed in sec. 5 illustrates this process. In the early stage of its development the right was not clearly defined, and petitions for provisional disposition involving sunlight rights were often denied. Gradually, however, some courts began to grant provisional relief; and they were followed by other courts that awarded damages for infringement of these rights. As precedents developed, the courts then began to enjoin constructions obstructing sunlight. Although a right to sunlight has not attained the level of an explicit statutory entitlement, it is now a solidly protected interest.

2.2.2   Permanent Injunctions

The Code of Civil Procedure also recognizes two subcategories of permanent injunctions. Courts have distinguished orders to take affirmative action (adoption of specific control technology) from other orders to refrain from action (termination of an offending polluting discharge). Mandatory orders must be specific to be enforceable. Although the courts cannot punish failure to observe an order, there are various means of compelling compliance. A court can authorize another party to perform the desired measure at the polluter's expense. Thus if a contractor defies a court's order and erects a new building, the plaintiffs can request the court to have the house demolished by a third party at the expense of the construction company, or the party might ask the court to require the polluter to post a bond to cover all potential damages stemming from violation of the injunction.

To assist the reader in following the doctrinal developments described, this chapter is divided into two parts. Section 3 introduces and analyzes the principal judicial developments in the civil area and sec. 4 examines the most important administrative suits. For both classes of litigation the hybrid Osaka International Airport case marks the end of the second generation of decisions and the beginning of the third and present era of cases. Table 1 illustrates the analytical scheme adopted.

Section 3   Creation of New Rights and Remedies in Civil Litigation

3.1   Judicial Recognition of a Right to Sunshine

3.1.1   Background

Although the Japanese historically have been concerned about assuring access to sunshine,[22] sunshine disputes, particularly sunshine-related lawsuits, increased significantly in the late 1960s and early 1970s. The immediate cause of conflict was the construction of high rise buildings that often obstructed the path of sunlight to traditional two-story wooden residences. Around 1965, residents in many of Japan's large cities began militating against street building, rapid transit, rubbish burning, noise, odors, and other nuisances; and by the early 1970s disgruntled communities had begun to confederate into powerful leagues dedicated to preserving the environment.[23] The sunlight issue was part of a larger citizen reaction to the loss of the urban amenities sacrificed during Japan's frantic program of postwar reconstruction.[24]

The Western reader may at first underestimate the desperation of these citizens' campaigns. They were neither whimsical nor poetic, but rather a last struggle to maintain a minimum standard of life. Unlike in the West, the occupant of the traditional Japanese home is dependent on the sun for light, heat, and other basic necessities.[25] A concern for health first prompted the judiciary to recognize the victims' claims in disputes over sunlight.

Although most of the early suits were defeated, their number multiplied, and gradually in the lower courts plaintiffs began to secure a few scattered victories. The legitimacy of a legal entitlement to sunlight, however, was uncertain until 1972. In that year the Japanese Supreme Court in *Mitamura v. Suzuki*[26] upheld the plaintiff's claim, thereby providing a powerful moral and legal endorsement for the protection of sunlight.

139

Table 1
The three generations of postwar pollution cases in civil and administrative litigation.

First Generation (1953–1973) Era of the Four Pollution Trials	Second Generation (1973–1974) Judicial Exploration of New Rights and Remedies	Third Generation (1975–1979)
Civil Cases:		
12/23/53 ... (Kōfu Dist. Ct.)	5/11/73 Hanshin Auto Expressway (Amagasaki Br., Kōbe Dist. Ct.)	9/9/77 Tagonoura Port Pollution (Tokyo High Ct.)
9/10/62 Otsu Human Waste (Otsu Dist. Ct.)	10/13/73 Kintetsu Fujidera Baseball Ground (Osaka Dist. Ct.)	10/27/77 Shizuoka Sunshine Demonstrations (Supreme Ct.)
9/18/69 Toba Bay Reclamation (Tsu Dist. Ct.)	2/2/74 Ikata Nuclear Power Contract (Matsuyama Dist. Ct.)	6/2/78 Hibiya Park Sunshine (Tokyo Dist. Ct.)
1/22/70 Toba Bay Reclamation (Nagoya High Court)	2/27/74 Osaka Int'l Airport (Osaka Dist. Ct.)	
5/20/71 Hiroshima Hygiene (Hiroshima Dist. Ct.)	5/23/74 Kōchi Vinyl Pollution (Kōchi Dist. Ct.)	
6/30/71 Itai-Itai Disease (Toyama Dist. Ct.)	5/30/74 Tagonoura Port Pollution (Shizuoka Dist. Ct.)	
9/29/71 Minamata Disease (Niigata Dist. Ct.)	1/11/75 Chuba Prov. Pipeline (Chuba Dist. Ct.)	
	2/27/75 Ushibuka Human Waste Treatment Plant (Kumamoto Dist. Ct.)	
	11/27/75 Osaka Int'l Airport (Osaka High Ct.)	
4/1/72 Izumi Crematory (Kishiwada Br., Osaka Dist. Ct.)		
5/19/72 Kokubu Human Waste Treatment Plant (Kagoshima Dist. Ct.)		
6/26/72 Mitamura v. Suzuki (Supreme Ct.)		
7/24/72 Yokkaichi (Tsu Dist. Ct.)		
7/31/72 Chiba Pipeline (Chiba Dist. Ct.)		
8/9/72 "Itai-Itai" Disease (Toyama High Ct.)		
10/19/72 Toshikawa Steel Works (Nagoya Dist. Ct.)		

Table 1 (continued)

	12/26/72	Daiwa Denki Steel Works (Kobe Dist. Ct.)
	1/18/73	Tokyo Five-Story Mansion (Tokyo Dist. Ct.)
	2/14/73	Hiroshima Hygiene (Hiroshima High Ct.)
	3/20/73	Minamata Disease (Kumamoto Dist. Ct.)

Administrative Cases:

6/23/68	Matsuyama Airport (Matsuyama Dist. Ct.)	5/31/73	Kunitachi Pedestrian Bridge (Tokyo Dist. Ct.)	7/17/75	Ikata Nuclear Power Plant (Takamatsu High Ct.)
7/20/71	Usuki Cement (Oita Dist. Ct.)	7/13/73	Nikkō Tarō Cedar Tree (Tokyo High Ct.)	12/8/75	Musashino City Guidelines (Tokyo Dist. Ct.)
12/.../72	Shinkansen Bullet Train (Tokyo Dist. Ct.)	1/14/74	Date Electric Power Plant (Sapporo Dist. Ct.)	3/14/78	Juice (Supreme Ct.)
		2/2/74	Ikata Atomic Power Plant (Matsuyama Dist. Ct.)		

Mitamura and Suzuki were neighbors in the Setagaya ward of Tokyo then designated by the National Construction Standards Law as an open space residential area. In 1960 Suzuki began to add a second story to his home even though the addition raised the ratio of floor space to surface area beyond the prescribed limit. Suzuki also failed to comply with construction certification procedures and ignored orders from the Tokyo governor's office to stop work and remove the illegal construction. Suzuki's construction had completely obstructed all midday sunlight and air ventilation to Mitamura's home. Because his family's health was threatened, Mitamura sold his home (in 1964 for a loss) and moved.

Mitamura sued Suzuki for damages of ¥1,000,000 (approximately $3330) for the loss in value of his home and for his family's pain and suffering. Unwilling to rule that obstruction of sunlight and air ventilation gave rise to compensatory liability under private law, the Tokyo District Court dismissed his claim. The Tokyo High Court, however, reversed the decision. On appeal, the Supreme Court upheld Mitamura's claim and rejected the defendant's contention that he could enjoy his property with no regard to consequences. The defendant's building, the court ruled, was analogous to a nuisance like smoke, noise, or odors. Although the court's opinion did not explicitly recognize the plaintiff's interest in sunlight and air as a legal right, it did hold that air and light were necessary to a comfortable and wholesome life and therefore deserved legal protection.

3.1.2   Opinion of the Supreme Court of Japan (Third Petty Bench) in Mitamura v. Suzuki[27]

Concerning arguments numbered one through seven raised by the appellant's counsel, Yasunobu Narutomi, the High Court, taking note of the fact that

141

the appellant's construction of the second floor illegally prevented sufficient sunlight and air from reaching the appellee's house, ordered the appellant to pay damages. The [free passage of] sunlight and air surrounding one's residence is a necessity if one is to live a comfortable and healthful life. Even if it must traverse the air space of another, it is an object which may be legally protected. It is proper to recognize the injured party's demand for the payment of damages where the person who has caused the obstruction of the sunlight and air has done so by abusing the neighbor's rights. Of course, it can be argued that interference with sunlight or air is a passive characteristic of land use and therefore it is unlike the expulsion or intake of noise, smoke, and odors which actively create nuisances. But when one who has a right to use land uses it in such a way as to receive sunlight and air while preventing them from reaching a neighbor's building, the nuisance created by the exercise of this right is not different from the emission of noise and there is no reason why the injured party's interests should not be protected.

In the present case, the High Court, relying on the evidence mentioned above, found that the appellant's construction of the second story section prevented an amazing amount of light from reaching the appellee's house and garden. As noted in the High Court's decision, the sunlight which reached the rooms in the house and garden changed slightly depending on the season but, except for a short period, almost all of the sunlight was blocked out during the entire day. In addition to the loss of almost all of the sunlight during the day, the ventilation from the south, compared to the situation that existed prior to the construction of the second story, became very poor. It can be easily inferred from the above that this situation had affected every aspect of the appellee's daily life.

Yet, the mere fact that the construction work on the house to the south interfered with the sunshine and ventilation of the house to the north is not enough [in itself] to make the construction an illegal act. However, the exercise of a right is something which may only be done within the scope of generally accepted concepts with due attention paid to the circumstances and the desired result. When it is found that a person has exercised a right without considering its social consequences and has caused injury to another to a degree which exceeds the proper limit, it must be said that the exercise of the right exceeded its proper limit, that it was an abuse of the right, that the action taken pursuant to the right was illegal, and that the responsibility for the illegal act must be borne.

In the present case, according to the decision of High Court, in addition to the act of constructing the second floor which was a violation of the [National] Construction Standards Law, the appellant ignored the stop-work order of the Tokyo governor's office and the removal of illegal construction order and completed his work. At the very least, the appellant's violation interfered with the sunlight and air surrounding the appellee's house. If the appellant had constructed the second floor in accordance with the provisions of the Construction Standards Law and if the construction manager's certification procedures had been followed, the appellee's expectation for normal sunshine and ventilation would have been protected. Even though the appellant's construction took place in a housing area, it greatly reduced the sunlight and air ventilation, thus causing the appellee to lead an unpleasant life and eventually forcing him to move his residence in order to escape the situation.

It cannot be simply said, however, that the appellant's violation of the Construction Standards Law was wrong as against the appellee. Nevertheless, it is proper to hold that the appellant's act, as described above, was an abuse of gen-

erally accepted concepts of the proper exercise of a right and as such was illegal. Therefore, it must be said that the appellant cannot escape liability for the damage created by his illegal acts. It is his duty to pay damages. The decision of the High Court, which required the appellant to pay damages, was correct. The appellant's arguments cannot be accepted. (28)

3.1.3   The Significance of Mitamura

Despite its significant influence on later judicial decisions, it is difficult to decipher what is the precise legal basis for the Supreme Court's holding. One theory used by the Court is abuse of rights.[28] According to this theory,

When a person exercises a right without considering the social consequences and injures another to a degree which exceeds the proper limits, he has abused that right. The action taken pursuant to the right is illegal and the responsibility for the illegal act must be borne.

Some analysts suggest that because Suzuki's action "was an abuse of a generally accepted concept for the proper exercise of a right" it was illegal; thus a finding of liability was proper.

The Court's decision also emphasized the defendant's violation of the National Construction Standards Law, for had the defendant observed the law, Mitamura's expectation for normal sunshine and ventilation would have been protected. Scholarly writings also discuss the personal and property rights theories, art. 709 of the Civil Code, and local custom as additional bases for liability.[29]

Although its doctrinal basis is uncertain, the contribution of the Mitamura decision is undisputed. It abolished the requirement dating from the prewar period that plaintiffs have to prove malice.[30] It unequivocally recognized the plaintiffs' legal interest, if not absolute right, to receive sunshine. Perhaps most important, for the first time in the postwar era the Supreme Court ruled unequivocally in favor of a victim's struggle to safeguard the environment.[31]

3.1.4   The Judicial Development of Sunlight Rights After
Mitamura

Like the four pollution trials, the Mitamura decision encouraged residents throughout Japan to place greater hope in the courts and, as a result, new sunlight suits substantially increased in the short period immediately after the Supreme Court's decision. Indeed between 1972 and 1975 the law of sunlight underwent an important transformation.

The most significant development is the courts' refinement of the interests protected by sunlight rights. Although the central concern continues to be health, courts have begun to link the spiritual and aesthetic benefits of sunlight to the idea of human well-being. The opinion of the Tokyo District Court in the Tokiwadai Maeda Mansion case[32] is typical.

In our country sunlight and ventilation in buildings where people live, and the enjoyment of the benefits of nature, brighten an individual's way of life. As absolute necessities of life for the maintenance of health and the assurance of a wholesome living environment, the possession of this interest shall be recognized as the subject of legal protection.

Despite the expanded judicial interpretation of the interests protected, the courts have not yet unequivocally affirmed a plaintiff's absolute *right* to sunlight.

A second trend is the relaxation of the requirement that plaintiffs prove the il-

143

legality of the defendant's conduct. As noted in Mitamura, Suzuki had acted in clear violation of a national construction standard and the Supreme Court weighed the illegality of his conduct heavily in assessing the harm the plaintiff suffered. Subsequent courts have granted relief in sunlight cases where defendants were actually complying with all existing laws and regulations.[33]

A third, perhaps even more significant, trend is the development of remedies for sunlight obstruction. In addition to increasing damage awards, the courts have granted injunctions, mandated changes in building designs, reduced the height of some buildings, increased setbacks, and angled upper stories. For example, in *Motozawa v. Fujisawa Construction Co.*[34] eleven residents attacked the construction of a six-story building because it totally obstructed sunlight from their one- and two-story homes. The building was located near the border of a third class combined use zone, and the plaintiffs' residences were just beyond the boundary in a second class residential zone. The defendant contended that the building would satisfy existing building requirements and insisted that any change in design would result in lost profits. Several units in the cooperative had already been sold. The Tokyo District Court held that rights involving relations between neighbors may be expanded or limited depending on circumstances. It ordered the north side of the building angled to permit the sun's rays to reach the plaintiffs' homes for at least part of the day.

In a second case, the Japan Public Highways Corporation (Nihon Dōro Kōdan) faced similar difficulties in completing the Chūō Expressway, a major highway running through Tokyo. For three years it had suspended construction because of neighborhood opposition to noise pollution. Just when agreement seemed in sight (the company agreed to construct five-meter high noise barriers on either side of the expressway), residents objected to the barriers, alleging obstruction of sunlight. Only after the corporation finally proposed to relocate the complaining residents to a new building (at a cost of approximately ¥200 million) could construction be resumed.

A third controversy involved a planned enlargement of the Russian Embassy's Trade Representative Office in Tokyo. The Embassy eventually had to pay approximately ¥1 million in damages and agree to angle the building before the neighborhood permitted the continuation of the development.

A final trend is the increase in negotiated settlements.[35] A critical factor has been the mediatory role of the courts. Indeed the doctrinal uncertainty of sunlight rights has somewhat facilitated this development. In ordinary petitions for temporary dispositions, where clearly delineated statutory or contractual rights are in issue, the plaintiffs usually prevail. But in sunlight cases, the interests, albeit important, are still legally less certain. Thus a judge can exert great leverage over the plaintiffs. Conversely the Supreme Court's Mitamura decision and the increasing number of other court decisions recognizing sunlight rights[36] also encourage builders to offer concessions.

The following are two typical examples of settlements negotiated by the parties under the court's supervision.

Sample Settlement Agreement A

Provisions of Settlement Agreement
June 18, 1973, Tokyo District Court
Article 1. *Defendants who are in the process of constructing an apartment in the Shinjuku District may not and will not build the portion of the eleventh floor of the building which is marked in red on the attached sheet.*

Article 2. The defendants recognize an obligation to pay damages to the plaintiff(s) of ¥3,000,000. They will deliver this sum by mail or in person to the plaintiffs' attorney's office.

Article 3. Defendants must set up a large scale antenna to which each of the neighbors may connect their facilities. Defendants will also attempt to minimize in the appropriate ways the problem of interference with television reception.

Article 4. The parties recognize that they have no rights or obligations other than those explicitly set forth in this concession agreement.

Article 5. Each party will pay his own litigation expenses.

Sample Settlement Agreement B

Bunkyōku District, Tokyo University Area
X (Construction Company) v. Y et al., neighborhood group
Tokyo District Court, July 12, 1975

Article 1. Y recognizes and agrees that X is entitled to build the house shown on the attached sheet. Y will not disturb the construction of this building.

Article 2. If Y consents to X's construction of a third floor, X may later enlarge his building by constructing a fourth floor.

Article 3. X will install screens on the windows of the north face of the building.

Article 4. If the stone wall owned by defendant Y is destroyed by X's construction, X must repair the wall at his own expense, and if Y suffers other damages to his property, X will be liable for these costs.

Article 5. The parties recognize the boundaries of the property shown on the attached sheet and each must respect these boundaries.

Article 6. Each party will pay his litigation expenses.

3.1.5   The Administrative Protection of Sunlight Rights

The continuing conflicts over sunlight, the Supreme Court's opinion, and the increasing body of lower court decisions eventually caused the government to pass the "sunshine" amendments to the Building Standard Law in 1976.[37] These amendments added light, plane, and setback standards to minimize the obstruction of sunlight to adjacent buildings[38] and authorized local governments to apply these standards to designated areas within their jurisdictions. The amendments require a developer or contractor to obtain a construction permit from a "building review council" under the appropriate government authority. In Tokyo the developer must furnish residents with the dimensions of the proposed structure, the date anticipated for construction, and other details.[39]

In practice the developer, architect, and other parties often meet the residents to discuss the environmental impact of the proposed building, and sometimes a report on the meeting must accompany the application for a building permit. When the parties still cannot reach an accord, local officials often mediate the disputes.[40]

3.1.6   Assessment

The development of sunlight rights in Japan is best viewed as a compromise between the individual residents' interest in sunlight and the nation's compelling need to deploy its meager land resource as productively as possible.[41] Although the question of future judicial interpretation of the doctrine is at present uncer-

145

tain, the courts' recognition of a right to sunlight has been salutary on balance. Although it may not have helped residents who are already locked in shadows, it affords many others reasonable assurance of at least some sunlight.

The development of the doctrine, however, has at times caused significant dislocations and therefore requires refinement. Although in theory negotiations should reduce the total sum of costs (damage, abatement, avoidance, transaction costs), leading thereby to more efficient resource allocation,[42] in practice the outcome of many sunlight negotiations is unsatisfactory. For example, because of the time and expense of negotiating with large numbers of people, and because settlements do not always insulate builders from liability,[43] some developers have abandoned otherwise beneficial projects.

The courts' recognition of sunlight rights apparently has also skewed the development of the city's inner core. Entrenched low-storied neighborhoods in the center of cities have forced large buildings to the fringes. This has necessitated longer commuting times that are especially onerous for low- and middle-income families. Finally, the piecemeal recognition of sunlight rights has frustrated comprehensive land use planning and has aggravated urban sprawl.

3.2   The Development of a Legal Right to Environmental Quality

Several factors fostered the development of a "right to sunshine" by the courts, local governments, the Diet, and the executive. First, the Civil Code provisions governing relations between neighboring property owners provided a traditional basis for this new concept. Second, the idea of a "right" to sunshine was conceptually simple and sufficiently defined so that the courts were able to develop needed remedies for its protection.

Advocates of an environmental right, however, advance a more sweeping, constitutional claim,[44] for they assert that the citizen's environmental rights require the state and industry to establish substantive and procedural safeguards to protect and preserve the environment. The new environmental right embraces not only air, water, land, and other natural resources but also man's cultural environment. It is enjoyed collectively by the living; it is the birthright of the unborn. No Japanese court has yet squarely recognized this complex, almost revolutionary proposition.

Despite the difficulty of securing judicial recognition, the plaintiffs in over twenty-five suits have argued that infringement of environmental rights justifies damages, injunctions, and other relief. And although environmental rights are not yet legally enforceable, the concept today exerts a powerful political influence.

The idea of an environmental right was originally proposed by the Japan Federation of Bar Associations in the 1972 Niigata Symposium on Environmental Protection, and subsequently has been refined in draft legislation. The following is an excerpt from the Bar Association's original proposal.

3.2.1   Proposal for an "Environmental Right" (Japan Federation of Bar Associations, May 1972)

Fundamental Principles of an Environmental Right

Principle that the Environment Is Common to All People

The fundamental principle in proposing an Environmental Right is the idea that the environment should be common to all people. Until now the natural envi-

ronment of human beings, such as air, water, sunshine, quietness, and natural scenery has been considered "natural property" which is rarely protected as a right. However, our environment is being destroyed rapidly and extensively and now it is quite expensive to obtain pure air and water. The environment which has been considered as natural property should now be recognized as valuable property. One important economic issue is how to assess the value of a proprietary interest in utilizing air and water and how to incorporate this value into existing economic theory.

In the field of law, the environment has not been treated as an object of an independent right. A privilege to utilize the environment has been considered as part of the ownership of real estate or as a benefit incidental to a property right to make use of real estate. The wear and tear or destruction of natural property resulting from utilization of real estate has been considered something to be tolerated for the convenience of such utilization. The principles that ownership is almighty and omnipotent and that the owner of land is completely free to use his land limitlessly from the sky to the ground, on the one hand, and the principle of free enterprise, on the other, have been united to grant an unconditional license to enterprises to destroy the environment. In other words, it has been as if the Environmental Right belongs to industry.

This way of thinking must be changed drastically. The natural environment such as air, water, sunshine and natural scenery is an indispensable and valuable resource for human beings. It must be completely separated from the ownership of the real estate. The right to utilize the environment must be a common property to be distributed equally to everyone regardless of whether or not he is an owner of real estate. The environment should belong not to the owner of real estate, but to all people.

> *Principle that the Environmental Right Is the Community's Collective Right*

Since public nuisances always damage a great part of some particular region, they must be taken to concern the environment of that entire region. It must be a question between the enterprise causing the public nuisance and the whole community in the region. However, traditional legal theory has treated public nuisances as an issue between the enterprise and a particular individual suffering from such nuisance and has not weighed the interests of the enterprise and the individual. Thus, the individual suffering from the public nuisance has been in an extremely weak and powerless position vis-à-vis industry. This situation is similar to the relations between management (employer) and labor (employee) before the rights of labor (rights to organize, to have collective bargaining, and to strike) were legally established. With labor rights, the laborer has been able to stand in a substantially equal position with management. The Environmental Right, therefore, must be established as a community's collective right, giving the people a position substantially equal to that of industry.

3.2.2   A Comprehensive Legislative Proposal for Environmental Protection

The Bar Association's recommendation was subsequently incorporated into a three-part proposal for legislation; "A Basic Law for Environmental Protection," "The Planning Law for Environmental Protection," and "The Law to Ensure the

147

Participation of Residents and Others in Determination for Local Development."
The following is a brief excerpt from the first draft.[45]

The Basic Law for Environmental Protection

Chapter 1. General Provisions

Article 1. Purpose. *The intent of this law is to declare a national policy to protect and to restore a healthy environment for human beings by enumerating the principles of protecting a healthy environment, essential protection measures, and procedures to implement this policy, in order to promote affirmatively the policy of environmental protection for the health and welfare of present and future generations.*

Article 2. Environmental Right. *Every citizen has a right to a secure and healthy environment. According to the provisions of the law, every citizen is entitled to exercise the environmental right against the State, its agents, public corporations, and other private persons.*

Article 3. *In recognition of this most fundamental national policy to ensure an environment healthful for man's physical and spiritual life, to protect natural ecological systems from harmful human activities, and to restore the quality of the environment from its present state of deterioration, the State hereby declares the following:*

(1) The State shall give precedence to the policy of environmental protection and restoration by improving land use planning and economic planning in cooperation with local governments and public corporations, and by maintaining a balance between human activities and ecological life to the extent financially and administratively practicable.

(2) The State shall preserve and continue to improve a safe and good environment for the people. In order to achieve this, the State must stop pollution by protecting the natural environment before pollution adversely affects human health.

(3) The State shall preserve for present and future generations, the air, water, land, animals, plants, and minerals contained in the biosphere through careful control and planning, and seek to improve the quality of recycled materials, and to make the maximum use of limited resources.

(4) The State shall secure a good environment and establish regulations for the wise use of land, for those measures necessary for population control, and for the control of large-scale consumption of resources.

(5) The State shall seek to promote technical innovations that protect the environment. The State shall promote technological developments and studies in order to preserve the environment.

(6) The State shall redevelop overcrowded and unhealthy areas, and develop social, cultural, and educational facilities in underpopulated areas.

(7) The costs of preserving and restoring a good environment shall, in principle, be borne by those who caused pollution.

Article 4. Definitions. *A "good environment" in this law denotes the environment in which people in the present as well as future generations can maintain healthful minds and bodies, and enjoy safe and pleasant lives.*

The good environment defined in the preceding paragraph shall satisfy each of the following conditions:

(1) Present and future generations shall benefit from a natural, cultural, and so-

cial environment and from environmental resources that secure a healthy, safe, and pleasant life.

(2) A good ecosystem shall be preserved, in which plants and animals can grow.

(3) Necessary and sufficient safeguards shall be undertaken so that a living environment that is healthy, safe, and pleasant for human beings is maintained (including property and those public establishments and facilities that are closely related to human life).

(4) Historical and cultural sites of importance shall be preserved.

Article 5. Relations to Other Laws and Regulations. *Regulations and policies of the State and public corporations shall be set and enforced in accordance with the policy declared by this law.* [46]

3.2.3   Yoshida Town Waste Treatment Facility Case[47]

Facts of the Case. Summary of District and High Court Opinions

[The appellants (defendants) are public corporations[48] engaged in waste treatment established by seven towns in Hiroshima Prefecture and Yoshida Township. Hiroshima Prefecture planned to build a human waste treatment facility in Yoshida Township. One hundred and thirty-two appellees (plaintiffs), who live in and own properties within 2.5 kilometers of the proposed construction site and along the banks of the Eno River, brought an action to enjoin construction. The plaintiffs contended that building and operating these facilities would produce offensive smells, water and air pollution, and would adversely affect public health and agricultural crops. They sought to enjoin this construction because, they alleged, it would infringe their personal right to preserve the environment and other rights based on ownership or possession of real property, protected by the Civil Code provisions governing relations between owners of adjacent lands. Two additional plaintiffs were owners of lands immediately adjacent to the proposed site.

The defendants argued that: (1) the facilities were needed to serve public needs; (2) the appropriation for the construction had been approved by the town council; and (3) expected pollution would be limited to the standards set by the Ministry of Health and Welfare.

The Hiroshima District Court granted a temporary injunction on May 20, 1971,[49] after recognizing the standing of all 134 plaintiffs. The court held that the plaintiffs had standing to sue if they lived in the affected area, and that even though it would not be possible to determine the extent of injury to each plaintiff, since the action was to enjoin future injury, a temporary injunction was appropriate because prompt relief was needed. The court made the following findings.

(1) It was foreseeable that the operation of the treatment facilities would adversely affect the plaintiffs' health through air and water pollution expected to continue for many years. Impairment of health could not be remedied by monetary compensation.

(2) The proposed facilities were to serve public needs and were obviously not intended to harm the plaintiffs.

In deciding whether to grant the injunction, the court considered not only the extent and nature of harm but also factors like site selection and the defendants' examination of alternative, available methods of controlling pollution. The court also stressed the importance of local townships providing residents with an op-

149

portunity to express their opinions and negotiating with them matters such as compensation and the introduction of a pollution-monitoring system. Finally, the court held that every feasible alternative site had to be considered. In granting the injunction the court balanced the advantages and disadvantages to the parties.

(3) The court found the defendants' attempt to negotiate with the residents unsatisfactory. After the construction plan for the proposed site had been made public, the residents in the area had filed several petitions with the town council to reconsider the plan and suggested two other sites as alternatives. The town officials insisted on the proposed site and sought to persuade the council individually. The proposed site was approved by the council without adequate explanation and discussion. Although the mayor (who was also manager of the cooperative) met with the residents a few times, he merely repeated his belief that the facilities would be pollution-free and that a minority of residents should endure whatever inconvenience was accepted by the majority. He neither proposed nor discussed any concrete plan to alleviate the residents' fear of pollution. In short he attempted to bulldoze acceptance of the proposed site over the strong opposition of nearly half of the area's residents.

The defendants appealed and the Hiroshima High Court sustained the District Court's injunction. The High Court made these additional findings.

(1) The treatment facilities in the instant case were designed to meet effluent standards set by the Waste Disposal and Public Cleansing Law,[50] *despite the fact that other facilities already in existence had not been able to comply with estimates of operational capacity and were malfunctioning, especially in times of peak demand. At times, overloads had caused major public health problems because of the overflow of untreated waste.*

(2) The design capacity of the proposed facilities was small when compared to the expected development of the area. No emergency capacity had been installed to deal with mechanical failure or an increased demand after holidays. When the facilities became overloaded, it would be difficult to maintain their operational effectiveness. At such times it was probable that they would release offensive odors and their discharges would exceed the legal standard.

(3) In view of the difficulty of obtaining water from the river (because some owners of lands between the proposed site and the river had refused to let the defendants build a pipeline), it was doubtful that the present water supply was sufficient to operate the facilities. Similarly the alternative of underground water was not readily available.

(4) River water itself was inadequate to dilute the facility's discharges. There was a risk that downstream wells from which many people drank would be contaminated.

(5) It was likely that a refuse treatment facility of the proposed capacity and size would be overloaded because the area's refuse was increasing. It surely would become a source of air pollution because the proposed site was in a "V" shaped valley that made smoke diffusion difficult and created an inversion layer.]

Opinion of the High Court (Extract)

Generally, an injunction should not be granted where persons merely suffer some emotional discomfort or other inconvenience from polluting activities which one should be expected to endure as a part of daily life. Yet, an injunction should be granted where there is a high probability of harm to the health of many people and to their living and residential environment. Pollution causing activities, even those activities serving public purposes, should be enjoined, in the absence of other special circumstances, because physical harm cannot be remedied by monetary compensation.

In the instant case, there is a high probability that the health of the residents would be adversely affected by water and air pollution because there are at the present stage no practical plans for pollution control measures such as monitoring of water and air quality, construction of an independent water supply system, the development of safety procedures for the operation of the facilities and planned measures for the improvement and enlargement of the facilities in response to the expected increases in demand. . . .

It is understandable that the defendants urgently need to construct the treatment facilities to meet public needs and that the defendants have chosen the site after some consideration of the circumstances of the case. Yet it is highly probable that the health of many residents in the area would be adversely affected by the establishment of these facilities, and, as a result, it may become impossible for these residents to enjoy the use of their lands and houses for daily necessities. In this case the construction of such facilities can be enjoined. In other words, we will permit those residents whose health may with high probability be adversely affected to seek an injunction against the construction of such facilities.

In the instant case the defendants have brought this situation upon themselves by failing to undertake an appropriate course of administrative conduct. We find no reason to impose the burden of such administrative incompetence upon these residents and to ask them to make any sacrifice in the present case. Moreover, defendants have not proved the absence of any alternative better site or more effective control technology. We do not find other special circumstances to convince us of the necessity of denying the plaintiff's claim for injunctive relief.

Therefore, the defendants are expected either to modify their proposed plan or to change the site so that the level of risk to these residents' health is unchanged. (33)[51]

3.3   Judicial Consideration of New Environmental Protection Remedies Before the Osaka International Airport Case

3.3.1   The Hanshin Highway Case[52]

Summary of the Facts of the Case and of the Kōbe District Court's Opinion

The petitioners reside in houses (one operates a health clinic) within 85 meters of National Highway 43. At the time of the suit, the defendant, Hanshin Public Highway Corporation[53] *had been constructing an elevated new highway (between Osaka and Nishinomiya) above and along Highway 43. Hanshin planned to maintain control of the new highway until its completion.*

Highway 43 is 50 meters wide and has ten lanes and pedestrian walks on both sides. It is capable of handling 90,000 cars daily. In 1972 the number of cars using it exceeded that capacity. Another major national highway between Osaka and Kōbe is Highway 2. It is also overloaded. Fifty-nine thousand cars traverse it daily, an amount greatly in excess of its capacity. The median noise level during a 24-hour period along Highway 43 in Amagasaki City exceeds 70 phons., and at peak hours the noise from traffic easily exceeds 80 phons. Local residents consequently suffer from noise, road vibrations, and air pollution from car exhausts.

In May 1969 the Minister of Construction decided to build the new highway at issue as a part of its urban planning for the area. And in December 1969 the governor of Hyōgo Prefecture approved the routing of the new highway and designated it exclusively for the use of motor vehicles. In October 1971 the Minister of Construction granted the defendant a permit to build the new highway, and construction began promptly thereafter.

The petitioners seek a temporary injunction to prohibit the construction of the new highway. They contend that if the new highway's construction is completed, the amount of automobile traffic would double, intensifying the injuries already noted. They also assert that in addition to noise, vibrations, and air pollution, the new elevated highway structure will obstruct sunlight, circulation of air, and radio and television reception. Petitioners argue that their living and residential environment will be destroyed permanently and that they will suffer irreparable damage.

There are several bases for this action. First, the plaintiffs allege that the State and the defendant have a duty to introduce controls against existing pollution around Highway 43, and to take necessary precautions against anticipated pollution from the proposed new highway. They also assert that the defendant has proceeded with the construction of the new highway without satisfying its duty of preventing pollution and, therefore, its actions infringe the petitioners' rights to enjoy a peaceful environment and a healthy and pleasant life. The plaintiffs argue that their environmental interests are protected by their constitutional right to life (seizonken), their personal right (right to human dignity, jinkaku-ken), and their environmental right (kankyōken).

The defendant emphasizes the importance and necessity of constructing the new highway and its compliance with all required legal procedures. The defendant also emphasizes its promise to adopt pollution control measures, such as constructing green belts, using sound-arresting boards, soundproofing the windows of school buildings, and resurfacing Highway 43, that, it alleges, will protect the petitioners from additional pollution resulting from the completion of the new highway. Finally, the defendant contends that the highway corporation is part of an administrative agency and because its action is performed in "the exercise of public authority," it cannot be contested or enjoined by an injunction in a civil suit.

[In denying the plaintiffs' request for a temporary injunction, the court focused mainly on three issues:
(1) the nature of the legal right to be protected by the injunction;
(2) the degree of anticipated pollution; and
(3) the meaning of an action in the exercise of public authority and a civil injunction.]

Opinion of the Court

(A) The Nature of the Legal Right to be Protected by the Injunction

People need to establish a home as a center for daily life. The conditions of one's residence directly affect the drive and motivation for a family. Therefore, people usually try to maintain good and pleasant surroundings in their residential area, including their natural environment. It is human nature to do so.

A human habitat is benefited in part or whole by the following environmental amenities: (1) sunlight, (2) flow of air, (3) quietude, (4) view, (5) clean air, (6) pri-

vacy, and (7) freedom from the general sense of being overwhelmed by imposing structures. Naturally it is desirable for homeowners to enjoy these environmental interests indefinitely. Yet, in the long run, the residential environment may inevitably change in accordance with the progress of society. Therefore, it is difficult to give judicial recognition to a right to enjoy these environmental interests at a certain point of time and to make these rights permanent.

The rights pertaining to one's residential environment exist for the purpose of maintaining amenities currently enjoyed by residents. It is similar to a possessory right that is recognized to protect the existing state of possession of property. The nature of this right may change in response to one's changing environment. When the environment that is the basis for the environmental right itself disappears, then the right is also extinguished. Unlike a possessory right on which one can predicate an action for the recovery of one's lost environment, it is difficult to construct a right to environmental quality that could have such an effect.

Whatever benefits a resident derives from his so-called "neighborhood environment" are merely reflective; such benefits are not yet protected by law. One cannot claim a personal right over one's neighborhood's environment. A resident cannot bring a civil suit to preserve a neighborhood environment that is not directly connected with his own residence.

Changes in the residential environment mainly result from human activities that often give rise to unnecessary and meaningless disturbances. It is a serious matter for residents to be deprived of such environmental amenities as sunlight, quietude, and clean air without reasonable justification, and to be forced to accept the deterioration of their environment. Their tangible, as well as intangible injury, is especially serious when environmental deterioration affects their health. Therefore, it is legally necessary to recognize and protect environmental amenities. Those who unduly infringe on the residential environment of others may be held liable for injuries arising from such infringement. Yet it is difficult to evaluate environmental interests in monetary terms; and it is impossible to preserve those interests by monetary compensation. A damage award does not serve to protect environmental interests. We need to devise effective means to prevent infringement of environmental amenities and to save the environment from deterioration.

These residents (who are owners or tenants) currently enjoying environmental amenities, have a right to protect these from the threat of encroachment where the impending encroachment is clearly unreasonable. We may call this right the "right to prevent unreasonable encroachment of environmental interests."

When the danger of unreasonable encroachment is present, residents can preserve their environmental amenities by seeking an injunction based on the above right to the extent necessary and adequate for the protection of such interests, provided that there is no special reason for them to refrain from doing so. One such case would be where compensation is deemed appropriate. (We believe that the exercise by residents of this entitlement must not be allowed to create an undue hindrance to other people's lawful and free activities.) . . .

This preventive right may be regarded as a part of the right to life viewed from the perspective of the importance of residential amenities as a basis for life. It may also be regarded as a part of a personal right distinguishable from a property right. (30) . . .

(B) Degree of Anticipated Pollution

[In the instant case, the court continued, there was a special reason for petitioners to accept a compensation settlement regarding their sunlight claims. The court held that because the disruptions of air flow and radio-television reception had not yet amounted to an "unreasonable encroachment," the petitioners had not perfected their contentions on these points.]

A small number of petitioners who live on the north side of the proposed elevated highway would be blocked from sunlight by the supporting structure of the highway. Although there is no evidence that the infringement of sunlight would adversely affect the petitioners' health, it would be grossly unfair if such serious injuries are imposed on them without compensation, especially in view of the constitutional provision that provides just compensation for property taken for public use.

Appropriate compensation is obviously in order for these petitioners. The defendant has proceeded with the construction without clarifying this issue of compensation with them. Their interests in having sunlight unobstructed have been clearly and unreasonably threatened.

Yet, we cannot grant an injunction based on petitioners' preventive rights where there are special reasons militating in favor of other forms of settlement. In the instant case, the defendant can neither alter the course of the new highway nor stop its construction altogether. No danger, however, will be posed to the petitioners' health by the resulting obstruction of sunlight. Under the circumstance, we hold that the defendant's payment of compensation to the petitioners for the loss of sunlight would remedy the "unreasonableness" of the defendant's action. (31)

The noise from Highway 43 is already unbearably high. Since the opening of a new highway would further increase such noise, the petitioners may seek injunctive relief in order to prevent the further risk of harm.

[The court then examined various available options. One possibility would be to enjoin the defendant from proceeding with construction entirely. The other would be to compel the defendant to limit its noise to a certain level, while allowing the completion of the new highway. The latter would provide a means to control the level of noise from Highway 43 and the new highway. An injunction against the construction of the new highway, on the other hand, would permit serious noise from Highway 43 to continue unabated.

The court next declined to find that the level of vibration was clearly unreasonable.

The court then pointed out that the petitioners would be entitled to injunctive relief to the extent necessary to prevent harm from the increased amount of exhaust fumes if the combined total amount of traffic increased, as was projected with the opening of the new highway. It emphasized, however, that if the number of automobiles using both highways were limited to 150,000 per day (yearly average), the level of air pollution at the petitioners' residences would not differ greatly.]

On the completion of the new highway, Highway 43 will become an eight-lane highway. The amount of traffic is estimated to decrease from 100,000 to 70,000 cars a day; 80,000 cars are expected to use the new highway daily (its capacity is 120,000). The combined total of vehicles using the two highways is estimated to be 150,000. Therefore an increase of 50,000 vehicles is expected. The amount of exhaust fumes from automobiles on the elevated highway is about one-tenth of those on the ground level road. Moreover, the new highway will be elevated at least 12 meters from the ground. In addition, a 3 meter fence is to be constructed on the elevated highway. Exhaust fumes would be widely dispersed and the effect on petitioners' residences near Highway 43 would be slight. (32)

If the Japanese version of U.S. Senator Edmund Muskie's law (Regulations on Automobile Exhaust Fumes, 1975 and 1976) is implemented, as the defendant contends, air pollution caused by automobiles would drastically be reduced. In the instant case, too, the anticipated increase in air pollution would thereby easily be arrested. It is, therefore, reasonable for the petitioners, while allowing the construction of this new highway, to impose on the defendant a limit of 150,000 cars daily for both highways and to lift the limit at such a time when the risk of increased air pollution disappears. (33)

[The court concluded that the petitioners' request for a flat prohibition on all construction exceeded the necessary preventive relief required in the case.]

3.3.2   Environmental Risk as a Basis for Contract Rescission: Ikata Atomic Power Plant Case[54]

Facts of the Case

[In July 1969 the plaintiff (Shikoku Electric Power Co.) bought two tracts of land from the defendants Tamura and Kubo. The plaintiff agreed to conduct borings in the land to determine (in its judgment) whether the land was suitable as a site for an atomic power plant. Both parties agreed and signed a contract containing the following provisions:
(1) The plaintiff shall pay the defendants ¥76 per square meter for tract no. 1 and ¥50 per square meter for tract no. 2 upon the completion of registration of the transfer of ownership.
(2) The defendants shall deliver the titles of the land and transfer the land to the plaintiff upon satisfaction of the condition stated above.

After the geological investigation on September 21, 1970, the plaintiff concluded that the land was suitable for the site of an atomic power plant and would meet the necessary requirements for obtaining government approval. On the same day the plaintiff so informed the defendant in writing.

The plaintiff brought this action to compel the defendant Tamura to transfer tract no. 1 by July 31, 1971, after removing the trees, and to compel the defendant Kubo to transfer tract no. 2 immediately.

The plaintiff asks the court to order the defendants to comply with the terms of the sales contract and to order a provisional execution of the judgment.

The defendants denied the facts alleged by the plaintiff and made the following contentions:
(1) Because there was a mistake in the material facts involved, the contract was invalid. Under the circumstances the defendants never intended to execute the sale of the land. *The defendants argued that they had signed the contract lightheartedly, trusting the statements of the officials of Ikata town who were acting as intermediaries that "the contract is only for boring explorations and investigations" and "you can rescind the contract later if you don't want to sell the land." They truly had not intended to sell their lands and were not aware of the fact that the land was sought for the purpose of constructing an atomic power plant. Only later did they learn of the intended use of the land. Therefore, the contract was void because the defendants signed it without intending to effectuate it and because there was a mistake in a material fact.*

155

(2) The construction of the atomic power plant was hazardous and contrary to public order and good morals. *The defendants noted that although it was stated in the contract that the purchased land was intended for the site of an atomic power plant, the establishment of an atomic power (reactor-generator) plant was itself extremely hazardous and for this reason was contrary to public order and the good morals of the society as contemplated by art. 90 of the Civil Code.*

In their argument the defendants placed particular emphasis on the fact that the plaintiff planned to build a PWR (pressurized water reactor) plant designed and manufactured in the United States. The PWR model, however, they noted, was still in an experimental stage. After one and a half years of operation, the generator of the first such reactor built at Kumihama, Japan, had developed trouble with its steam pipe system. Many cracks and holes had been found in the pipes. The danger to the residents from the radioactive vapor produced by the operation of the plant, the defendants argued, could not be ignored. Furthermore, because methods for treating and handling radioactive solid wastes had not been satisfactorily developed in Japan, the cooling water from the plant would adversely affect fisheries in the area. Moreover, the chosen site was near a zone of frequent earthquakes.

The defendants warned that the residents around the plant would be exposed to harmful radioactive substances if a serious accident at the atomic reactor should occur. Even during the normal operation of the plant, the discharge of a certain amount of radioactive substance, they alleged, could not be prevented. They also pointed out that many scientists in Japan, as well as in foreign countries, strongly opposed the construction of atomic power plants because of these dangers, and for all the above reasons the plant was contrary to the concept of justice, public order, and good morals.

(3) Defendants can rescind the contract because of a specific reservation. *The defendants argued that even if the court would not support their first two points, they had signed the contract relying on the statements of the officials that "you can rescind the contract later if you don't want to sell your lands." They claimed that this created a specific reservation entitling them to rescind the contract. The defendants in fact informed the plaintiff of their intention to rescind the contract in May of 1970.*

The plaintiff denied the defendants' allegations except for the part that stated explicitly that the purchased land was intended for the site of an atomic power plant.

The plaintiff argued that in Japan the establishment and operation of atomic power plants was legally recognized by statutes such as the Basic Atomic Energy Law,[55] the Law to Establish an Atomic Energy Commission,[56] the Law Regulating Atomic Energy Materials, Fuel, and Reactors,[57] the Electric Power Supply Law,[58] the Law to Promote the Development of Electric Power,[59] and the Law for the Compensation of Damages Caused by Atomic Power Accidents.[60]

As long as the establishment and operation of an atomic power plant complies with these laws, the plaintiff argued, it was legitimate; and there was no basis for finding it contrary to public order and good morals.

The plaintiff also pointed out that the PWR model adopted had been widely used in various parts of the world, representing 52% of all atomic power reactors. It was a practical and commercial reactor with operating experience and tested reliability. No operating accident of this model had yet occurred resulting in damages to the general public. The plaintiff also noted that the measures developed for treating radioactive wastes were adequate.]

Opinion of the Court

Having evaluated all the evidence submitted to the court, we find that the contract to sell the land in question was validly concluded. Tract no. 2 was registered as a rice field, but its present condition is fallow. It can be transferred without obtaining a permit as required by the Agricultural Land Law.

First, we examine the defendants' contention under (1). In the contract, the plaintiff made it a condition of the transaction that the land be judged suitable for the site of an atomic power plant. The defendants' alleged mistake was that they did not recognize the condition when they signed the contract. We find from the evidence that the defendants were aware of the intended use of the land as an atomic power plant site at the time of signing the contract. There was no mistake on the part of the defendants on this point. We do not believe the defendants' contention that they signed the contract believing that the site would be used only for boring.

We turn to the defendants' contention that the establishment of an atomic power plant on the proposed site is hazardous. We have already found that the intended use of the land was stated explicitly in the contract and that it was clearly a motive for concluding the contract for both parties. It is also clear that the plaintiff made it a condition of the transaction that the land should be judged suitable for an atomic power plant site.

Is the establishment and operation of an atomic power plant as a motive and a condition of a contract contrary to the public order and good morals required by art. 90 of the Civil Code?

The construction and operation of atomic power plants is, though strictly regulated by such statutes as mentioned above, not prohibited. Its existence is recognized under those statutes. Unless there is an extremely high degree of danger to the life and bodily safety of residents, and the danger is present and imminent, even though regulated, the construction and operation of an atomic power plant cannot be said to be contrary to public order and good morals.

In the case before us, the proposed PWR model has been commercially operated in the United States and other countries. It has also been in operation at the Tsuruga and Kumihama power plants since 1970. It cannot be said that its commercial operation has demonstrated beyond any doubt the safety of the reactors or has developed sufficient experience in controlling its dangers. The operation might still be accompanied by some hidden danger. However, we cannot say that the risk involved is extremely high and the danger is clear and imminent. Scientists are still debating the safety of reactors and the possible risks of radioactive wastes to the human body. Although all do not agree on the degree of safety of atomic power plants, there are some scientists who vouch for their safety. It was neither alleged nor proved in the case before us that the plaintiff would undertake the construction and operation of the atomic power plant while flouting the strict regulations. There are neither specific indications of a high degree of risk nor of a present and imminent danger. We are unable to find that the establishment and operation of the proposed plant is contrary to public order and morals.

We do not find the defendants' contention that they reserved the right to rescind the contract consistent with the evidence. It must be rejected.

The condition that the land should be judged suitable for an atomic power plant site is a part of the contract and the condition precedes the execution of the contract. According to the contract, the plaintiff, not a third party, can judge whether the condition was satisfied or not. The defendants' intention is not relevant to judging this point.

Therefore, we conclude that the plaintiff's contentions have merit and should be sustained. We render the judgment stated at the beginning of the decision. The cost of litigation shall be paid by the defendants according to art. 93 and 89 of the Civil Procedure Law, and the order for preliminary enforcement may be entered as provided by art. 196 of the same law.

3.3.3   The Kōchi Vinyl Pollution Suit[61]

Abstract of the Case

[This case received considerable attention in Japan because of its recognition of the liability of local governments and the State for injuries to fisheries resulting from the discharge of industrial waste. The plaintiff fishing cooperative possessed a right to draw nets on the beach of Nangoku City in Kōchi Prefecture. Twenty individuals within this cooperative had acquired rights to conduct net fishing in a designated fishing area and had obtained permission from the Governor of Kōchi Prefecture to carry on the business of net trawling pursuant to a cooperative fishing regulation.

The Ushiro River flows through the city of Nangoku and is designated as a first class river. (First class rivers are owned by the State, which delegates its powers of management to the prefectural governors.) Two water canals, through which the river flowed directly into the fishing area assigned to the fishing cooperative, had been erected along the river.

At the time the use of vinyl in agricultural cultivation and house construction had become so popular that every establishment along the river in Nangoku City used vinyl covering. Because vinyl sheets are replaced each year and the disposal of this material presented formidable difficulties for the local farmers, it was dumped into the river, or piled up in the rice fields on the ridges of irrigation fields and in the ditches. But during torrential rains, the vinyl thus accumulated passed out into the Ushiro River and from there through canals 1 and 2 to the fishing area of the fishing cooperative, where it piled up and floated around. Thereafter the vinyl became entangled in the nets of members of the cooperative and obstructed fishing operations. From around 1968 this refuse continued to accumulate in great amounts, clogging propellers and nets so that the fishing vessels themselves could not operate. Consequently, the members of the fishing cooperative had to suspend work to remove the vinyl. Much time was wasted also in drawing in the nets to land to remove the vinyl waste. Because of the intrusion of waste, the commercial value of the fishing catch itself declined and caused economic hardship to the members of the fishing cooperative.

In view of these injuries the members of the cooperative claimed that the city of Nangoku had been negligent in fulfilling its duty of waste collection and disposal under the Waste Management Law,[62] and that defendants, the Prefecture of Kōchi and the State, had been negligent in performing their duties of overseeing the city's waste disposal and in not providing technical or financial assistance to the city. The plaintiffs asserted that this dereliction of duty violated art. 1(1) of the State Compensation Law.[63] Finally, the plaintiffs argued that because the defendant, the State, was the owner of the Ushiro River, it also had the responsibility of maintaining the river through the Minister of Construction. Since the Minister of Construction had delegated his authority to the Governor of Kōchi by Cabinet Order, the plaintiffs asserted that both the State and the Pre-

fectural Governor, who had incompetently maintained and administered the river, should be responsible for the injuries plaintiffs had suffered. The suit was brought under art. 2(1) and 3(1) of the State Compensation Law (against the Prefecture of Kōchi) and under art. 2(1) (against the State). 64]

Opinion of the Court

(A) Responsibility of Nangoku City

According to the facts of this case, after farmers used vinyl, it was deposited in the rice fields and on the ridges and ditches of irrigation canals. Thereafter, it was taken up and carried away in a great rainfall, passing away from the area in the possession of these farming enterprises. It is thus unclear who had the responsibility (at this time) of disposing of this material and a situation developed that made it impossible to assign the responsibility of disposal to the original enterprises. Thereafter, this material began to float about and pile up in great quantities in the plaintiffs' fishing area, seriously interfering with their businesses.

The defendant city of Nangoku had the responsibility of disposing of these waste products under art. 10.2 of the Waste Management Law. Although the defendant city began to take some measures after the tenth typhoon (of the season) on August 12, 1970, but before the sluice-gates incident, 65 the defendant's measures for that period were incomplete and produced no observable effect. Only after the sluice-gates incident and the filing of this suit did defendant undertake a complete cleanup and remove a large amount of used vinyl sheets from the area, thus effectively reducing the injuries caused to the fishing grounds.

The following circumstances indicate that the defendant was particularly negligent in its retrieval of the vinyl material, and consequently inflicted injury on the plaintiffs. In 1970 the total yearly amount of vinyl used in the city was estimated at about 300 tons. In 1969 only 25 tons were retrieved. In 1970, 221 tons and in 1971, 407 tons of vinyl were retrieved. After the closing of the gates incident in 1972, the amount that was retrieved exceeded 1000 tons. From this we can infer that a considerable amount had been left unretrieved (before 1972). It is clear that the defendant city was negligent in not disposing of the used vinyl sheets and thus caused damages to the plaintiffs.

The defendant, Nangoku City, alleges that it was not negligent in its administration of the cleanup. The defendant argues that the area through which the Ushiro River flows is at sea level and that the ground level of the area at the mouth of the first and second water canals is particularly low. In an average rainfall, the rice fields in the area easily flood, and agricultural and straw materials, as well as the vinyl in question, naturally flow out from these areas into the sea.

During the period when the vinyl was first removed from the agricultural cultivation houses until the time of flooding, however, much vinyl continued to pile up. The city of Nangoku cannot escape responsibility for not having retrieved this substance during this time. Consequently, we hold that the city is responsible for and liable to the plaintiffs for their injuries under the State Compensation Law art. 1(1). (44)

(B) Responsibility of Kōchi Prefecture and the State

We hold that the defendant prefecture of Kōchi [although the State owned the river, it had delegated its administration and maintenance responsibilities to the prefecture] was negligent in the following respects. The prefecture of Kōchi was under a duty to employ appropriate measures, such as dredging the river bottom and setting appropriate wire nettings to remove the waste products from the river so that the wastes would not be carried into the fishing areas of plaintiffs. The Governor of Kōchi, nevertheless, did not take appropriate measures for the maintenance of the river. Consequently, until the sluice-gates incident, great amounts of mud were mixed with wastes, vinyl kept accumulating at the river bottom, and water flow was obstructed during rains. When the river flooded, vinyl and other materials were washed away. These were the consequences resulting from the governor's failure to maintain the river properly. Therefore, we hold that the State was liable under art. 2(1) of the State Compensation Law, and that the prefecture of Kōchi, as the party charged with the costs of the administration of the Ushiro River (45), was liable under art. 2(1) and art. 3(1) of the same law.

(C) Damages

[*The court then found that from March 1968 the fishing catch was reduced by approximately 20%. Members of the cooperative sold their catch to the cooperative, which in turn sold fish to wholesalers. The cooperative received commissions for 5% of the proceeds. According to the court's calculation, damage was estimated to be around ¥1,443,162. Although it was very difficult to assess the damage because various factors were involved, the court awarded ¥1,000,000 to the cooperative.*

Moreover, because it was difficult for members of the cooperative to prove the amount of economic loss resulting from this waste, the court held that compensation in the amount of ¥500,000 be paid to each member of the cooperative for pain and suffering, noting that these people were faced with future uncertainty due to the disruption of their fishing and the degradation of their fishing ground.]

3.3.4   The Tagonoura Port Pollution Taxpayers' Suit[66]

Abstract of the Case

[*The plaintiffs, nineteen residents of Shizuoka Prefecture, brought a citizens' (residents') suit under art. 242.2.1 of the Local Autonomy Law[67] against the prefectural government, the governor of the prefecture, and four paper-pulp companies. These companies were all engaged in the production of paper pulp in the prefecture and discharged their wastewater directly or indirectly into the neighboring rivers. Their discharges caused extensive damage to agricultural crops and fishing. Since the completion of Tagonoura Port in 1961, the prefectural government had dredged the port every year to remove the accumulation of sludge from the factories' wastewater. But implementation of national water quality regulations began belatedly. The rivers and port became heavily polluted by evil-smelling sludge, and the water turned a brownish color. Pollution extended into Suruga Bay. After April 1970 the dredging of the port was terminated because local residents began demonstrating against the prefectural government's dumping sludge into the ocean. As sludge accumulated, the port actually became more shallow and its functions were partially paralyzed. Boats were disabled as they sucked the sludge into their cooling pipes, ships' bodies were corroded by hydrogen sulfide, and large ships were prevented from entering* the port without first unloading their cargoes outside; also, several piers were subsequently rendered unusable.

The plaintiffs requested that the court issue:

(1) a declaration that the prefectural government's failure to compel the defendant companies to terminate waste discharges from pulp factories and the prefectural government's failure to prevent the pollution of the port from these discharges was illegal;

(2) an order requiring the governor to refund ¥10,000 million to the prefectural government as part of an illegal expenditure of ¥121,803,000 in 1969 for dredging the port;

(3) an order requiring the companies to refund ¥10 million to the prefecture as partial damages on the theory that the companies were jointly liable for the international tort of damaging the prefectural property (the port); and

(4) an order enjoining the companies from further discharging wastewater into the port in excess of 10 ppm of SS, 5 ppm of BOD and COD, and 0.3 ppm of sulfide.

In dismissing the plaintiffs' action because it did not satisfy the requirements of the citizen's suit provision in the Local Autonomy Law, the court made the following findings:

1. The prefectural government was not responsible for the control or maintenance of the rivers as prefectural public properties. Rather the governor had merely been delegated authority by the State to issue permits for rivers and to maintain rivers (River Law, art. 10; the Local Autonomy Law, art. 148-2). Furthermore, the prefectural government's right to control rivers was not the same as the superficies enumerated in art. 238-1-4 of the Local Autonomy Law. The court held that the rivers and the port were not property subject to a citizen suit.

2. Even if the port were a public facility and subject to a citizen suit, the court hesitated to find the governor negligent in administration. Moreover, regulations cited by the plaintiffs such as the prefectural pollution control ordinance, the ordinance relating to maintenance of rivers, and the prefectural regulations concerning fishing rights were not directly related to the administration of the port. Therefore, the court concluded that the governor's alleged failure to exercise the police power in these fields did not make the maintenance of the port unlawful.

3. The plaintiffs sought an injunction to enjoin the companies from discharging wastewater and further polluting the port based on the (governor's) right to control and supervise the port. The court held that it would be difficult to construe this right as a property right, necessary for injunctive relief. Moreover the governor's powers were limited to the maintenance of the port's functions. In the instant case, the functions of the port were not sufficiently impaired to warrant an injunction.

4. The plaintiffs contended that the polluting companies should be liable for damages to the port and that the governor should have made these companies bear the costs of dredging operations. By failing to do so, the plaintiffs argued, the governor should be held accountable for unjustified payments of public funds, and he should refund these payments. The court held that in order to hold the defendant companies liable for the dredging costs, their polluting acts must constitute a tort. Although injury had occurred (since functioning of the port was impaired by the defendants' discharge), their actions did not violate any existing regulation. The court found it inappropriate to apply a general, vague standard of "socially acceptable" conduct to the defendant companies in the present case because for a long time they had been left free to discharge wastewater in the ab-

161

sence of regulation. The court also found that the port was originally constructed on the assumption that discharged sludge from the factories would eventually be dredged. The court held that since the defendant companies were not liable in tort, the plaintiffs' claims against the prefectural governor and the companies could also be denied.]

3.3.5   The Ushibuka Human Waste Treatment Plant Case[68]

Abstract of the Case

[The plaintiffs are sixty-seven residents of three villages in Amakusa, Ushibuka City in Kumamoto Prefecture, who make their living by fishing. The defendant, the city of Ushibuka, planned the construction of a human waste treatment facility, and since 1971 has purchased land for this purpose near the residences of the plaintiffs. The plaintiffs opposed the construction of this facility for the following reasons.

1. Even under normal operating conditions, the planned facility could treat only about 97% of BOD;[69] the remaining untreated 3% would be discharged directly into the ocean. At times of overload, however, a considerably higher percentage of activated untreated sludge would be released. There was substantial evidence that overloads might occur.

2. After effluent was discharged, it would be carried by the tides along the coasts and into the harbors forming a single band by the villages where the plaintiffs resided and made their living. The anticipated harms were grave, and included economic and health injuries. In addition to adverse effects on fish and clams and injury to naturally growing sea grasses and seaweed fields, the plaintiffs also feared that the wastewater would also injure a fish processing production plant operating in "A" village and that the rise in BOD would trigger an outbreak of red tides and poisonous plankton in the area of each harbor. Moreover, the plaintiffs alleged, because they used sea water (due to the absence of daily fresh water) for washing rice and farm products, they too would be exposed to the contaminated water. In short the waste treatment facility constituted a major health hazard that would cause headaches, vomiting, eye trouble, and loss of sleep in addition to more serious maladies.

The plaintiffs argued that these injuries exceeded the limits of endurance and that the defendants' planned action would violate their fishing rights, possessory rights and rights of ownership, rights of human dignity, and environmental rights. Accordingly, they requested an injunction to prevent construction of the human waste treatment facility.

The court first investigated five facilities similar to that planned by the defendant, and operated under comparable circumstances. In many cases, the court found that waste treatment plants did not operate as originally planned. As a consequence the court took notice of the fact that 3% of BOD could not be processed, and that there was a high probability that overloaded unactivated BOD would be discharged.

Thereafter the court addressed the issue of the planned site's appropriateness. It carefully analyzed the amount of wastewater to be discharged into the surrounding environment, the wind currents, and the conditions of the tides. The court concluded that because there was a high probability of pollution in the adjacent waters, the villages, and the harbor, there was also a high probability that the alleged harm might result. Because the plaintiffs thereafter might not be able to continue to make a living, the court held that at least with respect to the twenty-three individuals living in "A" village, the facility's construction was illegal despite its obvious "public" benefit. Given the fact that there were no special countervailing circumstances that would militate in favor of permitting the construction of the plant, the court issued an injunction to stop its construction.]

Opinion of the Court (Extract)

. . . Next we examine whether there exist special circumstances militating in favor of the defendant.

1. According to the evidence, the defendant, Ushibuka City, had previously disposed of human waste by means of open ditch treatment. Previously it had paid compensation for injury to rice under cultivation, and recently it had purchased a special disposal area in the mountains and had disposed of the waste there. No harm has been reported. If a site were chosen without wells, rivers, or houses, and if the appropriate administration were given the disposal site, the risk of harm (from the open ditch method) would not be that great. Yet because it was considered difficult to guarantee or to safeguard the area, Ushibuka City concluded that such a practice could not continue forever; consequently since 1971 the city had planned the construction of a human waste facility. With this intent in May of the same year the city concluded a contract with Kubota Company in the amount of ¥87,000,000. Because construction was delayed, the city decided in a City Council meeting in March 1973 to allocate a budget of ¥143,070,000 to the project. In view of the fact that the open ditch disposal of human waste had not posed a grave danger to the residents' lives, if the construction of the facility were enjoined, the defendant would have no recourse but to continue handling the disposal of human waste by the open ditch method. Because the defendant is unable to realize the benefits of the contemplated contract, it will suffer considerable economic loss.

2. Despite the above, we believe that the defendant has not earnestly undertaken the following pollution prevention or avoidance measures. Since the purpose of the treatment facility is the protection of the living environment of the people of Ushibuka, and since it is related to public health and is supposedly in the public interest, its construction is of some public importance. However, the defendant proposed to construct the human waste facility in an area where there was the possibility of adverse effects on human health, and where these residents were already engaged in a livelihood dependent on clams and fish. In such a case special attention must be given to whether or not the contemplated facility will keep ocean pollution to a minimum level. The defendant has the duty of taking measures to explain the above facts to local residents.

At the very least, the defendant should have had an expert survey made of the tidal direction and speed in the vicinity of the contemplated site, and have forecast the dispersal of wastewater, and its stagnation. The defendant, moreover, should have undertaken an ecological investigation of the influence of wastewater on fish and clams living in this area and upon seaweed; the defendant should have made clear to local residents the existence or nonexistence of possible harm resulting from the construction of the facility; the defendant should carefully have investigated whether these results were evidence of greater harm than that resulting from the open ditch disposal method presently used. Upon convincing itself that there were no appropriate alternatives to the construction of the present facility, the defendant should have discussed compensation for spe-

163

cific harms predicted by their investigations and have discussed these and other measures with local residents.

The protection of the plaintiffs' lives and their environment and the planning of public health measures is the responsibility of the defendant, an administrative body. With the exception of plaintiffs, those receiving the benefit of the construction of the present facility are the people of Ushibuka, or put another way, the defendant itself, which is an administrative body. Since the defendant is to receive the benefit of such measures, it is only natural that it should have undertaken the above investigation. Because the defendant had the duty of making an appropriate investigation and because it neither did so nor came forth and initiated negotiations based on the results of its investigation, any resulting harm should be attributed to the administrative incompetence of the defendant. There is thus no reason why the plaintiffs should bear this situation without complaint. Consequently on contemplating all of these circumstances, we hold that there are no special circumstances militating in favor of the construction of the facility by the defendants in the face of the sacrifice of the plaintiffs. (27)
3. Accordingly, although the resulting harm to the plaintiffs' fishing industry and to their health from the facility may differ between the various plaintiffs, nevertheless we hold that there is a high degree of probability that the use of their residences or residential areas for their livelihood will be made difficult and that the injuries they will receive will exceed the bounds of endurance. Thus, even though plans for the present facility are on the verge of being implemented, we hold that it is necessary to grant this injunction. (27)

3.4   The Osaka International Airport Case: The End of the Second Generation

The Osaka International Airport is only twenty minutes by bus from the center of the city of Osaka, Japan's second largest metropolitan area. The Japanese government began operating the airport before World War II as a small suburban facility but after the war it grew into a busy international airport. Although noise and fumes around the airport had always been a nuisance, the introduction of jet planes in the late 1960's further disrupted the lives of those living nearby. When the government rejected their repeated appeals, two hundred and sixty-four residents finally filed suit in the Osaka District Court in 1969.

The suit differed from the four pollution cases discussed in the preceding chapter in three critical respects. First, the defendant was the national government. Although the courts in the major trials had hinted at the state's responsibility for pollution damages, this had never been made explicit.[70] Second, the plaintiffs argued that both their personal rights protected by the Civil Code and their environmental rights had been infringed and would continue to be violated unless an injunction was granted. Finally, the plaintiffs sought injunctive relief plus compensation for mental and physical injuries. The plaintiffs asserted that the airport was too small and lacked adequate facilities for international use, and the government's negligent maintenance of the facility had injured them. Individually, the plaintiffs requested: (1) ¥500,000 in damages for past injury and attorney's fees, (2) an injunction against the use of the airport between 9:00 P.M. and 7:00 A.M., and (3) payment of ¥15,000 per month until such time as the government reduced the airport's noise level to 65 phons.

The Osaka District Court's decision, issued on February 27, 1974, proved less than satisfactory to either party. Although the court found that the airport's noise exceeded levels set by Cabinet Order and that the plaintiffs suffered mental anguish from the resulting disruption of their lives, it failed to find a causal relation between the plaintiffs' injuries and the alleged harms. Although the court awarded the plaintiffs a nominal sum for past damages, their claim for future damages was rejected as too speculative; the court noted that the government's planned noise pollution abatement measures would probably remedy the problem.

At the same time the court rejected the defendant's argument that ordering the government to adopt a new set of airport regulations would violate the separation of powers doctrine. The court noted that the injunction would only prohibit the defendant from using the airport during certain hours (10:00 P.M. to 7:00 A.M.) and, therefore, the intrusion on executive prerogative was minimal. In fact the injunction merely formalized the status quo ante because, by the time of the court's decision, the government had already terminated all flights after 10:00 P.M. under mounting pressure from residents. Finally, the court rejected the plaintiffs' request to enjoin the use of the airport from 9:00 P.M., finding that the public interest in the airport's use (both domestically and internationally) outweighed the plaintiffs' interest in question. All parties appealed.

The Osaka High Court's decision of November 27, 1974, substantially expanded the District Court's decision on the plaintiffs' behalf. The High Court increased the sum of monies awarded for past harm and ordered the defendant to pay future damages of ¥10,000 per month per person until such time as the injunction was observed, and ¥6,000 per month until the government agreed to decrease the number of daily flights. The court also extended the injunction until 9 A.M. and awarded attorneys' fees.

The High Court's decision differed from the District Court's opinion in several other important respects. This time the court found that aircraft noise was the cause or at least one cause of the plaintiffs' alleged hearing impairments, stomach pains, heart conditions, high blood pressure, menstrual irregularities, and other difficulties. Where the District Court held that the plaintiffs must individually demonstrate proof of causation, the High Court ruled that individual proof was unnecessary since the plaintiffs as a group had demonstrated observable symptoms presumably caused by aircraft noise.[71] In granting the injunction, the High Court weighed the cost of the plaintiffs' substantial injuries from aircraft noise against the benefits to the public of the facility. Whereas the District Court had explicitly rejected the plaintiffs' environmental rights theory, the High Court simply found the theory unnecessary because the defendant's violation of personal rights was sufficient to support injunctive relief. The court thus left open the question of the violation of the plaintiffs' environmental rights.

3.4.1　　　Decision of the District Court (Extract)

Concerning a Right to Human Dignity and an Environmental Right

First, we consider the plaintiffs' claim in their request for compensation in this suit; that rights to human dignity and a [wholesome] environment have been infringed, and moreover that these rights provide a legal basis for the claim in this suit for injunctive relief. The essence of the plaintiffs' claim is that the right to human dignity and the environmental right are exclusive controlling rights that embody the rights of art. 13 of the Constitution, to wit: the right to life, lib-

erty, and the pursuit of happiness, and the right of art. 25 of the Constitution, to the maintenance of minimum standards of wholesome and cultural living. Moreover [it is alleged that] no infringement of these rights will be tolerated irrespective of the reason and that illegality is determined only by the existence of harm without balancing the public interest or other factors. In order to assure protection of the outer limits of the right to human dignity, an environmental right is a strong basis on which to frustrate such specific harms at the outset and to put an end to widespread environmental destruction which has violated the rights of residents.

We believe that unreasonable invasions of an individual's life, liberty, honor, and so on, or of the benefits of one's life as an individual should not be permitted. Such individual benefits are esteemed as legally protected in themselves. They can be called the right to human dignity, by analogy to the property right.

In the present case of noise from aircraft there exists no remedy but compensation for injuries already suffered from a continuing and uncompromising interference with human life, entailing an extremely serious influence on an individual's daily life and health. Because the deficiency in the safeguards from harm to the individual is apparent in this case, we hold that we should recognize the request to terminate those acts which have given rise to the harm. . . .

In this case one may assert, as a legal basis for a request to terminate such harmful acts, the right of ownership or other property rights which are recognized as grounds to permit termination of such interferences. However, because we value the benefits of an individual's life [and believe they should] be protected in the same way as property rights, it is reasonable to recognize the claim to terminate such harm even in the case of one who does not possess such property rights. Since the (remedy of) termination of such harm is implied in the right to human dignity, we hold that termination of such acts can be based on this right.

On the other hand, we doubt whether an environmental right has been given [explicit] recognition in legislation. Articles 13 and 25 should be construed as general guidelines establishing duties for all the people, and this view must be faithfully reflected by a national policy in legislation and in its administration. However, this is not to suggest that these provisions provide each individual citizen with some kind of direct and specific claim against those who inflict harm upon them.

As the plaintiffs point out, recently the destruction of the environment by pollution has become awesome; it has become necessary to maintain and preserve the environment from destruction. These are precepts that no one can deny. Moreover, it is obvious that the government and local communities, in order to protect the environment, are establishing measures to prevent pollution, have duties to carry out these measures, and that enterprise and citizens must also labor to prevent pollution. However, to spring from these facts directly to the conclusion that an environmental right be recognized as a private legal remedy for pollution is premature. Moreover, when an individual's interest is invaded by the destruction of the environment, he may request compensation for this tort.

When judging the question of the legality or illegality of the act, even if we consider the destruction of the environment as the essence of the injury claimed, we are not concerned with the problem of whether there is an invasion of an environmental right. And even when considering a claim for injunctive relief, one might equally well assert property rights and individual rights to human dignity as a legal basis for such claims. In any event, although we do not here recognize an environmental right, we believe that this does not mean that the individual's interest is left unprotected. According to the plaintiffs, pollution and environmental destruction, which is injuring individuals' legally protected interests, could be curtailed before the outbreak of a specific harm by the exercise of an environmental right. However, for us to endow an environmental right with this function would be going beyond the zone of individual legal remedies, and it would be necessary to have a clear legislative provision as a basis. It is thus difficult to assent to the theory of an environmental right that holds that illegality must be recognized merely from the existence of harm without any balancing of interests. Even in the case where freedom from injury to an interest is an absolute right, we must decide whether to recognize the existence of the illegality only after considering all the circumstances in a given case. By beginning with an assessment of interests, it is possible to identify the fair remedy in a specific case. In this connection, although the plaintiffs have emphasized that the general environment in their residential areas has been destroyed by the noise of the airport, this after all has infringed on the interests of the lives of each individual plaintiff. Thus, since plaintiffs have also alleged separate, specific, individual harms to life, to their health, and to the benefits of life, failure to adopt the concept of an environmental right as a vehicle for private legal relief would not compel us to abandon protection of the plaintiffs' claims.

3.4.2   **Opinion of the High Court[72]**

I. Facts Pertaining to the Airport, Adjacent Areas, and
Aircraft Noise

[The court summarized the factual determinations of the District Court and made the additional findings:
(1) The airport is located at the point where the boundaries of three cities in metropolitan Osaka converge. It was established in 1937 and has since been controlled and operated by the State. In 1959 it was made an international airport and, thereafter, expanded on several occasions; the B runway was completed in 1970. The airport now has an area of 3,043,600 square meters and two runways (A runway, 1828 × 45 meters and B runway, 3,000 × 60 meters). The size of the airport is one-third that of London's Heathrow Airport, although the number of flights at Osaka is fifty per cent greater than those at Heathrow.
(2) Most of the plaintiffs live within 2,000 meters of the ends of the runways and directly under either the trajectory of landings or takeoffs. Most of these districts have been zoned as residential under both the former City Planning Act (1919, Law No. 36) and the present City Planning Act (1968, Law No. 100).
(3) The number of airplanes using the airport has greatly increased over the years. In 1964 jets were introduced. In 1965 there was an average of 188 landings and takeoffs per day, including those of 30 jets. By 1972 this number had increased to 418 including 248 jets. Planes were landing and taking off once every 2 minutes during the day (a jet every 3 minutes 31 seconds); and every 3 minutes 36 seconds during the night (every 4 minutes 5 seconds for jets). In May 1975 ten flights were scheduled between 9:00 and 10:00 P.M. From 10:00 P.M. until 7:00 A.M., eight flights, all propeller mail planes, were operating when the District Court delivered its decision [on November 27, 1974].
(4) The plaintiffs have been exposed to levels of noise between 80 to 100 phons, or

85 to 95 WECPNL (weighted equivalent continuous perceived noise level) in their residential districts. The environmental quality standards for airplane noise set by the Environment Agency on December 27, 1973, in compliance with art. 9[73] of the Basic Law for Environmental Pollution Control (1967), required that the level of noise be below 70 WECPNLs in residential areas and 75 WECPNLs in other areas, except those areas designated as industrial or commercial. In residential areas around class 1 airports, the Environment Agency directive provided that the standards be achieved with all deliberate speed sometime after ten years. Meanwhile, 85 WECPNLs was to be achieved in residential areas within five years (by 1978), and 75 WECPNLs was to be achieved within ten years (by 1983).

(5) According to a noise contour map provided by the Ministry of Transportation in 1972, 38,000 persons in 12,500 households were living in a zone where noise exceeded 90 WECPNLs; 111,000 persons in 33,200 households lived in the above 85-WECPNL zone. Itami City drew a similar map in 1973 that showed 1,302,000 persons in 377,900 households living in an area where the WECPNL exceeded 70 (the environmental standard set for residential districts)].

II. Injuries

1. Preliminary Remarks

In judging the plaintiffs' contentions with regard to various injuries, the court regards the following four points (A through D) as essential to its evaluation.

(A) The court undertook two on-site inspections for a total of 15 hours to view the actual situation around the airport. These investigations were important to corroborate evidence necessary in formulating basic points for judging the plaintiff's injuries. We summarize some of our observations below:

(1) Noise. *Noise is extremely strong in those areas directly under the landing patterns. The impact of noise from low flying aircraft is intense in the area within 1,000 to 2,000 meters from the southeast end of the B runway (Kotobukichō, Nishimachi, Yamasan, Toshikurahigashi, Toshikura), and the area within 1,600 meters of A runway and adjoining the fence alongside B runway and taxi ways leading from it (Katsube and Hashirii). Houses in these areas, including those of the plaintiffs, are mostly wooden structures and can hardly be called sound-resistant. We agree with the trial court that "in contemporary daily life, there is nothing like the noise produced by jet planes." We found that noise in the area where we visited at night was devastating when accompanied by the overwhelming and imposing image of a huge airplane descending with full landing lights on. We felt strongly that the longer people are exposed to such noise day in and day out, the more uncomfortable they will become since their discomfort increases and amplifies as time passes. Their discomfort never decreases because it is not possible for them to get accustomed to such noise. We assume that the situation at night in other areas that we viewed only during the day is not much different. Those areas are about the same distance from the airport as the spot that we visited at night, and planes approach for landing from only a slightly different direction and altitude.*

We obtained a similar impression from the situation (of the takeoff course) in Takashiba and Mutsumi. We regret that because of an unexpected and sudden change of landing and takeoff patterns on the day of our visit, we could not actually observe the impact of planes taking off over Setsuyo Hill and other districts. Setsuyo

District is a densely populated area on a hillside directly under the takeoff course. The planes gain altitude by turning to the left of the hill after flying over Takashiba and Mutsumi. Considering the fact that these districts are on a hillside, along with our impression obtained from other districts, it is reasonable to assume that the district is exposed day and night to unusually severe noise from the planes.

By going to a district in the vicinity of highways, we also compared noise produced by jet planes with noise from automobiles and other urban activities. We found jet noise to be far more severe. The defendant contends that jet noise is intermittent and momentary. We doubt, however, that we can ignore the noise problem for this reason. It has been several years since the opening of the B runway and the introduction of many jet planes. During the intervening years, people have been exposed repeatedly to the noise of jets almost every day, from early morning into the night, at the frequencies mentioned above (see I(3)).

(2) Exhaust Fumes. *The court viewed successive landings of two large jet planes from the bank of the Senri River at the south end of B runway. The planes came in as if to knock our heads off. As soon as they touched down on the runway, they reversed their jet engines. We were indeed surprised by the ferocity of exhaust fumes that blew squarely against us because of the direction of the wind at that time. In addition to the fumes discharged from takeoffs and landings of the planes, we saw a large amount of exhaust emitted by planes waiting to take off. Because of the short distance and topographic features of Katsube and Hashirii districts, we can easily assume that airplane fumes are often blown toward Katsube and Hashirii districts on the east side of the airport.*

During the day we often noticed a long, black stream of fumes trailing from the jet planes. The amount of exhaust was far greater than that of automobiles. We have no doubt that a considerable amount of exhaust was blown into the districts under and around the landing course.

(3) Loosening of Roof Tiles. *In Takashiba, Mutsumi, and Setsuyo districts, we observed many houses with loosened roof tiles. Some houses had bracelets attached to the roof tiles to prevent them from becoming loose. This is an unusual sight in ordinary city districts. It has been argued that this is caused by vibrations from the planes. The defendant has not refuted this contention. We find no reason to deny it.*

(4) On the Problem of "Coming to the Nuisance." *There are a considerable number of houses built by various developers between 1962 to 1965 in the affected areas. It has not been contended that the people who have since moved into these houses did so with the knowledge of the expansion of the airport or with the intention of joining in the opposition movement. We feel that it would be rather harsh to blame them for their not foreseeing the present situation at the time when they moved in.*

Except for the airport problem, the districts involved in the present case are at the outskirts of the metropolis and very suitable for residential areas. Even if the people find alternative sites to take up residence, obviously there are tangible as well as intangible difficulties in making a decision to move. We also noted that a number of families have moved out after receiving compensation and that their houses have since been removed. There are now many vacant lots left unattended and deteriorating, which causes additional problems for the remaining residents.

(5) We consider the intensive use made of this airport that is in close proximity to high density residential areas, and we rely upon our observations from on-site inspections. We note that the nature of this airport, originally serving only con-

ventional airplanes with a single runway, was drastically changed when jet planes were introduced in 1964 and when the B runway was constructed in 1970 to serve more and larger jet planes.

Since we do not have the experience of living there, we need to deliberate further on the issue of whether ordinary families, inevitably including older people, children, and sick people, can stand to live in these affected districts. In general, many people use air transportation and view it as an essential convenience in modern society. However, we feel that it is necessary to consider both the view of affected residents around the airport and the predominant and prevailing view of the general public and to compare and examine both factors in order to arrive at a judgment in the present case. (49)

(B) We feel that plaintiffs' 226 statements submitted to the court deserve careful examination before a judgment is made regarding their alleged injuries. This is because the alleged injuries consist of mental and physical suffering and various types of inconveniences in daily living. Plaintiffs contend that these injuries are interrelated and compounded in many ways. In order to judge the existence of such injuries, it is essential to understand accurately the experiences of the affected persons. For this reason, it is important to listen open-heartedly to their appeals in order to grasp the reality of their injuries. It is not possible to judge the full extent of their injuries solely from objective scientific observations. We must pay attention to their complaints, even though these are purely subjective evaluations. Of course it is extremely difficult to evaluate the contents of these statements. It is not certain that they accurately express their true meaning because they differ from statements made in court. Yet after examining these statements closely, we find them sufficiently persuasive, although some points have apparently been overemphasized. Judging from the way they were composed, these statements might not have been written by the affected persons themselves. Nonetheless, they contain detailed and concrete descriptions of each family, such as name, age, occupation, personal history, health, and the hobbies of its members. Some of this information is usually regarded as private. The statements also touch on: (1) the existence of various inconveniences of daily life caused by noise from early morning until night, and the difficulty of resting at night, not to mention disruption of the enjoyment of an evening with one's hobby; (2) sufferings of families with students or elderly, very young, or sick persons; (3) injuries from the exhaust fumes in Katsube and other districts under the landing course, particularly areas near the B runway; (4) the impact of vibration from planes on the structure of houses and on daily indoor living; (5) impairment of hearing; and (6) difficulties encountered in moving out even after a decision was made to do so, although the family had originally wished to live there permanently.

The statements well express the predicament of each family in a particular and individual way. We find no trace of influence by third persons on the content of these statements. They express feelings that we found to be true from our on-site inspection. (50)

(C) The defendant disputes the credibility of the questionnaire presented by the plaintiffs. In order to insure a fair and objective result from such a study, it is necessary to arrange questions so as to exclude any bias in the answers that might be created by prejudiced questioning. However, a similar result may be obtained irrespective of the method of polling employed when those questioned have strong feelings about a given subject matter. Professor Junichi Maekawa's study supports this point. It concluded that there was no meaningful variation

between the results of the direct and indirect questioning methods of polling used in Itami City.

Because of the particular nature of injuries caused by airplane noise described above, it is important to hear the victims' subjective complaints in order to understand the real situation. The questionnaire method is essential and effective for grasping the substance of their injuries and noting at least general trends. In the present case the results of various questionnaires, as cited below, show a clear correlation between degrees of exposure to noise and the frequency of complaints. We clearly regard these results as being far out of touch with reality. Yasuhio Nagata, who appeared as a witness, undertook at the request of the Ministry of Health and Welfare an analytical comparison of the results of various surveys of residents near airports, including a study of the Osaka International Airport in October of 1965 by the Kansai Action Council for Urban Noise, one of Yokota Air Base in July of 1970 by Tokyo Metropolitan Pollution Research Center, one of London's Heathrow Airport known as the Wilson Report, and one of Chitose Airport. He then compared those results with those of other questionnaires dealing with factory and highway noise. According to his testimony, the results of these studies show an extremely high correlation between the level of noise and the frequency of various complaints. The findings of these studies support each other and are sufficiently reliable. Furthermore, the environmental standards for airplane noise mentioned above were based on the results of questionnaires by the same witness and his comparative analysis of airport noise evaluation data from foreign countries. Nor can we agree with defendant's questioning of the reliability of the studies conducted by the Kansai Action Council for Urban Noise and by the Tokyo Metropolitan Pollution Research Center. The defendant's contention would amount to a denial of the objectivity of environmental standards set by the defendant itself. There is no problem, therefore, in adopting in this case the results of the questionnaires, including the ones just mentioned, as evidence supporting the substance of the plaintiffs' complaint. (50)

(D) The injuries which we examine below are of such a complex and severe nature centering as they do on mental distress, that they interact and lead to new injuries and compound each other. If the injuries are substantiated, then special attention should be given to the mental suffering caused by the various injuries. It is not enough to evaluate each injury separately, because each is an interrelated part of the whole destructive experience. Taken as a whole, it may be said that the air traffic of the airport in the present case has a great and severe impact on the plaintiffs. It is also important to consider the mutually intensifying aspects of plaintiffs' injuries when we evaluate each separate type of injury.

The following evaluation of each type of injury is based on premises A through D, even where we do not repeat these explicitly. (50)

2. Noise, Exhaust Fumes, and Vibration

[With respect to aircraft noise the court next cited the description in I. The court then described the pollutants in the exhaust fumes discharged from airplanes and compared the exhaust fumes from a DC-8 jet plane with those emitted from automobiles; carbon monoxide (equivalent amount to 341 cars with 1,600 cc exhaust capacity), hydrocarbons (1,611 cars), and nitrogen oxides (550 cars). The estimated amount of pollutants discharged at the airport in a year is as follows:

2,246 tons of carbon monoxide, 1,436 tons of hydrocarbons, 620 tons of nitrogen oxides, and 4,304 tons of total pollutants (including some aldehydes and cyanides). Although the average concentrations of these pollutants is no greater than that of downtown Osaka, the court recognized that a peak level concentration was observed in areas near taxiways passed over by jets and that air pollution in areas near the airport may have had an adverse effect on the plaintiffs' health. The court then repeated the findings of the District Court.]

3. Mental Suffering

(1) As we noted during our on-site inspections, the noise of airplanes, especially of jet planes, is extremely powerful and is beyond comparison with any other noise in contemporary life. It has a metallic sound and a high frequency. It is natural that anyone exposed to such noise would feel strong discomfort and irritation. The sufferings of the plaintiffs, who have had to live under the noise of airplanes every day and night for many years, must be beyond imagination.

[Medical research shows that the amount of blood flowing to one's brain significantly increases when a person is exposed to noise of 100 phon and at that level schoolchildren become emotionally disturbed and aggressive. The younger the children, the stronger the impact.]

According to several questionnaires, the nearer people live to an airport, the more they express emotional discomfort and irritation. And almost all persons who live in the plaintiffs' districts have voiced their discomfort. It is evident that the plaintiffs are suffering from an extraordinary degree of mental discomfort. In extensive parts of districts like Hashirii and Katsube, plaintiffs are suffering from exhaust fumes and odors as well as noise. (52)

(2) The plaintiffs also voice constant fear of a possible plane crash. Although statistically airplanes may be very safe, the possibility of accidents always remains. Were an accident to occur, it is clear that extensive damage on the ground would be inevitable. When we consider that most airplane accidents occur during landings and takeoffs and that the plaintiffs live very close to the airport and directly under the landing and takeoff patterns, we clearly cannot dismiss the plaintiffs' fears as groundless. . . . (52)

(3) Since they live under extreme stress and discomfort, it is understandable that some plaintiffs complain about their emotional state. [. . .] The development of their children's personalities is also adversely affected by the constant tension. (53)

(4) The psychological impact of noise on the mental and emotional state of an individual may differ, as the defendant argues, depending on the individual's sensibilities, psychological inclinations, or attitude toward the source of noise. Yet because of the particular characteristics of the location, residents have had negative feelings toward the operation of the airport from the beginning. They have naturally developed strong feelings of discontent and irritation about its activities. It is not possible to expect plaintiffs to be tolerant of or indifferent to airplane noise. In other words, we can presume the existence of the plaintiffs' mental sufferings as an objective fact by observing the nature and severity of noise to which they have been unwillingly subjected. We cannot say that their complaints are merely subjective. We cannot deny them legal relief. Nor do we expect that such mental suffering is alleviated by the plaintiffs' exposure to the noise for a long period. Plaintiffs have not become accustomed to it. The plaintiffs' mental suffering may itself eventually lead to or result in other physical

injuries and disorders, and we, therefore, find it to be an important and central element in the plaintiffs' injuries. (53)

4. Physical Injuries

(1) Hearing Impairment and Ringing in the Ears. [*The court next made these factual determinations. A high incidence of hearing impairment was observed among young and old plaintiffs and their families. The symptoms of hearing impairment and chronic ringing in the ears had been observed among residents since 1967. The situation, however, was getting worse. The court referred to several questionnaires and the medical examination of schoolchildren around Osaka Airport and the Yokota Air Force Base. The court found that the closer people live to the airport, the greater the incidence of hearing difficulty.*]

We find from the evidence that the plaintiffs, repeatedly exposed to the powerful noise of airplanes, suffered a temporary hearing impairment (TTS). As their exposure continued, some may have suffered a permanent hearing impairment (PTS). (54)

[*Some plaintiffs are suffering from TTS, and other plaintiffs may possibly be suffering from PTS; others besides plaintiffs also share the risk of suffering these same injuries.*

The court then explained how TTS is caused and how it develops into PTS. It referred to several noise standards proposed by organizations like the Japanese Association of Industrial Hygiene, the United States National Academy of Science, and others. It also referred to experiments on the impacts of noise on human beings.]

Results of audiometry examinations and the diagnosis of individual persons who have complained of difficulty in hearing were not submitted to the court. The hearing impairment of older people who were the majority of the complainants suffering from such ailments may have been caused by old age in addition to the noise. Old people may also suffer from hearing impairment in the future due to age. It is also not impossible that there are additional causes besides airplane noise for the hearing difficulties among the young people. It cannot be denied, however, that it is at least possible that aircraft noise is one cause of the hearing difficulty of which the plaintiffs complain. For the reason described in the summary it is not necessary to ascertain the individual causal relationship between the existence and degree of hearing difficulty of each defendant and the airplane noise.

It is true that there are individual differences in the degree of impact of noise on hearing. Yet when we discover injuries over such an extensive area, we must employ the standard of an average person in the area. We must also take into consideration the fact that within an individual's family there may be persons of weak health who are sensitive to noise. These people did not voluntarily accede to airplane noise by living in the area, and they have received little if any benefit from the airplanes. They should, therefore, be better protected from hearing loss than the worker for whose protection industrial hygiene standards for noise have been established. . . . Considering the above and keeping in mind the four items (A through D) noted during our on-site inspection, we find that noise, as measured and estimated around this airport is different from the noise problem of every other airport in the world; and that this noise causes by itself, or at the very least when coupled with other factors, the plaintiffs' hearing im-

pairment and ringing in the ears. The defendant's evidence to the contrary is not sufficient to refute this finding. (56)

(2) Other Adverse Effects On Health. [*The court refers to the plaintiffs' complaint alleging that noise caused the following physical disorders: (1) headaches, dizziness, irritation; (2) loss of appetite, indigestion, stomach ulcers and inflammation; (3) high blood pressure and heart irregularity; (4) miscarriage and menstrual irregularity; (5) frightening of babies and the retarded development of children. The court then reviewed the results of several medical studies mentioned in 4(1) of the impact of noise on human health and concluded that daily exposure to powerful aircraft noise is at least one of the causes of diseases that result from high mental stress.*]

The physical disorders complained of are in some way affected by airplane noise. We do not have data that show that the incidence rate in this area of serious physical disorders such as stomach ulcers, myocardial infarction, and miscarriages is higher than that in other areas where the noise is much less. These disorders and others such as stomach and heart disorders, high blood pressure, headaches, and menstrual irregularity, of course, are commonly encountered and result from various causes. Noise is one of many causes that may indirectly contribute to the alleged disorders. In contemporary society there are many causes other than noise involving human relations or one's employment that impose stress on people. It is not reasonable to place the blame for the plaintiffs' disorders solely on the noise of airplanes nor is it possible to define the extent of the contribution of noise to these disorders. These factors, however, are insufficient to prove that airplane noise is not one of the causes of the plaintiffs' disorders. It is enough to find that the physical effects of noise depend largely on the mental and psychological state of each individual. The plaintiffs are exposed daily to powerful noise and suffer from mental discomfort and inconvenience in their routine daily life. It is reasonable to think that their mental suffering may result in physical disorders; and physical disorders may in turn increase their mental sufferings. By this mutually amplifying effect, their disorders may become more extensive and serious. Under these circumstances, it is unrealistic for the court to recognize none of the defendants' physical injuries or to ignore the future risk of expansion and aggravation of these injuries. Therefore we find it reasonable to presume that airplane noise independently, or jointly with other causes, contributed in part to the injuries complained of and has aggravated injuries originating in other causes. Those plaintiffs who are not suffering from these injuries are equally exposed to the risk of similar injuries. (58)

[*The court recognizes that the impact of airplane noise and exhaust possibly affect sick persons by disrupting their rest and making their recovery difficult. The airplane exhaust fumes may cause nosebleeds, respiratory disorders, and eye pain among the residents.*]

5. Disturbance of Sleep

[*The court describes the results of experiments concerning the impact of noise on sleep. It points out that airplane noise disturbs the plaintiffs' sleep and prevents them from recovering from fatigue.*]

6. Disruption of Daily Life

[*Airplane noise disrupts the daily life of residents. It disturbs daily conversations, telephone conversations, and the reception of television and radio broad-*

*casts. It is presumed that noise seriously affects intellectual exercise like read-
ing, writing, and thinking. It lowers work productivity. Especially during the
evening hours, it disrupts family conversations and togetherness and creates adverse effects on family relations and associations with friends and relatives. The
noise drowns out the warning honks of automobiles or the sound of approaching
trains at railway crossings and increases the danger of traffic accidents. Vibration from noise is one of several factors that accelerate the dilapidation of housing. It is also likely that plants and trees around the airport have been affected by
the great quantity of exhaust fumes and gas from airplanes. Tar and soot contained in exhaust fumes seem to make spots on washed clothes, window panes,
etc. An increasing number of people have moved out of the area and their houses
are being torn down. The vacant lots are fenced temporarily but otherwise left
unattended, creating a secondary effect of airplane noise safety problems and an
impression of general decay of the residential environment.]*

7. Effects on Education

*[The court also finds that airplane noise affects education. It is presumed that
even after the school buildings were soundproofed (an action which reduced
noise by 30 dB), classroom work was still disturbed, and educational activities
outside the buildings were exposed to the same noise as before. One also cannot
deny that schoolchildren are disturbed when studying at home, and that students preparing for their entrance examinations have been seriously disadvantaged.]*

8. Summary of Findings on Injuries

*(a) The defendant argues that the plaintiffs have not proved their individual in-
juries. Yet it can be said that they suffer the same mental anguish and together
share disturbances of sleep and daily life even though the particular form of the
individual's complaints may differ. It may be possible to prove more clearly the
existing state of ill health and the extent of injury by requiring individuals to file
certificates of medical examination. We cannot say, however, that the plaintiffs'
claims are all spurious because of the absence of medical certificates. Also one
must realize that even with more extensive medical examinations it would not
be easy to determine individually and concretely the causal relationship between each plaintiff's existing physical harm and airplane noise and exhaust.
Furthermore, the possibility of these injuries exists equally for all residents in the
districts affected by airplane noise and exhaust, and the mere fact that some do
not yet evidence apparent symptoms does not mean that they are not being affected. There is, therefore, no reason to distinguish between those who are obviously affected and those who are not, since they are all subjected to the same
possibility of harm. The residents should be viewed collectively. If it is presumed
that some of them are suffering from symptoms caused by aircraft noise, it is
clear that the rest are subject to the danger of the same suffering. We must
provide for relief or protection for all residents of the area, since they are either
actually suffering from injuries or are at least exposed to the present danger of
injury. In this sense and in light of the particular nature of the damages in this
case, it is not unreasonable to argue, as the plaintiffs have, that mere exposure to
a noise level capable of causing the types of harm described should be sufficient
evidence of their injuries without going into proof of individual harm.
    Viewing the harm as affecting the area's residents as a whole, we naturally ex-*

175

pect that there are a number of elderly, very young, and sick persons who are particularly in need of quiet surroundings and who are less resistant to noise. There are also those who have particular life patterns (for example, those who work at night). We will determine the existence of injuries with full consideration to the special requirements of these persons for whom noise poses special problems.
(b) The defendant compares these phenomena with the ordinary noise of urban life and claims that there exist no cognizable injuries in this case. It is certainly true that contemporary urban life is usually beset by noise and other adverse conditions, that despite this some people live in urban areas for various benefits, and that most persons simply endure the discomforts. It does not follow, however, that exposure to airplane noise is one of the unavoidable phenomena of urban life; the plaintiffs would enjoy a relatively quiet life if there were no airplane noise. . . . The fact that there are several areas of Tokyo, Nagoya, etc., with noise pollution problems does not alleviate the plaintiffs' injuries nor, as set forth below, mitigate the defendant's liability. (63)

III. Illegality

1. Applicable Law

We have found that the plaintiffs have suffered severe injuries. We come now to the issue of what law is applicable to this case, Civil Code art. 709[74] and the State Compensation Act. art. 2(1) as urged by the plaintiffs or art 2(1) of the same act as applied by the trial court. To do so we must first examine the process of the airport's expansion, the measures taken by the defendant related thereto, and thereafter the public nature [of the defendant's activity].

2. The Process of Airport Expansion

(1) We found the following facts concerning the expansion of the airport and the responses of the area's residents and local governments.
(A) Process of Airport Expansion.
(1) In October 1951 a civil air transportation service started under the supervision of the U.S. military force.
(2) Thereafter Kansai business circles, for example, the Osaka Chamber of Commerce, expressed their hope of having an international airport in the area. In April 1957 the U.S. Commander for the Far East announced the early and unconditional return of Itami Air Base (involved in the present case), and that September the Aviation Bureau of the Ministry of Transportation drew up a plan to equip it for use as Osaka Airport. The plan, however, was not implemented because of budgetary considerations. Thereafter six organizations, the Osaka Prefectural Government, the Osaka Government, the Osaka Chamber of Commerce, the Hyōgo Prefectural Government, the Kōbe City Government, and the Kōbe Chamber of Commerce, decided to advance the ¥600 million necessary for the government to purchase land for the expansion of the airport. These entities also strongly requested the central government to designate it as an international airport. In August 1958 the Ministry of Transportation formally notified the concerned local governments of the airport's expansion and renewal plan which included building a new 10,000-foot runway.
(3) On March 18, 1958, Itami Air Base was formally returned to the Japanese government, which then became responsible for maintaining it as Osaka Air-

port. *The six organizations formed the Itami Airport Association, a corporate entity, and began activities such as the purchase of land on behalf of the central government, public relations for airport development, and lobbying of the area's local governments.*

(4) On July 3, 1959, the government designated the airport as a first class airport under the Airport Development Act and named it Osaka International Airport.

(5) In April 1960 the airport began serving international carriers. The number of planes using the airport increased each year and large airplanes with four propeller engines were commonly used.

(6) Thereafter, the Itami Airport Association did not make much progress in its expansion-related activities because of local government opposition. Around November 1962, however, the approval of Toyanaka City and Itami City was obtained and the association actively began buying land.

(7) On June 1, 1964, the present A runway began serving jet planes. At that time such planes as the Convair 880 and Boeing 727-100 were used. Many other types of jet planes have since been put into service. These have occupied an increasing share of the total number of airplanes using the airport.

(8) Toward the end of 1964, the government began construction to expand the airport even though some land had not yet been acquired. These lands were later voluntarily turned over to the government by their owners according to a memorandum signed in December 1966 (see hearing under (5) below) after the government instituted eminent domain proceedings against them. Thereafter the construction was completed.

(9) On February 5, 1970, B runway was put into use. From that time until 1972, the number of planes, especially jet planes, continued to increase as noted above.
(63)

(B) The Attitudes of Area Residents and Local Governments.

(1) The cities around the airport, Itami, Toyonaka, and Ikeda, had formed the League for Opposition to the Expansion of Itami Airfield in response to a plan to enlarge the U.S. air base. After the return of the base, the League initially maintained its position of absolutely opposing any airport expansion and petitioned the Ministry of Transportation to that effect. Partly due to the influence of the Itami Airport Association, however, the league gradually modified its position to one of conditional approval of the expansion. In November 1958 Itami City submitted a petition entitled "Petition for Compensation for Nuisance Accompanying the Expansion of Osaka International Airport" to the Ministry of Transportation, and in March of 1959, Toyonaka City submitted "Claims for Compensation Related to the Plan for the Extension of Osaka Airport." In these documents, the cities requested full compensation for appropriated lands and appropriate provisions for roads and waterways in the area. The cities specifically pointed out that damage from noise and vibration from jet planes was increasing as more planes were put in service. The city requested that soundproofing be added to schools and hospitals and that owners of ordinary houses be compensated for damages. They also demanded that other necessary measures be undertaken to forestall any disturbance of the daily lives of citizens.

[(2) The court referred to resident opposition around the Tokyo airport. In response the Ministry of Transportation banned jets from 11:00 P.M. to 6:00 A.M.

(3) Itami Airport Association's Noise Investigation Committee issued a report entitled "Antinoise Measures in Expanding Osaka International Airport." The court noted that, even though the report underestimated the level of jet noise, it

pointed out the necessity of soundproofing the schools within 3 km of the runways.]

(4) Despite repeated petitions from Toyonaka City and others, the Ministries of Transportation and Education took no action to soundproof the schools. They said there was no statutory authority for the government to undertake such a project. Itami Airport Association offered, however, to bear part of the costs of soundproofing and indicated that it was also prepared to provide compensation for other requested items. With these conditions, both cities then gave their approval to the land acquisition and the expansion of the airport. An agreement was made between Toyonaka City and the association in December 1961 entitled "Agreement Concerning the Expansion of Osaka International Airport." It provided that: (1) The association would pay up to ¥250 million to modify these schools. (2) The association would be responsible for compensation. (3) The above would also apply to ordinary houses, factories, and other facilities. In April 1962 the association concluded a similar agreement with Itami City. It was agreed that the association would pay ¥250 million to modify Kōzu Primary School and would cooperate with the city in lobbying for and implementing legislation to control noise and compensate injuries. (However, the association became inactive after it succeeded in acquiring the land, and the antipollution measures promised for the future under these agreements remained unfulfilled.)

(5) Itami Airport Association bought most of the land necessary for the proposed expansion. However, some owners in Katsube District refused to sell their land. In January 1966 the government, together with the Expropriation Committee of Osaka Prefectural Government, instituted eminent domain proceedings against them. The Committee gave approval in March and fifty-two owners in the Katsube Branch of the All Japan Farmers' Union filed a suit with the court to nullify the Committee's decision. When the government applied for and obtained from the Committee an order of execution by proxy, the owners built a "solidarity cabin" on their land where they sat in to stop the execution. But the Mayor of Toyonaka City intervened and, after negotiations, they decided to resolve the dispute. A memorandum was signed on December 21, 1966, by the seven parties involved, the representatives of resident-owners, the chairman of the Osaka Prefectural Federation of the All Japan Farmers' Union, Chief of the Bureau of the Ministry of Transportation, the governor of Osaka Prefecture, and the mayor of Toyonaka City. The memorandum centered on noise abatement measures. In this memorandum, the All Japan Farmers' Union criticized the Ministry of Transportation for having been idle and indifferent to the pollution caused by the operation of the airport. The union insisted that the ministry adopt affirmative policies such as the removal of the international airport, the imposition of a curfew on night flights, the denial of the use of the airport to jet planes noisier than those currently in service, the construction of soundproof study and community centers, the repair of damage to roof tiles and walls, and the reduction or exemption of television-radio fees. The Ministry of Transportation expressed its regret that it had not been possible to adopt affirmative measures to control pollution caused by the operation of civil airports, including the airport in the present case. It also expressed willingness to deal positively with the situation thereafter and to propose necessary legislation. It promised to effect some items demanded by the residents and in the future to meet readily with these individuals to discuss antipollution measures. In response to the positive attitude of the involved government agency, the resident farmers at Katsube District agreed to vacate their lands and to drop their suit.

(6) In June 1964, jet plane services were introduced and the noise intensified. On October 16, 1964, eight cities around the airport, Itami, Amagasaki, Kawanishi, Takarazuka, Nishinomiya, Toyonaka, Ikeda, and Minō formed the Osaka International Airport Noise Countermeasures Council (the so-called Eight Cities Council) to coordinate the activities of the member cities to investigate noise levels, collect information, promote noise abatement, and lobby for legislation to these ends.

More concretely, the council contracted with the Kansai Metropolitan Noise Control Committee to measure the noise level of the airport in April 1965 and to study, by means of a survey, the effect of airplane noise on the residents. The following October the council repeatedly petitioned the Ministry of Transportation and other agencies to propose antinoise legislation, impose a curfew on night flights, and regulate the use of large aircraft.

On August 1, 1967, an Act to Prevent Harm from Airplane Noise in Areas Around Public Airports (hereafter referred to as the Airplane Noise Prevention Act) was put into effect. The council found the act inadequate and began an active campaign to make it more effective. It demanded a large increase in the budget for compensation and financial aid under the act, restrictions on the number of jet flights, the immediate promulgation of environmental standards for airplane noise, and restrictions on any increase in flights or introduction of new types of aircraft made possible by B runway's completion. As the affected area was enlarged by the use of B runway, three more cities, Ashiya, Osaka, and Suita, joined the council on May 31, 1971 (now the Eleven Cities Council).

(7) On August 25, 1964, the Noise Countermeasures Council of the Southern Districts of Kawanishi City (soon to change to the Airport Countermeasures Council of the Southern Districts of Kawanishi City) was formed by thirteen neighborhood associations consisting of 5,000 households. Its activities had four aims: the removal of the airport, the completion of full soundproofing of public facilities, the compensation of the residents, and exemption from television fees.

The organization repeatedly petitioned the Ministry of Transportation and the airport authority. It also conducted a questionnaire of residents concerning damage. In addition to the types of activities undertaken by the Kawanishi group, communities in districts such as Takashiba and Setsuyo actively campaigned through neighborhood meetings, senior citizen clubs, and women's clubs for a prohibition on night flights, the redesignation and enlargement of the area where compensation was to be paid for relocation and the reduction of property taxes. From about 1965 their activities included petitions, demonstrations and pamphleteering directed at, among others, the Ministry of Transportation, the director of the airport, Hyōgo Prefecture, and Kawanishi City. Through this experience, however, the residents came to the conclusion that they could not expect any affirmative measure to be undertaken by the government, and from about the beginning of 1963 they began to realize that their only alternative was to institute a suit against the government. The Takashiba and Setsuyo districts' neighborhood associations held meetings and decided to file suit. Other neighborhood associations passed resolutions in support of the suit. The first group of 28 plaintiffs initiated the present suit in December 1969, and the second group of 126 filed the second action in this case in June 1961.

(8) Fear of increased noise was one of the reasons that the residents in Katsube District initially refused to sell their land for the extension of the airport. After signing the above memorandum, these Toyonaka City residents repeatedly re-

quested the Ministry of Transportation and the Osaka Civil Aviation Bureau to fulfill the promises of the memorandum and to proceed with antinoise measures. In October 1969, twenty-one neighborhood associations, including those of all the plaintiffs of this suit and consisting of 5,000 households in Toyonaka City, formed the Toyonaka Districts Action Federation against Airplane Noise. It petitioned the Minister of Transportation, the Secretary General of the Environment Agency, the Director of the Osaka Civil Aviation Bureau, the airport director, and the Governor of Osaka to take measures against the noise. In August 1971 the residents organized a litigation group and 122 plaintiffs brought the third suit in November 1971.

(9) Itami City adopted an environmental protection ordinance in April 1971 and declared that the mayor and airlines were under a duty to prevent harm from airplane noise and on March 25, 1972, drafted an environmental standard based on the 1971 ordinance. In April it developed its own monitoring and surveillance system separate from that of the Ministry of Transportation. (65)

(2) [The court then criticized measures that had been undertaken by the defendant in response to requests by residents and local governments. The measures were found inadequate. The court pointed out that the number of flights had not actually been reduced, that the noise monitoring system had proved ineffective, that soundproofing of schools and hospitals could not fully prevent noise, and that compensation payments were insufficient.]

(3) From the findings made concerning the process of expansion of the airport, the response of the residents and local governments to the expansion, and the measures undertaken by the defendant, we think that the following points should be noted.

(a) Although the airport expansion plan was inaugurated at the urging of Osaka Prefecture and the other five organizations, the possibility of resulting harm from noise was pointed out by local governments from the beginning of the plan. From these warnings and its experience at the Tokyo International Airport when jet planes were first introduced there, the defendant should have foreseen the impact of noise on residents around the present airport. Nevertheless, the defendant proceeded with the introduction of jet planes at the present airport without adequate study or analysis of the impact of these planes and without taking any measures to prevent or reduce potential harm. It is true that the Itami Airport Association paid ¥250 million each to Toyonaka and Itami cities for soundproofing schools there. However, the association has not fulfilled its other obligations under its agreements with these two cities to investigate noise pollution and to take countermeasures against it.

(b) The Airplane Noise Prevention Act was passed during the period between the beginning of jet plane service and the completion of B runway. The act was prompted by the campaigns of nearby local governments, such as the Eight Cities' Council, and other residents' movements. Under the act, financial assistance for soundproofing of schools and public facilities was implemented relatively early. Aside from the effectiveness of soundproofing, however, this assistance served only as partial relief from the various forms of constant discomfort caused by aircraft noise. Furthermore, although it may be regarded as partial compensation for the harm, financial aid for television reception fees does not reduce the harm itself. The defendant also contends that it imposed a first-stage restriction on the use of the airport, but this restriction was only to prevent additional jet plane service during midnight hours.

(c) After B runway was put in use, in addition to the continuation of the financial *assistance mentioned above, compensation for the relocation of families from the affected area belatedly began. There are intrinsic limits to the effectiveness of these measures, however, and they have not been adequate. A second-stage restriction on the operation of the airport was imposed when B runway became operative but this restriction has been very mild. Two years later a third-stage restriction was imposed based on the recommendation of the Director General of the Environment Agency, but the defendant implemented only a part of this recommendation. It has reduced noise slightly by modifying landing and takeoff patterns, but it is obvious that the defendant's measures have been inadequate and belated. As one example of the delays, one need only point out that it was after B runway began operation that the defendant built antinoise walls and banks at the end of the airport facing Katsube District.*

From these facts, we find that the defendant expanded the airport, introduced jet planes, increased the number of flights, introduced larger planes, and so on, without a prudent study of their predictable impact on neighboring residents. . . . We further find that the measures adopted by the defendant up to 1973 were makeshift and extremely inadequate. The plaintiffs are right in blaming the defendant for having neglected necessary antipollution measures in its rush to increase the use of the airport. It should be said for the officials in the Ministry of Transportation that they took considerable effort to secure the passage of the Airplane Noise Prevention Act and its subsequent amendments and to propose a budget for protective measures. However, their difficulty in securing an adequate budget is an internal problem of the government. It cannot be used as an excuse to reduce the defendant's responsibility to the plaintiffs. In sum, if we consider the whole process of expansion of the airport, we must agree with the merits of plaintiffs' contentions and hold against the defendant. (66)

3. Some Measures Proposed for the Future

(1) Among current or future measures proposed by defendant, especially important as a way to reduce the source of the noise itself is the introduction of a new larger type of aircraft (called by the defendant low noise, large aircraft, hereinafter called the airbus) and the restriction of the number of flights based on the use of the airbus. As for measures to be taken in the surrounding areas, the soundproofing of homes and compensation for those families moving out of the area is proposed. We shall examine each from the perspective of its effectiveness and appropriateness as a measure to avoid damage. (66)

(2) Antinoise Measures.

(A) The Introduction of the Airbus.

(a) The defendant proposes the introduction of low noise, large aircraft certified by ICAO (International Civil Aviation Organization) as meeting international noise standards. These include the Boeing 747-SR, Lockheed L-1011 and Douglas DC-10; they are already in use in domestic routes. An airbus can carry a significantly larger payload than a conventional airplane. The defendant hopes to put this type of plane in service at Osaka, but it has been unable to do so because it has been unable to get the consent of residents and local governments.

(b) Using the United States Federal Aviation Agency's official figures, the defendant contends that these airbuses can reduce noise by 10 phons compared to the noise from a Douglas DC-8. However, the noise level of airplanes may differ depending on flight conditions. According to the data of trial flights of these planes in Fukuoka and Osaka airports, it cannot be said for certain that the airbus

will reduce the impact of noise upon the plaintiffs' residential area to the level claimed by the defendant. By switching to the Boeing 747 or Lockheed L-1011, it may be possible to reduce noise by a few phons, but the plaintiffs will still be exposed to noise levels of 90 phons or more.

(c) It can easily be assumed that large airbuses will consume greater amounts of fuel and discharge more exhaust than conventional planes. According to the evidence, the Boeing 747 discharges less carbon oxides and hydrocarbons, but four times more nitrogen oxides than the Douglas DC-8. It also discharges more aldehydes.

(d) It is claimed that airbuses show a high level of design safety and a good in-service safety record. However, it cannot be said that the risk of an accident is nonexistent; and if it should happen, the damage to the ground would be greater than that caused by conventional aircraft. Therefore, we cannot dismiss as groundless the plaintiffs' fear of possible accidents, nor can we ignore their fear of imposing, low-flying, large aircraft.

(e) Therefore, although the introduction of airbuses might effect some reduction of noise, the defendant's contention and evidence is not persuasive enough to overcome the adverse effects of using airbuses pointed out by the plaintiffs.

(B) Restriction on the Number of Flights.

(a) Up to the present, the defendant has restricted the number of takeoffs and landings to below 410 a day (240 for jets), and in March and April of 1975, it eliminated 12 propeller plane flights and 4 jet flights from the February daily average of 389 scheduled flights (235 jets). If airbuses are introduced, the defendant intends to limit the daily total number of takeoffs and landings to 370 (200 jets) and in particular to reduce the number of post–9:00 P.M. domestic flights from 7 to 5.

(b) The defendant's plan is, for the time being, to reduce the number of jet plane flights by the introduction of airbuses. However, it is not clear what part of the reduced number of flights would be airbuses or how long they would continue to retain this share. According to some of the defendant's estimates, one-half of the YS 11s would be replaced by airbuses. But, as the plaintiffs pointed out, all YS 11s would eventually be replaced by airbuses because the YS 11 is no longer being produced. Therefore the introduction of airbuses would not necessarily result in the reduction of noise in the long run.

(c) The defendant's intention is clearly to use the airbuses to maintain the present level of load capacity while decreasing the number of flights. However, in view of the manner in which the defendant extended its airport operations, it will be impossible for the government to recover the trust of local governments and residents unless it hears the airbus question separately and expresses its intention of reducing the number of flights even if this means a reduction in the current load capacity. Even if we assume that the defendant could succeed in obtaining local consent to its plan [for airbuses], it remains doubtful whether this plan would effectively reduce the harm.

(d) The defendant further contends that the engines of the Boeing 727 and 737 are in the process of being modified to reduce noise. It is, however, not evident that the impact of noise on the plaintiffs' residential area could be reduced as the defendant claims. The defendant has already implemented one of its proposals by requiring certain planes to make a rapid acceleration of its rate of climb immediately after takeoff. Its effect on the plaintiffs' residential area is not clear. Other measures such as a cutback in jet planes and two-stage landings are proposed without specific implementation dates, and we cannot take them into

consideration. Even if all the above measures to reduce the noise at its source are combined, therefore, we are not persuaded that the defendant can reduce the total amount of noise by 7 to 10 WECPNL as the defendant claims. (67)

(3) Measures for the Surrounding Area.

(A) On March 28, 1974, an amendment (Law No. 8) to the Airplane Noise Prevention Act became effective. Under this amendment, the Osaka International Airport Neighborhood Redevelopment Organization (hereinafter the Organization) was established on April 19, 1974, with a fund of ¥1 billion contributed by the central government and the Osaka and Hyōgo prefectural governments. The Organization is empowered to carry out measures adopted in the Airport Neighborhood Development Plan such as creating a green-belt buffer zone, implementing neighborhood reconstruction projects, and securing land and housing for the relocation of area residents. The Ministry of Transportation has also delegated to the Organization the power, under art. 9 of the act, to handle matters concerning compensation for relocation and, under art. 8-2, the power to administer financial assistance for soundproofing of ordinary houses. The organization spent ¥11,100 million in 1974 and has a budget of ¥22,800 million for 1975. The central government will contribute ¥1,483 million as a subsidy, ¥4,545 million as an interest free loan, and ¥7,160 million as a special bond loan. In addition to these activities, the financial aid extended by the Ministry of Transportation for soundproofing, education, and community facilities was continued into 1975.

(B) Financial Aid for Soundproofing Ordinary Houses.

(a) The Organization administers the financial aid for soundproofing ordinary homes. The aid is given to houses in class 1 areas designated by the Minister of Transportation under the Airplane Noise Prevention Act with preference given to households with elderly or sick persons (all of the plaintiffs' residential districts fall within the class 1 area where noise exceeds 85 WECPNLs). It provides soundproofing and air conditioning for one or two rooms in the house depending on the family size. Of 1,245 applications in 1974, it completed the soundproofing of 457 houses including the houses of some plaintiffs. In 1975 there are plans to soundproof approximately 3,200 homes with a budget of ¥5,300 million.

(b) There are, however, 33,200 families living in the class 1 district in this area. Even if we assume that the construction will proceed as planned, it will take several more years to complete the soundproofing of all houses remaining in the area. Moreover it is doubtful that the noise level can be reduced by 10 phons by soundproofing one or two rooms in a wooden house, as the defendant contends; but assuming that this is accomplished, we still cannot say whether these measures are sufficient to alleviate the overall suffering of the plaintiffs in their daily life. It is not possible for several members of a family to live in one or two rooms all the time. The sick and aged may be adversely affected by living in a closed in and air-conditioned room. Soundproofed rooms, therefore, merely treat the symptoms not the cause of the plaintiffs' suffering.

(C) Compensation for Relocation.

(a) Compensation for relocation started in 1971. By the end of the 1974 fiscal year, 222 families had received compensation and moved out. When it was established, the Organization was delegated the authority to administer compensation payments under the program. As a result of the amendment to the Airplane Noise Prevention Law, the area where families are eligible for compensation for relocation has expanded and the area where noise levels exceed 90 WECPNLs has been designated as a class 2 area by the Minister of Transporta-

*tion. This area covers 2,000 meters extending into Kawanishi City and 4,000 me-
ters reaching into Toyonaka City from the end of the B runway. The 1975 budget
is ¥6,100 million, and there are plans to relocate 214 families (including the purchase of 6.4 hectares).*

*(b) Compensation for relocation provides a basic solution for those who are
willing to move out. However, the plaintiffs are dissatisfied with the prices offered in compensation for their property and, needless to say, are unhappy about
leaving places where they have deep roots. The compensation price is set according to the Official Guidelines for Compensating Losses Occasioned by Appropriation of Lands for Public Purposes. The guidelines take the price of nearby
land similar to the plaintiffs as a basis for the computation of the amount of
compensation. This amount has proved to be insufficient in most cases to buy
equivalent land in another area at the prevailing market price. It is difficult for
these people to buy land which can offer the same conveniences in transportation and other amenities as before. Many families have relocated at great expense to themselves. They may require a larger tract of land because of the restrictive zoning and building codes in the areas into which they are moving.*

*(c) In relation to the above, the defendant contends that the Organization has
bought 70,825 square meters of land for relocation and had already completed
43,697 square meters, or 208 lots, for housing sites by the end of the 1974 fiscal
year. It is not clear that these sites are to be offered at reasonable prices. Generally these substitute sites are not conveniently located and it is hard to predict
whether families will be satisfied with them or how many more lots can be se-
cured in the future.*

*(d) Compensation for relocation has primarily been paid to those who have
owned land. Those residents in the affected area who rent land or houses have
not been considered for compensation, and no project to develop multiple family
dwellings for them has been started.*

*(e) There are about 12,500 families living in the area where noise levels exceed 90
WECPNLs. It would be difficult, if not impossible, to remove all of them. In view
of the one-sided process of extending and increasing the adverse effects of the
airport on the residents, it is unreasonable to force them to move out of the area,
severing all of their attachments to places familiar to them. Moreover, relocation
of only some of the residents has created the secondary adverse effect of deterioration of the neighborhoods. The question of the new Kansai Airport is related
to this problem. Without knowing how the airport will be used in the future, it is
difficult for the residents to decide between moving with the assistance of compensation payments and staying and having a room soundproofed. When a policy
decision is made as to whether in the future this airport will be used in connection with the proposed new Kansai Airport, it will be made public.*

*(f) We conclude that compensation for relocation is not at the present stage a
fully effective means to avoid further damage. (69)*

4. The Public Nature of Air Transportation

*(1) In our modern society, airplanes offer the most rapid mode of both passenger
and freight transportation and serve important economic and social functions.
They are essential for international as well as for domestic long distance and in-
terisland transportation. There has been an increasing demand for air transportation and this trend will undoubtedly continue into the future. Both parties
have offered much evidence on the public nature of air transportation. Both have*

argued over basic economic policies of the government which may affect the future demand for air transportation. We consider it unnecessary to go into the details of these arguments. It is evident that air transportation is serving the public in the abstract terms described above, and there is no need for us to go beyond that judgment. The plaintiffs have contended that a large portion of the passengers are tourists whose use of air transportation is nonessential and induced by advertising campaigns by the airlines and that modern air transportation is a privately owned industry operated for profit. These points, however, do not refute the fact that air transportation serves the public.

(2) The Osaka International Airport serves the combined area of Kyōto-Osaka-Kōbe which has a population density second only to Metropolitan Tokyo. The area is a center of economic and cultural activities for Western Japan. There is a great need to establish and maintain a large airport in this area as an important terminal in the air traffic network. As for the actual situation at this airport, in addition to the findings regarding the number of flights, supra, we also adopt the findings of the district court opinion [citation omitted]. In 1973 this airport served 1,322,000 international and 10,685,000 domestic passengers. The amount of international cargo has increased from 4,400 tons in 1969 to 16,300 tons in 1971, and to 26,200 tons in 1973. From these figures it may be argued, as the defendant has, that a restriction on the number of planes at the airport will have a considerable economic and social impact.

(3) In discussing the public nature of air transportation and the operation of the airport, however, it is necessary to consider not only the social and economic benefits that they create, but also the costs they impose. We find that the impact of noise and other nuisances from airplanes using the airport has extended over a large area inflicting serious injuries on many people including the plaintiffs, and that the use of the airport has been continued without the adoption of other measures to alleviate injuries. Under these circumstances, there is a limit to the defendant's argument of the public nature, especially when airport operations are still inflicting injuries on local residents. It cannot be helped that restrictions on the use of the airport to curtail injuries may create some inconvenience to others. (69)

5. Defects in the Establishment and Management of the Airport

(1) The Osaka International Airport is clearly a public facility established and controlled by the government. The airport is cramped and adjoins residential areas. When such an airport is used by many large planes and jet aircraft, inevitably the residents suffer from the harmful effects of the noise and exhaust fumes. With such an inferior location, it was indeed unreasonable to expand the airport to introduce jet planes and designate it an international airport. The defendant did not adopt timely and suitable measures to abate injuries even after the risk of injuries from noise had been pointed out. Instead the defendant allowed the adverse effects to intensify by the introduction of jet planes, the addition of more flights, and the opening of B runway. The defendant has continued to inflict injuries on the plaintiffs from the use of the airport while procrastinating in undertaking the necessary abatement measures. The defendant is at present still unable to offer any basic solution.

We find, therefore, that there have been defects in the establishment and management of the airport and that the plaintiffs' injuries have been caused by these defects.

(2) The defendant argues that a defect in the establishment or management of a public facility as provided in art. 2(1) of the State Compensation Act refers to a facility that lacks ordinary safety measures. However, the safety of the facility is not limited to the actual users of the facility, but also includes the safety of persons other than users. In short, when the facility lacks the characteristics or equipment ordinarily provided and thereby presents a risk of damage to people, it should be considered unsafe. The defendant cites a decision of the Supreme Court (August 20, 1970, 24 Minshū 1268) in which the court found a defect in the government's maintenance of a road, which created a hazard to passengers on that road. The decision cannot be interpreted as limiting the concept of safety only to users of that road. In the case of an airport, necessary safety measures are not limited to its air traffic control system and security equipment, which are essential to reduce the risk of accidents in landings and takeoffs. In considering airport safety, the adverse effects of noise, and so forth, on the residents around the airport must also be examined. In other words, the use of a facility should not inflict arbitrary or indiscriminate injuries on other persons. It is this standard by which one should judge whether there is a defect or not.

The noise level inflicted on the residents may be affected by flight patterns and antinoise measures. However, the location of the airport in the present case is particularly unsuitable, and it was unavoidable that surrounding residential areas would be seriously affected when the airport began to be used by many jet planes. In such a situation, when the ordinary and intended use of the airport automatically inflicts damage on others, the airport does not have the characteristics necessary for a major international or domestic airport intended to serve numerous airplanes. It can be said, therefore, that there was a defect in the establishment of the airport itself and also that there was a defect in the management of the airport when the defendant allowed the heavy use of it without undertaking proper measures. The fact that the defendant has not succeeded in reducing injuries in any adequate degree despite its large investment in noise abatement measures for the surrounding areas indicates the defectiveness of the airport itself. Therefore, we find that the plaintiffs' injuries have resulted from defects in establishing and managing the airport and that the defendant is liable under art. 2(1) of the State Compensation Act. (70)

IV. Injunctive Relief

1. Validity of Injunctive Relief

(1) The plaintiffs seek an injunction to prohibit all landings and takeoffs at the airport between 9:00 P.M. and 7:00 A.M. daily. The defendant claims it would be legally improper for the court to grant this injunction. The defendant argues that (1) the official responsible for the establishment and management of the airport is the Minister of Transportation, not the defendant State; (2) the management of this type of facility involves the exercise of public and administrative authority. Within the scope of his authority, therefore, the Minister of Transportation can freely exercise his discretion in deciding the manner of airport operations and determining who will be allowed to use the airport; (3) it is the aircraft or airlines which have infringed the plaintiffs' rights, not the defendant or the Minister of Transportation; (4) the airlines from the start possessed a right to use the airport without any time restrictions. If the court should grant an injunction in this case, the Minister of Transportation would be in effect compelled to take

affirmative action to prohibit the airlines from using the airport during certain hours. In other words, he would be forced to exercise affirmatively his management authority over the airport; (5) the injunction in this case is actually not to order mere inaction, but to force the state to exercise its administrative authority to manage the airport. However, the defendant State as such is not capable of exercising such administrative control over the airport and it is not possible to demand that it do so; and (6) for a court sitting in a civil case pursuant to the Code of Civil Procedure to order what is equivalent to the exercise of administrative authority violates the principle against suits for affirmative administrative action, and in doing so violates the principle of separation of powers as well.

(2) However, we cannot agree with the defendant's contention.

(a) The State is the legal entity which has constructed and maintains the airport in question and the Minister of Transportation acts as an agent of the State in overseeing the airport. Essentially a public airport could be operated as a private enterprise. Even in the case of a State-established airport, it is not necessary to view the construction and operation of the airport as being entirely state action. Of course, when the State constructs an airport for the purpose of serving the public interest, it may regulate the use of the airport, and the Minister of Transportation as a supervisor may impose regulations on the users of the airport as a part of the exercise of his administrative power. However, when the relationship is not between the Minister of Transportation and the users of the airport but rather between the Minister and an unrelated third party, that relationship is not necessarily based on administrative authority. The request for an injunction in the present case assumes that the conditions caused by the defendant's defective establishment and operation of the airport are infringing on the private law rights of the surrounding residents and demands the elimination of the offending conditions. In this context the relationship between the State and the plaintiffs is exclusively a private law relationship, and the plaintiffs' demands should be considered as the exercise of their right to demand relief under private law principles.

(b) Even though it is the airplanes and airlines that produce the noise, fumes, vibrations, and so forth, these would not infringe on the plaintiffs' rights if the planes did not land and take off at this airport. That the defendant, being responsible for the management of the airport, has a duty to prevent pollution resulting from its use is beyond question at this point. The defendant could stop allowing the use of the airport for landing and takeoffs during the hours requested by the plaintiffs and by simply doing so it would effectively cease its infringement of the plaintiffs' rights. We need not specify in this opinion a concrete means of carrying out this step. The Minister of Transportation is capable of choosing among various alternatives. He could change the airport regulations, make a case by case disposition, or undertake administrative guidance. But the plaintiffs' action for an injunction in the present case does not ask the Minister of Transportation to undertake a specific administrative action. It is a private law action asking the defendant to terminate a certain action, and anyway it is irrelevant whether the defendant has the capacity to act or not.

(c) It is primarily within the discretion of the Minister of Transportation to decide how the airport is to be operated and who may use it. However, the issue in this case is whether the operation of the airport is an infringement of the plaintiffs' rights. It is not related to the exercise of the discretionary power of the Minister of Transportation over the users of the airport.

There is no doubt that the plaintiffs can request relief from a court. In this situation whether the court restricts the use of the airport is based on a legal judgment of whether an infringement of rights has occurred and whether an injunction is proper. Necessary and appropriate relief is determined by a balancing of interests from the viewpoint of private law; it is not a discretionary judgment based on administrative goals with the court's judgment replacing that of the administrative agency. Therefore, there is no substance in the criticism that the court has usurped administrative authority or violated the separation of powers.

(d) The defendant contends that third parties like the plaintiffs, who are not users of the public facility, do not have standing to bring any action [under art. 3 of the Law Concerning Procedures of Administrative Litigation] against the Minister of Transportation, based on the enactment or change of the regulations governing airport management, or on a particular administrative disposition, even if they are injured thereby. If we adopted the defendant's position, which would deny standing in a civil action because the suit involves the exercise of administrative action, we would effectively close the door of judicial relief to the plaintiffs. It is an unreasonable contention. It is also contrary to the spirit of the Constitution for defendants to argue for administrative independence and to assert that the prevention and relief of damages in this case should be left to the discretionary judgment of an administrative agency.

(3) The defendant contends that the use of the airport between 10:00 P.M. and 7:00 A.M. has already been stopped and relief is to that extent moot. Yet the measure taken by the defendant was conducted through administrative guidance and is not legally enforceable. It can be changed at the defendant's will. Furthermore the defendant has contended throughout this action that it is necessary to use the airport during these hours. Therefore the plaintiffs are entitled to have a judgment on the merits of their request for an injunction. (71)

2. Nature of the Right Infringed

Since we have found above a wrongful act committed under art. 2(1) of the State Compensation Act, we must now consider the nature of the right infringed and the appropriateness of injunctive relief sought.

(1) The plaintiffs contend that the actions which injured them in this case violate both their personal rights and their environmental rights. Based on these rights, the plaintiffs seek to enjoin the defendant's use of the airport during certain hours. The plaintiffs also all allege individual personal injuries. On the other hand, according to the plaintiffs' own assertions, the significance of an environmental right is that it defends the periphery of the personal right against environmental deterioration and thereby prevents individual injuries from resulting. Therefore we should first consider the propriety of the assertions based on the personal rights involved.

(2) Individual life, physical safety, and freedom of mind are fundamental to human existence and should be granted absolute legal protection. Also, as long as humans exist, a peaceful, free, and dignified life should be preserved with utmost care as provided in art. 13 of the Constitution. Article 25 of the Constitution also supports these rights. Personal rights are a composite of fundamental interests which relate to the physical as well as the mental well-being of individuals. No one can be permitted to infringe on such personal rights, and everyone has the power to act against any infringement of such rights. In other words, not only can a person request the elimination of conduct that causes him

physical harm, but he may also do so when it is a question of pronounced mental suffering or interference with the conduct of his daily life. Furthermore, even when such harm has not yet materialized, in cases where there is an imminent risk of such harm, one can demand the cessation of the dangerous activity. When based on an infringement of this type of personal right, the right to demand abatement of or protection against a nuisance forms the basis for a demand for a private law injunction.

The defendant contends that there are no explicit provisions on personal rights in substantive law and therefore there is no basis for an injunction. However, personal rights consist of those fundamental interests essential to human existence. They are no doubt to be construed as rights without any explicit provision and should indeed be recognized as fundamental rights. Previously, personal rights have primarily been discussed in the context of protecting one's reputation, name, likeness, privacy, or copyrights, that is, those personal rights which are most likely to conflict with other people's freedom of action and, therefore, where adjustments are most often necessary. This does not mean that personal rights do not include more fundamental human interests. To be sure it is difficult to define abstractly and definitely the extent of personal rights. Nevertheless, they include without doubt those fundamental interests described above. Viewed from another perspective, an action for an injunction based solely on property rights might not always suffice to protect these interests from possible infringement. So we find that there are substantive reasons for granting injunctive relief based on personal rights. We do not agree with the defendant's contention that it is premature to do so or that we should wait for theoretical developments and a systematic analysis of the subject.

(3) Applying this to the present case, noise and other adverse effects of airplanes using the airport have brought serious mental suffering and severe disruption to the daily lives of the plaintiffs, have already caused physical injury to some plaintiffs, and have exposed the others to risks of similar injury. In view of the seriousness of the injuries, we find that the plaintiffs' personal rights have been infringed and that compensation for past damages does not provide sufficient relief. Therefore, we consider injunctive relief to be necessary. (71)

3. The Extent of Injunctive Relief

(1) Had the plaintiffs asked for an injunction to prohibit the use of the airport entirely or during a substantial portion of the morning and afternoon hours, we would have to deal with a serious problem of balancing interests.

. . . According to the plaintiffs' testimony, they discussed a plan to seek a prohibition on the use of the airport from 8:00 P.M. on, but they finally agreed among themselves to limit their claim to a minimum, only requesting a curfew on the use of the airport after 9:00 P.M. At the conclusion of the court's second on-site investigation at the home of the plaintiff Shibahara Haru, the plaintiffs' attorneys stated that "Since we are willing to grit our teeth and endure the noise of planes from seven in the morning to nine in the evening, please grant us an injunction against the use of the airport after nine o'clock." The court was impressed with this brief statement which crystalized the seriousness and sincerity behind the plaintiffs' request to extend the injunction by one hour. Upon consideration of this action along with the extent and severity of injuries in the present case and the fact that the plaintiffs' injuries still remain largely unmitigated despite considerable efforts on the part of the defendant, it cannot be said that the defen-

189

dant has fully performed its duty to prevent pollution. It has been five-and-a-half years since the opening of B runway and the large scale introduction of jet planes. During these years the airport has been used continuously, and its use from 7:00 A.M. to late evening will inevitably continue into the foreseeable future. The hour between 9:00 and 10:00 in the evening is a precious time for family conversation, reading, thought, and rest. Some people have already gone to bed. It is quite natural for people to seek quiet at night. We, therefore, find no grounds to reject as unreasonable the plaintiffs' request for a curfew on use of the airport during the hours between 9:00 P.M. and 10:00 P.M. in addition to the prohibition on night flights already granted by the trial court.

(2) The defendant has already restricted the use of the airport between 10:00 P.M. and 7:00 A.M., and there is no evidence that serious damage has resulted therefrom. However, the defendant strenuously opposes the prohibition of landings and takeoffs between 9:00 P.M. and 10:00 P.M. and contends that it would be absolutely impossible to comply for the following reasons: (a) The demand on the airport is especially great during the hour in question because many businessmen making day trips use the airport during that time; (b) The needs of domestic cargo and mail service; (c) It would be impossible for arriving international flights to adjust departure times at the places of origin, land at intermediate airports, and arrange for connecting flights; and (d) There are no instances of curfews imposed at nine o'clock in the evening in other countries and the defendant has already received a protest from the International Air Transport Association for its restrictions on flights past 10:00 P.M.

We are not unmindful of the effects of an injunction for that hour in various areas including the international aspects. However, we cannot ignore the serious and extensive injuries that residents suffer as a result of the exceptionally unsuitable location of the airport, no matter how great the public interest involved in its use. We must grant the injunction sought, and we expect the best efforts of the defendant in dealing with the resulting problems. It was, after all, the defendant who constructed the B runway when he should not have, and who allowed the expansion of jet plane services in the face of mounting injuries.

(3) It is a matter of course that emergency landings and takeoffs necessary to avoid dangers to life are exempted from this injunction and is so stated in our judgment. Therefore, the defendant's objection based on this point is without substance. It might still be possible to allow a small number of light, quiet propeller planes different from those now in service which would not affect the neighboring residents to use the airport during the hours proscribed by the injunction. However, their use should be considered separately as a matter bordering on the limit of the injunction. It does not affect the court's prohibition against general use of the airport during the prescribed hours. (72)

4. Plaintiff Kondō

(1) One of the plaintiffs, Shimae Kondō, moved into her present house in June 1970 after the opening of the B runway. According to her testimony, she decided to move in without much knowledge of the noise problem in the area. She did so on the recommendation of a real estate agent and because the area was close to where her husband worked. She had seen the house only once for fifteen minutes before deciding to take it and came to realize the severity of the noise only after settling down. Given the present shortage of houses, she did not have many possibilities to choose from, and it was unlikely that the real estate agent or the

owner of the house would inform her of the existence of noise pollution. In these circumstances it is inevitable that new residents would not do a thorough investigation, and judging from our own on-site investigations, it is difficult for a stranger to comprehend the reality of airplane noise in fifteen minutes. Actually it requires living there to feel the intensity of noise pollution. In addition, since at least the time it was returned by the U.S. military forces and was designated as a first-class airport, the airport has been located close to a residential area with many inhabitants. It has since been expanded. It is natural to expect that some people would move into the essentially residential area around the airport, and the defendant cannot claim priority for the airport over the interests of individual residents who moved into the area later simply because of the prior existence of the airport. We find that the principle of coming to the nuisance has no application in the instant case as long as the individual did not move into the area intending to take advantage of pollution problems. The plaintiff Kondō's action for the injunction, like other plaintiffs' is granted. (72)
[(2) ...]

5. Environmental Rights

As we are granting the injunction and awarding damages based on our findings of infringement on the plaintiffs' personal rights, we find it unnecessary to discuss the environmental rights asserted by the plaintiffs. (73)

V. Damages

1. We have found that the plaintiffs have suffered severe mental and physical injuries and interference in their daily life. They seek compensation for all these injuries as nonpecuniary losses. This demand can be interpreted as a request for the payment of solatium, the amount of which is to be computed by consideration of all the injuries and suffering of the individual plaintiffs. Since all these injuries have been caused by the defects in the construction and maintenance of the airport, the defendant is responsible for their damages under art. 2(1) of the State Compensation Act. The plaintiffs claim separate damages for past and future injuries. We take the time of closing oral arguments before this court as the dividing point between past and future and consider both claims individually. (73)
2. Past Damages.
(A) In computing the amount of damages, we should consider the gravity of the plaintiffs' injuries and the process by which the infringement of the plaintiffs' rights occurred (including the inadequacy of the defendant's antinoise measures). The plaintiffs' injuries take various forms depending on the life patterns of the affected individuals. However, they are equally exposed to noise and exhaust fumes in each district. They suffer from mental anguish, risk of physical harm, and daily inconveniences, major elements of which are common to all residents in a given area. Therefore, we find it sufficient to consider the location of each plaintiff's house and the duration of his residency there. We find it unnecessary to consider other subjective elements of damage for each plaintiff.
(B) (1) A considerable number of the plaintiffs moved into the area after June 1964, when jet planes were introduced into the airport. However, on careful examination of their testimony and statements, we cannot say that they foresaw the aggravation of the noise pollution or that they moved in with the knowledge of the deteriorating situation. Also, in the light of the actual situation of this air-

191

port, we find it unreasonable to say that they should have realized what was happening at the time they moved in. Plaintiff Shimae Kondō moved into her present home after B runway had started serving airplanes, but she will not be excluded from damages for the same reasons as stated above regarding her request for an injunction. Plaintiff Yoko Tsune is in a similar situation. Therefore, we do not think it reasonable to distinguish these plaintiffs from other plaintiffs by the time when they started living in the area. We will simply compute the amount of damage according to the duration of residency of each plaintiff. We do not agree with the trial court's decision to adopt the time of the enactment of the Airplane Noise Prevention Act as a point for differentiating between plaintiffs with regard to the amount of damages to be awarded.

(2) Although a significant part of the harm suffered here was to the family life of the plaintiffs, the harm resulted from the infringement of their individual personal rights. Thus we see no reason to compute damages by household units as the trial court did or to adjust the award by the position of an individual plaintiff in his family or because another member of the family may also be a plaintiff. We will adopt uniform standards in computing damages for each plaintiff.

(C) (1) The plaintiffs request that the damages be computed as follows:

(a) For each plaintiff of the first and second suits who lives on the Kawanishi City side, ¥500,000 plus interest at the rate of 5% per annum for the period January 1, 1965, to December 31, 1969, and ¥10,000 per month for the period since January 1, 1970.

(b) For each plaintiff of the third suit who lives on the Toyonaka City side, ¥500,000 plus interest at the rate of 5% per annum for the period January, 1965, to December 31, 1971, and ¥10,000 per month for the period since January 1, 1972.

(2) We now examine the degree of damage in each district according to the data on noise levels admitted above and the number of landings and takeoffs. In the Districts of Takashiba, Mutsumi, and Setsuyo on the Kawanishi City side, severe noise existed by 1965. Thereafter with the increase in the flights and the loss of larger planes, it got worse and then further intensified with the opening of B runway. In the two districts of Katsube and Hashirii on the Toyonaka City side, noise similarly increased after 1965. A slight reduction in the maximum noise level resulted from the rerouting of large airplanes after the opening of B runway. However, in other respects, damage has intensified because of the increase in the number of planes and the noise and fumes from planes waiting to take off. Before the opening of the B runway, the other districts on the Toyonaka side suffered from a somewhat lower level of noise, but not one so low as to be ignored. Noise levels reached a maximum of 85 to 100 phons. In all areas, NNI exceeded 45. After the opening, it is apparent that damage has intensified. We find, therefore, that the plaintiffs in each of these districts have suffered about the same degree of injury since the opening of the B runway to the present. There is thus no reason to distinguish the plaintiffs by district in the computation of damages.

(3) Based on the above considerations, we award to each plaintiff in districts of Takashiba, Mutsumi, Setsuyo, Katsube, and Hashirii ¥8,000 per month; to each plaintiff in the districts of Toshikura, Toshikura Higashi, and Nishimachi ¥3,000 per month for the period from January 1965 to January 1970. . . . (However, the total awarded to plaintiffs from Katsube and Hashirii for the period January 1965 to December 1971, shall be limited by the ¥500,000 requested.) (74)

[The court then noted the application of 5% interest rate to the various totals.]

With respect to the plaintiffs who either moved in or moved out of this area during this period, damages shall be awarded with interest either after the month of moving in or until the month of moving out. (74)

(4) [The court then computed damages for each plaintiff.]

3. Future Damages.

(1) The plaintiffs seek compensation for damages which may result in the future. Their claims rest on the assumption that the wrongful act will continue to be committed and damage will result in the future. In this case, where the infringement of the plaintiffs' rights has continued over a long period of time, it can be presumed that the same situation will continue into the future unless the defendant proves that there is a probability that the wrongful act will cease or that the damage will be terminated in the near future. Therefore, it is possible to ascertain at present a factual basis on which an action for future damages would arise. Even if there is uncertainty about the degree or extent of damages arising in the future, the court can make a judgment on that portion of the damages which is accurately predictable. Furthermore, it would be unfair to punish the plaintiffs who have suffered continuously over a long period simply because one portion of potential damage cannot be ascertained at present. In the event of any new developments, the defendant can prove facts sufficient to show the mitigation of damages or the nonexistence of further injuries. Defendant could then ask the court to enjoin further execution of the judgment.

It is clear that the defendant is not about to impose regulations to drastically change the number of flights, types of planes, or the schedule of airport operations. The defendant has not shown that the measures taken to reduce the noise at its source will significantly decrease noise levels or that other measures such as soundproofing, although somewhat effective, are sufficient to prevent injury. It must be assumed, therefore, that the infringement of the plaintiffs' rights will continue for the foreseeable future. Needless to say, if some of the plaintiffs move out of the area on accepting compensation for that purpose, the infringement of their rights no longer exists and they cannot enforce the judgment against the defendant. The plaintiffs' action in the present case is for damages during the period that they live in the area. It is unthinkable that any plaintiff, after moving out of this area, will try to execute this judgment. Should that occur, however, the defendant may file an objection at that point.

Therefore, the plaintiffs' continuing right to claim damages as they arise should be recognized. Because the defendant has disputed these claims throughout this litigation, it is clear that there is a need for the plaintiffs to request it in advance.

(2) We now will examine the amount of damages for future injuries. Until this court's prohibition of night flights after nine o'clock is implemented, we assume that the same situation which existed at the conclusion of oral arguments will continue. We find that the plaintiffs in that situation suffer damages of ¥10,000 per month. When the prohibition is effectuated, a measure which the plaintiffs have desired as minimal and essential relief, and other measures such as soundproofing of their homes are implemented, their injuries will have been reduced to some degree. At such time each plaintiff will be only entitled to claim ¥6,000 per month, collectible at the end of each month.

(3) The plaintiffs' claim for future damages is, of course, valid as long as the infliction of injuries by the defendant continues. The plaintiffs proposed at the conclusion of the trial that their claim continue "until the prohibition of landings and takeoffs of all airplanes whose noise levels as measured in the plaintiffs' residential areas exceed 65 phons." That condition would make it impossible for

193

virtually all types of airplanes currently in service to use the airport. As a practical matter, we can hardly expect that the airport will be closed or its operation reduced to the level of a local airport in the foreseeable future. In effect, the plaintiffs must accept the fact that the airport will continue to be used by many jet planes from 7:00 A.M. to 9:00 P.M. for some years to come. We expect that the plaintiffs who continue to live in the area will make some concession for the public interests involved in the operation of the airport in view of the fact that their action for the injunction is now granted and that the ultimate purpose of their action is clearly not to obtain monetary compensation for damages. We would expect the plaintiffs to negotiate dispassionately with the defendant to realize reduction of the number of flights (including changes in the type of planes, reduction of noise levels, and shortening of the daily operation of the airport) in place of their monetary compensation for future damage (60% of the solatium for past damage). The defendant is also expected to reduce the number of flights to a number less than currently scheduled, or face an injunction and future damages. From past experience, the defendant should realize that the cooperation of residents in the area is essential for the operation of the airport. The defendant should make an earnest effort to reduce the number of flights and to negotiate a prompt and satisfactory solution with the residents.

Up to the present both parties have confronted each other squarely without modifying their positions. Now that the decision of this court is rendered, it is an opportune moment for both parties to make some concessions. Because the practical effect of the judgment extends beyond the litigating parties to all the residents in the neighboring areas, it is not enough to limit the negotiations to the parties themselves. The Pollution Committee is also working on the case, and it is expected that some good results will come out of its proceedings. In short, it is not possible to solve once and for all a dispute like the one before us with one lawsuit. It involves and affects the interests of many users of the airport including the residents around it and those from foreign countries. It is important for all the parties involved to reach a temporary agreement restricting the operation of the airport and to reestablish mutual trust by adopting suitable measures to improve the situation from time to time by taking full advantage of scientific and technological advances and by anticipating and adjusting to changes in circumstances.

The reason that the court has not set specific conditions for the termination of future damages in terms of the noise level, types of planes, or the number of flights is not because it would be difficult for the court with its limited technological knowledge to set specific numerical standards but also because to do so might impede a possible settlement by making the settlement conditions inflexible.

We sincerely hope that both plaintiffs and the defendant along with other concerned parties will make a sincere and positive effort to find an appropriate and practical solution to the dispute, thus ending at the earliest possible date the unfortunate continuation of the payment of future damages. In the happy event that the parties are able to reach an accord on the reduction of the number of flights, it would clearly affect the plaintiffs' cause of action. Since the circumstances underlying their cause of action can be foreseen only up to the time of such a compromise, the right to future damages in this case is granted with this limitation. (75)

4. **Attorney's Fees.** *It is clear from the record that the plaintiffs have retained attorney Yasuo Kimura and others to pursue these cases and have agreed to pay*

them 15% of the amount of past and future damages claimed and ¥75,000 per plaintiff for the injunctive relief sought.

It is clear that this suit was necessary to protect the plaintiffs' rights and that, in order to do so adequately, the specialized knowledge and techniques of an attorney were required. It was thus unavoidable that the plaintiffs had to rely on an attorney, and their payment of a reasonable attorney's fee is a part of ordinarily expected damages for the wrongful act committed by the defendant. Therefore, the plaintiffs are entitled to have the attorney's fee compensated.

Taking into account the substance of the plaintiffs' claim, the difficulty and length of the litigation and all other factors, we find it reasonable to order the defendant to pay the following attorney's fees in addition to the damage payments to the plaintiffs:

¥50,000 per plaintiff for whom injunctive relief was granted at the trial and on appeal.

¥150,000 per plaintiff [living in Takashiba, Mutsumi, and Setsuyo], ¥120,000 [for Katsube or Hashirii], and ¥110,000 [for other districts] for past damages awarded, amounting to a little more than 13% of the amount awarded, ¥1,000 or ¥600 monthly per plaintiff for future damages awarded, i.e., 10% of the award. (76)

VI. Conclusion

[In its conclusion, the court summarized the details of the injunction and the important points of the trial court's decision. The court permitted the plaintiffs to execute the judgment provisionally before the decision became final.]

3.4.3   Recent Developments in the Osaka International Airport Case and Related Matters

The defendants subsequently appealed the Osaka High Court's decision to the Supreme Court, the appeal being based principally on the following six points.

1. The High Court's attempt to compel (by injunction) the Minister of Transportation to perform various administrative acts violated the principle of separation of powers.[75]

2. Because the personal rights theory on which the court based its judgment was not defined or supported by any statutory provision, personal rights could not legally support the injunctive relief.

3. Because the airport is a public facility whose benefits redound to the public, an injunction will obviously cause serious economic and social dislocations. In deciding whether to prohibit the use of the airport, the Osaka High Court overemphasized the plaintiffs' injuries and underevaluated the public benefits of the facility.

4. The High Court's theory of collective harm, which it used to circumvent the problem of individual proof of causation, was not legally warranted. Indeed, there was no proof, the government argued, that noise had actually caused the plaintiffs' physical injuries or that the noise of the area even exceeded a socially acceptable limit of tolerance.

5. The airport's safety measures were not "defective" within the meaning of art. 2-1 of the State Liability Compensation Law nor could airplane noise be deemed a defect of the airport.

6. In its award of future damage, the High Court obviously ignored the fact that

the noise problem around the Osaka Airport would soon be substantially mitigated by the government's implementation of antinoise pollution measures.

Since the defendant's appeal in 1975, the First Division of the Supreme Court (consisting of five justices) has heard oral arguments from both parties. At the conclusion of a three hour argument, the Head Justice of the Division (Yasuo Kishigami) quite unexpectedly (and virtually without precedent) asked the parties if they would be willing to negotiate a settlement.[76] Although the plaintiffs' lawyers indicated that they would be willing to consider this option, if the contents of the proposed settlement reflected the intent of the Osaka High Court's decision, the government's lawyers postponed their answer stating that they needed time to discuss the matter with government officials. Subsequently the Supreme Court decided to review the High Court's decision en banc, and a final decision is expected in 1980.

On a more general level the Osaka International Airport controversy continues to exert an important influence on administrative and judicial thinking. Since December 27, 1973, when ambient standards for aircraft noise were established, the government has embarked on an ambitious program of countermeasures. These measures include improving aircraft equipment, adjusting flight frequency, establishing green buffer zones, rezoning areas for house construction, subsidizing the insulation of private residences, schools, and hospitals, and compensating residents for the costs of relocation.[77]

On April 12, 1978, the Diet strengthened the role of local governments in controlling aircraft noise by passing the Law Relating to a Special Provisional Arrangement for Countermeasures against Aircraft Noise around Specified Airports. Under this law, governors of affected prefectures may establish basic countermeasures against aircraft noise in consultation with heads of concerned towns and villages. The governors may also implement special land use plans and other countermeasures for airports in special noise control areas designated by Cabinet Order. The law exhorts the central government to provide financial assistance to local bodies initiating these measures.[78]

The Osaka International Airport suit has also served as a litigation model for residents in other areas. The most interesting current suit, initiated by forty-one members of the Association to Eliminate Noise at Yokota Air Base, seeks (1) an injunction terminating all takeoffs and landings at the base between the hours of 9:00 P.M. and 7:00 A.M.; (2) an order prohibiting noise over 5.5 phons; (3) ¥1,150,000 for past damages plus interest; (4) ¥243,000 per month as future damages until defendants satisfy condition (1); and (5) costs of suit.[79] From a doctrinal perspective the Yokota suit is significant because it seeks to enjoin the operations of a military facility, not a commercial airport, and because the base is under foreign, not Japanese, jurisdiction. The suit may also establish an important precedent on whether the American military must meet Japan's environmental standards and other controls.[80]

3.5   Assessment

Taken together the second generation civil cases suggest these generalizations. First, the courts have opened an alternative route around the barriers of the Administrative Litigation Law for deserving parties aggrieved by governmental action. They have, however, been extremely circumspect in expanding the scope of judicially protected interests. Indeed, from a doctrinal standpoint, the judiciary's recognition of a right to sunshine and its gradual acceptance of the personal rights

theory represent only cautious, albeit salutary, extensions of traditional principles; the courts' reluctance to recognize environmental rights attests to their continuing conservatism. Second, despite their reluctance to expand cognizable legal
interests, the courts have experimented with innovative, at times even radical
new remedies. Third, although the third generation cases are too few to permit assessment, the courts appear to be following their pattern.

3.5.1   Erosion of the Distinction between Public and Private Law

Perhaps the most distinctive aspect of some of these cases is that the plaintiffs
have survived the government's threshold attack that the action in issue was an
"exercise of public power," and therefore insulated from civil assault.[81] Although
this distinction may appear to the American reader senselessly technical, Japanese
courts have continued to endow it with great importance. In 1965, for example,
the Otsu District Court relied on the doctrine in refusing to enjoin the construction of a highway despite potentially significant environmental effects. Indeed
until the Osaka Airport case, a mistake in the choice of forums meant dismissal
or, perhaps worse, application of stricter public law principles.[82]

Although the government's subsequent appeal to the Supreme Court still
leaves the status of the doctrine uncertain, the Osaka High Court opinion has
helped clarify, incrementally, the bases for citizen environmental actions against
the sovereign. Indeed in some ways the court's distinguishing third party actions
over the infringement of personal rights from other administrative challenges to
public authority, recalls earlier American cases dealing with sovereign immunity
from tort liability. These early American cases granted proprietary acts only limited private law protection, distinguishing them from "public" acts that deserved
a higher level of protection. In Japan, however, the Osaka High Court rejected the
government's attempt to distinguish private and public acts, holding the government's conduct illegal where it resulted in serious public injury.[83]

Although the barriers to the environmental litigant in the Japanese cases may
be even greater than those faced by American counterparts, the policy considerations underlying the American decisions and commentaries still seem relevant. In both countries the courts have been primarily concerned about protecting
the innocent third party injured by government action and have seen no reason to
insulate governmental activities simply because they are government initiated.
To a lesser extent, they have also been concerned about creating incentives for
government to maintain a high standard of care.[84]

3.5.2   Environmental Rights Reconsidered

In fairness to the courts, the claims of the residents of infringement of environmental rights have raised almost intractable doctrinal and practical problems.
First, unlike sunshine rights, the concept of an environmental right is based on
the idea of a legally enforceable claim held by *all* members of the public. The
"collective" nature of the right, however, challenges a tradition dating from the
Meiji era that has considered rights to be individual, personal claims.[85]

Second, the concept is so open ended, that even its own defenders differ on its
proper interpretation.[86] Equally uncertain are the standards for measuring the extent of possible environmental destruction deemed necessary to trigger injunctive
relief for an infringement of environmental rights. Third, although some environmental rights champions concede that environmental destruction may at

197

times be an unavoidable price for living in modern society, the orthodox interpretation of environmental rights would not brook any environmentally hazardous development. Thus the point at which environmental rights must be compromised with other interests is still uncertain. Fourth, the critics note that recognition of a new concept may not even be necessary, since the objectives of an environmental right can be achieved by expanding interpretations of personal rights or limiting the "endurance theory" of torts. Finally, the theory's critics argue that the courts have no business recognizing a new right absent statutory authority. Only special legislation, they argue, can create a new environmental right.[87]

These considerations lend insight into judicial vacillations over the development of this new doctrine. Although the court in the Osaka International Airport case apparently was convinced that recognition of a special new right was not necessary, other courts seem to have equivocated. The courts in the Hanshin Highway and Yoshida cases, for example, in their efforts to find an appropriate remedy to safeguard the health of the plaintiff residents, at least seem to recognize the plaintiffs' legally protected interest in environmental protection. The Yoshida court, of course, did not explicitly rest its decision on the concept of environmental rights, and the Hanshin court evaded a definitive commitment by inventing a new term, "the environmental right to be free of unreasonable encroachments."

Although it is still uncertain whether the courts will someday fully support the concept of environmental rights, the 1972 proposal of the Japan Federation of Bar Associations has already had great political influence. It has at times served to rally residents' groups and has inspired politicians and bureaucrats to take a stronger stand on environmental issues. The idea of an environmental right has also focused the society's attention on the critical present need for legal reform in three areas: the expansion of citizen access to governmental decision-making, the development of procedures to facilitate class actions, and the formulation of new doctrine to address the long-term, cumulative impairment to health that virtually all members of the public suffer from the degradation of the environment. Ironically the vision that has given the environmental rights concept such sweeping force has at the same time constituted its principal stumbling block in attaining final judicial recognition.[88]

3.5.3   The Search for New Remedies

Although the general legal question presented in the four pollution cases was whether the plaintiffs were entitled to monetary relief, the courts were at least equally intent on reforming the chemical industry. But because the plaintiffs only sought compensation and also because of doctrinal limitations, the court's principal means of averting future ills was to blame past conduct. Virtually all the cases in sections 3.3 and 3.4 illustrate how the courts subsequently began searching for new remedies to intercept a polluting activity at an early stage before harm to humans or the environment resulted, and how, in this process, they have also attempted to influence the course of future governmental decisions.

The Hanshin Highway case is important because it is the first of a number of decisions that begin to develop a common law of supplemental remedies despite the absence of a separate and independent equity tradition in Japan. Essentially the Hanshin court used the possibility of an injunction to pressure the defendants to set a numerical limitation on the number of cars permitted to use the highway and to mandate the adoption of other traffic control measures. Despite its good intentions, however, the decision has been criticized as unenforceable. For although the court ordered the Highway Corporation to regulate traffic flow, the corporation possessed no legal authority of its own to do so. And the court failed to provide meaningful guidance. The unfortunate result of this initial ambitious effort is that most of the measures mandated have simply not been heeded. Indeed, in August 1975 another group of local residents again filed suit to block the highway's operation.

The Yoshida Town and Ushibuka decisions are significant for two reasons. First, they illustrate how the Japanese courts have begun to issue injunctions where there is a high likelihood of injury to health. Like the American courts, the two courts stressed the importance of granting relief where monetary compensation would be inadequate. Second, the two cases are interesting for the way the courts have used their injunctive powers to create an array of affirmative procedural obligations nowhere mandated by statute.

Of the two cases, the Ushibuka decision, some scholars argue, is perhaps the more explicit in its identification of at least five procedural obligations.[89] These scholars contend that the Ushibuka court reaffirms the earlier Yokkaichi court's emphasis on pre-siting studies, environmental impact assessments, and alternative plans. But they suggest that the court goes beyond the Yokkaichi decision in its analysis of the defendant's duty to consult. It refines this general obligation by four corollary requirements. The first element is the duty to publicize information; the second is the duty to provide residents with an adequate explanation of the implications of the development; the third is the duty to listen to local opinions, particularly unfavorable opinions, and to consider these feelings carefully in its final decision of whether to undertake the project; the final element is the duty to discuss the specific subjects of compensation, antipollution measures, monitoring, and factory inspection.

Although the Ushibuka case focused principally on procedural considerations, the Kōchi Vinyl court seemed less concerned with dispensing monetary relief and more intent on using the compensation award as a means of compelling local and central authorities to improve management of the river system. The Hanshin court's motive, however, was slightly different. Although it cited the requirement of tendering compensation, the court's principal purpose was to provide an incentive for pollution control that would avoid the harsh results of an injunction. The Kōchi Vinyl and Hanshin cases together truly set the stage for the Osaka High Court opinion.

The Osaka High Court's opinion is interesting from two perspectives: (1) the Court's balancing of the equities in deciding whether to grant an injunction; (2) the remedy of future damages. On the cost ledger, the court initially weighed the extent and severity of the victims' injuries, the long period they suffered, and the prospect that their injuries would continue unmitigated. The court then evaluated the benefits to the public (both domestic and international) of uninterrupted operation of the airport. The court next refined its analysis in two respects. It accounted for the harm to the many people residing in the surrounding exposed metropolitan areas, some of whom were not parties to the suit; it also noted that the community's injuries were not due to any contributory fault but solely to the government's failure to select a suitable site for the airport.

The basic question for the Western analyst presented by the court's opinion is: Was it appropriate to align the harms (costs) to the plaintiffs and the community against the (public) benefits conferred by the facility? Or rather, should the court have refrained from attempting any assessment of the harms (costs) to the general community? We believe the High Court's analysis was correct. The principal theoretical criticism of the court's approach is that whereas the defendant may be in a good position to present evidence on the benefits of the airport, only third parties not before the court can adduce reliable evidence on community injuries.[90] The court's approach risks becoming an expensive "fishing expedition." Although this danger was not entirely eliminated, the court's own on-site investigation, which gave the judges personal experience with the exposed areas, somewhat undercuts this argment.

The reader will recall that the High Court awarded "future damages" in the amount of ¥10,000 per plaintiff per month based on calculation of past damages that was "accurately predictable." The State was instructed to pay ¥10,000 until it met the court's order prohibiting night flights after 9:00 P.M. Thereafter the court ordered the defendant to pay ¥6,000 per plaintiff per month until the "concerned parties make a sincere and positive effort to find an appropriate and practical solution to the dispute."

The benefits of the court's approach are readily apparent. First, the order provided both an economic and "moral" incentive for the government to implement noise control measures.[91] This was necessary because the court probably had no power to enforce its own injunction. The award of future damages also exerted some "moral" influence in that it publicized the illegality of the defendant's conduct. Second, the award of future damages provided a source of individual relief. Third, perhaps most important, the remedy encouraged bargaining.[92]

At the same time there are several important difficulties with the court's approach. Although the continuous monthly payment of future damages may create a greater incentive for control than a lump-sum award,[93] it is clear that the remedy itself only reflects part of the social costs imposed by the defendant's activity. To the extent that marginal damage costs still exceed control costs the government will theoretically be operating inefficiently. This problem, however, is mostly alleviated by the additional payments the government has independently made through mediated and other settlements. Similarly it is uncertain whether the remedy of future damages averts further litigation because the plaintiffs are not barred from filing other suits to increase the amount of compensation. The court's exhortation to the parties to settle, the ministrations of the Central Committee on Dispute Settlement, and the government's accelerated program of antinoise measures, however, will all undercut litigiousness.

More complicated, however, is the court's reluctance to fix a termination date on future damages or to limit the airport noise level to a fixed standard. At the trial the plaintiffs asked that monthly future damages be awarded until the airport noise level was reduced to 65 phons. The court's rejection of this proposal was probably theoretically correct if, as the defendant alleged, the standard's attainment was technically impossible, or if the government's failing to achieve the standard would have forced the closing of the airport. The court's reluctance to set any other precise limitation is at least understandable since this technical decision seems more within the competence of the administrative agencies. At the same time the benchmark adopted by the court for terminating future payments ("sincere and positive efforts") must certainly seem to the Western reader unduly vague. Even in Japan this standard, although subtly preserving for the parties the greatest leeway of choice, is uncertain and invites misunderstanding.

Future damages also raise difficult and complex questions from a strictly economic perspective. Is the court's award of future damages on a *continuing* basis (as opposed to a lump-sum payment) inefficient because it discourages, or at least does not facilitate, the plaintiffs' leaving the area? Because a lump-sum payment would presumably be greater than the court's pro rata award, victim plaintiffs would at least be given the financial means to relocate.[94] The court's award, by providing a continuing incentive to remain, becomes in effect a license fee paid by the defendant to reimburse the sacrifice of the victims' health. The Osaka International Airport case will continue to serve as a tragic living laboratory to test the human applications of economic theory.

3.5.4   Recent Developments: The Third Generation

Although the number of third generation cases is still too few to permit assessment, some developments should be noted. In February 1978 a nine-member group filed suit in the Tokyo District Court demanding that a thirty-story office building opposite Tokyo's Hibiya Park be scaled down by over half because the building considerably marred the park's beauty by blocking sunshine. The plaintiffs, including environmentalists, local residents, housewives, and others, alleged that the "building would cause strollers the damage of being looked down on by people in the structure," thereby constituting an infringement of the "citizen's right to enjoy the park." The suit was the first judicial challenge of a building construction based on a citizen's "common" right to sunlight. On May 31 the Tokyo District Court rejected the plaintiffs' arguments holding that they did not have standing.[95]

Local residents in the Tagonoura case also scored an important victory when the Tokyo High Court ordered four major paper manufacturing firms to repay ¥10 million to Shizuoka Prefecture for work conducted with public funds to remove sludge from the port. Despite the fact that pollution of the port had preceded many water pollution laws, the court ruled that "the citizens of Fuji have a valid reason to seek damages from the companies in place of the prefectural government."[96]

In another suit, still unsettled in 1980, some 1,200 residents of the Kinki area (near Kyōto) have asked the Otsu District Court to enjoin the massive Lake Biwa Comprehensive Development Project. The plaintiffs' principal allegation is that the state and various local authorities have failed to conduct a proper environmental assessment and to provide various planning safeguards and, as a result, implementation of the project will endanger public health, pollute Lake Biwa irreversibly and destroy aquatic life, recreational areas, and invaluable archaeological sites. The Biwa case epitomizes the concerns of contemporary environmental litigation and the public's demand that government plan both wisely and well.

Section 4   Recent Judicial Developments in Administrative Law

4.1   Creation of New Environmental Rights and Remedies in Administrative Litigation (The Second Generation Suits)

As in civil litigation, the plaintiffs in administrative suits have asked the courts to recognize new rights and remedies as the primary means of upsetting or modifying environmentally hazardous governmental actions. But the objective is slightly different. These plaintiffs are more concerned with basic reforms. They seek minimum procedural guarantees of public hearings and access to information and on occasion have requested the courts to review the very substance of a ministe-rial decision. Although the reader will see that the plaintiffs in some cases have carved inroads into the vast traditional domain of administrative prerogatives, the way has been hard and success infrequent.

4.1.1   **Development of Due Process in the Environmental Field**

Matsuyama Airport Case[97]

Abstract of the Case

[*The governor of Ehime Prefecture granted a permit to A to reclaim land in order to construct a runway for the Matsuyama Airport.*[98] *The Ōkaga Yoshida-hama Fishery Association, the appellant, consisted of seventy-three fisher-men who possessed fishing rights to the reclaimed area and to other areas.*[99] *The association alleged that over twenty of its members and their families based their livelihood principally on fishing in the area of this reclamation and nearby areas. Because the reclamation would seriously and irrevocably damage their fisheries, the association sought suspension of the execution of the reclamation approval.*

The Matsuyama District Court ordered the suspension of the approval, holding that art. 31 of the Constitution[100] *required that the governor give notice and an opportunity to be heard prior to granting approval of the reclamation. Because he had failed to do this, the court found the approval illegal.*]

Opinion of the Court

. . . That there is a dispute over whether art. 31 of the Constitution should be in-terpreted to apply to administrative procedures and to the divestiture of property is common knowledge.[101] *However, this court thinks it appropriate to affirm the applicability of art. 31 to these cases. Although art. 31 applies both to adminis-trative and criminal procedures, criminal sanctions generally impose greater hardship on the person suffering sanctions than do administrative acts. [On the other hand], the business of administration requires speed and dispatch. Thus, the substance of due process mandated by art. 31 itself will differ between the two (types of) procedures. In other words, ordinarily, observance of the due pro-cess mandate in administrative actions can be more relaxed than in cases where punishment is imposed. Also, with respect to due process guarantees relating to the divestiture of property, rational distinctions can be drawn depending on the degree of benefit to be divested, the urgency for administrative action, and the specific substance (of the action). Accordingly, during emergencies, in cases of temporary divestiture of property of minor value, due process guarantees can on occasion be considered unimportant.*

There are many interpretations of the content of due process. However, it is the opinion of this court that when an action is adverse to someone, it is basic that notice and an opportunity to be heard be given. (The content of the hearing is of course an additional issue. But there can be no doubt that at a minimum a hearing should allow the party to state an opinion orally or through a memoran-dum.) In this way the people are given some opportunity to protect their rights, while the just conduct of administrative decisions is assured. The Supreme Court too has made the following comments regarding a quasi-criminal penalty that resulted in the confiscation of the possessions of a third party during a judi-

cial proceeding: "Absent notice, or any opportunity for an explanation or defense to the owner of the property, the deprivation of ownership rights by confiscation is extremely unreasonable, and must be said to be constitutionally intolerable." (Supreme Court, Decision of November 28, 1962, 16 Keishū (No. 11) 1953)

It is clear that the instant case involves an administrative action materially disadvantaging the appellant owners of fishing rights in the reclamation area. The license (approval), granted under the Public Surface Waters Reclamation Law,[102] does not in itself directly destroy the public's [rights in the] surface water or extinguish fishing rights. But because the fishing rights to the surface water are extinguished by the execution of the reclamation based on this law, the license (approval) action becomes an important administrative act that leads to grave results such as the extinction of fishing rights. We cannot say that fishing rights are not property rights requiring due process guarantees against their deprivation, especially in the instant case where these fishing rights attach to a wide stretch of water surface and where over twenty members of a fishermen's union and their families derive their livelihood principally therefrom. From this perspective, due process guarantees should attach to their deprivation as a matter of course. We have insufficient data demonstrating a situation of urgency such that the purposes of the administrative action would be frustrated if the governor had undertaken these procedures prior to granting the license.

We have here a problem of due process guarantees with respect to the license under art. 4(3) of the Public Surface Waters Reclamation Law[103] because this action was based on laws and regulations entirely lacking in procedural protections for the rights of the interested parties. That is, since the action was taken under conditions where no rules provided for notice or an opportunity to be heard, appellant holders of fishing rights have been materially disadvantaged by this administrative action.

It is easy to understand how the provisions of the Public Surface Waters Reclamation Law are deficient when we consider the following: The Land Expropriation Law[104] provides scrupulous provision for the protection of the rights of interested parties. Similarly these other laws contain provisions of notice to the affected party. (See permit actions under art. 40(1)(1) of Riparian Law[105] which bear a close resemblance to license actions under art. 4(3) of the Public Surface Waters Reclamation Law. Riparian users under 40(1)(1) include fishing rights holders; compare these to Riparian Law Enforcement Order art. 21;[106] also actions under art. 39(1) of the Fisheries Law[107] that provides for alteration, cancellation, or suspension of fishing rights upon public necessity; also compare the legal provisions in law for prior notice to interested parties and an opportunity for the presentation of views in arts. 38 and 39 of the Riparian Law[108] and arts. 34 and 39 of the Fisheries Law.)[109]

As set forth above, due process guarantees need not be recognized in administrative proceedings on as high a level as in criminal proceedings. Consequently, when an administrative action is undertaken in circumstances where there does not exist a specific statutory guarantee, the demands of art. 31 of the Constitution will be met even when notice and an opportunity to be heard are given only to those materially disadvantaged by the action. However, in the instant case, there is insufficient data to conclude that notice and an opportunity to be heard were given to the appellants.

According to the evidence in this case, the appellants did have an opportunity to conduct negotiations for compensation for their fishing rights with the responsible officials of Ehime Prefecture, Matsuyama City, and the no. 3 Harbor Con-

struction Bureau, in connection with the construction of the Matsuyama Airport runway. We recognize that such negotiations were actually conducted numerous times. Nevertheless, according to the evidence, it is hard to say that these negotiations were conducted in connection with the licensing (approval) action under art. 4(3) of the Public Surface Waters Reclamation Law. Rather, we consider them no more than contacts undertaken to solicit the cooperation and agreement of the appellants, who possessed fishing rights in the surface waters in order to implement the landfill construction smoothly. It is hard to see these contacts as providing the appellants with notice and an opportunity to be heard in connection with the action in this case consonant with due process. Therefore, we suspect that the action in this case violates art. 31 of the Constitution.

4.1.2   Environmental Considerations in Administrative Decision Making at the Local Level

Usuki Cement Case[110]

Plaintiffs: Hideo Sudo and fifteen others.
Defendant: Governor of Oita Prefecture.
Intervenor: Usuki Cement Company.

Judgment

(1) The defendant's directive issued as Fishing-Port Directive No. 889 on December 25, 1970, permitting the intervenor to reclaim 7,579 square meters of the public water surface adjacent to the shore line from Dojiri 85-5 to Nishinoyama 477(1) of Usuki City is hereby reversed.
(2) The defendant shall pay the costs of this litigation and the intervenor shall pay the costs of intervention.

Opinion of the Court

1. Administrative Disposition and Standing to Sue. *The parties do not dispute the fact that on December 25, 1970, in accordance with the Public Surface Waters Land Reclamation Law (hereinafter Land Reclamation Law), the defendant granted intervenor a permit to reclaim land under the public surface water for a site for a cement manufacturing factory. It is also not disputed that the plaintiffs are fishermen and are members of the fishermen's union, which holds a cooperative right to fish in the public waters in question, and that according to union regulations on the exercise of this cooperative fishing right, the plaintiffs are entitled to fish in the fishing grounds which include the public surface waters in question. Since the plaintiffs are entitled to fish in these waters, they have standing to sue for the cancellation of the permit issued for the reclamation of public surface waters. (42)*
2. Persons Who Have Rights to the Public Surface Waters. *The Usuki fishermen's union was established in accordance with the Cooperative Fishery Union Law and was granted by the defendant on January 1, 1964, rights of Types #1, 2, 3 to fish in the public surface waters in question.*

The first question is whether the cooperative fishing right granted by the defendant was partially waived as to the public surface waters in question.

Waiver constitutes a partial abandonment of the fishing right. According to art. 50(1) of the Cooperative Fishery Union Law,[111] *a resolution to relinquish the right at the general meeting of the union is required. Over one-half of the regular members must be present and a vote of more than two-thirds of those present must be recorded. The right to fish belongs legally to the fishermen's union. But the union cannot itself engage in fishing (art. 11(1) and 17 of the Cooperative Fishery Union Law). It merely holds the right for the individual members of the union who actually engage in fishing. These members enjoy the benefit of the right because of their status, defined by the union's regulations on exercising the fishing right. A member of a fishermen's union must be a fisherman who engages in fishing for more than 90 days a year (art. 18(1)(i) of the Cooperative Fishery Union Law). The status of individual members is defined in the union regulations and is important in determining who enjoys the benefits derived from the fishing right. It takes a special resolution to change the union regulations regarding the exercise of a fishing right (the Cooperative Fishery Union Law, art. 50(5)). This provision alone, however, cannot serve as an adequate safeguard of the interests of the minority members in the case of the majority's abridgement of their rights by amendment of the regulations. The risk of domination by the majority increases when the union's size increases by merger or otherwise. In order to minimize the risk to the minority, the Fishery Law requires, in the case of amendment of the regulations, the approval of the administering agency (art. 8(4) of the Fishery Law) and the written consent of at least two-thirds of the union members who engage in type 1 fishing and who live in the adjacent beach areas. The law is designed to protect the interests of individual members who actually engage in fishing. A waiver of the fishing right may be a different matter from amendment of the union regulation. However, by waiving the fishery right, the individual members of the union lose their right to fish (Fishery Law, art. 8(1)). In this respect it has the same effect as an amendment of union regulations. Moreover, if the fishing right is once waived, its effect is more lasting and certain than amendment of the union regulations because it terminates the fishing right altogether. (43)*

In the case before us there were 129 members who engage in type 1 fishing and live in the adjacent beach area. We cannot find from the records that more than 86 (or 2/3) of 129 members consented to the waiver in writing. (Although it is possible for clearly expressed oral opinions to serve in lieu of written consent, we are unable to find in the entire records such an unmistakable expression of opinion by the members.) As we cannot find that the decision to waive the cooperative fishing right (Kyō No. 30) was supported by more than a third of the qualified members, we hold, without going into other issues, that the right was not extinguished and remains as to the entire area of the designated fishing grounds.

Therefore, unless requirements provided in art. 4 of the Land Reclamation Law are satisfied, the permit should not have been issued. We examine the requirements to determine whether they have been satisfied. (44)

3. The Consent of the Fishing Right Holders. *On March 23, 1970, Terutoshi Sasaki, a board member of the fishermen's union drafted a document and submitted it to the defendant through the intervenor, indicating that the union approved of the proposed land reclamation plan for the public surface waters in question.*

According to the Land Reclamation Law, when a fishing ground is filled with earth, the cooperative fishing right in the affected water is also extinguished, and the rights of individual members of the union are extinguished. Therefore, when the union gives its approval, as required by art. 4(1) of the Public Surface Waters Land Reclamation Law, it must do so by a special resolution in accordance with art. 50 of the Cooperative Fishery Union Law, and the resolution must be carried by more than a two-thirds vote of the members who engage in type 1 fishing and live in the adjacent beach areas as provided in art. 8(5) and 3 of the Fishery Law. In other words, the same procedure required for the waiver of the fishing right discussed under 3 must be followed. As we have already found, these procedures were not followed by the union. The approval submitted by the board member, therefore, does not have the effect required by art. 4(1) of the Land Reclamation Law. (45)

4. Benefits and Injuries Arising from Land Reclamation. Next, we determine if the issuance of the permit by the defendant was in accordance with the require-ments of art. 4(2) of the Land Reclamation Law. The land in question is intended for the site of the intervenor's cement manufacturing factory. Article 4(2) provides that a permit to reclaim land in public surface waters may be granted, even if there is someone who holds the right to use those public surface waters, when "the benefit arising from reclaiming land would exceed by far the injuries therefrom." This provision specifies a case where a permit may be granted for re-claiming land even if the holder of the right to use the public surface waters does not give his approval and the land is not intended for an enterprise which is primarily in the public interest. From the wording and context of art. 4, it is ap-parent that the requirement must be strictly interpreted. When a private com-pany applies for a permit, it is not enough to compare the economic benefits ex-pected of the reclaimed land and the production of factories built on it with the economic loss of the holders of rights to use the public water. An examination must be made of the existence, degree, and extent of adverse effects of the con-struction of factories upon the life and environment of residents in the area. [em-phasis added] *Also, the direct as well as indirect loss to holders of rights must be accounted for accurately. The permit may be granted only in cases where, in light of comprehensive land use plans and national economic needs, it is objec-tively clear to everyone that benefits arising from the reclaiming of the land far exceed losses arising therefrom and that it is not at all unreasonable to terminate the existing rights and incur such losses. In the case before us, it would cost ¥17 billion to reclaim the land which has an estimated value of ¥13 billion. The intervenor plans to build a cement factory at a cost of ¥170 billion which is ex-pected to produce 20,000 tons of cement per month for the first two years and 25,000 tons thereafter. The factory will require 120 workers for the first two years and 150 thereafter. The intervenor expects to gain annual profits of ¥400,000,000 for the first two years and ¥800,000,000 thereafter. The proposed site is close to good limestone mines. The factory operation is expected to stimulate economic activities in related industries in the area and produce a good ripple effect in the economy. The factory is also expected to be a source of tax revenues for the local governments.*

On the other hand, the factory may create a labor shortage for other industries in the area. Many questions as to the extent of benefits from the factory operation for the area still remain to be examined. It is foreseeable that the construction, operation, and related activities of the factory will also produce air and water pollution and adversely affect the life and environment of the area. The inter-venor has concluded a pollution prevention agreement with the city and prefec-tural government. The agreement provides that the intervenor shall make every effort to prevent air pollution from smoke and SO$_2$, water pollution, and noise, *and that the company shall faithfully respond to and negotiate with the parties about compensation if damages for which the company is deemed responsible should occur. However, the agreement alone cannot offer adequate measures for preventing pollution.*

As we have determined, the union possesses a cooperative fishing right to the public surface water proposed for the factory site. Many union members have engaged in fishing in the designated water. However, the defendant, when granting the permit, grossly underestimated the amount and market value of fish caught in the designated water, and considered only that it was used as a bathing beach in the summer. There is no evidence that the defendant ever made any investigation as to the direct and indirect losses the fishermen might incur. The defendant offered an annual figure of ¥16,570,000 for fish caught in that water. The defendant submitted no evidence to support its estimate except one document which was prepared after the litigation began and merely listed the figure with no indication of the sources from which it had been adduced.

As we have found above, the intervenor stands to gain from the reclaimed land and the production of cement, and related industries in the area may also profit. On the other hand, this development will require a large capital investment and will predictably produce adverse effects on the living environment of residents and decrease the income of fishermen in the area. This is hardly a case where it is objectively clear to everyone that benefits from the reclaimed land will far outweigh losses therefrom. There is also insufficient evidence to hold the contrary. We find the defendant's contention that the agency has satisfied the requirement of art. 4(2) of the Land Reclamation Law to be without substance.

5. Conclusion. The permit was issued on the erroneous assumption that the cooperative fishing right of the fishermen's union had been duly waived and contrary to the requirement of art. 4 of the Land Reclamation Law. Therefore, we hold that the permit is null and void. (45)

4.1.3   "Ripeness" of Administrative Action for Judicial Review

Shinkansen Bullet Train Case[112]

Abstract of the Case

[*In late 1971 the Ministry of Transportation published its basic plan for the Narita Extension of the Japan National Railway System. Thereafter the Japan Railway Construction Corporation drafted its construction plan and on February 8, 1972, presented its implementation plan to the minister, who approved it two days later. Plaintiff residents, living within a corridor designated by the plan, challenged the minister's action as an infringement of their property rights and personal rights in tort, alleging that they would most likely be forced to surrender their land; and that even if their land escaped actual condemnation, they would suffer severe physical and psychological harm from noise and vibrations.*]

Opinion of the Court

[*The court held that the plaintiffs' action was not ready for judicial review.*] *At the stage of the approval of a Construction Implementation Plan, it is not yet concretely confirmed who will become an interested party when the plan is executed. In this sense, a Construction Implementation Plan and its official approval must be considered as abstract in nature. In other words, an approval here is unlike a concrete disposition directed at a specified individual. Furthermore, there is no provision that requires its publication and it itself has no effect whatsoever on citizens' rights and duties.*

The approval in this case can be viewed not only as the defendant's endorsement of the . . . fundamental provisions concerning the construction of the Narita line, but also as authorization given to the Japan Railway Construction Corporation to proceed with construction based on the plan. It is therefore internal behavior directed at the above corporation and cannot be said to be a concrete disposition directed at a private citizen or to have any influence whatsoever on citizens' rights or duties. As a legal case, therefore, it lacks the ripeness necessary for a controversy.

[The court later explained that dismissal of the plaintiffs' action was without prejudice to any later suit after their land was actually condemned. This would be adequate legal redress, the court held, for any "concrete infringement of their rights."][113]

4.1.4   Standing in Administrative Suits

Kunitachi Pedestrian Bridge Case[114]

Abstract of the Case

[*The defendant, Tokyo Metropolis, decided to construct a pedestrian bridge over a street known as University Avenue in Kunitachi City. The purpose of this bridge was to provide a safe crossing for students pursuant to the Law Concerning Emergency Measures for Traffic Safety,[115] which governs the installation of public safety transportation facilities. The city hired an appropriate contractor and the bridge was completed in February 1970. It is already in general use.*

The plaintiffs are seven residents of Kunitachi City. Of these, five live within a radius of 1,000 meters from the bridge. Two, who live at a greater distance, have been selected from a group concerned about the future of Kunitachi City. The plaintiffs assert that the act of construction of the bridge violates the Emergency Measure Law and violates the plaintiffs' environmental rights. The plaintiffs request revocation of the decision to construct this bridge according to art. 9 of the Administrative Litigation Law.[116]

The basis of the plaintiffs' claim of standing to contest this matter is that they have a legally protected interest to traverse the street at grade, that such crossing is frustrated by the construction of the bridge, that they are forced to go up the bridge and then down the other side, that these acts have impinged upon legally protected interests, and that these interests form the basis for the action to cancel the act of construction. With respect to their claimed environmental right, the plaintiffs assert that although University Avenue possesses unusual urban beauty, construction of the bridge has disrupted the beauty of the environs, has stimulated an increase in traffic, will give rise to traffic accidents, and will produce air pollution from the exhaust fumes of the automobile traffic.]

Opinion of the Court

The plaintiffs' claim is dismissed.

The decision concerning the issue of the plaintiffs' standing is as follows:

Certainly, even if we recognize that we cannot ignore the complaints of the pedestrians prevented from a level crossing and forced to go up and down this bridge, and even if such individuals are deemed to have a legally protected interest to contest an obstruction where the inconvenience exceeds the limits of human tolerance, [this does not suggest] that ordinary citizens who do not actually experience these circumstances have a legally protected right to contest these matters. The plaintiffs allege that because they live in the vicinity of University Avenue they must cross over this road for work, shopping, and walking in their daily activities. However, they do not argue nor have they proved that it is a specific daily necessity for them to traverse this road at or near the point where the bridge was built.

[The following is the court's decision with respect to the claim of an environmental right.]

We recognize that the maintenance of the human environment, i.e., sunlight, air, water, and quietude, in a wholesome state is an absolute necessity for a healthy, comfortable life. And when these (amenities) are destroyed beyond the limits of tolerance, human life and health may be also infringed. Even if we recognize a legally protected interest that the environment not be so degraded, we cannot infer that one has a legal interest in environmental protection in the sense of the term "legal interest" as used in art. 9 of the Administrative Litigation Law, so long as the degree of degradation does not exceed the limits of tolerance.

[From this basic premise the court then specifically analyzed the plaintiffs' claim of a violation of their environmental rights and held that "the question of whether a pedestrian bridge, average in scale, size, and condition, violated the aesthetic surroundings was perhaps a subjective and emotional one." Moreover, the court held that it did not believe that the bridge had "destroyed the environment beyond human tolerance." Consequently, the court also refused to recognize the plaintiffs' claim that the introduction of the bridge would also increase the risk of traffic accidents and air pollution from car exhausts. In refusing to recognize the legal basis for the plaintiffs' claim for cancellation, the court held that the plaintiffs lacked standing and dismissed their complaint.]

Date Electric Power Plant Case[117]

Abstract of the Case

[Hokkaidō Electric Power Company (intervenor) decided to construct facilities for an intake water cooling system, a water channel, cargo facilities, a breakwater, and an outfall in the city of Date. Upon application by the Hokkaidō Power Company, the governor of Hokkaidō (defendant) on June 25, 1973, granted a landfill permit in the vicinity of Chōwa Town in preparation for the construction of the proposed facilities.

The plaintiffs, who are members of the Date fishermen's union and the Usu fishermen's union, challenged this permit as illegal, alleging that it would cause irrevocable harm, and demanded that the permit be voided. The bases of the plaintiffs' claims are as follows:

1. The fishermen's union possesses a cooperative fishing right to the superjacent waters. A prior waiver by the union of part of its fishing rights was invalid.

2. The union was not given the opportunity of a hearing.

3. Adequate scientific investigation of the site had not been made.

4. The environmental rights of residents, including the plaintiffs, would also be violated.

In response to the plaintiffs' allegations, the defendant made the following assertions:

209

1. The plaintiffs did not have standing.
2. The instant case was neither one of irrevocable injury nor of extreme emergency requiring prevention (see Administrative Litigation Law, art. 25(2)).[118]
3. The great impact on public welfare asserted by plaintiffs was questionable.]

Opinion of the Court

The plaintiffs' request for suspension of the intervenor's construction works is dismissed.
(A) The Issue of Plaintiffs' Standing. *The issue of whether one may ask for cancellation of an administrative ruling (a suit for cancellation) is dependant on whether one has a legally protected interest to make such a request for cancellation (Administrative Litigation Law, art. 9). Those who have such legally protected interests must be those whose rights or legally protected interests are violated by the ruling. Where an infringement of such rights results from administrative ruling we do not question whether the individual affected was originally a party to the ruling or a third party. Consequently the plaintiffs have standing to raise the issue of cancellation. In the present case the plaintiffs were not parties to the present ruling, and, as we will describe below, because the plaintiffs' rights were infringed by the ruling, we hold that the plaintiffs have standing to raise the issue of cancellation of the ruling and consequently have standing to petition the court to enjoin implementation of the ruling. [4] A fishing right is an absolute right to operate a fishing business in a designated fishing zone. It is created by a permit granted by the prefectural governor (Fishery Law, art. 10).*[119] *It is in the nature of a real [property] right (Fishery Law, art. 23(1)).*[120] *The fishing right belongs to the fishing union or to the fishing union federation. Fishermen who are members of the union have only those rights which conform to the regulations pertaining to the exercise of fishing rights as adopted by the union or federation (Fishery Law, art. 8(1), 8(2)).*[121] *Consequently, none of the fishing rights belong to the individual fishermen. Theirs are derivative rights. But these rights are protected in their entirety; they must be distinguished from de facto or reflective rights.*

If we consider the relationship between the scale and nature of the area that has been designated for reclamation in the present case and the marine zone in which the plaintiffs conduct their fishing operations one cannot deny, based on our common sense, that pollution produced by the land reclamation in the area under construction will not be limited to this marine area alone but rather will disperse and spread out to adjacent marine areas. Similarly we are unable to judge whether such pollution can be kept from entering the forty-eighth fishing zone, which is adjacent to the marine area under construction. (The forty-eighth fishing zone is that zone in which the Date plaintiffs conduct their fishing operations.) We also cannot say that there is absolutely no risk that pollution from the area under construction will not extend to the proximate forty-seventh fishing zone, which is the area operated by the Usu plaintiffs. Since if pollution should occur, it would affect the fishing operations of plaintiffs in each area, adversely affect fishing operations, and cause a reduction in the catch, we hold that the right of the plaintiffs to conduct fishing operations could be infringed by the ruling in the present case. (5)
(B) The Issue of Irrevocable Injury.
1. Damages to Fisheries from Construction work. [*The court examines the evidence of potential damage to fisheries from such construction.*]

210

In sum, after our comprehensive consideration of the degree, time period, and ease or difficulty of recovery from the impact of reclamation works, the adverse impact, if any, will not be irrevocable. For this reason it cannot be said that the harm from water pollution would be as serious as provided in art. 25(2) of the Administrative Litigation Law. (7)

[2. Safety of Fishing Operations. . . .]

3. Injury from Outfall and Thermal Effluent. *When the Date Power Plant is completed and operations begin, cooling water will be drawn into an intake pipe constructed on the reclaimed land. Thereafter it will be discharged in a heated state from outfalls constructed along the coast about 1 kilometer to the southeast of the reclamation area. These facts are clearly indicated on the materials presented in this case. If we compare this thermal water with the temperature of the surrounding marine waters, we can see that in summer it will be higher by 5°C and in winter by 7°C; it is predicted that the outfall will be 22 cubic meters per second in amount. It is clear that there will be some influence from this thermal water on the fishing operations in the marine areas adjacent to the outfall (these areas are those zones currently fished by the Date plaintiffs). It must also be said that there is risk of adverse impact from the effluent and thermal water upon those areas currently operated by the Usu plaintiffs, although those areas are somewhat distant from the outfalls.*

The harm caused by the construction and by the facility to be constructed on the reclaimed land can be characterized as injury directly resulting from the administrative ruling in the instant case. Nonetheless, we cannot say that the harm was anticipated from the instant administrative ruling. It may appropriately be regarded as not within the harm contemplated by art. 25(2) of the Administrative Litigation Law. Of course, one might say that if there were no reclamation, no intake or effluent would occur even if other facilities of the Date Electric Power Plant could be built. Thus in this sense the intake and discharged waters can be related to the administrative ruling. However, intake and effluent do not usually accompany the reclamation of marine areas. The administrative ruling in this case is only a means of granting authority to the intervenor to conduct reclamation. It does not directly relate to the possibility of intake and discharge of water. Administrative regulations have already been established elsewhere with respect to the prevention of pollution in marine areas resulting from such effluent. The Water Pollution Prevention Law[122] regulates water pollution generally. Article 23(2) of the Electric Power Enterprise Law,[123] however, provides that power facilities, which are designated in art. 2(7), do not have to comply with the administrative regulations in the Water Pollution Prevention Law but they are regulated by the Electric Power Enterprise Law. The establishment of power facilities is regulated through permits for construction plans under art. 41(3). In addition art. 48 and 49 of the same law provide that electric power companies shall have a duty to follow standards regarding the maintenance of the facility. Should the power company fail to comply with this duty, the Ministry of International Trade and Industry may order the termination of the company's use of the facility and is empowered to regulate its use.[124]

The adverse impact resulting from the intake or effluent would seem to be closely related to these administrative regulations. In the final analysis, we cannot say that this ruling was the direct cause of the injuries anticipated from the intake and discharge of water. (8)

[4. Injury from the Reduction of the Fishing Area. . . .]

5. Infringement of an Environmental Right. *The plaintiffs allege that as a result of*

*the reclamation in the present case the physical beauty of the land to be re-
claimed and neighboring marine areas and the natural environment will be destroyed, and that plaintiffs' environmental rights will be infringed. However, the
content of the plaintiffs' so-called environmental right has not been defined.
Apart from this point, as noted above, the reclamation will be on a comparatively small scale. Moreover, if we compare the natural environment of the
marine areas and the coastal areas adjacent to the land to be reclaimed with
other areas along the Funka Bay coast, the former cannot necessarily be held to
be better because human waste facilities, factories, and railroad tracts have already been built in this area. Nor can we necessarily say that the breakwaters to
be constructed on the east and west and the reclamation necessarily will destroy
the scenery or natural environment. For the above reasons we cannot say that
the reclamation in the present case will produce irrevocable harm to the scenery
and living environment. Moreover, with regard to possible harm resulting from
air pollution accompanying the operations of the Date Electric Power Plant, we
hold that the administrative ruling in the present case bears no direct relation to
such harm in much the same way as we have explained earlier in the context of
thermal water from the plant. (8)*

4.1.5   The Incorporation of Environmental Values in Administrative Decision Making at the National Level

Nikkō Tarō Cedar Tree Case[125]

Abstract of the Case

[*On May 22, 1964, the defendant Minister of Construction approved an application by the Governor of Tochigi Prefecture, the second defendant, for a public
works project that involved an expropriation of land owned by a religious
cooperative, Nikkō Tōshōgū Shrine. The purpose of the project was to improve a
national road running through the Nikkō area. The minister's approval was
given in accordance with art. 20 of the Land Expropriation Law (1951). On May
26, in accordance with art. 33 of the law, Tochigi Prefecture posted a public
notice identifying the land to be expropriated. The Nikkō Tōshōgū Shrine filed
suit to revoke the actions taken by the aforementioned defendants. However, the
Expropriation Committee of Tochigi Prefecture, the third defendant, decided on
February 18, 1967, to execute the expropriation. The shrine filed a suit to revoke
this action. The Utsunomiya District Court consolidated the shrine's suits and
rendered a decision in 1969 holding that the minister's approval was unlawful
and, therefore, that subsequent actions by the prefecture and the Expropriation
Committee were also unlawful.*

*National Road No. 120 narrows in width from 16 meters to 5.7 meters at the
point in question where the road is sandwiched on the left by the Shinkyō Bridge
(God's bridge), which crosses the Ōtani river and is designated as an important
cultural property, and the Tarō cedar in the hillside on the right. The proposed
project was to widen this bottleneck in the road to 16 meters, so that traffic
could flow without congestion. To do this, however, the Tarō cedar (1.75 meters
in diameter, 40 meters high, and estimated to be 600 years old) and fourteen
other trees of similar caliber in the area would have had to be sacrificed.*

*On appeal the Tokyo High Court sustained the District Court's decision and
decided for the plaintiff.*

The plaintiff contended that the public works project in question was in viola-

tion of art. 20(3) of the Land Expropriation Law which provides that the "work project must contribute to an appropriate and reasonable use of land."[126]

The court held that the requirement of the article is satisfied when public benefits outweigh losses (public as well as private) resulting from the utilization of land by a public works project. The court admitted that the Minister of Construction has a discretionary power to decide whether or not a public works project contributed to an appropriate and rational use of land. Yet it was illegal, the court held, for the minister unreasonably not to take into consideration the most important factors in reaching his judgment. The court's reasoning in weighing these factors was as follows.]

Opinion of the Court

1. The project was approved by the Minister of Construction as one that would "contribute to an appropriate and rational use of land." In reaching this judgment, however, the defendant underestimated such significant factors as various cultural values and the environmental beauty of the land. The land in question is situated at the entrance to the Nikkō National Park. It is an awesome composition of man-made beauty with the Shinkyō Bridge painted in red, shrines, the natural beauty of the surrounding forest of huge cedars, and the crystal stream of the Ōtani river. The area has been designated as a special scenic preserve.[127] It is also known as a place of religious and historical significance, since it is the site of the Nikkō shrine and one of the original places of religious worship. The Tarō cedar tree and other huge cedar trees are considered to be of as great cultural value as those cedar trees along the old Nikkō trail, which has been designated a special historical site.[128] The existing state of scenic beauty surrounding the special preserve area in the National Park should be maintained and preserved with utmost care as a national cultural heritage. The cultural heritage observable in Nikkō is the creation of a long natural and historical process. It would be practically impossible to restore the area once it were artificially changed. Although the plaintiff has private ownership of the land, this land also has scenic, religious, and historical value which should be shared and preserved for all the people as their common cultural heritage.

Should the road expansion plan be implemented, a considerable portion of the hillside opposite the bridge will need to be cut away, thereby destroying a dozen huge cedar trees and the site of a shrine. A two-level stone wall of 3 and 5 meters high and 40 meters long will be constructed in their place. Obviously the existing state of scenic, historic, and cultural value of the area would be drastically and seriously altered. The defendants' (proposed) artificial restoration plan to be initiated after the completion of construction cannot remedy such destruction nor restore the natural and scenic beauty of the area.

It is also easily predictable that the increasing amount of traffic which would result after the completion of the expansion would inevitably further violate the serenity of the area.

2. Therefore, it is not sufficient for the Minister of Construction to identify some public benefits deriving from the project such as the accommodation of increased traffic. He must also find the project to be necessary even at the sacrifice of the scenic, historical, and cultural value of the area, and worth the cost of such environmental deterioration and destruction. The preservation of the environment and these values should be given the utmost consideration by the administrative agencies involved because these are factors that provide the people with a healthy and culturally satisfying life.

A road may be constructed by human initiative when necessary. It is replace-able in the sense that one can build it anywhere and at any time by expending money and time. In the instant case, the Governor of Tochigi Prefecture had four plans, each proposing a different route. He decided on plan A in question because it was the least expensive, technically easiest, and required the shortest con-struction period. The existence of alternative plans demonstrates that it was pos-sible to construct a road meeting the increasing demand of traffic flow without disrupting the precious values in the area. Needless to say, there would be time and budgetary limitations to be considered. According to the estimate by the governor, the most expensive plan C would cost ¥1,351,000,000, but the adopted plan A cost only ¥43,000,000.

We wonder if plan C would be too expensive from the point of national econ-omy if the benefits of the preservation of the cultural value of the area were con-sidered. We do not agree that the defendants have no alternative but to adopt plan A for economic reasons. For example, it might be possible for them to oper-ate the road in plan C as a toll road.

3. The defendants have argued that the proposed expansion would still be neces-sary even if plan C or other alternate routes were constructed, since tourist and business traffic to the shrines in Nikkō are projected to increase in any event.

The road in question is obviously unsuited for further expansion for the pur-poses of industrial development or tourism. It runs through the area which is considered a gateway to the Nikkō National Park and should be left undis-turbed. If some other road were to be constructed to serve those purposes, the necessity of expanding the present road should certainly be reconsidered. If the preservation of cultural values and the environment in the area were given the highest consideration, the view that automobiles should be excluded from the road and the road converted into a pedestrian walk might prevail. It is fool-hardy to destroy the very resources of tourism by expanding the road in the name of developing tourism. (37)

The decision to expand the present road takes on added significance when considered in light of the fact that construction of other main roads was also proposed. The Minister of Construction should review the necessity of the project after a clear formulation of the policy concerning whether other roads should be constructed in the near future for industrial development or tourism in the unde-veloped area behind Nikkō and after a consideration of the significant value of the preservation of the environment of the area. The defendant Minister of Con-struction has approved the project without formulating a clear policy or plan for the construction of a separate road and without adequate consideration of scenic, historical, and cultural values. As a consequence, the minister has significantly underestimated these values and their importance. We hold that there was an error in the procedure by which he exercised his discretion. The defendant did not undertake a full deliberation of (all) possible means to recon-cile the needs of solving traffic congestion with the importance of the preserva-tion of the environment.

4. One of the reasons for the road's expansion noted by the governor in this appli-cation was the need to cope with the anticipated increase of traffic in an Olym-pic year. The Minister of Construction listed as the justification for his approval of plan A, among several contemplated plans, the short period required for the construction. We infer from the defendant's contention that the anticipated in-crease of traffic from foreign tourists during the Olympic year to be one of sev-eral factors considered by the Minister of Construction in his examination of plan A. However, from the perspective of the cultural and historic values of the area and the significance of the environment as a national heritage, a temporary increase of traffic anticipated in the Olympic year is a short-term and transient phenomenon. It should not be a factor to be taken into consideration in deciding the appropriateness of a given land use.

5. Before the project was approved by the Minister of Construction, the National Park Council voted for the expansion of the road with the condition that the cutting down of the cedar trees be limited to a minimum and that necessary restorations would be performed upon the completion of the road. The council chose plan A after examination of all plans. However, their considerations were influenced by the desolate state of the area after a storm in March of 1963 in which many trees had been blown down. In making their choice, they apparently presumed that the desolate state would remain unchanged, and this perception obviously influenced the judgment of the Minister of Construction regarding the project. However, we doubt from our own on-site investigation that the council's presumption has proved to be right. Moreover, in view of the necessity of preserving the natural state of the environment and the significant value of the area, factors such as the risk of falling trees in storms (and the risk of their blocking the road) and the declining state of old trees should not be overestimated in deciding whether a project would contribute to a rational use of the land. (38)

6. In deciding the issue of the rational use of land, the Minister of Construction significantly underestimated the importance of cultural values of the area and of preservation of the environment. Consequently, the defendants failed to consider all possible means of reconciling the need of preservation of the area with the need for expansion of the road (1 to 3). The defendant considered irrelevant factors such as the anticipated increase of traffic during the Olympic year (4), and overestimated such factors as the risk of falling trees and the state of old and declining trees (5). We hold that there was an abuse in the exercise of defendant's discretion, and the defendant's decision is, therefore, deemed unlawful (voided).

4.1.6   Public Access to Government Information

Ikata Nuclear Power Plant Case[129]

Abstract of the Case

[After the Shikoku Electric Power Company had been granted permission by the prime minister to construct a nuclear power plant at Ikata, about thirty local residents filed a kōkoku appeal demanding a retraction of the license on the grounds of procedural illegality and for safety reasons. In the court proceedings, the plaintiffs asked the court, on the basis of art. 312(2)(3) of the Code of Civil Procedure,[130] to order the prime minister to submit documents that had been submitted by the power company in connection with its application for a construction permit.]

Opinion of the Court

1. Article 312(3) of the Code of Civil Procedure, concerning the relationship between the "holder of documents" and the "person with the burden of proof," is not limited exclusively to direct legal relationship between the two parties. It also deals with facts that are relevant in a more general sense, including mate-

rials prepared in the negotiation process. Applying this interpretation to the kōkoku appeal under consideration here (in which it is contended that the administrative agency's decision was illegal and ought to be retracted), it is clear that the article refers both to documents drawn up in the course of the procedure but prior to issuance of the administrative decision, and also to those documents which constitute the material on which the decision was based.

Because, in principle, administrative decisions ought to be made on behalf of the public through fair and open procedures and because many of the rules relating to the process of administrative decision-making are designed to guarantee its reasonableness and impartiality, the above interpretation does not unfairly disadvantage the administrative agency that holds the documents. Moreover, the party that is contesting the legality of the administrative agency's decision (the people) does not usually have access to the documents on which the decision is based.

Essentially this is an issue of offsetting the disadvantage of the party with the burden of proof (the people) who lacks the necessary documents for such proof. In this sense our interpretation goes a long way toward the discovery of the substantial truth demanded in the kōkoku suit. Because there is a substantial legal relation between the residents and the prime minister (created by the circumstances), the residents have a substantive right (legal standing) to appeal the decision granting a license. Thus because the documents on which the licensing decision was based are recognized to exist, the principles embodied in art. 312(3) of the Code of Civil Procedure are applicable to the documents that create the legal relation between the prime minister and the company.

2. The "official secrets" mentioned in arts. 272 and 273[131] of the Code of Civil Procedure refer to matters that become known by government workers in the course of official employment but which, if made public, would endanger either the public welfare or national security. However, even if the documents containing industrial secrets were clearly submitted on the premise that the prime minister would not publish them, their publication is merely intended to make public those secrets relating to the business of private industry whose main purpose is the seeking of profits. It is difficult to understand how this would endanger either national security or the public welfare, particularly when a business matter (the obtaining of the agency's permission) produced the necessity in the first place of submitting the required records.

Consequently, the refusal to publish the above materials, based on the pretext of industrial secrecy and the duty of nondisclosure, cannot be allowed under the principles of fairness and good faith.

4.1.7   **Validity of Development Guidelines**

Musashino City Residential Development Guidelines
Case[132]

Abstract of the Case

[A developer failed to comply with two provisions of the Musashino Residential Development Guidelines in constructing a condominium in Musashino City, a suburb of Tokyo. The city invoked the measures prescribed by the guidelines for dealing with violations and cut off the condominium's municipal water supply and sewer service. The guidelines provided:

*1. The owner of a building must receive the consent of the residents of the neigh-
borhood as to the effect the building will have on their exposure to sunlight.
2. Where more than fifteen family units are involved, the owner of the building
must donate a school site to the city, conforming to standards set by the latter, or
bear the costs of acquiring the site and constructing the facilities.*

*In response, the developer sought a temporary injunction against the city's re-
fusal to supply water and sewer services. The developer claimed that according
to art. 15, paragraph 1 of the Water Supply Law, the city cannot deny an application for water service unless it has "reasonable cause." Measures in the
guidelines restricting water supply exceed the city officials' authority to issue
administrative rules and are illegal. The developer contended that the Sewer
Law does not give the sewer administrator authority to restrict the use of public
sewers.*

*The city argued in response that (1) because usage of the public water supply
and sewer system is governed by public authority, the court may not issue a
temporary injunction under art. 44 of the Administrative Litigation Law;*[133] *(2)
the guidelines express a norm of natural law essential to the adjustment of the
interests of the residents and the developers concerned. Violations therefore
qualify as "reasonable cause" under art. 15, par. 1 of the Water Supply Law, and
the city may deny water service. The residents concerned in this case joined the
side of the city.*]

Decision

[*The court issued a temporary order to supply water from the waterworks.
However, the court rejected the rest of the claims.*]

Opinion of the Court

*1. Regarding the legal status of use of the public water supply: . . . The goal of
the welfare state is the advancement and improvement of the lives of the inhabitants and not the exercise of public power by an administrative body imposing its will on the people. Since art. 14, par. 4, item 1 of the Water Supply Law
provides that water be supplied according to "cost principles" (there is to be a
charge for water), water supply transactions can therefore be seen as having the
character of private law relationships and as nothing more than bilateral
agreements in private law. Furthermore, art. 15, par. 1 of the same law uses the
term "water supply contract," thus postulating equal positions for the water
utility and the user. It is therefore reasonable to view the legal status of water
use relationships as agreements between parties at private law. [Consequently,
the court held that art. 44 of the Administrative Litigation Law did not apply.]
2. Regarding the legal status of use of the public sewer system: It is undeniable
that there are certain characteristics analogous to the situation of the public
water supply. . . . However, since the public sewer system is under the sole control of the local government (Sewer Law, art. 2, par. 1), the residents do not need
the consent or permission of the administering local government to use the sewer
facilities; on the contrary, residents are forced to use them by virtue of being
residents (Sewer Law, art. 10, par. 1). It is thus clear that use of the sewers is not
based on a contractual relationship between the utility and the user. Taken altogether, it is apparent that the sewer service and water supply operations are entirely different in nature. The legal status of use of the sewers is rather like that of*

a public road offered for the use of general traffic. . . . The residents may freely use the sewer system subject to the limitation that they cannot prevent others from using the facilities. Usage of the sewers falls under the so-called general usage of public facilities. Its legal status is that of a public law relationship. It is natural to interpret the actions of the local government in restricting use of the sewer system as being an exercise of public authority. [Consequently, the court held that the developer could not seek a temporary injunction.]

3. As to whether refusal to make a supply contract with those who have not complied with the guidelines meets the "reasonable cause" standard in art. 15, par. 1 of the Water Supply Laws: The guidelines are not formal laws such as regulations or ordinances, or based on legal authority. They are therefore no more than expressions of policy regarding the enterprises concerned. They do not have legal binding power. Consequently, the rejection of the supply contract in compliance with the aforementioned provision in the guidelines does not directly satisfy the "reasonable cause" standard of the Water Supply Law. It is therefore necessary to determine in their actual context whether or not the facts that promoted the rejection meet the "reasonable cause" standard. . . .

As to the second provision of the guidelines: Since the local government may not obtain donations other than through the free will of the donor, noncompliance with this provision does not directly constitute "reasonable cause." Moreover, where an enterprise willfully violates the guidelines and undertakes construction, all the circumstances must be taken into consideration. The nature of the violation, what it consisted of, and the situation and the particular events leading to the violation must be evaluated. In cases where in view of these factors the developer's coercion of a water supply contract would constitute an abuse of rights, refusal of consent may of course be regarded as meeting the "reasonable cause" standard. However, in this case there is no such abuse of rights.

4.2   Assessment

The decisions presented in sec. 4.1 provide evidence of a parallel "common law" development in the administrative law field. Without clear statutory guidelines, in several cases the courts have imposed procedural and substantive obligations deemed necessary for environmental protection on the administrative agencies. Although some cases break new ground, the judicial approach has generally been unswervingly traditional. The courts have taken pains to identify environmental protection guarantees within the context of existing statutes and have endeavored, at least outwardly, not to trespass on administrative or legislative prerogatives.

4.2.1   Procedural Considerations

Standing

Although the courts have not hesitated to develop new procedural safeguards, once the issue of a plaintiff's standing to sue has been decided, standing still remains the critical procedural barrier to administrative environmental litigation in Japan. The courts have looked principally to the nature of the plaintiffs' interest, and to the type and extent of harm; generally the interest at stake has been regarded as the important issue. When plaintiffs claim infringement of a statutory right, the courts have recognized standing under conditions where the alleged harm was still only speculative. And they have even intercepted large developments far in advance of the anticipated injuries. But when the plaintiffs' interests were less clearly defined, the courts have not recognized standing despite the prospect of harm. In such cases they have explained that the risk was insufficient to justify special consideration.

From the perspective of traditional doctrine, the plaintiffs' position in the Nikkō Tarō Cedar case was the strongest. Because the Tōshōgū shrine itself held title to the Tarō cedar and the other trees in issue, there was no question that the shrine possessed a property right recognized by the Civil Code. The planned construction of the road would undisputedly have destroyed the trees.

Other cases suggest how the courts have begun to expand traditional doctrine. For example, the Usuki Cement case indicates that absolute ownership may not be necessary because in Usuki the plaintiffs merely possessed a usufruct recognized by the Fishery Law. And in Usuki the possibility of injury was less certain than in the Nikkō Tarō Cedar case. The Date Electric Power case represents a comparable holding based on similar considerations.

Scholars have differed sharply in their interpretations of the problematic Kunitachi decision. Some view the case as merely reaffirming the traditional doctrine that those without a right protected by statute cannot sue. Others, however, read the court's decision as actually expanding traditional doctrine by its statement, "and even if such individuals are deemed to have a legally protected interest to contest the obstruction, where this inconvenience exceeds the limits of human tolerance. . . ." By implication, they argue, the court is suggesting that at least those plaintiffs who actually used the road would have standing, if they could have demonstrated that the construction of the bridge might have injured them beyond human endurance.

This second interpretation is significant because the court appears to be expanding the bases for plaintiffs' "legally protected interest." The plaintiffs, of course, alleged the infringement of their "environmental rights," but the court rejected this concept. Rather the court seemed to rely on the "personal right" (right to human dignity, jinkakuken), an interest that a number of important judicial opinions and scholarly articles have recognized. According to some scholars, the Kunitachi court provides a judicial "hook" for other courts to expand the standing doctrine by broadening the zone of interests included in the right to human dignity.

The Kunitachi decision, however, leaves many issues unanswered. The meaning of "exceeding the limits of human tolerance," remains extremely vague, and the status of the second group of plaintiffs concerned with the future of Kunitachi is left uncertain. Ostensibly these individuals failed to demonstrate a requisite interest and risk sufficient to justify judicial attention.[134]

More recent decisions affecting standing are difficult to assess. On the positive side, on April 25, 1978, the Matsuyama District Court recognized the standing of thirty-three local residents to challenge the construction of the no. 1 nuclear power plant of the Shikoku Electric Power Company at Ikata. The plaintiffs alleged that the reactor was structurally defective, safety precautions and preliminary environmental assessments were grossly inadequate, and the defendant state's approval of the reactor under the Basic Law for Atomic Power[135] violated their constitutional rights to health and the pursuit of happiness.[136] In recognizing the plaintiffs' standing the court placed special emphasis on the risks of radiation to public health. From the perspective of the law of standing the Ikata decision is important because of its concern with health risks. It has generally been interpreted to signal the courts' willingness to countenance other administrative challenges to the government's nuclear development program, at least until a clear national consensus is reached.[137]

More restrictive, however, is a recent decision of the Supreme Court holding that the Federation of Housewives (Shūfuren) lacked standing to challenge a Fair Trade Commission (FTC) ruling upholding soft drink manufacturers' failure to notify the public that their product contained little or no natural juice. The housewives' federation did not have standing, the court held, because the appellants were not "directly" injured by the FTC action.[138] Although the "Juice Case" dealt with consumer rights, the Supreme Court's decision has deeply concerned environmentalists and others who interpret it to signal the beginning of a judicial retrenchment.[139]

The Right to a Fair Hearing

Although the courts have continued to interpret standing restrictively, they have been more aggressive in expanding procedural safeguards once the standing issue has been resolved. One important example of this trend is the Matsuyama Airport decision, which recognized the local fishermen's right to present their views before the prefecture approved the airport's construction. The case is important because the court established the plaintiffs' right to a hearing on independent constitutional grounds even though the Public Surface Waters Reclamation Law at the time did not explicitly authorize public hearings. Of course the substance of the public's due process rights in this early decision was carefully qualified. The plaintiffs were holders of fishing rights to the superjacent waters around Matsuyama, so standing was not an issue. The court left uncertain whether persons not holding recognized rights should receive comparable protection. The court also noted that the plaintiffs might not be entitled to a hearing in emergencies involving minor property claims. Finally, it did not elaborate in any detail on the substance of the "minimum hearing requirement," apart from noting that the "contacts" between the fishermen and the Ehime Prefectural authorities were inadequate to satisfy the plaintiffs' right to be heard.

The Matsuyama case is historically significant because it occurred in 1968, several years before the judgments in the major pollution trials. In part the court's progressive decision can be attributed to the strident public demands of the period that the government adopt aggressive environmental protection measures. The court may also have been influenced by the scholarly and judicial opinions at the time calling for expanding fair hearing procedures.[140]

The Ikata court seems to extend the due process requirement to the public's right to obtain information held by the government. The decision suggests how the courts have labored to stretch the provisions of the Code of Civil Procedure, since Japan, like many other countries, has not enacted a Freedom of Information Act. Although the decision stresses the need for some qualifications on the public's right to receive information in sensitive areas like security, it expresses an evolving stern judicial attitude toward government efforts to conceal all meaningful information merely because disclosure might prejudice commercial interests.

The courts' attention to minimum procedural safeguards has also stimulated legislative changes, as in the civil area.[141] For example in 1973 the Public Surface Waters Reclamation Law was amended to include the following provisions:

1. When there is an application for a land reclamation permit, the governor of the prefecture shall give notice without delay of a summary of the proposal and shall make public for three weeks from the day of notice documents containing matters relevant to each item of the previous article . . . and related books and, fixing a deadline, shall solicit opinions of the heads of the local cities, towns and villages. However, if the application is dismissed, this law is not applicable.

2. When the governor of the prefecture gives notice pursuant to the previous paragraph, he shall inform the concerned governors of this fact without delay.

3. When the notice in paragraph 1 has been given, people who have interests in the land reclamation can submit written opinions to the governor up until the expiration date of the three weeks set out in paragraph 1.

4.2.2   Substantive Considerations

Despite incremental progress in procedural matters, the courts have remained extremely conservative on the merits. The one major exception is the Nikkō Tarō Cedar case. This case illustrates how, as early as 1973, at least one court stressed the independent obligation of the administrative agencies to develop rational policy guidelines and planning criteria for environmentally hazardous projects. Although the court analyzes the problem in procedural terms (that is, the agency did not follow appropriate procedures, and hence abused its discretion), we interpret this case to be a virtually unprecedented judicial inquiry into the very substance of the planning process (or lack thereof) of the Ministry of Construction.

In voiding the approval of the Ministry of Construction, the Utsunomiya District Court relied primarily on the "appropriate and rational use" requirement in art. 20(3) of the Land Expropriation Law. The court interpreted this provision to warrant judicial scrutiny of the defects in the ministry's cost-benefit calculations. Its principal finding was that the ministry so arbitrarily overvalued certain factors and undervalued others, particularly environmental and cultural values, that the procedure itself was defective.

The ministry's assessment of the public benefits of the project was misguided for several reasons. First, the court held it had been overly concerned with finding the cheapest, technically easiest alternative, requiring the shortest time period for construction. Second, the ministry had overvalued short-term benefits such as the road's capacity to accommodate tourists during the Olympic games.

The nub of the court's opinion, however, is that the ministry failed to calculate accurately the costs of the destruction of the cedars to the Tōshōgū Shrine and the nation. And the court is fairly explicit about the values represented. The first is the economic interest of Tōshōgū itself. The second is the value to the nation of the resource in question. Although part of this cost is reflected in the anticipated loss in tourism to the area, another part is the cost borne by the nation because of the loss of this irreplaceable part of its environmental and cultural heritage. By its emphasis on the concept of heritage, the court indicates its concern not only with the interests of the living but also with the interests of unborn generations.

The court's effort to place some numerical value on the environmental-cultural factor is especially interesting. By its calculations the court effectively is saying that in its opinion the value of the cedars to be sacrificed at least exceeded the difference in the costs of plan A, the cheapest option elected by the ministry, and plan C, an alternative that would have avoided the destruction of the cedars. In other words, since plan A would cost ¥43,000,000 and plan C ¥1,300,000,000, by implication the court suggests that the cedars' value, at least in its judgment, exceeded ¥1,257,000.

Although the court does not develop this point in detail, its analysis is more than a catalogue of costs and benefits, for it suggests that environmental protection is important enough to constitute a fundamental constraint on other activities and is to be given "utmost consideration by the administrative agencies." Because these values "provide the people with a healthy and culturally satisfying life," the court holds that the executive must demonstrate that a given project justifies the incidental cost in environmental deterioration.[142]

In addition to art. 20(3) of the Land Expropriation Law, the court's decision rests on another ground. The Nikkō footpath and the surrounding area were within an area designated as a nature preserve under the Natural Parks Law, but the cedar trees in issue were not, even though the Tarō cedar and other trees were contiguous to the designated area. Although it avoids suggesting that the cedars should have been designated, the court does note that based on its own on-site investigation, public opinion, and other factors, it was unreasonable under the circumstances for the Ministry of Construction to have treated the two sets of cedars (those within the designated area and those outside of it) differently.

Recent administrative decisions after Nikkō Tarō Cedar have been distinctly less generous on the merits. Perhaps the most troublesome is the Matsuyama decision in the Ikata nuclear power plant case. In accepting the government's assertion that radiation emitted from the plant was infinitesimal and therefore harmless, the court essentially placed its imprimatur on the entire nuclear development program.[143] Subsequently, the Fukushima District Court also rejected a suit brought by residents requesting cancellation of a land reclamation permit granted by the Fukushima prefectural government to the Tokyo Electric Power Company, which intended to construct a second atomic power plant.[144] Finally, the courts have continued to reject other environmental rights suits along with parallel actions involving "rights of access to beach areas" (irihama ken).[145]

In order to reach a balanced assessment, let us now integrate the developments in the civil and administrative law this chapter has discussed. Because Japanese courts have not been able to rely on specific "action forcing" provisions in the environmental laws as have American courts, they have had to predicate decisions essentially on extensions of tort doctrine, various obligations appearing in statutes unrelated to environmental protection, and constitutional prescriptions. Perhaps erratically, the courts may be groping for a format for government decisions somewhat analogous to the environmental impact assessment requirement of section 102(c) of the United States National Policy Act (NEPA) of 1969.[146] Thus, the Matsuyama Airport, Usuki Cement, Ikata Nuclear Power Plant, and Ushibuka Waste Treatment Plant decisions affirm the agencies' responsibility to prepare a memorial of decision-making and disclose this record to the public. The Nikkō Tarō Cedar and Ushibuka cases also underline the importance of an ecological impact assessment, the identification of unavoidable or irreversible adverse effects, and the scrutiny of alternatives—all obligations mandated by the American statute.

The Western reader may view the obligations created by the Japanese courts without explicit statutory authority as officious meddling in the affairs of government. Indeed the civil and administrative decisions considered together express a judicial willingness to confront substantive questions exceeding most American courts' penchant for procedural solutions in the environmental field.

Although cases like Nikkō Tarō Cedar or Ushibuka raise serious questions as to how future courts and administrative agencies will refine and implement these obligations, we applaud these early efforts. For as long as Japanese statutes fail to provide clear and specific duties for the government to protect the environment and as long as the administrative agencies refuse to embrace this responsibility on their own, the nation's cherished cultural and environmental heritage will be mortgaged mindlessly for short-term commercial gains. Although environmental values need not always prevail, they at least deserve scrupulous attention.

Section 5   Role of U.S. and Japanese Courts in Environmental Litigation

The courts in Japan and the United States have served important common purposes. They have strengthened the average person's bargaining position with powerful industries and the government; they have focused national attention on neglected or ill-understood problems and on interests needing legal protection. In both countries the courts have also been a place of refuge because legislators and administrators have been unwilling or unable to protect the environment; and the courts have responded by delineating minimum standards. Finally, judicial intervention has been necessary to open governmental decisions to public scrutiny and to establish fair procedures.

But the differences in the courts' function in Japan and the United States are also very great. In Japan the courts have focused on problems of damages or injunctive relief in civil litigation. The review and reform of government's decision-making has not been a primary objective, although it has been a secondary consequence of some decisions. In the United States, however, environmentalists and industry have asked the courts primarily to scrutinize the conduct of the executive.[147]

The attention of the United States courts to governmental decision-making has influenced their role and contribution. Generally they have focused on procedural questions and have been chary of substantive review.[148] More often they have deferred to the comparative expertise of the administrative agencies in environmental matters and to traditional presumptions supporting administrative action. At times also environmental statutes have explicitly limited judicial discretion.[149] As a result, the courts have come to view themselves as "partners" with the agencies in implementing the intent of Congress.[150]

These differences in role and orientation suggest that Japan's contribution lies principally in the area of civil private litigation. Let us first consider doctrinal questions of liability, causation, and cognizable interests, and then examine briefly the legal and institutional aspects of remedies.

Perhaps the most interesting Japanese case is the Yokkaichi court's treatment of joint and several liability. Although comparable American cases in the environmental field are scant, Michie et al. v. Great Lakes Steel Division, National Steel Corp.[151] is one important precedent. In Michie, thirty-seven residents living near La Salle, Ontario, Canada, sued three corporations operating sewer plants in the United States immediately across the Detroit River from Canada. The plaintiffs alleged that the defendants' factories had discharged pollutants that were carried by air currents onto their premises in Canada impairing health and damaging property. The plaintiffs claimed damages between $11,000 and $35,000.

The plaintiffs asserted that the defendants' actions constituted a nuisance for which all three corporations should be jointly and severally liable. The Sixth Circuit Court of Appeals ruled that when polluters concurrently cause an indivisible harm, the defendants should be held jointly and severally liable, where plaintiffs were unable to apportion the harm between the wrongdoers.[152]

Although Michie seems directly on point, there is one important difference. In Yokkaichi one defendant was able to establish that its discharges alone were insufficient to cause the plaintiffs' injuries, whereas the court in Michie does not attempt to distinguish the harmful effects of the defendants' individual emissions. As noted in Yokkaichi, the court handles the problem of minimal discharge by a theory of attribution that to the writers' knowledge has not been explored widely in American environmental law.[153]

But outside environmental law, in tort cases dealing with hospital accidents, United States courts apparently have ventured even further than Yokkaichi. In Anderson v. Somberg,[154] for example, the plaintiff recovered damages for injuries sustained during a back operation. The plaintiff sued the doctor for medical malpractice, the hospital for negligence, the medical supply distributor on a warranty theory, and the manufacturer for strict liability in tort. On appeal the Supreme Court of New Jersey held that the burden of proof shifted to the defendants; and that the jury *had* to find at least *one* primary or third party defendant liable.

Although the Yokkaichi court's analysis of negligence is paralleled by a few American decisions, Yokkaichi offers a useful and important precedent on the joint liability of multiple pollution sources. The contribution of the five major pollution trials (including the Osaka International Airport case), however, is even more distinct in the less well charted jurisprudence of pollution disease causation.

Perhaps the most significant aspect of the Japanese decisions is the courts' attitude toward the problem of scientific uncertainty. The defense of scientific uncertainty, they implied, had been long abused by polluters to subvert the victims' cause; courts must deal with scientific precision differently from the scientific world because other policies and objectives are at stake; and in law, justice at times must prevail over the disinterested quest for abstract truth.[155] On a more mundane level, the decisions are significant because of: (1) the way the courts integrated statistical (epidemiological), clinical, and experimental data within a legal analytic framework;[156] and (2) the way they applied this framework to a variety of diseases, many of which were chronic, some unrelated to pollution,[157] and some attributable to multiple factors and multiple sources.

Three kinds of cases in the United States (and perhaps other countries) merit particular attention from the perspective of these decisions. The first is the situation of people who have been exposed to Kepone poisoning around Hopewell, Virginia, PBB poisoning in and around Grand Rapids, Michigan, and PCB poisoning in the Hudson River Basin. Although (to the authors' knowledge) no cases of cancer induced by these substances have been verified to date, these examples of known exposure to toxic substances deserve careful attention. Indeed these cases may be easier analytically than the Niigata, Toyama, and Yokkaichi examples, because the source of the toxicant is uncontested.

A second, harder case is the "cancer belts" in New Jersey or the Mississippi River Basin that appear to be "linked" with industrial activity in other areas. In a hypothetical suit by cancer victims, an American court might profitably consider a two-part analysis like that adopted by the Yokkaichi court. The court could first ascertain whether the defendants' activities taken as a whole caused the plaintiffs' injuries, and upon an affirmative finding use presumptions and evidentiary burdens to address the problems of individual causation. Although this would represent an extension of the Yokkaichi court's analysis, the approach deserves at least careful judicial scrutiny.

The third, most complex situation is illustrated by (what has been euphemistically called) the case of Roger Diamond versus the World.[158] Diamond was a class action on behalf of all residents of Los Angeles County against 292 named industrial corporations and municipalities and 1,000 additional unnamed defendants, to recover damages for air pollution-induced injuries and for injunctive relief. The Court of Appeals held that the suit could not be brought as a class action and that injunctive relief was beyond the court's capabilities.[159] Although we do not suggest that Diamond itself was procedurally feasible, it is important that American courts start treating causation for what it often is—a collective phenomenon. American courts might begin by considering the applications of the Yokkaichi rationale to complex cases of injury from "mixed" mobile and stationary polluting sources.

Another area deserving more careful American attention is the Japanese courts' development of sunlight rights.[160] The case for granting legal protection to this interest rests on several grounds. First, as in Japan, sunlight is important to health; second, sunlight is valued for aesthetic reasons; third, perhaps most important, United States law and policy now encourage the development of solar energy technology essential for capturing sunlight's vast economic and energy potential.[161] The current state of the law, however, poses formidable obstacles to technological change and development.[162] In the absence of comprehensive federal and state schemes for assuring adequate solar access (e.g., solar zoning), judicial intervention seems necessary.

The courts have several doctrinal options. Sunlight interests, for example, could be predicated on a state constitutional or legislative declaration of environmental policy or implied from local, state, or federal laws encouraging solar energy use. Common law doctrines for the protection of property rights, like riparian rights, trespass, and nuisance, also provide bases for legal protection. Of these, nuisance offers the greatest chances of success.

Like the Japanese sunlight doctrine, nuisance law requires careful balancing of the plaintiff's, defendant's, and community's interests and a determination that the challenged conduct is "unreasonable." Under the Restatement,[163] intentional invasions of another's interest in the use and enjoyment of land are presumed unreasonable, unless the utility of the invasion outweighs the harm. It is precisely regarding these more subjective determinations of "reasonableness" and "utility" that Japan's experience can best aid American courts in protecting the justified expectations of solar energy users.

Three aspects of the Japanese courts' approach to remedies also deserve attention. The first is the judges' mediating local controversies like sunlight disputes. The second is the courts' willingness to conduct on-site investigations on their own initiative or at the request of the parties. As described, the on-site investigation in the Osaka International Airport case significantly influenced the High Court's decision. The third area is the substantive remedies themselves, particularly the use of future damages imposed as an alternative to injunctive relief. As noted, the Osaka High Court's approach may represent a more effective and efficient solution to a difficult resource conflict than that adopted by a comparable American precedent.[164] The field of remedies seems particularly suitable for a continuing judicial dialogue between the two countries in the years ahead.

III   Environmental Protection
Legislation and Its Administration

5   Formulation and Implementation of Environmental Policy: A Comparative Assessment of the Regulatory Process

Japan's 1976 report on the state of the environment submitted to a special meeting in Tokyo of the Organization of Economic Cooperation and Development (OECD) offered the world an impressive record of achievement. In many parts of the country pollution had declined to a remarkable degree.[1] Elsewhere it had been arrested. And in several areas Japanese industry had met the world's strictest environmental standards.[2] From an economic perspective the most striking result of Japanese pollution control policies, analyzed under three economic models, was that GNP and employment were practically unaffected.[3] Foreign observers recalling the dreary Japanese report to the 1972 U.N. Stockholm Conference on the Human Environment may find this transition startling, for Westerners have come to think of Japan, perhaps as a result of the conference, as a veritable cauldron of pollution.

In large measure Japan's successes in the environmental field (and to some extent also her failures) can be traced to the regulatory process, the subject of this chapter. Our principal objective is to describe the administration's formulation of pollution countermeasures and to analyze the positive and negative incentives used to induce industry's compliance. In analyzing the "compliance problem" we are particularly interested in the contribution made by structural factors, particularly the innovative role of local governments, and by "metalegal" institutions such as the Japanese bureaucracy's reliance on "administrative guidance." The discussion is supplemented by three case studies dealing with the development of desulfurization, automotive emission controls, and the administration's adoption of environmental impact assessment procedures.

Japan's experience is particularly interesting for several reasons. First, pollution control efforts have proved at times more effective than measures conducted in the United States or other Western countries. Second, there is some evidence that countermeasures in Japan have been executed more efficiently, and also more equitably, than comparable Western initiatives. In view of the increasing interest in the United States and other countries in strengthening the enforcement process, in this chapter we attempt to identify ideas and approaches deserving of more careful attention by the West.

Section 1   Structural and Organizational Aspects of Environmental Protection in Japan

1.1   The Role of the Bureaucracy

To grasp the strengths and dysfunctions in the administration of environmental policy in Japan, the reader must be aware of the historically dominant influence in the society of the bureaucracy. During the nineteenth century the bureaucracy served as the guardian of the imperial institution and the moral order that it embodied.[4] Between the promulgation of the Meiji Constitution of 1890 and World War II, the bureaucracy possessed independent legislative powers (meirei) in the name of the Emperor that were constitutionally equivalent to parliamentary enactments. Throughout this period judicial review of administrative action was carefully circumscribed by the jurisdictional limitations of the prewar Administrative Court.[5]

Quite naturally the bureaucracy enjoyed all the accouterments of an elite class. In the Tokugawa era (1600–1868) officials were ranking samurai, the apex of the military establishment.[6] And after the Meiji Restoration, officials, perhaps even more than the army, were the samurai's legitimate heirs. Often, government services made high ranking officials wealthy men. They sported fancy buttons and braids, and wore swords on ceremonial occasions. Salaries were high, so that even lower ranking officials reportedly visited geisha. A bureaucrat's education was the best to be had since virtually all trained at the prestigious imperial universities, particularly Tokyo University. In comparison to their countrymen, bureaucrats also enjoyed considerable career mobility. For example, Craig notes that "in the late Meiji period, a Tokyo University graduate might enter the Home Ministry, then become the assistant chief of police in a small city, move back to a different bureau in the ministry, then to a prefectural government post, and so on." From the peaks of lofty superiority, the bureaucracy surveyed the rest of society with vast disdain—the maxim, "officials honored, the people despised" (kanson minpi), epitomized the prevailing attitude.

During the early period of the Occupation, the SCAP authorities attempted in various ways to remedy the more egregious aspects of the prewar bureaucratic system. Some key wartime officials were purged,[7] while the bureaucracy as a whole was stripped of many of its symbolic perquisites.[8] Under the 1947 Constitution, bureaucrats were redefined as "servants of the people" (kokumin no kōboku)[9] and the principle of separation of powers was explicitly reaffirmed.[10] Also, the prewar Administrative Court was abolished and replaced at the suggestion of the SCAP authorities by the new Administrative Procedure Code.[11] Finally, to oversee the bureaucracy's continuing reform,[12] as well as to introduce outside citizen advice on its operations, a system of commissions was established.[13]

Despite such efforts, the basic prewar bureaucratic pattern continued. In part this is due to the fact that while the SCAP authorities chopped off the apex of the bureaucratic pyramid, they left the base and middle undisturbed. Thus, much of the prewar system remained intact. Although bureaucrats are now perhaps more heterogeneous as a class, educational backgrounds are still almost invariably similar.[14] Moreover, the processes of recruitment, promotion, and retirement still adhere closely to the prewar pattern. Indeed, some scholars argue that the "consciousness of power" (kenryoku ishiki) of today's bureaucrats may even exceed the status consciousness of the ultranationalistic and feudal periods.[15]

What is critical for the present study is the continuity of the bureaucracy's virtual monopoly over the legislative process. The success rates of governmental and individual member bills, the declining rate of amendments, and the singular lack of success of opposition-sponsored bills all support this thesis.[16] In addition, the bureaucracy still possesses powers to issue orders virtually of equal legal dignity to statutes themselves. As Pempel notes: "the emerging picture is of an increasing proportion of serious political policy making in Japan taking place outside the public arena of the Diet and under the increasing control of a democratically unresponsive bureaucracy."[17]

Let us consider the implications of this arrogation of law-making powers by examining the decision-making process within the bureaucracy, as viewed from the outside, and with particular reference to the regulation of industry. Perhaps the critical issue in decision making within the bureaucracy is jurisdiction. Historically, although broad statutory grants of authority provided a mandate for innovative administrative action and engendered feelings of loyalty, pride, and solidarity within the ministries, bureaucratic territorialism at the same time also greatly impeded interministerial collaboration.[18] Since the 1950s, jurisdictional conflicts have seriously hampered the implementation of environmental policies. Such conflicts delayed by almost a decade the formulation of the Basic Law and contributed to weakening it. Jurisdictional rivalries even today deprive the Environment Agency of the support of the traditional ministries and continue to frustrate that agency's initiatives.[19]

Conflicts over jurisdiction are tempered and usually resolved through consensus. Japan's concern for consensus can be traced to the earliest political writers,[20] who extolled the virtues of concord and harmony, and today the establishment of consensus continues to be a primary objective. Consensus invokes several interrelated ideas. It expresses an affirmative state of mind and also implies that once concord is achieved, the "right" conclusion to a problem will emerge. The number of people from whom assent must be obtained will generally depend on the importance of the issue.[21]

Consensus can be built in many ways. The most customary means is by frequent, at times protracted, consultation.[22] There is no predetermined order of consultation, and consultations may be formal or informal depending on the issue. Informal consultations ("nemawashi," literally "binding the roots of a plant before pulling it out") are most frequent; they are perhaps best compared to the American practice of "buttonholing," although the Japanese version may often be far more subtle, indirect, and painstaking. A second means of building consensus is by more formal meetings (kaigi). For example, during the preparation of a bill, a ministry's section chief may discuss the matter at a weekly meeting.[23] Thereafter, a meeting of the ruling Liberal Democratic Party's (LDP) Policy Affairs Research Council will clear a bill and at some point before its submission to the Diet, the vice-minister and bureau chiefs will discuss the bill. A final important means of building consensus is the "ringi" system, wherein documents (ringishō) are drafted at lower levels of an organization and then circulated to various superior units for approval.[24] Although in a sense the imprint of an official's seal reflects a fait accompli, the process of exchange and deliberation itself generates momentum that influences the formation of consensus.[25]

From the perspective of the implementation of environmental policy, the system's emphasis on consensus is double edged. As the case studies in this chapter demonstrate, the opposition of prodevelopment ministries has delayed or blocked many ambitious environmental protection proposals, and by conditioning consent, these ministries have seriously weakened many other measures. Yet, because of the importance of consensus, the Environment Agency has at times been able to exert influence on some decisions beyond the apparent scope of its meager explicit statutory powers.

Political factors also influence decision making within the bureaucracy. Historically, the Japanese bureaucracy has cherished a self-image of political neutrality, and there is historic support for its belief. The laws that established the bureaucracy during the Meiji period encouraged independence as a bulwark against the ambitions of the nascent political parties, and the pre- and post-war structure of the Japanese ministries has fostered insulation from political concerns. While the American department is open at the bottom, middle, and top (the latter to political appointees) the Japanese agency is a closed body. It recruits only at the bottom; employment is for life and promotion progresses ineluctably by seniority. Indeed the three top nonbureaucratic posts (i.e., the minister, a cabinet member, and two "political vice-ministers") deal with the agency only through its bureaucratic vice-minister (the highest career post in the bureaucracy) and do not penetrate the organization.[26] The bureaucracy has often pointed in its claim of political independence to these structural factors and also to its practice of consulting all the opposition parties during the drafting of legislation and on other important policy matters.[27]

Despite these indices of neutrality, the LDP does possess a number of strong controls over the bureaucracy.[28] For example, the party controls promotions within the bureaucracy so that it is virtually impossible for an unacceptable individual to rise above the level of bureau chief. To a lesser extent the party influences a bureaucrat's prospects for retirement (amakudari) either by facilitating or interfering with the retention of a choice post. The party also may significantly affect the legislative process because a bill must be approved by its Policy Affairs Research Council (Seimu Chōsakan), its Executive Council (Sōmukai), its Diet Policy Committee (Seisaku Shingikai), and finally by the Cabinet. Discussions between party and ministry officials at each of these stages also exert a powerful, although perhaps more subtle, influence. By the time a bill is presented to the Diet, or an important decision ready to be made, a consensus has already been established among the key bureaucratic and political actors. One is therefore not surprised that the bureaucracy has rarely opposed an LDP directive and has collaborated closely with a series of conservative administrations.[29]

The strong personal ties between LDP members and bureaucrats obviously also constitute a continuing source of influence over the implementation of environmental policy. With the exception of Buichi Ōishi, the first director of the Environment Agency, the LDP has never appointed a man with strong environmentalist sympathies and technical competence to head the Environment Agency, although certainly the party includes such individuals within its membership.[30] The party has also helped defeat the passage of environmentally progressive legislation,[31] and continues to influence the speed of implementation of other measures.

Yet, there are also checks on the extent of political interference. As might be expected, the press, and the public generally, would be highly critical of blatant LDP influence-mongering and it is certain that the opposition parties would take political advantage of attempts to subvert environmental policies. Some commentators indeed suggest that the bureaucracy, if so disposed, could rebuff political interference, for they assert its present powers and influence have never been equalled even in the pre–World War II era.[32]

The public's capacity to influence internal ministerial affairs is extremely limited. As discussed in ch. 4, although citizens' groups have scored some recent successes in administrative litigation, decision making within the bureaucracy remains a closed process. Public participation is generally by administrative grace, not by right.[33] The law does not require formal adversarial hearings on important rule making decisions,[34] and agencies are not even legally required to maintain a record, although most agencies do keep records for internal purposes.[35] Japan has not enacted a Freedom of Information Act, and a recent trend favors even greater governmental secrecy.[36] Finally, the almost overwhelming legal presumption upholding the broad discretion of public officials preserves the integrity of the system from public intrusion.

The bureaucracy's insulation is reinforced by the scope of its administrative powers. Not only is it the drafter, it is also the principal interpreter of law.[37] Moreover, the bureaucracy controls the implementation of statutes by administrative enactments, cabinet orders (seirei shikōrei), ministerial orders (shōrei), and various rules and regulations (shikō kisoku).

All administrative acts of course must have some statutory basis.[38] Japanese legal scholars generally subdivide administrative acts into two types. The first type is comprised of official acts that must accord strictly with law, with nothing left to discretion. The second is comprised of discretionary forms of administrative actions.[39] In addition to the statutory limitations on discretion, official action is also bridled by the consensus system and tempered by abstract principles of common sense (jōshiki), reason (jōri), goodwill (kōjō), and benevolence (jin).

Although thus far the discussion has focused primarily on the formal aspects of the Japanese administrative process, the reader should be aware that the greater part of decisions are based on negotiation, discussion, and consultation. This informal, nonauthoritative process by which government seeks to induce industry's voluntary cooperation and compliance is commonly termed "administrative guidance" (gyōsei shidō).[40] Administrative guidance is distinguished from administrative acts (gyōsei kōi), administrative legislation (gyōsei rippō), and administrative enforcement (gyōsei shikō). It may take the form of directions (shiji), requests (yōbō), warnings (keikoku), suggestions (kankoku), or simply encouragement (kanshō). Although these actions often resemble legally sanctioned directives (kamei kōi), administrative guidance ultimately has no legal force or effect.

The development of administrative guidance is historically tied to the government's relationship to industry. In the early Meiji period (1858–1880) government, not the private sector, developed the railways, ports, roads, and urban services; and, following the well-established Tokugawa practice, government continued as industry's patron. Indeed men like Kaoru Inoue[41] and other farsighted Meiji leaders were chiefly responsible for cajoling reluctant businessmen by subsidies, loans, and at times by outright gifts, into the new and risky fields of industrial development. Between the Matsukata retrenchment of 1880[42] and World War II, the Japanese economy developed its "dual" structure of gigantic financial industrial combines (zaibatsu) on top, and a large number of small enterprises, usually of less than five workers, at the base of the economic pyramid.

Technically, it would probably be incorrect to employ the term "administrative guidance" to describe the government's relation to industry during this period, for to a large extent the zaibatsu was more powerful than the prewar bureaucracy. Indeed, collaboration with the zaibatsu proved more expedient than attempting to force its submission. Conversely, the bureaucracy's conduct toward small and medium enterprises was more imperious than the gentler practice of administrative guidance. Despite these differences, the government's prewar relationship with industry established a framework for cooperation essential to the later development of administrative guidance.

Many government officials today believe that administrative guidance proved most effective during the early postwar period when industry and government shared the common objective of economic rehabilitation.[43] Indeed, between 1945 and 1960, giant strides were made. Recently, however, the government-industrial relationship is changing. It has grown increasingly adversarial, as government has had to address more diversified, and at times conflicting, concerns over environment, energy, consumer safety, or trade. Whereas industry's compliance with administrative recommendations was virtually assured during the early postwar period, today this is not so. The government is more cautious, makes its objectives more explicit, and usually identifies a broader range of options for industry.

But despite the occasional tensions of the relationship, administrative guidance still remains the government's primary means of controlling industry. It can soften the effects of a rigid statute or enhance a regulation. By permitting industry to contribute to identifying and implementing its own solutions to a problem, administrative guidance contributes significantly to the effectiveness and fairness of the administrative process.

At the same time, an increasing number of commentators note a darker side of guidance—the absence of institutional review,[44] the danger of arbitrary or discriminatory application,[45] its vulnerability to political influence,[46] the potential for conflict with statutory objectives,[47] and its possible incompatibility with international commerce and foreign relations generally. We will address these charges in the concluding part of this chapter and in later chapters.

1.2   Environmental Protection at the Central Level

1.2.1   The Environment Agency: Its Organization and Performance

Japan's unitary governmental system resembles more the British parliamentary model than the American federal system. At the apex of apparent power[48] is the prime minister, who is elected by the House of Representatives from among the members of the Diet.[49] Chosen by his peers and replaceable by them, the prime minister has substantially less personal power than an American president.[50] In fact, he serves more like a chairman of his affiliate party.[51] One main function of the prime minister is to choose a Cabinet. The Cabinet now numbers twenty persons, a majority of whom must be members of the Diet.[52] Twelve of the Cabinet members head the major ministries (fig. 1), while the remainder direct some of the more important subsidiary governmental bodies.[53]

The Environment Agency is an executive body established under the Prime Minister's Office and headed by its director general. The director general is a member of the Cabinet and holds the rank of state minister. As set forth in fig. 2, the agency is constituted into four major bureaus or departments in addition to the Secretariat of the Director General.[54] These are the Planning and Coordination Bureau[55] that expanded many of the tasks of the Headquarters for Environmental Pollution Countermeasures, the Nature Conservation,[56] Air,[57] and Water Quality[58] bureaus and the Environmental Health Department.[59]

Despite its ostensible role as the principal administrative guardian of the environment, the Environment Agency has been generally regarded within official circles as a parvenu and its contribution continues to remain uncertain. Many of its present problems can be traced to the political circumstances of its creation. The establishment of a new and independent agency meant that some of the prerogatives of other traditional ministries or agencies would need to be curtailed. In fact, many departments were stripped of regulatory powers,[60] while others were entirely abolished.[61]

The traditional ministries and agencies fought this usurpation of their powers in three ways.[62] First, they placed their own members in key positions of influence within the newly formed agency. For example, the Ministry of Health and Welfare (MHW) sent two hundred and eighty-three officials; the Ministry of Agriculture and Fisheries (MAF), sixty-one; MITI, twenty-six; and the Economic

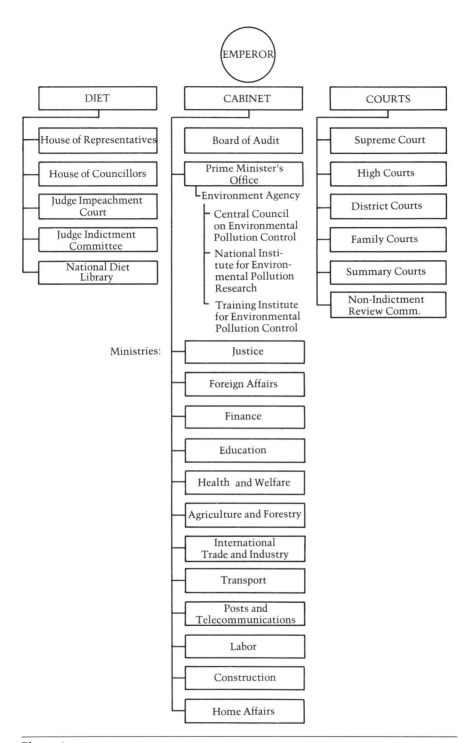

Figure 1
The government of Japan.

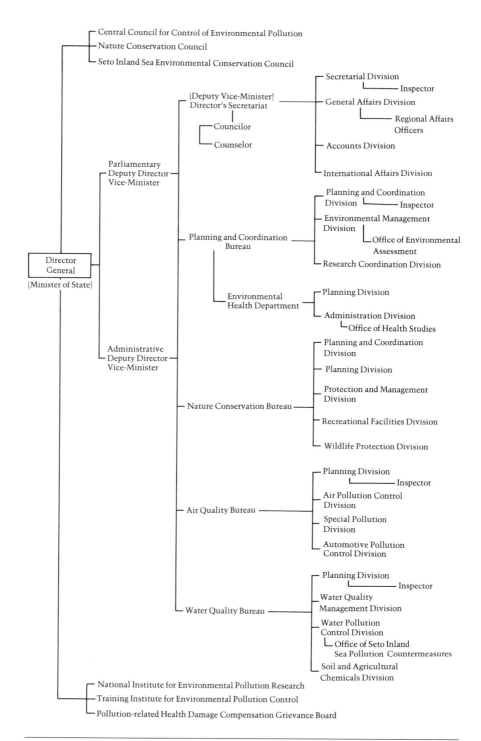

Figure 2
Environment Agency (March 1976).

Planning Agency (EPA) twenty-one. In the competition for high posts, MHW won a deputy directorship (vice-minister), secretary general of the secretariat (deputy vice-minister), and two bureau chiefs; the Ministry of Finance (MOF) and MAF each took a bureau chief; MHW won seven section chiefs, MITI, EPA and MAF two, and the other ministries grabbed one apiece. Generally, former officials of industry-oriented ministries gained key posts in the general affairs, accounting, planning, and coordination sections of the Environment Agency.

Second, the traditional ministries attempted to protect jurisdiction by limiting the Environment Agency's enforcement powers. Apart from its statutory authority to establish ambient, emission, and related standards in the air, water, noise and other fields,[63] the agency's principal function is to coordinate, rather than to implement, environmental protection policies.[64] Table 1 suggests that the

Table 1
Laws and administrative machinery for the protection of the human environment.

Item	Area of Responsibility	Relevant Laws	Jurisdiction
1. Pollution control measures in general	Planning of pollution control measures	Basic Law for Environmental Pollution Control	EA
	Establishment of environmental quality standards	do.	EA
	Formulation of the Environmental Pollution Control Programs	do.	EA, HCEP, EPA, MITI, etc.
	International relations		Ministry of Foreign Affairs, EA, MITI
2. Air pollution	Establishment of standards for emissions from factories and the enforcement thereof	Air Pollution Control Law	EA
		Electric Power Industry Law	MITI
		Gas Industry Law, Mine Safety Law	MITI
	Establishment of automobile emission standards and the enforcement thereof	Air Pollution Control Law, Vehicles for Road Transportation Law	Ministry of Transport (MT), EA
		Road Traffic Law	National Police Agency (NPA)
	Measures for control of smoke and soot emitted from household heating systems, etc.	Air Pollution Control Law	EA
3. Water pollution	Establishment of quality standards for effluent from factories and the enforcement thereof	Water Pollution Control Law	EA
		Electric Power Industry Law	MITI
		Mine Safety Law	
	Control of water pollution caused by effluents from sewerage systems	Sewerage Law	Ministry of Construction (MC)
	Control of marine pollution caused by wastes (including oil) from vessels	Marine Pollution Prevention Law	MT

Table 1
(continued)

Item	Area of Responsibility	Relevant Laws	Jurisdiction
	Control relating to river waters	River Law	MC
	Conservation of fishery resources	Fishery Resources Conservation Law	Ministry of Agriculture and Forestry (MAF)
	Establishment of potable water quality standards	Water Works Law	MHW
4. Noise	Establishment of factory noise standards and the enforcement thereof	Noise Regulation Law Electric Power Industry Law, Gas Industry Law, Mine Safety Law	EA MITI
	Establishment of standards for noise emanating from construction sites and the enforcement thereof	Noise Regulation Law	EA
	Establishment of automobile noise standards and the enforcement thereof	Noise Regulation Law Vehicles for Road Transportation Law Road Traffic Law	MT EA NPA
	Measures for control of aircraft noise	Law Concerning Prevention, Etc. of Disturbance Caused by Aircraft Noise in the Vicinity of Public Aerodromes	MT
		Law Concerning Adjustment, Etc. in the Environs of Defense Facilities	Defense Facilities Administration Agency (DFAA)
	Control of other kinds of noise	Minor Offense Law	NPA
5. Ground subsidence	Basic measures to prevent ground subsidence		EA
	Control of pumping of ground water for industrial use	Industrial Water Law	MITI
	Control of pumping of ground water for use in building	Law Concerning Regulation of Pumping of Ground Water for Use in Buildings	EA
	Measures to prevent subsidence of agricultural land		MAF
6. Offensive odors	Control of offensive odors emanating from plants, etc., processing dead animals	Law Relating to Dead Animal Processing Plants, Etc.	EA
7. Soil pollution	Soil pollution control and measures for cleansing polluted soil	Agricultural Land Soil Pollution Prevention, Etc., Law	MAF
8. Waste disposal	Disposal of industrial and non-industrial wastes	Waste Disposal and Public Cleansing Law	MHW, MITI
	Poisonous and deleterious substances control	Poisonous and Deleterious Substances Control Law	MHW

Table 1
(continued)

Item	Area of Responsibility	Relevant Laws	Jurisdiction
9. Agricultural chemicals	Establishment of standards and registration system for agricultural chemicals	Agricultural Chemicals Regulation Law	MAF
10. Control of land utilization and construction of facilities	City planning	City Planning Law	MC
		Building Standard Law	MC
	Control of new and/or additional construction of factories	Law Concerning Restriction on Industries, Etc., in Built-up Districts in the National Capital Region	Commission for the Development of the National Capital Region (CDNCR)
		Law Concerning Restriction on Industries, Etc. in Build-up Districts in the Kinki Region	Commission for the Development of the Kinki Region (CDKR)
	Surveys of conditions affecting location of plants	Factory Location Law	MITI
11. Improvement of pollution control facilities and conservation of nature	Regional development planning	National Capital Region Development Law Law for the Conservation of Green Belts around the National Capital Region	NLA
		Kinki Region Development Law	NLA
		Law for the Development of Conservation Areas in Kinki Region	NLA
		Chubu Region Development Law	NLA
	Development of new industrial cities and industrial development of special areas	Law for Promoting Development of Special Areas for Industrial Consolidation, Law for Promoting Establishment of the New Industrial Cities	EPA, MAF, MITI, MT, MC, Ministry of Home Affairs (MHA)
		Factory Relocation Law	MITI
	Construction of sewerage systems	Sewerage Law	MC
		Law Concerning Emergency Measures for Sewerage Construction	
	Construction of buffer zones	City Planning Law	MC
	Conservation of the environment under the natural park system	Natural Parks Law	FA
	Conservation of forestries	Forestries Law	MAF
	Protection of wildlife	Law concerning Wildlife Protection and Hunting	MAF
	Protection of cultural properties	Law Concerning Special Cultural Properties	Culture Agency
		Law Concerning Special Measures for Preservation of Historical Natural Features of Ancient Cities	Prime Minister's Office (PMO), MC

Table 1
(continued)

Item	Area of Responsibility	Relevant Laws	Jurisdiction
12. Research and development	Coordination of research and development activities		Science and Technology Agency (STA)
	Implementation of research and development activities		STA, EA, Ministry of Education (ME), MHW, MAF, MITI, MT, MC, etc.
13. Settlement of disputes and relief	Settlement of environmental pollution disputes	Law Concerning the Settlement of Environmental Pollution Disputes	PMO
		Mining Law	MITI
		Temporary Law for Compensation of Damage Caused by Coal Mines	
	Relief for the patients affected by environmental pollution		EA, MITI
14. Cost bearing and incentive measures	Determination of the entrepreneurs' share of the cost of public pollution control works	Law Concerning Entrepreneurs' Bearing the Cost of Public Pollution Control Works	PMO
	Loans by the Environmental Pollution Control Service Corporation	Environmental Pollution Control Service Corporation Law	MITI, MHW
	Loans for the modernization of small and medium enterprises	Law for Loans for the Modernization of Small and Medium Enterprises	MITI, MT, MAF
	Special measures concerning taxation	Corporate Tax Law, Special Taxation Measures Law, Local Tax Law	Ministry of Finance, MHA
	Subsidies for noise control measures in the vicinity of public aerodromes	Law Concerning Prevention, Etc., of Disturbance Caused by Aircraft Noise in the Vicinity of Public Aerodromes	MT
	Subsidies for noise control measures in areas surrounding defense facilities	Law Concerning Adjustment, Etc., in the Environs of Defense Facilities	DFAA
	Subsidies for pollution control facilities of schools		ME
15. Punishment of crimes relating to environmental pollution	Punishment of crimes and offenses relating to environmental pollution	Law for the Punishment of Environmental Pollution Crimes relating to Human Health	Ministry of Justice (MJ)
	Protection of human rights		MJ
16. Other	Regulation of toxic substances, etc., used in factories	Labor Standards Law	Ministry of Labor
	Education for environmental conservation		ME
	Toxic Substance Act		MITI

traditional ministries have in fact succeeded in retaining a greater part of regulatory power.[65]

Third, the influence of the Environment Agency has been hamstrung by niggardly budgetary allocations. In fact, the environmental protection budgets of many of the traditional ministries, often a small fraction of the monies appropriated for these agencies, have consistently exceeded the entire budget of the Environment Agency.[66]

Despite budgetary limitations, jurisdictional rivalries, and a shortage of trained officials, the first director general of the Environment Agency, Buichi Ōishi,[67] ably demonstrated the capacities of a powerful man. From the outset, Ōishi sought to project the novel concept of a government agency established to defend the weak or afflicted. For example, upon his appointment, Ōishi declared that the pollution control administration "should be conducted from the viewpoint of the people, who are the victims."[68] Ōishi first dramatically demonstrated this resolve by "freezing" construction of tourist roads in the Ozenuma and Daitsetsuzan districts because, he claimed, the construction might destroy parts of these national park areas.[69] Thereafter, Ōishi urged the Minister of Transport to impose stricter noise pollution standards than those recommended by the Central Council for Environmental Pollution.[70] This action has remained virtually unprecedented because in almost all cases the Environment Agency has substantially adopted the council's recommendations.[71] In a final case, Ōishi gave new hope to victims of Minamata disease by ordering the governors of Kumamoto and Kagoshima Prefectures to reconsider their rejection of the certification of some patients.[72] By these acts, Ōishi demonstrated the powers inherent in the Environment Agency's mandate to serve as a coordinating agency[73] and at the same time revived the hopes of pollution victims and conservationists that the government might become more sympathetic to environmental concerns.[74]

Despite the public applause accorded Ōishi's efforts, the new agency has exerted little influence on the actual administration of pollution control. One reason for this has been its continuing conflict with MITI. An early example of the rivalry between the two agencies was MITI's opposition to an Environment Agency proposal that polluters be subject to strict liability.[75] After some struggle, the Environment Agency ultimately retracted this idea.[76]

A second battle with MITI took place in September 1971 when MITI's Pollution Safety Bureau attempted to prevent the publication of the names of companies discharging poisonous heavy metals into the nation's waters, despite the fact tht the polluting plants reportedly included one company that exceeded the existing discharge limitation for chromium by 320 times. MITI maintained that disclosure was improper, given its prior promise of secrecy to the concerned prefectures and enterprises. Ōishi insisted, however, that all data acquired by MITI must be publicized, and that if MITI failed to do so, as director general of the Environment Agency, he would assume responsibility for its publication. The dispute was eventually settled four days later when MITI published the names of the violators.[77]

In recent years public doubts about the agency have increased in tandem with its inability to promote effective environmental protection legislation and other measures. For the most part, Ōishi's successors have not shared his early pioneering spirit. Indeed many of these individuals have evidenced little understanding, sympathy, or knowledge of environmental problems.[78]

A good example was the tenure in office of Shintarō Ishihara. A writer by avoca-

241

tion with an avid interest in marine ecology, Ishihara had little practical experience with environmental problems prior to his appointment. After assuming office, Ishihara announced that the agency's prior "open door" policy of easy access to pollution victims and environmentalists dating from the era of Ōishi was at an end.[79] Henceforth, he declared, some meetings with lobbyists and victims would be closed to the press, their content "prescreened," and their number substantially restricted.[80]

As a result of these and other actions by the agency's ministers and other high-ranking officials, there has been a substantial loss in confidence and trust in the agency.[81] Morale within the agency reportedly is low, for most officials lent to the agency on a temporary basis have not developed new loyalties. Few junior officials have been promoted, and most officials express little pride in the agency's contribution to the national welfare.[82] The traditional ministries still consider the Environment Agency to be an official upstart, and the public now seriously doubts its willingness to act in the environment's behalf.[83]

1.2.2   The Central Council on Environmental Pollution Control (Chuō Kōgai Taisaku Shingikai) and Other Affiliates of the Environment Agency

The Central Council

The establishment of the Central Council concluded a series of governmental reorganizations spanning several years. Prior to the enactment of the Basic Law, the most important governmental advisory body was the Pollution Commission (Kōgai Shingikai) established in 1965 and attached to MHW. This forty-member body, together with MITI's Council on Industrial Structure, played a critical role in determining many of the important policies embodied in the Basic Law. When the Basic Law was passed in 1967, the major functions of the Pollution Commission were transferred to the Central Council, initially located within the Prime Minister's Office.[84] With the establishment of the Environment Agency in 1971, the Central Council was thereafter reorganized as the principal advisory organ to the new agency's director general.[85]

The majority of council members are professors and scientists, although other members include ex–governmental officials, representatives from local government, business associations, other private organizations, and labor unions.[86] In addition to the regular council members, who now number ninety,[87] council subcommittees have also included about two hundred expert members. As shown in fig. 3, the council is divided into twelve subcommittees (bukai) most of which have their own expert subcommittees dealing with specific problem areas. Regular council members and experts work on a part time basis.

As has been observed in other studies of Japanese government practice, the bureaucracy has tended to dominate council affairs.[88] Perhaps the clearest example of this general trend is the bureaucracy's control over the council's investigations. For example, in virtually all cases the questions put to the chairman of the council are narrowly defined,[89] and indeed the posture of the questions in many cases anticipates (and influences) the direction of an eventual recommendation.[90]

The bureaucracy in some sense also manages the council's research. Because the council, like other advisory committees, does not possess an independent research staff, most research is conducted by Environment Agency staff members, and research conducted by the council itself is always closely supervised. Fur-

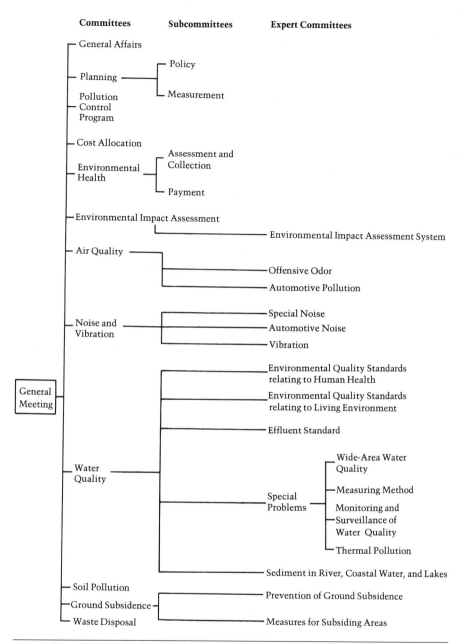

Figure 3
Central Council for Control of Environmental Pollution (September 1976).

thermore, most technical data and other information studied by the council is made available to it by the agency.[91]

The Environment Agency bureaucracy is also in a position to influence the drafting of the council's final report. Under the usual practice, Environment Agency officials will directly supervise the drafting of the expert subcommittee's report. Even in cases where experts are chosen to draft a report, government officials are usually involved. Often a draft report is circulated to other agencies, and it is not uncommon for officials of these agencies to attach comments. Thereafter, the experts' report is submitted to the subcommittee where council members representing various interests debate its merits and weaknesses. Unless a matter is particularly controversial, a decision of a subcommittee is usually accepted.[92] The final report is thereafter submitted to the director general of the Environment Agency or to the appropriate official(s) who initiated the request to the council.

Because the Central Council is only an advisory body, its recommendations are not binding. In theory, a minister is free to modify or to reject the council's report. However, modification or rejection is rather unusual. If an intended recommendation of the council appears to the Environment Agency to be unacceptable, the agency will settle the troublesome aspects of the report in advance.

Although the Environment Agency in many ways dominates the affairs of the Central Council, it should be pointed out, in fairness, that the council has played a more active role in policy making than is usual for such advisory bodies. The council has also performed a valuable service in synthesizing, collating, and analyzing highly technical issues at times beyond the scope of the competence of the Environment Agency staff. As a result, most important policy decisions of the agency have been conducted in close collaboration with the council.

In addition to these formal proceedings of the council, Environment Agency bureau chiefs (kyokuchō) may at times establish ad hoc expert committees. Although the functions of these committees parallel those of the council, these ad hoc groups are often formed to circumvent the restrictions on the number of official advisory organs permitted each ministry.[93] Usually these informal councils are called "study groups," and their expenses are defrayed by the agency's research budget.

The National Institute for Environmental Pollution Research

The National Institute for Environmental Pollution Research was established on March 15, 1974, as an affiliate of the Environment Agency. Some of its divisions are concerned with air, water, or soil pollution, while others deal with public health and the environment in general; a consultative body advises its director.

In 1976 the Environment Agency appropriated a total of ¥2,856,320,000 for one hundred environmental research projects, to be undertaken by various national institutes. These projects included an analysis of complex air pollutants in urban areas, the development of low-pollution cars, the assessment of new toxic pollutants such as PCBs, the improvement of waste water disposal systems, and a study of pollution control in the Seto Inland Sea.[94]

The Training Institute for Environmental Pollution Control

This Institute, established in March 1973, processes 1,500 to 2,000 trainees annually. For the most part, individuals are personnel of various central and local agencies who supervise pollution control programs. Recently, the Environment

Agency has also conducted technical training courses designed to assist middle-level officials from other countries. Participants from Southeast Asia, the Middle East, Africa, and Central and South America have attended these seminars.[95]

1.3   Development of Local Environmental Activism

1.3.1   Introduction

Unlike the loose association of states that predated the founding of the republic of the United States in the 1780s, the present forty-six prefectures of Japan (including the Tokyo Metropolis) were created only after the collapse of the Tokugawa Shogunate in 1868 and the establishment of a central government. The present prefectures are essentially artificial units, carved by the new Meiji government from the territories of the feudal clans.[96] During the Meiji period, and extending throughout most of the prewar period, the benefits of local administrative autonomy were poorly understood. Until World War II the powerful Home Ministry appointed prefectural governors, controlled the police, and exercised close supervision over other local affairs.

Despite the ambitious reforms instituted by the Occupation authorities,[97] some commentators argue that the SCAP efforts to promote broad local autonomy have generally not proved successful.[98] For example, after a short period of experimentation, the central authorities reconsolidated local police systems within the National Police Agency in 1954, local elective school boards returned to the supervision of the Ministry of Education in 1956, and the abolished Ministry of Home Affairs was partly reanimated in 1960.[99]

In addition to retrenchment, the weakness of local tax powers impeded decentralization. Even today local governments may retain only 30% of the amounts collected in taxes to cover local expenditures, with the remaining 70% diverted to the central government.[100] This has permitted the central government to maintain tight control over local fiscal affairs. As much as four-fifths of the work of local government is performed on behalf of the central ministries, and the central government actually employs many local officials whom it pays to oversee local affairs.[101]

The one outstanding exception to the central government's domination of local governments during the postwar era has been the local initiative in the area of pollution control, although even this development has occurred only recently. During the 1950s local governments obediently followed the national targets for economic growth, eager to attract large factories and hoping for increased tax revenues and new jobs. Thus, local governments permitted factories, ports, and highways to be hastily constructed. Often they granted subsidies and special tax incentives.

Yet after only a short while local governments began to feel the strains of the economic growth juggernaut. The construction of large factories often failed to increase local employment, because the use of modern technology tended to eliminate the need for all but the most highly skilled labor. And the expenditures of local governments in maintaining industrial sites often greatly exceeded the tax revenues generated by the large factories.

These strains helped radicalize many local governments and influenced local attitudes toward pollution. But other factors were also responsible. First, local communities suffered severe adverse health effects from industrial pollution. And because local politicians were usually the first to hear complaints about pollution, these individuals naturally had the greatest political incentive to marshal relief. Second, as pollution became more serious, local communities began to establish more stringent standards than those set nationally, and to impose other relief measures. Third, the problems of pollution control served for many local communities, especially those dominated by the opposition parties,[102] as a convenient platform to raise other political and social grievances. Finally, as discussed in ch. 2, local regulatory efforts began to reflect the emerging social ethic that asserted the rights of the average citizen to a healthy environment, and the right of local communities to remain integrated and undisturbed.[103]

1.3.2   Forms of Local Control

Local Ordinances (Jōrei)

Although local governments began sporadically to pass pollution control ordinances as early as 1962, the Tokyo Metropolitan Environmental Pollution Control Ordinance of July 1969 provided the greatest impetus for local legislation.[104] In many respects the Tokyo Ordinance established a number of important precedents. Unlike earlier measures that were limited mostly to the control of noise, the Tokyo Ordinance sought "to prevent and to eliminate all environmental pollution" that would "disturb the wholesome, safe, and comfortable life of the people of this large city." Moreover, the preamble granted pollution control priority over economic growth, while national policy still stressed "harmonization" of these two objectives. Where the national law at the time required merely a report of a factory's construction, art. 23 of the ordinance required a permit.[105] Although the national law regulated only the density of pollutants,[106] the Tokyo Ordinance established the important principal of regulating pollutants in terms of the total volume of discharges.[107] Finally, the Tokyo Ordinance empowered the Tokyo Municipality to impose draconian measures against polluters—such as shutting off the industrial water supply.[108]

As might be supposed the passage of the Tokyo Ordinance excited considerable political and legal controversy. Its principal defender was Michitaka Kaino. In response to the central government's argument that the prefectures had historically never enjoyed independent rule-making powers, Kaino argued that since 1868 the prefectures had been "platforms" for local citizens to express their grievances. In support of his assertion he pointed to the famous Fukushima incident of 1882, wherein the central government even under the old constitution had conceded some measure of autonomy to local citizenry.[109]

Second, Kaino contested the government's assertion that local governments (and their governors) were merely agents of the central authorities, unauthorized to enact legislation that exceeded the scope of national laws.[110] Kaino questioned the national government's monopoly over pollution control and maintained that art. 2.3 of the Local Autonomy Law authorized independent local action to protect citizens' health and safety.[111]

Third, Kaino pointed out that the object of the Tokyo Ordinance was the factory or workplace as a whole, whereas national laws were limited to factory discharges of smoke, soot, or effluent. He urged that although the subject and means of regulation differed, the Tokyo Ordinance did not conflict with existing national laws.

Fourth, Kaino insisted that the central government's failure since 1949 to object to Tokyo's regulation of public hazards had vested the Metropolis with the right to control these hazards.

Finally, Kaino urged that in light of the great risks to public health a uniform national standard was inappropriate. Tokyo, he emphasized, must have independent powers to enforce its own regulations.

Many other scholars quickly supported Kaino.[112] Their defense of the ordinance was based on an interpretation of art. 94 of the Constitution, [113] that permits local governments to enact regulations (jōrei, kisoku) within the scope of law, and art. 14 of the Local Autonomy Law, that authorizes local legislation not "inconsistent" with national law.[114]

Under traditional theory, local legislation is preempted when a statute is silent. Local legislation is also preempted where nationally uniform regulations are deemed (by the central government) more desirable. In the debate with Kaino, the central government adopted the orthodox position that the water and air pollution laws had preempted local legislation, except for those local regulations pertaining to nondesignated areas.[115]

The defenders of the Tokyo Ordinance counterattacked with three arguments. First, they urged that because local governments possessed both a right and a duty to protect the health and welfare of local residents, the Tokyo government should be able to impose more stringent regulations than those already set by the central authorities.[116] Second, they emphasized that local governments possessed "intrinsic authority" to protect local residents, and urged, therefore, that local pollution control legislation could not possibly be "inconsistent" within the meaning of art. 14 of the Local Autonomy Law.[117] Finally, they concurred with Kaino's assessment that although the Tokyo Ordinance differed in the subject of its regulation (i.e., factories rather than discharges), it was not inconsistent with national law. Although these arguments may not appear to the Western reader entirely logical, they were widely applauded by the general public at the time and thereby acquired considerable political weight.

Thus, increasingly after 1969, the demand for the more effective control of pollution became identified with the issue of local autonomy and the merger of these two issues influenced the course of subsequent legislation. In 1970, when the Basic Law was amended in response to public criticism of the "harmonization clause," the defenders of the Tokyo Ordinance hailed this change as a symbolic victory, for the Tokyo Ordinance had already recognized the principle of environmental supremacy. The 1970 amendments to the Basic Law also permitted local prefectures to establish stricter emission standards that those set centrally,[118] an idea also incorporated in the 1970 amendments to the air and water pollution control laws.[119] Finally, in 1974 the government amended the Local Autonomy Law to explicitly accord local bodies the right to control pollution.[120] Although local governments were still not permitted to pass ordinances that contradicted a statutory provision, the amendments were deeply symbolic. This was the first time that local governments had achieved some real measure of autonomy after many years of domination.

The trend toward decentralization of pollution control accelerated to the point that by 1975 every one of Japan's forty-six prefectures had passed some variety of pollution control ordinance; seven prefectures had passed ordinances establishing a comprehensive approach to environmental protection generally following the Tokyo Ordinance. For example, the number of pollution control, environmental protection, and other related ordinances increased from 79, 7, and 48 in 1971 to 426, 144, and 65, respectively, in 1975.[121]

Pollution Control Agreements (Kōgai Bōshi Kyōtei)

Local governments have also attempted to control pollution by executing pollution control contracts directly with individual factories. Until the amendments of the Basic Law and the air and water pollution laws, local governments possessed few independent regulatory powers and were not even permitted to enter and to inspect a polluting facility. The invention of pollution control contracts served as an imaginative device to circumvent inhibition by the central authorities.

Among the first recorded postwar contracts was the 1964 agreement between Yokohama City and Kanagawa Prefecture and the Tokyo Electric Power Company over the city's sale of reclaimed land to the company. In the contract, the company promised to meet strict standards for dust, SO_x, and noise, to install stipulated pollution control equipment, to use low sulfur oil and coal, to permit city officials to inspect its facilities, and to observe all future municipal instructions for pollution prevention. In cases of violation, the city was authorized to undertake pollution abatement at the company's expense. As a result of Yokohama City's success, other municipalities began to require that factories conclude pollution control agreements, and some municipalities concluded agreements that were wholly unrelated to the sale or lease of land.

Several factors influenced this development. One was the weakening of industry's bargaining position. As the adverse health effects of air pollution became known after 1970, the central and local authorities introduced many controls on factory emissions and siting.[122] These controls made industry's task of finding suitable factory sites, already onerous because of the dearth of land, even more difficult.[123] As a result, the companies' negotiating position with local governments was substantially undercut. Public pressure also served as a primary motivating force for both local officials, sensitive to their political consituents,[124] and for the companies. As discussed in ch. 2, the surge of antipollution feeling constituted a moral imperative for enterprises to be sensitive to the environmental consequences of their actions. Indeed, many companies soon deemed negotiating a pollution control agreement with local officials to be more desirable than a direct confrontation with irate residents.

The early pollution control agreements were rather abstract because their legal status was uncertain, and because local governments lacked technical expertise to insist on more specific provisions. Essentially, many contracts were simply "gentlemen's agreements" (shinshi kyōtei). For example, many early agreements declared that a company had an abstract duty to prevent pollution, to supply information, or to cooperate with local officials in a factory inspection. The company simply promised to take feasible and "appropriate" measures to prevent pollution. Such contracts in most cases did not contain provisions to handle violations.

Recent pollution control agreements, however, are drafted more precisely, and are richer in content. For example, many agreements require companies to meet emission or effluent standards stricter than national standards; others stipulate that a factory use special low sulfur fuel; some require the use of the most advanced pollution prevention technology, contain stipulations on factory operations, or provide for inspection of the factory. In some cases these contracts also authorize drastic enforcement measures like stop-work orders, emergency enforcement by proxy, strict liability for damages, fines, cancellation of a contract (where a sale of land between local government and a factory is involved), and as previously mentioned, interruption of the municipal water supply upon notice of violation.[125]

Tables 2–4 record the basic statistics on pollution control agreements concluded between 1971 and 1975.[126]

The increase in the number of pollution control agreements reflects a recognition of their wide range of uses. For example, one important recent use is to forestall conflicts between national law and local ordinances. Thus between 1970 and 1975 many local governments wished to set stricter SO_x emission standards for factories than the national emission standard,[127] but the central authorities probably would have opposed this action. Through the contract device, these difficulties were avoided.

Second, even where a local ordinance has set stringent standards, the pollution control agreement often proves useful because it can specify even more detailed controls, such as a specific pollution prevention device or operating procedure. In short, the contract allows factories to tailor controls to different geographical and technical conditions. Neither national laws nor local ordinances achieve so fine a tuning.

Third, the negotiating process often provides local governments with the opportunity of assisting a factory in developing a pollution prevention plan.[128] Because local governments these days are storehouses of information on all aspects of pollution control, they can instruct firms on suitable types of pollution equipment, operating procedure, or manufacturing process. In some cases local governments have even provided small- or medium-size factories with financial assistance. (Sec. 2.3.4)

Finally, many recent agreements have been used by the public as a powerful tool to promote the "democratization" of corporate and government decisions by securing citizen control over a factory's management and operations. Many arrangements now entitle the public to inspect a factory and obtain information relevant to pollution control or abatement.

A substantial number of these agreements have already been reached between factories, local governments, and citizens' groups either on a tripartite basis or with citizens' groups participating as observers[129] (table 5).

The legal validity of pollution control agreements has been the subject of a scholarly debate that has not yet been satisfactorily resolved.[130] Although vague or abstract agreements are generally regarded not to be legally enforceable as contracts,[131] they may at times remain deeply significant from a social and institutional perspective.[132] Some scholars, however, even question the status of precise and detailed agreements on the grounds that local governments do not possess statutory authority to impose their administrative powers over private parties by use of a contract. This argument has prompted some local governments to authorize pollution control agreements by passing enabling ordinances. Despite such actions, the debate continues, because some scholars and others insist that local governments still may not impose fines and other penalties by contract.

Another problem arises from the failure of most pollution control agreements to provide an enforcement mechanism in cases of their violation. Although it has been suggested that a prefecture or municipality may obtain an injunction on the grounds of violation of contract,[133] some provisions, such as a specific emergency low-sulfur fuel requirement, may simply not be enforceable under present civil procedure.[134]

Guidelines for Development (Shidō Yōkō)

Another early device that local governments used to enhance their powers was the promulgation of "guiding principles" for land development. The first effort

249

Table 2
Number of firms regulated by pollution control agreements (as of Oct. 1, 1975).

	General Regulation	Type of Fuel	Soot and Smoke	Water Discharge	Noise and Vibration	Noxious Odor	Other	Total
Number of firms regulated	7,135	2,473	3,908	4,947	4,169	3,010	3,856	8,923

Source: Kankyō Hakusho (White Paper on the Environment) 394 (Environment Agency, 1976).

Table 3

Provisions of pollution control agreements.

Year	1971	1972	1973	1974	1975
Total number of firms regulated	2,141	3,202	5,097	7,096	8,923
Agreements providing for suspension of operation and liability for damages	804	1,104	1,777	2,412	2,808
Agreements providing for absolute liability	1	10	37	113	510
Agreements providing for on-site inspection	1,509	2,271	2,603	4,863	6,062
Agreements providing for penalties in cases of breach	240	410	945	1,390	1,985

Source: Quality of the Environment in Japan, 1976 228 (Environment Agency, 1976).

Table 4

Number of private firms which have signed pollution control agreements, by industry (as of Oct. 1, 1975).

Type of industry	Number of firms	Type of industry	Number of firms
Agriculture	570	Rubber, Leather	88
Mining	268	Ceramics	677
Construction	350	Iron, Steel	372
Foodstuffs	619	Nonferrous Metals	221
Clothing, Textiles	383	Metal Products	1,016
Timber, Wood Products	331	Machinery	1,025
Paper, Pulp	353	Electric	177
Chemicals	942	Other	1,348
Petroleum, Coal Products	183	Total	8,923

Source: Quality of the Environment in Japan, 1976 226 (Environment Agency, 1976).

Table 5

Public participation in pollution control agreements.

	Number of Firms				
	1971	1972	1973	1974	1975
Agreements with Local Governments:					
Groups of residents participating as parties	18	40	45	67	76
Groups of residents participating as observers	175	231	276	315	337
Agreements with Groups of Residents:	223	395	796	1,113	1,394

Source: Quality of the Environment in Japan, 1976 227 (Environment Agency, 1976).

was Kawasaki City's guidelines that conditioned the city's approval to land developers on their undertaking various environmental preservation measures.[135] Recently, with the support of local residents, administrative "guidance" of this sort has become widely used, and like pollution control agreements, takes many forms. For example, some recent guidelines require that a developer dedicate a park or a school; others impose various restrictions on building design or construction; or authorize a municipality to turn off the water supply to a development not complying with the guidelines.[136]

Yet, the legal status of these guidelines is also uncertain. Although there is some support for the view that the Land Development Law and the City Planning Law implicitly authorize the issuance of guidelines as part of the permit powers accorded local governments, more recent initiatives appear to have exceeded the original intention of these laws. Indeed in response to the Musashino City decision,[137] a number of local governments have begun to rely more extensively on the legally stronger option of pollution control ordinances.

Section 2   The Formulation and Implementation of Environmental Policy

To help the reader follow more easily the discussion in this section, it may be useful to distinguish central from local, and formal (i.e., strictly "legal" in a traditional sense) from informal measures and institutions. The Environment Agency, the Central Council, and the National Training and Research institutes, as described, are central institutions established by statute; ordinances are formal, legal enactments by local bodies. Pollution control agreements, on the other hand, are examples of less formal, local institutions. Essentially, they are hybrids (i.e., some are formal contracts, others are not). Local administrative guidelines are the least formal institutions thus far discussed.

In the following discussion, "central planning" should be viewed as a quasi-formal (i.e., hybrid) central institution; standards are set at both the central and local levels and some types of standards (i.e., emission controls) are more formal than others. Enforcement of environmental policies occurs at both the central and local levels and combines formal measures (i.e., statutorily based administrative orders) with less formal procedures (i.e., administrative guidance). Similarly environmental pollution control programs, "pollution prevention service corporations," and the special enforcement program for small and medium enterprises are all hybrids: they function at both the central and local levels and have formal and informal aspects.

2.1   The Development of Centralized Environmental Planning

Although the Basic Law in some sense established a "master plan," the Japanese approach to environmental protection has been more piecemeal than comprehensive. From the start it was a reaction to the crises of Minamata, Toyama, and Yokkaichi. And the initial remedy was direct. Factories believed to be the major sources of mercury, for example, were forced to seal off effluents, to convert processes, or simply to discontinue production.[138] Stringent measures for some air pollutants like sulfur and nitrogen oxides were also implemented.

Yet, at the same time whole other fields were neglected. For example, water and marine pollution received scant attention and noise pollution was virtually ignored until relatively recently. There was no real attention given to environmental planning.[139]

Japan's unwillingness to address the possible need for comprehensive environmental planning contrasts with the emphasis the government placed on centralized planning in other areas.[140] For example, since the early 1950s, the government has formulated plans for economic growth, energy supply, and industrial development. And while comprehensive land use planning has at times touched on the issue of environmental protection, land use measures, particularly those conducted under the ambitious National Land Use Planning Law of 1974,[141] have not succeeded in bringing the field of environmental protection into the mainstream of central decision making.

With the establishment of the Environment Agency in 1971, the government began, albeit gradually, to address the need for centralizing its various environmental programs. The first effort was the 1971 "Long Term Prospectus for the Preservation of the Environment," an interim report prepared by the Planning Committee of the Central Council. Although general, the report attempted to examine various projected pollution levels, to analyze these with reference to other variables (i.e., increasing public demand for a better environment, projections of economic growth, etc.), and to set forth a series of long-term objectives and "tasks" for environmental policy.

A second early effort was the 1972 Cabinet Resolution that introduced environmental impact assessment procedures into the government's anlysis of public works projects. This resolution has served as a basis for reviewing a substantial number of projects and has stimulated greater national sensitivity to environmental planning.

The most ambitious measure to date was the government's announcement of a ten-year "Environmental Protection Program" in 1977.[142] The program's introduction outlines its four purposes:[143] (1) to refine existing policies and to set forth measures needed for the attainment of environmental quality standards; (2) to develop long-range and comprehensive administrative guidelines for the conservation of the environment accounting for economic growth, industrial and regional developments, and other environmentally disruptive factors; (3) to develop a framework for analyzing the costs of achieving pollution control and other objectives; and finally, (4) to outline the general direction of future central and local initiatives for preserving, maintaining, and enhancing environmental quality. The plan also attempts to establish a framework for coordinating controls on air, water, solid waste, and noise pollution, and addresses a wide range of additional subjects—future pollution damage relief measures, the promotion of environmental science, the establishment of environmental education programs, new pollution control technologies, critical environmental issues in land use planning, natural park conservation, wildlife protection, the protection of amenities, and finally the direction of international cooperation. Although the program is extremely vague and cursory in its treatment of these many subjects, it represents one of the first efforts anywhere to integrate conceptually the entire range of environmental programs and policies.

An equally impressive effort is also being made to establish an overall econometric model to accompany the "Long Term Environmental Preservation Plan."[144] In its present formulation, the model addresses the relation between (1) economic activity and pollution discharges; (2) future emission limits and "targeted emission limits"; (3) targeted emission limits and required pollution control investment (e.g., capital stock); (4) the "demand inducing effect" of pollution control investment; (5) the demand reducing effect of pollution control expenses; and (6) the influence of pollution control investment on prices and other variables. The model assists the administration in coordinating pollution control measures with other national policy objectives and provides a relatively objective method of analysis.

Yet, despite these advances, centralized environmental planning is still inchoate. Legally the status of the plan is uncertain, for unlike economic growth, industrial, or land use planning, the Environment Agency's action was ad hoc and not based on explicit statutory authority.[145] For some time the attainment of emission standards, and the implementation of various local pollution control programs, will continue to occupy the government's attention.

2.2   Standards

2.2.1   Types of Standards and Their Role in Environmental Protection

As in many countries, standards in Japan constitute an important part of the government's environmental protection program. The word "standards" refers to two sets of prescriptions. The first, "ambient" or "environmental quality standards" (kankyō kijun) represent levels of pollution that an area may not exceed. These standards are not by themselves legally binding and serve essentially as policy objectives.[146] "Emission standards" (haishitsu kijun), the second type, fix the quantity of pollutants (or their concentration in effluent) that a source may discharge per unit of time. These standards are legally binding and are enforced by administrative and criminal sanctions.

Ambient standards themselves are divided into standards designed for the protection of human health, and standards established for the conservation of the living environment.[147] In some areas like air pollution, the government has established only the former, health-related standards, reasoning that these standards would also protect the living environment.[148]

Japanese environmental quality standards also may vary by geographical area.[149] For example, different rivers may have different quality standards and at times the standard varies even within the same body of water.[150]

As Table 6 indicates, Japanese ambient standards for some pollutants are the strictest in the world. The reader should not conclude, however, that because environmental quality standards serve only as "administrative targets," they are not taken seriously. Indeed, the case studies in sec. 3 attest that environmental quality standards serve as critical reference points, and, because these standards are backed by consensus, their attainment is considered at least as important as in Western countries.

Unlike environmental quality standards, emission standards are set by both the central and local governments. The central government has established emission standards for air and water pollutants and noise applicable to stationary and mobile sources.[151] Often prefectural and municipal governments have established supplementary emission standards.[152]

Although there are many types of emission standards, a few generalizations are possible. Most standards are expressed in concentrations of pollutants, not in absolute quantities. In theory, this permits a polluter (particularly in the case of water pollution) to increase the amount of a pollutant discharged by increasing the quantity of emissions or effluent. The cases of emission standards for SO_x and

Table 6
Air quality objectives,[a] Japan and selected OECD countries, 1975.

	SO_2 (ppm)	Particulates (mg/m³)	NO_2 (ppm)
Japan	0.04	0.10	0.02
Canada	0.06	0.12	0.10[b]
Finland	0.10	0.15	0.10
Italy	0.15	0.30	n.a.
United States	0.14	0.26	0.13[c]
Germany	0.06	n.a.	0.15[d]
France	0.38	0.35	n.a.
Sweden	0.25	n.a.	n.a.

Sources: For SO_2 and for particulates: Werner-Martin and Arthur C. Stern, The Collection, Tabulation, Codification and Analysis of the World's Air Quality Management Standards, School of Public Health, University of North Carolina at Chapel Hill, N.C., Oct. 1974; for NO_x: R. Kiyoura, International Comparison and Critical Analysis of NO_2 Air Quality Standards, paper presented at the 69th Annual Meeting of the Air Pollution Control Association, Portland, Oregon, June 27–July 1, 1976.
[a] All figures are average daily values.
[b] The figure is for Ontario; the figure for Saskatchewan is much lower: 0.01.
[c] For the United States, the NO_2 objective is set in terms of average yearly value (0.05 ppm); the figure given is therefore an equivalent open to criticism.
[d] The German standard is 0.05 ppm for "long-term exposure" and 0.15 ppm for "short-term exposure."

NO_x, however, are somewhat more complex. Under the currently employed "k" value system established by the Air Pollution Control Law of 1968, the standard for sulfur oxides is defined not only by concentration, but also in terms of quantity, and the speed and temperature of the exhaust.[153] Although most emission standards do not vary with the location of a facility, the value of the coefficient k for SO_x may. To date, more than 100 areas, grouped in sixteen k categories, have been so designated. Because the value of k is about six times higher in the less-polluted areas than in the most polluted areas, a given facility may be permitted to pollute about six times more in Northern Hokkaidō than in Tokyo or Osaka.[154]

Whereas centrally established standards are usually geographically uniform, locally set standards are not. Indeed local ordinances and pollution control agreements often prescribe even more stringent standards[155] and these differences, at times, can be great.[156]

Although emission standards are theoretically established as a means of attaining ambient standards, the two types of standards are viewed somewhat independently. In the case of SO_x emissions the more industrial an area, the higher the k value. It does not follow, however, that for each of the possible eight areas for which k = 8.76, the resulting ground concentration will be the same in all these areas simply because all stacks in the areas meet the standard. For although national emission standards are uniform for each pollutant throughout the country, the density of polluting facilities is not. Consequently, despite the uniformity of the national emission standards, the ambient quality of an area will necessarily depend on the extent of its industrialization.[157] Although ambient and emission standards are not strictly linked in Japan as in the United States or other countries, this difference may not be significant because of the many supplemental standards and other controls established by local governments.

One innovation developed by the government in 1974 to assist in attaining ambient standards is the concept of "mass emission" controls. The intent of this measure was to remedy the defects of the k-value system by restricting the total emission of pollutants in a given area to a limit deemed scientifically acceptable.[158] Like the environmental quality standard, the mass emission control is an objective of environmental policy and not per se binding on polluters. It also must be translated into a set of emission standards. Although the government must still devise an appropriate "set" of emission controls (since many different sets can provide a "solution" to a given mass limitation), the idea of fixing a limit on total emissions has the advantage of imposing controls on individual firms. Each plant must then allocate its emissions between various polluting facilities within its confines. Consequently, the system encourages efficient resource allocation.[159]

2.2.2   The Standard-Setting Process[160]

The Environment Agency is entrusted with the responsibility of promulgating ambient quality standards that are issued by the director general after consultation with the Central Council for Environmental Pollution Control.[161] Under the usual practice, the Central Council delegates its responsibility to a subcommittee of about ten to twenty persons,[162] that in turn usually appoints an ad hoc committee of scientific and other experts.[163] The expert committee initially assesses the existing evidence on a pollutant's environmental effects. Although the committee is supposed to ignore implementation costs in its recommendations, the experts often consider these costs when discussing the issue of technical feasibility.[164]

After the experts' report is completed, it is sent to the subcommittee of the Central Council. At times the subcommittee will weaken the standards proposed by the experts or will extend the prescribed period for their attainment. Although the subcommittee's deliberations are closed to the public, officials of other ministries may observe the proceedings, and these officials are permitted to suggest modifications to a subcommittee's recommendation. In this way industries with close ministerial ties can indirectly influence a subcommittee's decision.

Unless the Central Council wishes to review the subcommittee's recommendations,[165] it will send the subcommittee's final report to the director general of the Environment Agency.[166] Should the council review the report, industry, government agencies, and environmental groups will then have an opportunity to communicate their views formally to the council and informally to individual council members. Because of public pressure,[167] however, the politically sensitive Central Council is not likely to weaken the subcommittee's proposals, although it may propose a delay in the implementation period. The director general of the Environment Agency can implement the council's report or modify it after considering additional information. Industry and other interested groups thus may have another opportunity to comment on the proposed ambient goals. Because of the council's prestige and broad base of public support, however, the director general is usually inclined to accept its recommendations.

The procedure for setting emission standards differs in several respects. With the exception of regulations controlling automobile exhausts, emission standards are set without consulting the Central Council.[168] Separate bureaus within the Environment Agency for air, water, and noise have responsibility for developing the regulations. When new standards are needed or old ones revised, the appropriate bureau chief usually consults an advisory group of experts to help prepare the regulations, or at times he may devise the standards without outside help.

If an advisory group of experts[169] is consulted, its task is to study medical, meteorological, and other technical data and to identify an emission standard that polluting facilities can meet. Within the advisory group, the emphasis on technical feasibility is great because a facility's failure to comply with emission standards can lead to civil or criminal liability.[170] Discussions on the issue of technical feasibility also often raise the question of economic feasibility. Once the advisory group has completed its work, the bureau drafts emission regulations for approval by the director general. Throughout the process the agency also generally collects outside opinions through direct consultation with interested parties or by discussion with other ministries.

The procedures for promulgating environmental standards thus give important economic interests ample opportunity to communicate their views to the government. Although it is not certain that implementation costs are actually considered when ambient and emissions standards are set (for it is conceivable that government officials give industry interests a hearing but ignore what they say about costs), the frequent communications between officials and industrial leaders suggest that the cost of compliance is a factor the government takes into account. It is thus highly unlikely that the standards that are set are entirely noneconomic.[171]

The history of the establishment of ambient standards for SO_x and NO_x well illustrates the interplay between politics and science thus far described. In 1969, MHW's Council on Living Environment called upon its expert committee to review the existing data and to offer its opinion on the appropriate standard. Pursuant to the ministry's request, the committee examined the available epidemiological and other data collected principally from Yokkaichi and Osaka and advised the ministry that an extremely strict standard (i.e., average concentrations of SO_x should not exceed 0.1 ppm, and the daily mean of hourly averages should not exceed 0.05 ppm) was warranted. The committee's opinion was thereafter duly transmitted to the Subcommittee on Pollution Control, the great majority of whose members were not scientists.[172]

The petrochemical, power, and other industries' fierce attack on the standard began at the subcommittee level. Industry contended that the proposed standard was scientifically unwarranted, that feasible technology for its implementation was not available, and that the establishment of so stringent a standard would disadvantage Japanese industry internationally. These views were embodied in a letter submitted by Keidanren (Federation of Economic Organizations) on October 29, 1968, to MHW; in this letter Keidanren requested that the SO_x ambient standard be set at a technologically feasible level until a more comprehensive scientific study of the issue was completed.[173]

Industry's attack won the day. The subcommittee and thereafter the council quickly capitulated, weakening the newly adopted standards by a factor of three and extending the period for the standard's attainment.[174] Heavily polluted areas such as Tokyo or Osaka were permitted ten years to achieve these standards, while rapidly industrializing—and therefore, increasingly polluted—areas, such as Chiba and Mizushima, were given five years to achieve the standard. Only the most heavily polluted areas, such as Kashima or Kinura, were expected to achieve the standards immediately.

The debate continued, however, at the ministerial level. Championing industry's case, MITI proposed an even more lenient intermediate hourly standard[175] to be met by heavily polluted areas within five years. Since government plans for the development of desulfurization technology were targeted for 1971, MITI requested that industry be permitted at least five years to adjust to the new regulations. (Sec 3.1) On February 12, 1969, after seven months of negotiations within the government, the Cabinet adopted the council's recommendations with some modifications. Specifically, these modifications incorporated MITI's proposal of intermediate standards and exempted the application of the standards to areas zoned exclusively for industrial sites.

Yet, politically as well as medically, the 1969 ambient standard soon proved unacceptably low. In the ensuing months, epidemiological studies of areas that had already met the SO_x ambient standard reported extremely high rates of pulmonary and other disorders associated with air pollution. The publication of these statistics caused the public to demand that the ambient standard be strengthened.

In 1971, immediately after the Environment Agency began operations, Director General Ōishi requested the Central Council to consider the question of the reassessment of the standard. After nearly two years of deliberations, the council finally recommended that the SO_x standard be returned substantially to the level originally proposed by the expert committee.[176] During its deliberations on the SO_x standard, one commentator notes,[177] the council multiplied the safety factor by two in an attempt to address the possibility of adverse synergistic effects with other pollutants.

While the debate over the SO_x ambient standard progressed, an even fiercer struggle raged over the NO_x standard.[178] Not only was the scientific community radically divided over the issue of the toxicity of NO_x, but also at the time of the publication of the standard there existed no feasible technology for the control of NO_x.

These factors greatly complicated the task of the expert committee when it was asked by MHW to propose an appropriate initial NO_x ambient standard. Although the committee possessed some data on the results of exposing laboratory test animals to NO_x, and other data based principally on clinical observations of the acute effects of high NO_2 concentrations on workers, the committee was unable to acquire any information on the effects on human health of low NO_x concentrations or the processes that form photochemical oxidants.[179]

Despite the inadequacy of existing scientific data, the expert committee was deeply concerned about the health effects of NO_x. Reasoning that the synergistic effects of NO_x might pose an even greater health hazard than SO_x, and bearing in mind the public cry for some affirmative government action, the committee concluded that the standard for NO_x should be one-half that for SO_x.[180] The expert committee's assessment was thereafter transmitted to the subcommittee and thence to the Central Council.

Although industry bitterly disagreed with the expert committee's recommendation,[181] a number of factors militated against the weakening of the standard. First, an episode of photochemical smog in July 1970 had frightened the public, so that any move to weaken the standard would be fiercely opposed. Second the Yokkaichi court's decision strengthened public support for control of NO_x. Third, MITI itself had reason to accede to the Central Council's recommendation, because MITI hoped to persuade other ministries that large-scale industrial complexes should be permitted an additional grace period of eight years. The NO_x standard was subsequently adopted as recommended without modification.[182]

Although the Cabinet's decision was strongly backed by public opinion, the decision shunted many problems to the future. Thus industry and others continued to criticize the NO_x standard even during the emotional period of the four pollution trials,[183] and the controversy was violently renewed when a team of WHO experts visited Japan in September 1976, to discuss the Japanese NO_x criteria. This event gave industry its opportunity to counterattack and it at last succeeded in forcing the government to begin the process of reassessment.

By December 1977, a subcommittee of MITI's advisory organ, the Industrial Structure Council, appeared with an overwhelming indictment of the standard. The subcommittee's report pointed out that there was little hope for the standard's attainment in the immediate future because of industry's lack of success in meeting the standard to date.[184] Also the report cited the standard's scientific unreliability, noting that no solid evidence linked NO_x pollution with various respiratory diseases, and that the existing Japanese standard was substantially higher than that adopted by WHO. Finally, the report pointed out the great difficulty of controlling NO_x because of the diversity and number of emitters[185] and the huge amounts of money industry had already (wastefully) invested in seeking to control NO_x pollution.[186]

The Industrial Structure Council's report placed the Environment Agency in a difficult position. Although at least on the surface the council's arguments appeared to have merit, many strongly opposed relaxation.[187] Behind the controversy, however, also lurked a deeper issue. The establishment of the NO_x standard had embodied the government's philosophy of the early 1970s that held that industry, not the public, must bear the burden of scientific uncertainty. Indeed, at the OECD meeting in 1976, Japan proudly named her overriding concern with public health the "noneconomic approach" to environmental decision making. Thus in a sense the issue of weakening the NO_x ambient standard posed the larger dilemma of how long Japan would continue this ambitious policy. In July 1978, the Environment Agency announced that the NO_x ambient standard would be substantially weakened.[188]

2.3   Implementation

2.3.1   Administrative Enforcement vs. Criminal Sanctions

In Japan the implementation of emission and ambient standards and the attainment of many other policy objectives depends primarily on administrative enforcement, not criminal sanctions. Let us examine the regulation of factories to see how the administrative process envelops location, design, and many other aspects of an enterprise's operations.

Although centralized environmental planning may still be in its infancy, industrial site planning is not. Under the Industrial Relocation Law,[189] the government has introduced a program of guiding factories to specific areas (i.e., "removal promotion areas" and "inducement areas") designated by Cabinet order and pursuant to the Comprehensive National Development Plan of 1962.[190] "Removal promotion areas" are defined as those areas where industry is so highly concentrated that factory relocation may be necessary; "inducement areas" are targeted for factory relocation because they are less heavily concentrated, and because population growth is projected to be low.

A variety of subtle measures are employed to "induce" relocation. For example, the central government subsidizes both the municipal government and the factory when the factory relocates from a removal area to an inducement area;[191] a relocating factory may also obtain loans at favorable conditions until the point of sale of the previous location; various tax incentives aid the factory in installing its new facilities; and finally, the government may itself construct an industrial site for the company or subsidize the municipality's construction of the site.[192]

259

A factory's location is strictly controlled by a variety of legal and administrative restrictions. For example, factories are legally required to assist the government in its environmental impact assessment of large-scale industrial development areas by providing data on local topography, wind direction, tides, and other natural conditions and land use patterns and related sociodemographic factors.[193] These reports assist the government in formulating a general strategy for administrative guidance.

The Factory Location Law also grants the Minister of International Trade and Industry broad powers over the factory's specific location within an area. Ministerial regulations designate the amount of space to be allocated to a specific factory site, those areas that must be used for green zones and those that need not, and the general kinds of pollution control equipment to be used.[194] By notification (kokuji), the government also designates the kinds of goods to be produced, the size and arrangement of manufacturing facilities, and the pollution prevention measures to be undertaken. After the above notification is delivered, a factory is enjoined from beginning operations for 90 days,[195] during which time MITI may impose various additional "conditions" or recommendations on operations.[196]

Other laws impose more specific obligations on an enterprise to be fulfilled before commencing operations. For example, art. 5 of the Water Pollution Control Law[197] requires every operator of a facility (or a person planning to operate or to modify a facility) to submit a report on its location, type, structure, method of operation, and plans for waste treatment. The law provides that if the prefectural governor to whom the report is submitted "deems that the effluent at the place of discharge does not satisfy the established effluent standard," he may order "the person who submitted the report to change the reported structure, the method of use, or the program of treatment."

Like factory siting, actual operations are also subject to constant surveillance. One interesting method of control is the requirement that designated factories appoint pollution control managers and "superintendents."[198] These persons are responsible for supervising the technical tasks of pollution prevention—inspecting fuels, measuring smoke and other emissions, and the like. The superintendents must pass a national examination administered by MITI, and factories are legally required to notify the prefectural governor of their appointment. Although managers and superintendents are employed by the factory, they also serve the government as a valuable check on heavily polluting industries.

An impressive network of monitoring stations also assists in the surveillance of individual factory operations. Since each large factory in an industrial area is required to install monitoring equipment that automatically sends collected data to a computerized central station, the system not only facilitates regulation of individual factories, but also contributes to the area's overall pollution abatement program.[199]

The principal responsibility for seeing that a factory complies with emission standards and other regulations is delegated, as is monitoring, to local governments.[200] As the case studies presented in this chapter demonstrate, protracted discussions, negotiations, and warnings represent the principal means of inducing compliance. Local governments also delicately measure an array of incentives—accelerated depreciation, loans, and procurements—to reward a firm's positive environmental performance.

Local governors also possess sterner powers. For example, under the Air Pollution Control Law, a prefectural governor can request a factory owner to supply data deemed "needed" by the government, and prefectural officials are now empowered to inspect factories on a regular basis.[201] When a factory's operation violates existing emission standards *and* poses a risk to human health or the environment, the governor is also empowered to order the polluter to improve or to suspend operations.[202] And persons who fail to comply with the governor's order are subjected to imprisonment and a criminal fine.[203]

For the Western observer, one of the most striking aspects of the Japanese administration's approach to enforcement surely must be the apparent reliance on negotiation and guidance and the disinclination to employ directly "coercive" measures.[204] Indeed, there have been very few orders to modify operations[205] and even fewer orders to comply with standard fuel requirements.[206]

Perhaps even more striking is the absence of criminal prosecutions for violations of pollution laws. This is not because Japan's pollution control laws fail to include criminal sanctions, for many do. Indeed, one statute, the Law for the Punishment of Crimes Relating to Environmental Pollution[207] is so conceptually innovative that it deserves special mention. First, the law punishes not only the act of injury but also the separate act of a polluter's negligently endangering public health.[208] In effect, the imposition of risk is criminalized. Second, the law establishes a presumption of causation where a polluting discharge creates an imminent danger to public health.[209] Third, the law punishes both the principal actor and also the corporation in cases where a corporate representative is responsible for the offense.[210]

Despite its inventiveness, the Law for the Punishment of Crimes Relating to Environmental Pollution, like criminal provisions in other statutes, has been seldom used.[211] One reason is that referral of a pollution case to the procuracy is unattractive from the administration's perspective. Generally administrators believe that guidance is a more effective and flexible tool than a clumsy, lengthy criminal trial. Because virtually all enterprises prefer guidance, the administration understandably also feels that the chances of successful administrative enforcement are greater. Moreover, by surrendering a matter to the procuracy, the administration loses its opportunity to attach additional conditions that can supplement existing statutory provisions and regulations.

Even from the procuracy's perspective, prosecution of a "pollution crime" is not viewed as a congenial matter. Although the police, of course, have the independent power to investigate a pollution-related offense (instead of waiting for a formal citizen complaint), they have generally chosen not to do so. At the same time, there have been a few pollution-related criminal complaints filed by the public. From an evidentiary perspective, prosecution of a pollution offense is also comparatively difficult because, although the police have strengthened their capacity to investigate pollution violations, police training and technical competence in the pollution field have been grossly deficient. Consequently, in many areas (e.g., air pollution) the prosecutors have had difficulty in marshaling concrete evidence. Finally, practically all pollution cases (97.3%) have been disposed of by means of summary trials, and polluting industries, even those responsible for extensive property damage, have often escaped punishment.[212]

2.3.2   Environmental Pollution Control Programs

Article 19 of the Basic Law[213] requires prefectural governors, on the instruction of the prime minister, to formulate "environmental pollution control programs" for areas where pollution is already a serious concern or likely to become so. The primary purpose of these programs is to facilitate the attainment of special ambient standards[214] applicable to the targeted program area.[215] Public works projects (the construction of monitoring stations, sewage systems, waste treatment plants, green buffer zones, the dredging of rivers and harbors, the removal of houses)[216] constitute the most important component of these programs, although other measures are also undertaken.

From the outset the central government has attempted to dominate the programs. During the preparation of the Basic Law, some politicians argued that the programs should be completely centralized, hoping that the national government would thereby be motivated to invest substantially in their implementation. These individuals also feared that without central aid the programs would soon drain meager local financial resources. Others, particularly members of the progressive political parties, insisted that local governments should assume the entire responsibility of formulating and implementing the programs. These individuals felt that the central authorities would capitulate to industrial pressure and therefore would inadequately respond to local needs. The programs' present structure reflects a compromise between these two views. Although the central government continues to maintain supervisory authority,[217] a special law[218] now authorizes the national government to subsidize the costs of many projects.

Based on ten years of experience with implementation, local pollution control programs on balance have proved useful, although a number of problems have now surfaced. First, although substantial progress has been reported in the reduction of SO_x[219] for some areas, the achievement ratio for NO_x standards reportedly has been low for all areas. Second, the programs have been criticized for their dependence on the central government.[220] Allegedly, over-reliance has led to programmatic uniformity and encouraged subservience to industrial interests. Third, the present system has apparently focused entirely on short-run measures, thereby neglecting long-range land-use, urban, and industrial planning.[221] Fourth, because the overriding goal of the program has been the attainment of a limited number of existing ambient standards, many other polluting substances as well as synergistic effects are ignored. Finally, the critics assert that the program has permitted relatively unpolluted designated areas to degrade to the level of existing ambient standards.[222]

2.3.3   Direct Government Intervention in Pollution Control and Abatement: The Environmental Pollution Control Service Corporation (Kōgai Bōshi Jigyōdan)

As noted in ch. 1 the government began to search with increasing concern in the mid-1960s for effective pollution controls that would not upset economic growth. The Environmental Pollution Control Service Corporation (Kōgai Bōshi Jigyōdan) was originally the brainchild of high-ranking officials within MHW. The basic idea behind the creation of this body was that the government itself would undertake the job of pollution control, shifting the costs over time back to the polluter. Through its management of these public works, it was thought, the government could assure its full control over pollution control programs, while avoiding dislocations from the imposition of stringent emission or other controls.

The Environmental Pollution Control Service Corporation is a multipurpose, government-owned and -operated entity.[223] It selects environmentally sensible sites, purchases land, builds green belts, installs pollution control and abatement equipment, and thereafter conveys title to the property to the concerned enterprises.[224] The corporation also builds and soundproofs housing for workers.

In addition to its direct involvement in pollution control, the corporation greatly facilitates administrative guidance of industry. Its most successful operation to date has been its direct loans, particularly to small- and medium-size enterprises, for the installation of pollution control equipment and waste treatment plants. It also provides its client enterprises with scientific and technical information and facilitates financing for pollution control.[225] By negotiating joint contracts with a number of enterprises, the corporation also encourages these firms to share information, pool funds for pollution control, and collaborate in a variety of other ways.[226] Finally, the favorable tax treatment accorded facilities constructed or installed with the help of the corporation provides an economic incentive for some enterprises to reduce pollution.[227]

2.3.4   Treatment of Small and Medium Enterprises

Small and medium enterprises constitute an important sector of the Japanese economy, but at the same time, as in many other industrialized as well as developing countries,[228] these firms can cause serious environmental problems. Because they usually eke out a marginal economic existence, such enterprises are usually not in a position to acquire sophisticated scientific information or to install expensive pollution control technology. Consequently, the imposition of stringent pollution controls often bankrupts smaller companies.[229] Yet, from the government's perspective, exempting these industries from pollution regulations can seriously weaken the overall effect of a pollution control program.

The Japanese government has sought to respond to the problems of small and medium enterprises in a number of affirmative ways.[230] First, the problem of know-how is addressed by a special counseling and guidance system. In each of Japan's nine largest cities and in many prefectures, the government has established a network of counselors[231] who offer advice on a wide range of matters from the most technical questions to simple referrals to appropriate lending and other institutions. Many local chambers of commerce also offer legal, technical, and financial assistance,[232] and the central government has established two specialized agencies, the Small Business Production Corporation and a Small and Medium Enterprise Agency to assist small firms in complying with pollution controls.

Second, small firms have access to a variety of special loan programs and are granted preferential tax treatment. One means of facilitating financing is the Environmental Pollution Prevention Service Corporation. This corporation is authorized to make loans of up to 80% of the cost of pollution control, and these loans are often used for business reconversion, the purchase and installment of pollution control equipment, equipment modernization, factory relocation, and pollution prevention insurance. Small and medium enterprises also benefit from special tax treatment, such as accelerated depreciation of pollution prevention equipment, exemptions from real estate acquisition taxes, and various incentives designed to encourage the purchase of waste-recycling equipment.

Finally, Japan has established at public expense a number of national testing and research institutions mandated to assist small firms in developing pollution control technology. These institutions grant subsidies to small enterprises, and at times, themselves conduct the research and development for an industry.

The assistance accorded small and medium enterprises should be viewed as another dimension of local administrative guidance. It is as beneficent as the stringent controls imposed by local ordinances, pollution control contracts, or various guidelines are severe. The overriding purpose is again to induce industry's compliance in as fair and effective a manner as is possible, and in large measure the Japanese practice in this area has been successful.

Section 3   Case Studies

3.1   Case 1: Implementation of Controls on SO$_x$

3.1.1   Major Legislative Developments

As Japan became increasingly reliant on petroleum during the early 1960s, SO$_x$ (sulfur oxides) replaced soot and dust as the principal air pollutant.[233] The government's first effort to control sulfur oxides was the Smoke and Soot Regulation Law of 1962,[234] but this was a piecemeal effort. The law applied only to designated air pollution control districts; owners of smoke-emitting facilities had simply to notify the prefectural governor before installing or modifying a facility; where a factory violated an emission limitation, the governor's only recourse was a "clean up or desist" order.[235] There was no idea of nondegradation, and undesignated areas were left unregulated.

Implementation of the new law was also dilatory. For example, the designation of Yokkaichi City was stalled until 1964 and implementation postponed for two more years. By 1966, the city had already certified three hundred and fifty-five patients of air pollution-induced pulmonary disease.[236] The government's establishment of emission standards was similarly halfhearted. Because emission standards were set in terms of density of pollutants at the stack, by increasing the number of stacks, using larger stacks, or diluting concentrations with fresh air, polluters could circumvent the law. Moreover, the government usually acceded to industry's request by setting low emission limitations.[237] Although large firms could easily pay the costs of meeting these standards, small and medium industries, like local porcelain manufacturers, suffered heavy economic burdens.[238] The increase in heavy oil consumption[239] coupled with the government's diffident attitude contributed greatly to the sharp rise in SO$_x$ concentrations during the same period.

As noted in ch. 1, the government began the task of formulating the Basic Law during the mid-1960s. The enactment of this statute in 1967 was important from the perspective of the regulation of SO$_x$, because it mandated the adoption of ambient standards and emission controls,[240] land use regulations,[241] and pollution control programs.[242] The Basic Law also directly influenced the passage of the Air Pollution Control Law in 1968[243] that replaced the ineffective earlier statute.

The Air Pollution Control Law of 1968 was a major advance beyond the Soot and Smoke Regulation Law for several reasons. First, it replaced the ineffective earlier approach to emission controls with the k-value system.[244] Second, the Air Pollution Control Law applied to all areas, even areas not yet polluted.[245] Finally, in emergencies, owners of polluting facilities above a certain size were required to submit abatement plans,[246] and the government was authorized to set especially stringent standards for these facilities.[247]

Despite these positive changes, the SO$_x$ concentrations in many cities did not decrease markedly, and the public began to demand the amendment of the Basic Law as one step toward a remedy.[248] As noted in ch. 1, the extraordinary "Pollution Diet" Session of 1970 amended both the Basic Law and the Air Pollution Control Law.

The amendments continued the general trend of strengthening the government's pollution control policies. For example, the harmonization clause in the Air Pollution Control Law was eliminated;[249] regulations were extended to all areas of the country;[250] the power of prefectural governments in standard-setting was augmented;[251] regulations of fuel use were introduced for the first time;[252] and new, more stringent measures against polluters were authorized.[253] Finally, in 1974 the government adopted the idea of mass emission controls that greatly strengthened its management of SO_x pollution.

3.1.2   Administrative Disposition of SO_x Pollution Cases

The first major administrative initiative to address the problem of SO_x pollution was the Kurokawa Investigation Task Force (Chōsadan), organized by MITI and MHW in 1962. The task force was an ad hoc research group composed principally of scholars from a variety of disciplines; Mutake Kurokawa, director of the task force, had headed the Industrial and Technological Research Institute and was an expert on energy policy. The team visited Yokkaichi and also the Mishima-Numazu area, and after a year submitted its recommendations to MITI. The Kurokawa group's report prescribed various remedies for the increased pollution anticipated in intensely concentrated industrial areas. These remedies included desulfurization from fuel and stack emissions, the use of pollutant-dispersing high stacks, and the creation of green buffer zones.

While the government discussed the recommendations of the task force, local governments began to take the job of SO_x control into their own hands. In 1964, the Mishima-Numazu demonstrations began and in the same year, Yokohama City conditioned a sale of land for the Tokyo Electric Company's power plant on the company's construction of high stacks and its use of low-sulfur oil. The city's virtually unprecedented action shocked the central government, because it demonstrated the impact that local governments could have on the nation's energy supply.

In 1965, the central authorities responded to these events by substantially expanding research and investigatory efforts in the pollution field. For example, between 1966-1972 the government began meteorological and health surveys in major trouble spots, e.g., Numazu and Yokkaichi, and also established a system of monitoring stations in Yokkaichi and other areas. At this time MITI also began to develop pollution simulation models to aid in impact assessments and future planning.[254]

In 1966 as public pressure mounted, MITI commenced administrative guidance of polluters in a few heavily industrialized areas,[255] despite the fact that the preparation of the Basic Law had not yet even been completed. Accordingly, MITI recommended that factories in the major industrial areas substantially curtail SO_x emissions, use fuel significantly lower in sulfur content, reduce the number of operating chimneys, and construct new and higher stacks.[256] These measures were also loosely designed to correspond to the prevailing ambient standards. MITI's recommendations themselves were aided by its newly developed simulation model.

It soon became apparent that these piecemeal measures were inadequate to deal with the nation's increasingly severe air pollution problems, and that a comprehensive approach was required. The government's first official declaration of this policy change was the statement released in February 1967 by the Comprehensive Energy Investigation Committee (Sōgō Enerugi Chōsakai), an advisory body to MITI.[257] The committee's announcement stressed the importance of a comprehensive approach, emphasizing the importance of smoke diffusion, the development of desulfurization technology, and the need for industry to expand efforts in using desulfurized oil. The committee suggested that a 1.7% reduction in the sulfur in oil be achieved by 1969.

In preparing its final report, the committee deliberated upon a wide range of possible options to promote desulfurization. These included the direct and indirect desulfurization of heavy oil, the adoption of stack desulfurization technology, the import of desulfurized heavy oil, desulfurization of gas, desulfurization by the conversion of crude oil and by naphtha, the alteration of sulfur-using industrial processes, greater use of LNG and LPG, greater emphasis on atomic energy, relocation of heavily polluting industries, and the transformation of the basic structure of many of those industries. Finally, the committee decided to place particular emphasis on direct and indirect desulfurization of heavy oil and the expedited development of new stack desulfurization technology.[258]

Several factors influenced the government's ultimate decision to emphasize desulfurization of oil and stacks.[259] First, given Japan's dependence on Middle East oil supplies and the infancy of the atomic energy industry,[260] direct desulfurization of heavy oil was the most sensible choice. Second, because the two options selected appeared to impose the least onerous economic burdens on industry, they seemed to entail the greatest chance of success. Third, from the perspective of evolving air pollution policy, adoption of the two methods seemed the most effective approach. Fourth, based on existing econometric models, the government estimated that this technology could be developed to a high level of efficiency.[261]

Implementation of the government's desulfurization policies began on an industry-specific basis even before promulgation of the committee's final report.[262] For example, as early as 1966 the government launched a ¥2.6 billion "large scale research project" (ōgata itaku kenkyū)[263] to develop on a pilot basis desulfurization technology in the power industry. Two types of technology were selected. The first such project, constructed under contract with Mitsubishi Heavy Industries (Mitsubishi Jūkōgyō) and Chūbu Electric (Kenryoku), developed a 150,000 m³/h large-scale manganese ("wet") desulfurization process[264] at the Yokkaichi City power plant. Under the terms of the agreement, the government indicated its willingness to support the wider development of this technology if the pilot project proved successful. The second contract was with Hitachi and the Tokyo Power Company for the development at the latter's plant of a 6,000 m³/h charcoal ("dry") desulfurization process. Because this facility operated at 90% efficiency, the government supported the use of a larger model (150,000 m³/h) in the following year.

The government's handling of the petroleum industry differed from its guidance of the power companies in that the former was left far more to its own resources in developing the appropriate desulfurization technology. Beginning in 1967 the government began "rating" refineries by the type of desulfurization process used. Permits and allocations to the refineries were then issued according to a company's ranking. Companies were also asked to agree to specific administrative "targets" negotiated with the government.[265] Active government assistance to the petroleum industry was limited in scope and duration and confined to the development of catalytic processes.[266]

In 1970, after the national ambient standards were substantially revised, the implementation of desulfurization policies was conducted even more aggressively and brought more in line with the new standards.[267] One important tool used by MITI in regard to the petroleum industry was the energy supply planning process. Under art. 3 of the Petroleum Enterprise Law,[268] the government is obligated to prepare a five-year energy supply plan that establishes inter alia the kinds and amounts of oil to be imported, and the capacity of waste oil disposal equipment to be used. From this broad grant of authority, MITI implied a further basis for strengthening its hand in negotiations with industry.

After 1970 the government also initiated a program of direct financial assistance to the steel industry. The steel industry's technological difficulties in developing desulfurization equipment were somewhat less onerous than those of the power industries, for the steel industry, unlike the power industry, possessed the basic technology even before World War II. Yet, in order to stimulate its further development, the government established a cooperative project with about ten companies affiliated with the Iron and Steel Institute to develop a small-scale ammonia desulfurization process.[269] The project proved successful and has led to the development of several further applications.

Direct regulatory measures, such as standards and permits, backed by administrative guidance were also carefully coordinated from the start with a complex of tax and other financial incentives. Tax incentives included a special tax credit of one-third the initial costs of installation of stack and oil desulfurization equipment, accelerated depreciation, and a two-thirds exemption during the first three years of property taxes relating to pollution control facilities. In addition, after 1974 the government reduced the tariff on imported desulfurized heavy oil and on low sulfur oil. Also of great importance was the contribution of the Environmental Pollution Control Service Corporation that made special loans to environmentally progressive small and medium enterprises, retrofitted plants with scrubbers, assisted the relocation of polluting facilities, constructed green belts, and planned low-pollution industrial estates.

After 1970 local governments also began to play a more important part in the regulatory effort. For example, heavily industrialized areas such as Chiba, Ōita, Mizushima, and Kashima, began to insist that new refineries, power plants, steel companies, and other polluters negotiate control agreements. In a number of cases, these agreements contained explicit provisions requiring the use of smoke desulfurization technology, low-sulfur fuel and other antipollution measures.[270] At the same time, the central government began to encourage local areas like Hokkaidō to take more aggressive measures to curtail SO_x. And to facilitate the harmonization of desulfurization policies at the central, prefectural, and local levels, MITI encouraged a number of major industrial areas, e.g., Chiba, Ōita, and Kashima, to conclude interprefectural compacts.[271]

The case of the development and implementation of desulfurization policy affords some valuable insights into the Japanese administrative process. First, the study demonstrates how the Japanese administration was able to conceive, formulate, and begin the implementation of a specific policy on desulfurization without any explicit statutory authorization. Initial administrative action actually preceded the passage of the Basic Law in 1967 and the Air Pollution Control Law of 1968; indeed, neither statute directly addressed desulfurization. Eventually the administration developed the entire program out of its implicit administrative powers.[272]

Second, this study demonstrates the general administrative preference for "guidance" over law. Indeed, laws such as the Smoke and Soot Regulation Law of 1962 and the Air Pollution Control Law of 1968 appear to have served primarily as political gestures rather than as the primary means of controlling air pollution. Although this legislation did establish emission standards and did authorize some useful regulatory tools such as emergency warnings and abatement orders, its principal purpose seems to have been to legitimize or justify actions already undertaken by the administration.

Indeed, in some cases law and the administrative process actually appear to have been in conflict. The reader will recall that MITI lobbied against statutory controls on the content of sulfur in fuel oil. Having just developed a plan for the importation of high-sulfur Middle East oil, the government insisted that the problem of reducing sulfur could and should appropriately be addressed through administrative guidance, and not through direct statutory controls. The administration viewed guidance as a more flexible and effective means of control.

This attitude was especially apparent in the case of ambient standards. Although extremely important to the government's desulfurization program, the ambient standards established in 1970 served more as administrative targets than as procrustean legal requirements. In virtually all cases, the job of selecting and developing a particular technological option was left to the free choice of the industry concerned.[273] Administrative actions sought more to encourage, guide, and facilitate, than to mandate a given course of industrial decision making.

Third, the study reaffirms the principle that administrative guidance tends to be industry-specific. Thus, although the government assumed a major part of the research and development costs of scrubbers within the electric power industry, and subsidized part of these costs for the steel industry, it preferred to adopt a more passive stance with respect to the petroleum industry. This study also demonstrates the government's willingness to stimulate intraindustry collaboration, and hence to tamper with competitive market forces, when such collaboration is deemed useful to central planning and other objectives.

Finally, the case of desulfurization affords insight into industrial attitudes toward various forms of administrative action. Most officials interviewed in the steel, power, and petroleum industries felt that the various tax and other financial incentives provided by the government, although beneficial, did not influence decision making significantly. Rather, all regarded direct regulatory controls (such as permits to new refineries) backed by administrative guidance to have been most effective. When asked what primarily motivated their compliance with the government's (mostly MITI's) recommendations, all emphasized the authorities' pervasive powers, especially in light of fierce competition within the industry. For example, petroleum company officials expressed fears that if they were not to comply with government suggestions, the government might deny permits, interfere with production quotas or distributions, influence users, or somehow otherwise favor competitors. Yet, beyond these specific sanctions, virtually all attested to a sense of being completely enveloped by the governmental presence, and expressed real concern about "displeasing." As one official put it: "It is as if we were children. . . . If we do not behave, our parent—next time when we want something—will refuse to give it to us. This is our chief fear."[274]

3.2   Case 2: Auto Emission Controls

During the 1960s, air pollution in Japan's largest cities intensified proportionally with the rapid increase in the number of automobiles. Because automotive exhausts were essentially unregulated, the Diet passed a series of resolutions exhorting the government to initiate remedial action.[275] In response to the

Diet's requests the Chief of the Automobile Bureau in the Ministry of Transportation began administrative guidance of the industry. Accordingly, in July 1966 the Automobile Bureau issued a circular requesting the industry to limit CO (carbon monoxide) emissions in all new car models to less than 3%. With the enactment of the Air Pollution Control Law of 1968, this standard was incorporated into the ministry's new automotive emission regulations,[276] limited further in 1969 to 2.5%, and extended to cover used cars in 1970.

In September 1971, jurisdiction over the setting of automotive emission standards was transferred from the Ministry of Transportation to the newly created Environment Agency. The agency immediately set to work in consultation with the Central Council[277] on the formulation of long-term targets. There was cause for considerable haste. During the summers of 1970 and 1971 photochemical smog struck Tokyo and other metropolitan areas, and there were frequent reports of smog-induced health injuries and related complaints. The immediate and effective control of automobile-induced air pollution was now a political necessity.

In its deliberations, the Central Council made use of an important foreign precedent. In June 1971 the council received reports that the U.S. Clean Air Act of 1970 (the "Muskie Act")[278] had established extremely stringent standards for CO, hydrocarbons, and NO_x, the standards to take effect in 1975 and 1976. The council quickly began to debate the question of what standards Japan should adopt. Several factors militated against the option of simply copying the U.S. standards. First, like its U.S. counterpart, the Japanese automobile industry in 1971 did not possess the requisite technology to meet the 1975 standards.[279] Second, it was unclear whether this technology could be developed by 1975. Third, the costs of this undertaking were unknown.

Despite these considerations, the Japanese government decided to establish emission targets identical with those of the Clean Air Act.[280] An important basis for this decision was the belief that the U.S. government would never establish a standard for an industry of this importance without firm evidence that the industry's compliance was possible. Because Japanese automobile exports to the United States were projected to increase, Japan's decision makers reasoned that Japanese automobile manufacturers would in any event have to comply with U.S. emission standards.

The government's new policy was first manifested in the Central Council's interim report to the Environment Agency submitted on October 3, 1972. The report recommended that the United States standard be adopted for CO, hydrocarbons, and NO_x (tables 7 and 8). In May 1973,[281] the Environment Agency commenced a series of hearings[282] on the 1975 standards and despite strong opposition from the industry it succeeded in issuing a notification that CO, hydrocarbons, and NO_x emissions must be reduced by 89%, 91% and 45% respectively from 1973 levels.

The automobile industry's attack on the Environment Agency over the 1976 standards was even more intense. A powerful new weapon was the fact that the U.S. EPA administrator, William Ruckelshaus, had suspended enforcement of the 1975 emission deadlines of 1973 on the grounds of technological infeasibility and, in the following year, the Congress had amended the Clean Air Act.[283] Armed with this information, Japanese auto manufacturers argued that compliance with the 1976 standards, particularly the NO_x limitation of 0.25 g/km, was not feasible and that more lead time was necessary. In any case, the industry pointed out, there was no longer a necessity for Japanese automobile exporters to hasten to meet the earlier, more stringent U.S. standards.

Table 7
Passenger-car exhaust limits in Japan and the United States (g/km).

Pollutant	Country	1973	1974	1975	1976
CO	U.S. Federal	24.2	24.2	9.3	9.3
	California	24.2	24.2	5.6	5.6
	Japan	18.4	18.4	2.1	2.1
Hydrocarbons	U.S. Federal	2.11	2.11	0.93	0.03
	California	1.99	1.99	0.56	0.56
	Japan	2.94	2.94	0.25	0.25
NO_x	U.S. Federal	1.86	1.86	1.93	1.93
	California	1.86	1.24	1.24	1.24
	Japan	2.18	2.18	1.2	0.86, 0.6[a]

Note: Japan employs the fiscal year and the United States the calendar year. Figures are average values. Measurement in Japan employs the so-called ten-mode procedure; in the United States, measurement is per LA-4C for 1973–1974 and LA-4CH for 1975 and thereafter.
[a]The first figure is for vehicles with an equivalent inertia weight exceeding 1,000 kg, and the second for vehicles weighing less.

Table 8
Automobile emission standards[a] for Japan and selected countries (g/km).

	CO	Hydro-carbons	NO_x
Japan (for 1976)	2.10	0.25	0.60[b]
Japan (for 1978)	2.10	0.25	0.25
U.S., Federal Government (for 1975)	9.30	0.93	1.93
U.S., California (for 1975)	5.60	0.56	1.24
Canada (for 1975)	15.62	1.25	1.94
Canada (for future)	2.13	0.25	1.94
Sweden (for 1976)	24.20	2.10	1.90

Source: OECD.
[a]Testing methods vary from country to country and comparisons must be made with great care.
[b]0.86 for passenger cars with equivalent inertia weight of more than 1,000 kg.

On August 3, 1974, the Environment Agency submitted the hearing record to the Central Council along with a request that it reconsider the 1976 standard. The Central Council thereon convened its Expert Advisory Committee on Automotive Production and began the process of reexamination.

The government's deliberations, however, quickly sparked a major political controversy. On August 28, the mayors of seven major cities (Tokyo, Kawasaki, Yokohama, Nagoya, Kyōto, Osaka, Kōbe), who opposed weakening the standards, launched a team of seven scientists and sixteen staff members to conduct a counter-investigation of the automobile industry. Between September 13 and 14, the team held hearings with representatives of nine auto manufacturers and the staff of the Environment Agency. The team thereafter published its conclusions in an interim report that upbraided the industry for its greed and called upon other cities to take independent action against high polluting automobiles.[284]

The director general of the Environment Agency gave the team's final report a chilly reception and the chief of the Air Quality Bureau denounced it at a subsequent Diet subcommittee hearing.[285] Learning of the chief's testimony, the Tokyo metropolitan government's Pollution Survey Commission, that had been assisting the seven cities' investigation, challenged the Central Council to hold a public debate with the seven cities' investigation team. The council naturally refused, basing its objection on the grounds that no precedent could be found for a public debate under such circumstances. Yet, after negotiating with the investigation team, the council at last agreed to a series of meetings. The upshot of these discussions was that the council agreed "in spirit" to the team's assessment that the industry possessed the requisite technology to meet the 1976 standards, but adhered to its original position that a postponement of implementation was necessary. The council maintained that the industry had not had sufficient lead time, that low-pollution cars had proved less easy to drive, and that these cars consumed more fuel.

After sixteen meetings, the expert committee submitted its report to the Air Quality Subcommittee of the Central Council. The report concluded that because of technological infeasibility, implementation of the 1976 emission standards for NO_x should be postponed until 1978.[286] The expert committee's report was accepted by the Air Quality Subcommittee and referred by it to a plenary meeting of the Central Council. The council could not, however, reach consensus at its December 17 meeting, because of the strong stand taken by representatives of consumer groups. Finally, on December 27, the council's General Affairs Subcommittee, staffed by the chairman of the other subcommittees and some key council members, overcame all opposition, and recommended postponement. At the same time the subcommittee urged the government to provide the industry with tax incentives to develop low-pollution cars and to implement drastic measures to reduce automobile traffic in heavily populated areas.

But the battle was not quite over. Soon after, the media began to publish accounts of how secret council debates had been leaked to the industry by a pro-industry council member. The media were quick to pounce on this new controversy, highlighting the problem of the conflicts of interest of council members affiliated with the automobile and petroleum industries. The onslaught greatly impaired the credibility of the council and forced one automobile industry representative, alleged to have been principally responsible for the leak, to resign. Despite severe public criticism,[287] the Environment Agency adhered to the council's recommendations, and on February 24, 1975, the agency issued a notification establishing the permissible maximum NO_x emission standard for 1976 at a less stringent level.[288]

With the conclusion of the controversy over the 1976 NO_x emission standards, the Environment Agency turned to the problem of obtaining the automobile industry's compliance with the 1978 limitations. This time the chief of the Air Pollution Bureau assembled a special unofficial Study Group on NO_x Emission ·Control[289] as a means of bypassing the political complications now attending recommendations by the council. During 1975, the study group itself held hearings in order to gather data on new technological developments and to ascertain whether technology needed to meet the 1978 standards could be developed. The study group's interim report, published in December 1975, was inconclusive on this issue.[290]

The entire debate changed in January 1976, when two of the smaller auto manufacturers, Mitsubishi Automobile Industry and Fuji Heavy Industry, released reports that they could produce cars meeting the 1978 NO_x standards.[291] Thereafter, Toyo Industry and Honda also announced that with favorable tax treatment their cars would clear the 1978 threshold.[292] Despite the progress of these minor manufacturers, the two major firms, Toyota and Nissan, continued to procrastinate, relying on an earlier extension permitting them to market older-model, heavily polluting cars.[293]

Yet, with Nissan's announcement[294] in March, 1976, that it too would endeavor to meet the 1978 controls, the battle was almost at an end. In August, 1976, five of the nine domestic manufacturers revealed at a hearing held by the study group that they would clear the NO_x standard (0.25 g/km) and on October 20, 1976, the study group submitted its final report to the director general of the Environment Agency, confirming the industry's technical and economic capability of meeting the 1978 standard.[295] On December 18, 1976, the Environment Agency issued a notification to this effect.

Only Toyota remained. Soon after, however, the company retracted an earlier request for a two-year extension.[296] Toyota simply capitulated to the accumulating economic, social, and administrative pressures.

In retrospect, the progress achieved by the Japanese automobile industry can be attributed principally to the actions of the central authorities who were themselves subjected to strong pressure from local governments, the general public, and the Diet.[297] The government employed a variety of positive and negative incentives. First, as we have noted earlier, it viewed the emission standards more as administrative targets than rigid, inflexible, legal requirements. Indeed, the emission standards served essentially as an initial reference point for negotiations between the government and industry. Within this framework the government exhorted compliance by a variety of means. For example, since 1972 the Environment Agency and several ministries had met on numerous occasions with automobile industry leaders to discuss the technical and economic difficulties of the industry, to offer recommendations, suggestions, and advice, and to hold hearings.

On some occasions, administrative guidance would also take a sterner form. When Toyota began distributing pamphlets criticizing the government's pollution control efforts as "unscientific" and unduly stringent, Michio Hashimoto, chief of the Air Preservation Bureau, summoned the vice-president of Toyota and other Toyota officials to the agency for a public reprimand. As a result, Toyota withdrew the pamphlets from distribution.[298] The intent of all these measures was to create an environment conducive to the industry's compliance. There was little expectation that such measures in themselves would effect compliance.

A further incentive was the various tax measures initiated since 1975 that were designed to reward companies producing low-pollution cars and to penalize other less progressive firms. These tax incentives included reduction in the commodity and motor vehicle acquisition taxes for passenger cars meeting the 1975, 1976, and 1978 emission standards, and a further reduction in ownership taxes (i.e., the light motor vehicle tax) between 1976 and 1977 to encourage the use of passenger cars meeting the emission standards.[299] Unlike the case of the development of desulfurization technology, the government did not finance or subsidize the development of automotive emission technology. Environment Agency officials explain that the principal reason for this was the prohibitive expense. For example, in 1972 the major companies, Toyo Kōgyō, Toyota, Nissan, Fujita, Mitsubishi, and Honda invested approximately ¥15.1, 45, 4.2, 4.6, 8.4, and 19 billion, respectively.

Environment Agency officials also emphasized the importance of the Study Group on NO_x Emission Control. The contribution of this body was threefold. First, it provided a flexible means of bringing automotive experts and auto industry officials before the agency to discuss frankly the technical and economic problems of compliance.[300] Second, the study group's final report containing a detailed comparison of each company's performance record (i.e., test results on air-fuel ratio, types of emission controls and their effectiveness, fuel economy, drivability) had a more subtle purpose. By means of this report the agency was able to put the full force of public support for control of the automobile industry to decisive use. Copies of the report were made available to all distributors who then proceeded to exhort prospective customers to buy low-pollution car models. Third, the report was also used widely by the automobile industry in its advertising campaigns.

In order to appreciate why the government's use of information was so effective during this period, the reader must grasp both the dynamics of public opinion in Japan and the economics of the industry. Chapter 2 has already described how public attitudes toward environmental destruction changed radically during the 1960s and how during the 1970s a powerful citizens' movement, imbued with a strong environmental ethic, challenged many environmentally destructive industrial projects. This same movement also directed public attention to the complaint of the automobile industry that compliance with the 1975 and 1976 emission standards was economically and technically infeasible. The industry's position was widely disbelieved and vociferously challenged as self-serving. Because it was generally felt that the industry's failure to meet the stricter emission standards would seriously jeopardize public health, the industry's procrastination was considered to be immoral.[301] Public attitudes were also strongly influenced by the enactment of the Clean Air Act of 1970 in the United States, because the U.S. statute was seen as evidence that the emission goals were obtainable.[302]

The citizens' movement and the press throughout actively pressed their views on the industry. Dilatory companies, particularly Toyota, were excoriated in the press. As a result, the corporate name was impaired. In the Japanese setting this had two adverse consequences. First it entailed a substantial loss of face because the preservation of good corporate name was a primary concern for Japanese companies. Second, many companies feared that if they became known as "bad" environmental citizens, this would affect sales.[303]

The peculiar effectiveness of this combination of administrative guidance and public pressure can only be understood after further consideration of the industry's position. First, the automobile industry in Japan is fiercely competitive, a fact intimately understood by the administration. As a result, every advantage a firm gained, either by a boost in corporate image, from tax incentives, or by other means, could be used to increase its market share. This is why Nissan and Toyota, the two major companies, were particularly distressed by the smaller firms' announcement of their successful development of emission technology in 1976. Both Nissan and Toyota recognized that the government and public could quickly use the technical progress of the smaller firms to drive a wedge within the industry that could result in the smaller firms gaining a dominant market position.

Many of these factors also explain why the automobile manufacturers did not challenge the implementation of the Environment Agency's regulations through litigation as did their American counterparts.[304] In addition to the probability that such a suit would not have been successful given the strong legal presumption in favor of the Environment Agency's action, the industry, by resorting to the courts, would effectively have invited violent public attack.[305] Again, corporate image would have been seriously affected and the companies would have been exposed to further public and governmental sanctions.

Local governments also served as a third important source of pressure on the automobile industry. Although many local governments severely criticized decisions of the Environment Agency, in effect the strong stand taken by local governments at the same time gave great force to guidance imposed by the central authorities.

The actions of local governments took a variety of forms. The action of the mayors of the seven major cities has been noted. As a result of such agitation, a number of cities and prefectures began to implement specific measures of their own. For example, Aichi Prefecture announced that it would purchase only the low-pollution cars of "progressive" companies;[306] Hiroshima Prefecture introduced a system of tax rebates for cars meeting the central government's NO_x standards;[307] and a number of other cities adopted restrictions on the operation of high-pollution cars in times of photochemical smog emergencies, strengthened parking regulations, and set up bus lanes.[308] These actions combined with public and central governmental pressure apparently provided a decisive incentive for the industry.

The case of the implementation of automotive emission controls illustrates some further general principles about the regulatory process. First, it demonstrates the subtleties of administrative guidance by showing how such guidance can be fashioned to the specific characteristics of a given problem. In the present case the automobile industry was significantly more competitive than the petroleum, power, and steel industries.[309] Thus a central objective behind guidance of the automobile industry was to capture the benefit of decentralized market forces. Essentially, Environment Agency officials manipulated public opinion to its most administratively advantageous end. In fact, the public reprimand of Toyota was timed exactly so that public pressure could be brought to bear most forcefully on the company.

The economics of the industry also played a critical role in determining the extent to which the government decided to become actively involved in the process of research and development. In comparison to the petroleum industry, the automobile industry was a far less critical sector. Moreover, there was evidence as early as 1974 that some companies had already made progress toward developing the requisite technology for 1975 and 1976 model cars. The government thus felt it unfair to provide general assistance to all firms when this differential existed. These factors, in addition to the prohibitive cost, militated against direct government involvement in the development of emission technology.

Finally, the present case illustrates the importance of compromise during negotiations between government and industry. An outer limit on agency action is the fact that Japanese administrations will rarely, if ever, attempt to force a company to carry out an action that is either technologically or economically infeasible. Yet, within this limit, compromise may be necessary. The purpose of such a compromise may be first to give the industry a "break" based on notions of fairness or economic expediency. But, once the government concedes in one instance, this compromise can be used to strengthen its position during the next round of negotiations. This was precisely the effect of the government's agreement to a two-year extension in 1976 on the implementation of NO_x limitations for that year. By acceding to industrial pressure in 1976, the government acquired overwhelming leverage to push through the industry's acceptance of the NO_x standard in 1978.[310]

3.3 **Case 3: Development of an Environmental Impact Assessment System**

In 1965 MITI began conducting rudimentary environmental impact assessments in areas where industrial development was either taking place or anticipated. The ministry's action was primarily motivated by a desire to avoid the reoccurrence of a major public health catastrophe like that in Yokkaichi or a serious political disturbance as had just taken place in Mishima-Numazu. The government's experiment with environmental impact assessments was initially limited to an analysis of SO_x concentrations in the case of air pollution[311] and to the effects of COD (chemical oxygen demand), BOD (biological oxygen demand), and oil in the case of water pollution. As noted, MITI developed a simulation model during this period that served to predict future pollution levels and provided a basis for administrative guidance of the petroleum and other industries.[312]

After the extraordinary "pollution control" Diet session of 1970, the government gradually began also to focus greater attention on the problems of environmental planning that had until then been neglected. This new sensitivity was first manifested in the following statement released by the Cabinet on June 6, 1972.

In view of the importance of environmental conservation, the government shall hereafter adopt the following measures in implementing various public works projects (through either government agencies or local public bodies to which the central government has granted subsidies, or has licensed, or has approved) so that problems of environmental conservation do not occur in the implementation of the projects.

1. Hereafter, when the central government agencies or ministries are planning to implement various public works, such as highways, harbor facilities, and land reclamation and so forth, they shall design and effect construction in a manner that contributes to the preservation of the environment and avoids the creation of pollution.

2. In furtherance of the purpose stated in par. 1, administrative agencies of the central government shall have the appropriate public works organization carry out, in advance where necessary, research and study in regard to public works within their respective jurisdictions, including the nature and extent of the effects on the environment, proposals to prevent environmental destruction, and a comprehensive investigation of alternative plans, and so forth. Based on the results (of the study), each agency shall provide guidance regarding the necessary measures to be taken by the organization.

3. Local public bodies should also be called upon to adopt the necessary measures prescribed above. [313]

The Cabinet resolution provided the basis for the administrative development of an environmental impact system.

In the following month the need for establishing an environmental impact assessment system became even more apparent because of the political crisis fostered by the Yokkaichi opinion. The Yokkaichi court had sharply criticized the government's precipitous development of new industrial areas and its insensitivity to environmental concerns. A dramatic demonstration of official sincerity was therefore politically essential. The government's response was to amend the Factory Location Law,[314] the Public Water Reclamation Law,[315] and the Ports and Harbor Law[316] to include various provisions for impact assessment on all major projects or developments.[317]

Yet the public was not to be put off by a piecemeal solution. Shortly after the amendment of the Factory Location and other laws, a strong public movement began to develop for the government's preparation of a comprehensive statutory approach. The opposition seized this opportunity. In January 1975, the Kōmeitō (Clean Government party) submitted a bill to the Diet that required certain industries to prepare comprehensive assessments of the environmental and social impacts of their activities. The Kōmeitō bill proposed the establishment of an administrative commission charged with reviewing the environmental consequences of industrial plans and obligated to reject unsound projects. The bill also developed procedures to assist the public in obtaining information, in presenting opinions on draft assessments, and in challenging decisions of the commission. In October of the same year, the Socialists introduced a counterproposal that, while containing explicit and detailed procedural provisions for the preparation of environmental impact statements (EISS), eliminated the cumbersome permit provisions of the Kōmeitō draft.[318] The Japan Federation of Bar Associations also joined the fray and submitted a proposal similar to the Kōmeitō bill, although it required the assessment of a broader range of acitivities.[319]

The initiative of the opposition parties prompted the government at last to begin formulating an omnibus approach.[320] The draft ultimately prepared by the Environment Agency was modest in comparison with the ambitious Socialist and Kōmeitō proposals, for it required assessments only on developments entailing the possibility of substantial impact. In most cases an EIS was required only after a development plan had already been formalized, and public participation in the assessment process was extremely limited.

Despite its limitations, industry strenuously opposed the Environment Agency's draft. Industry was concerned that the EIS law would be used by militant citizen groups to delay development, and that, rather than resolving conflicts, the EIS law would engender them. In support of industry's concerns, MITI, the Ministry of Construction, and the Ministry of Transportation all launched an attack on the Environment Agency's effort. The agency's initial draft was consequently shelved. In March 1977 the agency redrafted the EIS bill, offering even greater compromises than before.[321] Despite its conciliatory attitude, the Environment Agency could not obtain the consent of the other agencies before the adjournment of the regular 1977 Diet Session. Sensing defeat, the opposition parties also permitted their proposals to be tabled. The Environment Agency's third attempt to lobby its EIS bill through the goverment was defeated in May 1978.

Despite the failure of the Environment Agency and the opposition parties to establish a comprehensive statutory environmental impact assessment system, the government began to integrate impact assessment procedures into its administrative approach to pollution control based on its 1972 Cabinet resolution. The first and most controversial project considered under the new administrative procedures was the superindustrial complex development at Eastern Tomakomai City in Hokkaidō. The complex was part of a comprehensive plan for Hokkaidō's development formulated initially in 1950. One part of this program contemplated the construction of Tomakomai Harbor, the second part projected the development of an industrial base in Western Tomakomai,[322] a third contemplated the industrialization of the entire area of eastern Tomakomai. The size of the proposed development was gigantic, encompassing about 13% of all land allocated for industrial purposes in Japan.[323]

In late 1972, the Hokkaidō Environmental Pollution Control Council prepared and filed a report on the environmental impact of industrialization on eastern Tomakomai Harbor. Pursuant to the 1972 Cabinet understanding this report was thereafter transmitted with comments to the Environment Agency. The agency, however, refused to assent to the report, citing a number of its deficiencies. These included the report's analysis of only a few pollutants, and its failure to address the possible health effects of these substances on the town of Yūfutsu that was sandwiched between the eastern and western industrial bases.

In June 1972, Hokkaidō Prefecture responded to the agency's objections by amending its earlier assessment. The revised version was thereafter approved by the agency and forwarded by it to the Council on Ports and Harbors, with the recommendation that a supplementary investigation of the effects of the entire plan, not just the harbor portion, be undertaken.

The negotiations between the agency and Hokkaidō Prefecture, however, angered many environmental and other activist groups. These groups charged that the assessment had been prepared without citizen participation, and that Hokkaidō Prefecture and Tomakomai City had never released their impact statements for comment. To respond to these criticisms, Tomakomai City in July, 1973, held "open hearings" in order to persuade residents of Yūfutsu that the environmental assessment was adequate. The residents remained unconvinced.

In October 1973, Tomakomai City recommended its planned development on a more cautious "step-by-step" basis. City officials reasoned that if the more egregious environmental problems could be brought under control, the remainder of the huge industrial park might be completed without local protest. Accordingly, Hokkaidō Prefecture revised its EIS in December 1973, and resubmitted the document to the Environment Agency. Finally, the agency approved the project.

Despite the compromises achieved, local residents continued to express great concern over the course of decisions by the Environment Agency, the Hokkaidō Prefecture, and Tomakomai City. Their basic contentions were that the assessment was unnecessarily limited, did not support its conclusions, failed to address ecological impacts, afforded no opportunity for meaningful citizen participation because of inadequate notice and hearing procedures, and took account of only a small range of expert opinion (principally in the engineering and physical sciences).

Aware of these deficiencies, the Japan Federation of Bar Associations dispatched an investigatory team to Tomakomai. At the annual meeting of the Association in March 1974, the team urged the government not to approve current plans for construction.[324] The Bar Association's report had little effect. In January 1974 the Ports and Harbor Council approved the eastern harbor construction plan. Thereafter, the Ministry of Transport also approved and subsequently promulgated the plan.

Despite official approval of the Hokkaidō plan, the government's critics continued to attack the weaknesses of the EIS procedure. In May, 1974, a Diet subcommittee questioned Takeo Miki, then director general of the Environment Agency, as to why the government had approved the further industrialization of Tomakomai when air pollution around Yūfutsu already exceeded existing ambient standards. After these hearings, the Environment Agency pressed Hokkaidō Prefecture to revise its EIS, and the Prefecture responded by preparing a supplementary assessment of the expansion of fossil fuel power stations in Tomakomai.

Partially in response to the criticisms suffered over the Tomakomai episode, the Environment Agency began to give more serious attention to formulating new environmental assessment procedures. In 1974 the agency established a special office, the Office of Environmental Assessment, to oversee environmental impact analyses, and thereafter expanded its operations to include the drafting and implementation of impact assessment "guidelines."[325] These guidelines were first applied in September 1976 to the huge Mutsu-Ogawara petroleum refinery and industrial park in Aomori Prefecture.[326] And in 1977, the Environment Agency also developed similar guidelines for a planned bridge to link Honshu and Shikkoku, Japan's two major islands.[327]

As in other areas, the central administration approach to environmental impact assessments was influenced and supported by the actions of local governments. The first local enactment was the Kawasaki EIS ordinance of September 1976, which required land developers to file impact statements with the mayor. Based on this report, the municipal government was empowered to order changes in a development project.[328] Public disclosure and participation were encouraged, and the ordinance also established procedures for the public to request a formal hearing on a given project. Violations of the ordinance were punished by a fine of ¥50,000. The Kawasaki ordinance has subsequently served as a model for the establishment of similar provisions by other local governments.[329]

The case of the development of an environmental impact assessment system demonstrates several further, critically important aspects of the administrative process. First, it illustrates the difficulty of attempting to reform administrative decision making in Japan by forcing traditional ministries to alter present decision making to account for "new" environmental (or other) interests. Whereas a statute forcing interagency cooperation could have some chance of success under the more flexible U.S. administrative structure,[330] the fate of the Environment Agency's legislative proposal demonstrates that it was not possible under the more vertically rigid Japanese administrative framework.[331] Indeed, MITI and some of the other mission-oriented agencies were not the only ones to resist the idea of modifying traditional jurisdictional prerogatives. The Liberal Democratic party and the power industry, as well as elements within the Environment Agency itself, were opposed to, or at best unenthusiastic about, this effort at reform.

Second, this case demonstrates how the administration has been able to "generate" authority for guidance by creating within itself a new administrative division. After interministerial consensus was achieved, the Environment Agency's Environmental Assessment Office was able to expand its original mandate of simply supervising environmental impact assessments to include quasi-regulatory functions; today (1979) this office supervises many projects. As explained, this rule making does not require explicit statutory authority.

Third, this case underlines the importance of consensus in the administrative process. Here the Environment Agency was not able to obtain MITI's and other ministries' support for its draft legislation; consequently this legislation failed to pass. Conversely, MITI's incorporating impact assessment procedures into its own guidelines for construction permits to power plants and other facilities reflected its willingness to cooperate to some extent.

Fourth, this case illustrates yet again the rich interaction between the judicial and administrative processes. For example, the Yokkaichi decision stimulated greater administrative interest in the problems of industrial plant siting and provided a legal basis for subsequent guidance. The Ushibuka case and other administrative decisions have influenced local decision making along similar lines. Conversely, the expansion of guidance in the impact assessment field has itself influenced a number of courts in their determinations of the environmental soundness of developments.

Fifth, this case suggests that administrative guidance applies not only to the private sector; governmental and quasi-governmental bodies may also be the subjects of administrative instruction.[332] Examples of this process are the Environment Agency guidelines issued for the Tomakomai and Mutsu-Ogawara developments, and the Cabinet understanding of 1972 pertaining to public works projects.

Sixth, the present case further illustrates the subtleties of administrative action. For example, one form of guidance (i.e., that requiring companies to gather data on their emissions) can be used by another agency to conduct other forms of guidance (i.e., suggestions, advice, recommendations on factory siting, or as a condition for a permit). Often guidance is also employed to supplement regulations, thereby filling in gaps, easing the burdens of complying with a particularly harsh standard, and creating incentives. An agency's draft legislation itself can at times also be transformed into guidance. Thus, although the EIS statute proposed by the Environment Agency was defeated, many of the ideas contained therein have already been incorporated by the Environment Agency, MITI, and the Ministry of Construction into various guidelines or recommendations.

Finally, the case is instructive because it provides insight into the important emerging issue of administrative reform. Thus many citizens, such as those involved in the Tomakomai case, feel that the administrative process is arbitrary, does not fairly or adequately account for public opinions, and is extremely biased in industry's favor, and that the environmental assessments have been perfunctory, and in many cases decisions based upon them substantively wrong. The following section considers this indictment.

Section 4   The U.S. Regulatory Process From the Perspective of Japanese Experience: Contrasts and Possibilities

This section explores how the United States might use Japan's legal and institutional inventions to develop more effective environmental protection measures. Our principal thesis is that the United States has underestimated the potential contribution of institutional arrangements in solving complex scientific and technical problems. Furthermore, we believe that the U.S. approach to regulation can be modified and that cultural, political, social, institutional, and legal barriers to a more flexible, perhaps "Japanese" approach, are not insurmountable. Indeed, to some extent, Western commentators in very different settings have already celebrated the merits of some of the measures discussed, although their proposals have not for the most part been seriously heeded. Without detailing specific recommendations, the discussion seeks to motivate further inquiry.

First, let us draw broadly the basic institutional differences between the two approaches. Perhaps the most critical distinction lies in the relationship between government and industry. In Japan, although recently there have been increasing problems and conflicts, the relationship has been primarily a cooperative one. Both the central and local governments have considered the task of securing industry's compliance with pollution regulations to be essentially a shared responsibility.[333] For this reason the government has sought more often to be helpful than coercive, and its approach to regulation has been generally extremely flexible. Local governments have behaved similarly, although local environmental pollution policies have tended to be more radical than those of the central authorities.

Industry also has collaborated with government more often than it has opposed regulation. Of course, industry has at times fiercely remonstrated against governmental policies that it deemed scientifically unsound or economically onerous. Yet basically the interchange has been in spirit a partnership. Indeed, this cooperative pattern is also evident within and between industries, despite keen economic competition.

In the United States, industry and government have viewed each other more as adversaries than allies. History affords some explanation for this perhaps unfortunate situation. From its earliest days, American industry has prided itself on its achievements, accomplished independently and often unaided by governmental patronage. The conviction that the fruits of labor should honor the tiller, whether historically justified or not, underlies American industry's present resentment of encroaching federal controls, and its general unwillingness to assume any responsibility for achieving society's overarching objectives.

Conversely, feeling the pressures of environmentalists and other special interest groups, both the legislative and the executive branches have continued to believe that government's principal function, at least in the environment field, is to be industry's taskmaster. Indeed, the prevailing social ethic, now critical of industry, has naturally tended to be fundamentally suspicious of initiatives that would bring government and industry into even more intimate fellowship.

The use of the courts in both countries also reflects the industrial-governmental relationship. In Japan, strong social pressures have inhibited industry from judicially challenging the broad statutory authority of the administrator. And while citizens' groups have attempted to stimulate reforms through administrative litigation, such cases, although increasing, are still few in number, and only marginally influential. This has not been the experience of the United States where industry, the environmental groups, and even state governments have defied federal mandates. Indeed, the intrusion of the courts into agency decision making has set back some ambitious environmental protection proposals and programs for years.[334]

Finally, the U.S. Congress has influenced the development of environmental policies far more than its Japanese counterpart, the Diet. Whereas the Congress has been the principal source of environmental legislation and has generally maintained its influence through oversight hearings, the bureaucracy has virtually dominated the legislative process in Japan.

Recalling these basic differences, let us now consider how the implementation of automotive emission controls under the U.S. Clean Air Act of 1970[335] might have been facilitated and difficulties avoided. Suppose the United States in 1970 could have benefited from our present understanding of how Japan succeeded in controlling auto emissions. We wonder whether the real barrier to the American automobile industry's performance was technology. Although American industry has won the courts' sympathies by technological arguments,[336] the Japanese industry's success in meeting an even stricter standard suggests that institutional factors may be as important as technological considerations, if not more important.[337]

Let us assume the following: (1) that the Clean Air Act granted the EPA administrator wider discretion over standard setting and enforcement; (2) that the administrator is willing to experiment with a richer arsenal of enforcement tools and strategies; and (3) that the government has decided to intervene more actively and constructively in the automobile market.[338]

These guidelines are then translated into the following program of implementation. First, the approach to emission controls is changed. Under the revised approach the administrator, not the Congress, establishes the strictest standards warranted by the available scientific data.[339] These standards are legally binding on the industry, and also serve, as in Japan, as administrative targets. If the administrator deems it technologically or economically necessary, the EPA may also establish interim standards at the strictest levels considered technically "reasonable."

Second, the government alters its approach to enforcing the emission controls along the following lines. (1) The administrator imposes a variety of economic incentives as auxiliaries to direct regulation.[340] Some possible options that have already been proposed are a series of fines for the industry's failure to meet interim or final standards,[341] a smog tax,[342] or perhaps some other form of pollution tax. Like Japan's approach to the control of stationary sources, the principal intent of these and similar measures is to provide an additional economic incentive for the industry to develop technology and thereby to avoid paying these charges. (2) The administrator at the same time negotiates a compliance schedule with each of the major automobile manufacturers. One appropriate analogy here is the Japanese government's use of compliance agreements with the petroleum industry. In these agreements the manufacturers also promise not to litigate EPA's emission control program in exchange for the government's promise of constructive assistance to the industry as noted below. (3) In a procedure similar to the Japanese Environment Agency's treatment of the NO_x study group's report, the administrator widely publicizes the contents of the agreements and supplements its original public announcement by periodic reports on the progress of each company.[343] The government in a variety of ways rewards[344] progressive companies and rebukes and otherwise admonishes recalcitrant firms.

Third, as part of its compliance agreement, the government itself participates with industry in developing the necessary emission control technology. One possible approach is Jacoby and Steinbruner's proposal that firms not meeting interim standards pay a penalty and that this penalty be collected and earmarked under a special Technical Development Fund. These monies would then be used to contract research with the automobile manufacturers and others. The program could seek to develop: (a) engines with low emissions; (b) engines that require little or no maintenance; (c) engines with low fuel consumption; (d) improved safety design suitable for mass production.[345] The Technical Development Fund of course might also be supported by revenues from a smog or pollution tax or even various charges on stationary sources. There are many possibilities.[346] Finally, the government as part of its agreement with industry would facilitate intraindustry collaboration in the development of pollution control technology.

There are several reasons why this policy would be sensible. First, it avoids wasteful duplicating efforts; second, it reduces the costs to the individual manufacturer; third, it creates economies of scale.

Intradindustrial collaboration, of course, raises various questions under the antitrust laws and would represent a departure from present policy.[347] Yet, enforcement of the antitrust laws has been suspended in other instances. For example, carriers are permitted to negotiate prices and schedules under both CAB and ICC auspices and exemptions have been granted under the Federal Trade Commission Act when otherwise prohibited activities are deemed beneficial to national defense or security.[348] Under the proposed scheme, the EPA would sponsor and monitor cross-licensing agreements between the manufacturers,[349] and provide other assistance along the general lines of MITI's guidance to the steel and power industries in the development of desulphurization technology.[350]

The problems of stationary source pollution in the United States may also be attacked along the line of the regulatory approach contemplated for the automobile industry: broad administrative discretion supported by an arsenal of flexible enforcement tools, and tempered by constructive governmental action. Japan's experience here, too, offers a number of ideas that could usefully be incorporated within a state's implementation plan under the Clean Air Act. For example, the concept of mass emission controls, although untested, appears worthy of consideration because it encourages firms to limit pollution in the least costly way.[350a] Similarly, the Japanese institution of pollution control associations also could prove effective. Although at present there is no analogous institution in the environmental field, both the federal and state governments are routinely involved in public works projects, and such projects could easily be expanded to incorporate many of the activities that pollution control associations perform. Of course, there would be a need for appropriate enabling lesislation, as in Japan, to reallocate some of the costs of such projects to industry.[351] Yet, industry itself might support this legislation. If effectively implemented, it could prove cost-saving, protect some industries from litigation and other harassment by environmentalists, and possibly also contribute to greater regularity in the government's administration of environmental protection measures.

In other areas, an institutional development parallel to Japan's administrative approach is already evident. One example is Connecticut's enforcement program.[352] Connecticut has imposed a civil penalty on violators of pollution laws calibrated to the economic benefits garnered from noncompliance.[353] Reportedly, since the inauguration of this program, the number of recalcitrant polluters has dramatically declined, and a large number of violators have been brought into compliance. Although Japan has yet to adopt a civil penalties scheme,[354] the Connecticut program shares much of the spirit of the Japanese regulatory scheme. It is simple and administratively flexible, inexpensive, easily understood, and creates a powerful incentive for compliance with pollution regulations.

Japan's innovative use of pollution control contracts also deserves careful attention in the United States.[355] Although the authors are unaware of any comparable state or municipal effort, there is some precedent in the private sector. For example, recently two long-standing enemies, environmentalists and the coal industry, have announced their agreement on two hundred ways of exploiting the nation's coal resources while minimizing environmental damage.[356] Also, in a somewhat different context, EPA has recently concluded an extrajudicial pollution control "pact" with the United States Steel Corporation's Clairton Coke Works, modifying a 1972 court decree.[357]

The transition to a broader, more flexible administrative approach of course will not be easy and necessarily will give rise to legal, political, cultural, and institutional problems. Perhaps the threshold legal issue is whether this delegation of authority is constitutionally permissible. An affirmative answer is suggested by recent court decisions in the environmental and other fields.[358]

Of course from industry's perspective, legal questions of delegation aside, expanding administrative discretion can be burdensome, costly, often frustrating, and at times arbitrary, unjust, and oppressive.[359] Moreover, this approach consumes executive time, requires private consultants or outside counsel, forces the compilation of masses of economic and scientific data, and as a process, exhausts and debilitates, and from many accounts seems interminable.[360] Yet, despite this powerful indictment, some commentators still insist that a discretionary approach is to be preferred over strengthening judicial review or expanding congressional control over the executive. This is especially so where enforcement is fair and flexible, rewards responsible corporate behavior, and gives industry adequate lead time for compliance.[361]

Would expanding administrative discretion in the environment field conflict with the implementation of polices in other areas? As we have seen, this is always a danger when administrators feel less constrained by statutory mandates.[362] As the above discussion of automotive emission controls suggests, the benefits of a tailored solution in one area (pollution control) may need to be set off against the costs extracted from the attainment of other policy objectives (antitrust). Balancing or harmonizing environmental policies thus dictates greater rationalization, if not centralization, of these policies. This, however, is precisely the course that a number of U.S. commentators have begun to advise.[363]

The expansion of the administrator's authority raises a final, as yet unresolved, concern. In large measure, the successes of Japanese environmental policies between 1970 and 1976 can be attributed to the bureaucracy's ability to mobilize quickly, to control the formulation of pollution laws, and to direct their implementation, all without fear of judicial interference. Yet, accompanying these positive attributes is a darker side of the Japanese administrative process. Environmental policies can be abridged or subverted as easily, rapidly, and efficiently as they have been created. Already we have some hint of a trend in this direction. Whether the weakening of the NO_x ambient standard, for example, is merely "rationalization" as the government asserts, or whether this change portends a broader pattern of deregulation, time alone will tell. The 1973 pollution victim compensation system discussed in the following chapter continues our inquiry into how well a framework of expanded administrative discretion can adequately protect the public.

6   The 1973 Law for the Compensation of Pollution-Related Health Injury: Theory and Practice

In recent years, many industrialized nations and some developing countries have begun to recognize that chemical substances and other industrial by-products pose hazards to man and the environment. Only now are we learning that these substances are not naturally degraded, that they concentrate in the tissues of plants, fish, animals, and humans, that they are carried far from the sites of production by the winds, tides, rivers, and other agents of the natural world. Awareness of the problem has recently been jarred by episodes of pollution-induced catastrophies in Seveso, Italy, Hopewell, Virginia, Grand Rapids, Michigan, Minamata, and Yokkaichi. Today many nations have begun to respond through special toxic-substance control legislation.[1]

Despite the current international movement for control of toxic substances, only Japan has comprehensive legislation to aid the victims. Only Japan has recognized that serious injury to human health will accompany industrialization, that the misfortune of the victim must be addressed along with the control of the pollutant.[1a] This is the essential idea behind the Law for the Compensation of Pollution-Related Health Injury enacted in 1973.[2]

The 1973 Act grew out of the exigencies of the 1950s and 1960s. As described in earlier chapters, Japan was totally unprepared for the pollution diseases that began to appear during this period. The central government was absorbed with economic recovery, the scientific community was largely uninterested, and local authorities simply did not possess the requisite scientific and technical skills to be of assistance. An official response came only after years of protest.

This chapter analyzes the cultural and institutional foundations of the 1973 Act, and traces the transition from sporadic local relief to the beginnings of a comprehensive approach in the Basic Law, through the 1969 relief system, strict liability, and finally to national compensation; it also describes how the compensation system operates and assesses its performance during the past four years. Although we do not believe that other countries will wish to pass identical legislation, the 1973 Act serves as a useful paradigm. For this reason, the concluding portion of this chapter establishes a framework to assist efforts in the United States, and perhaps other countries, to redress the calamity of pollution's victims.

Section 1   From Local Relief to the Establishment of a National Victim Compensation System

1.1   Local Relief

Local governments were the first to assist.[3] The initial effort began in Kumamoto Prefecture in 1958 and was followed in 1960 by Niigata Prefecture's aid to Minamata disease victims. In 1965, Yokkaichi City established Japan's first medical payment allowance system for air pollution victims and the Yokkaichi system thereafter served as a model for comparable programs inaugurated by Nanyō Town (Yamaguchi Prefecture)[4] and Takaoka City (Toyama Prefecture) in 1967. Finally, in 1968, Toyama Prefecture set up a relief program for cadmium poisoning victims, analogous to that of Kumamoto and Niigata Prefectures.[5]

These early programs differed in several respects. Although all covered payments for medical expenses, some, like Niigata, extended loans with low or no interest; others reduced local income taxes (Toyama) and property taxes (Takaoka City). The financial burdens of relief also were allocated differently. In Nanyō Town and Takaoka City, the municipal governments bore all the costs of assistance; together Niigata City and Niigata Prefecture shared that system's expenses; the central, prefectural, and city governments supported the medical treatment of Kumamoto Prefecture's Minamata victims;[6] in Yokkaichi and Toyama, the polluting companies contributed to a relief fund jointly administered by the central, prefectural, and municipal authorities.

In retrospect, these were only stopgap, essentially propitiatory measures.[7] For example, no program reimbursed lost earnings. From the start Kumamoto Prefecture's relief effort was linked to Chisso's (then the Shin Nihon Chisso Fertilizer Corporation) mimaikin payments. Patients officially certified as Minamata victims by a prefectural council[8] became entitled to an annuity under the mimaikin contract as well as to relief under the prefectural system. As might be expected, the local programs were only marginally helpful. For example, during the ten years preceding the establishment of a national relief system in 1969, the Kumamoto Minamata Disease Council certified only thirty-seven patients[9] and other programs were similarly circumscribed.

Yet, despite their many limitations, the early programs must not be completely discounted. By "officially" recognizing pollution-induced diseases and "certifying" victims of these diseases, they, perhaps unwittingly, focused the government's, industry's, and the public's attention on the link between pollution, disease, and human suffering; they forced local governments to take responsibility for relief; they set in motion an administrative process that led gradually, but inevitably, to a nationwide program of compensation.

1.2   Development of a National Administrative Relief System

Several forces carried the issue of victim relief to national attention. First, the victims' protests, particularly those of Yokkaichi's asthma victims and fishermen, were extensively televised and reported by the press. Second, the success of the Mishima-Numazu residents' movement profoundly shocked industry and the central government, for the movement demonstrated that unless the grievances of victims and local residents[10] were addressed, the construction of large-scale petrochemical and other industrial complexes might soon be politically unacceptable. Third, the government's formulation of a comprehensive environmental policy motivated more critical scrutiny of the issue.

Curiously the movement favoring victim assistance also gained support from the defeat of an early proposal to impose strict liability on polluters. During the preparation of the Basic Law, victims' and residents' groups, the media, and many local governments had widely endorsed this idea. But industry rebuffed it, arguing that the government's failure to conduct wise land use planning was the chief cause of pollution, and that strict liability was inappropriate when industry was faithfully complying with all administrative regulations.[11] The deletion[12] of the provision in the final version of the Basic Law demonstrated to some "progressive" officials[13] the even more desperate need for some remedial action.[14] Many subsequently gave their full support to the abstract idea of national relief embodied in art. 21 of the Basic Law.[15]

Shortly after the enactment of the Basic Law, the Ministry of Health and Welfare began a serious study of various approaches to relief. The most carefully considered option was the Yokkaichi system.[16] Another idea was to tie the new program into the general framework of social insurance.[17] Under this approach, the government would bear the entire costs of relief payments and the relief system would be free of any notion of industry's liability. Another option was to establish a system of liability insurance.[18] Under this scheme, the fund would advance payments to victims whose interests would thereafter be subrogated to the fund. A fourth option was to require industries to post a relief bond.[19] A fifth plan contemplated the government's advancing relief payments on behalf of the responsible industries. A final proposal established a centralized administrative agency to dispense relief payments. This approach advanced an unusual idea: the fund would be generated from charges levied on polluters and would replace existing systems of donations from parties responsible for the victims' injuries.

The ministry's initial response limited aid to reimbursement for health damages (medical and nursing expenses) suffered by certified victims living in designated areas, and excluded property damage. The ministry also proposed that five-eighths of medical payments be financed by the contributions of factories operating in designated areas, proportionate to the share of the area's pollution.[20] The central, prefectural, and municipal authorities would shoulder the burden of the remaining costs of the system. Yet even this timid proposal was unacceptable to the development-oriented ministries. At length the Ministry of Health and Welfare's proposal was shelved and the matter remanded to the Central Council for more careful consideration.[21]

After the expert committee submitted its report to the council in June 1968, a subcommittee began considering the details of specific draft legislation. Although the establishment of dispute settlement procedures posed few problems, victim relief reportedly was intensely debated.[22] Representatives of local governments urged that industry bear three-fifths of the costs of compensation,[23] that compensation payments should include a living allowance, and that damages to property should also be reimbursed. Industry representatives, however, advocated separating the system from legal liability, and limiting it to health damages; they opposed compulsory collection.

The subcommittee's final report to the council (September 30, 1967) essentially was a compromise between these different positions. It proposed the establishment of a provisional administrative relief system limited to health injury and recommended that industry shoulder a part of its costs on the theory that industry bore "moral responsibility to maintain society."[24]

On October 18, 1968, the Central Council approved the subcommittee's report and forwarded its own recommendations to the government.[25] After further debate on the issue of cost allocation, the government submitted its proposal for victim relief—along with a companion dispute settlement bill—to the Diet, which passed the relief proposal at the end of the session.[26]

1.3   The 1969 Law for Special Measures for the Relief of Pollution-Related Disease[27]

Like earlier local relief efforts the new law was a limited effort. Modeled principally on the victim assistance program developed by Yokkaichi City, the program only reimbursed medical expenses (including a small nursing allowance) not covered by social insurance. It did not cover pain and suffering or disability. The number of compensable pollution-related diseases was also extremely limited. In fact only diseases already included in local assistance schemes—the four pulmonary diseases under the Yokkaichi plan,[28] mercury poisoning, cadmium poisoning, and chronic arsenic poisoning[29]—were included. The system's administration (including patient certification) was relegated to local governments.

Although the central, prefectural, and municipal governments and industry shared the financial burdens of the 1969 system,[30] industry's contribution was collected in a somewhat peculiar manner. In the hope of avoiding a direct confrontation with the victims and any suggestion of individual corporate liability, the link between relief payments and industry's contribution was severed.[31] The concerned industries collectively established a nonprofit corporation, the Pollution Measures Cooperative Foundation (Kōgai Taisaku Kyōryoku Zaidan), to which they all contributed voluntarily. Each company's donation was determined by consensus.[32] The monies collected were thereafter transferred under contract to a semigovernmental Pollution Control Service Corporation (Kōgai Bōshi Jigyō dan).[33] The government also deposited its contribution to the fund at the same time.[34] The Pollution Control Service Corporation then released the fund monies to the local governments responsible for dispensing relief payments.[35]

From the victims' perspective, the 1969 Law was patently inadequate. For example, art. 24 permitted prefectural governors to terminate medical payments and under some circumstances even to ask victims to make refunds. Finally, only victims below an established income level qualified.[36] Although the 1969 Law established a unique precedent that attracted world attention,[37] it was in reality a stopgap measure. Industry's leaders were the principal persons to applaud it.

1.4   From National Relief to National Compensation

The year 1972 marks the transition from national relief to national compensation. Two events outside Japan influenced this change. The first was the U.N. Stockholm Conference that publicized, to Japan's embarrassment, the agony of the victims.[38] The horrified outcry of many private groups at Stockholm lent some impetus to subsequent action within Japan. A second influence was Japan's endorsement of the OECD's "polluter pays" principle.[39] The original intent of the OECD declaration was to mitigate dislocations in international trade by requiring polluters at the outset to bear the costs of pollution abatement and control. The Japanese government expanded this idea into a political means of persuading recalcitrant industry leaders to pay the costs of pollution damage.[40]

These events directly influenced the passage of strict liability legislation.[41] Strict liability was an important conceptual precursor to national compensation, because it focused attention on the responsibilities of industry, and derivatively on the need for supplementing victim relief. The government's attitude to the new legislation, however, was less than enthusiastic. By spring 1972 the government had already prepared a draft bill under an earlier pledge by Prime Minister Satō,[42] but the draft was sharply criticized and its passage by the Diet was uncertain. During the Diet debates in June 1972, the government fought aggressively for as narrow a construction of strict liability as was possible, urging that strict liability represented a significant departure from Civil Code tort principles based on negligence; that the new legislation should be limited to injuries caused by air and water pollution;[43] that a broad statutory presumption on causation should await informal recognition of this principle;[44] that available injunctive relief and other remedies were adequate;[45] and that retroactive legislation violated general legal principles. In the face of violent criticism from the public and the opposition, the "polluter pays" principle provided the government with a decisive political weapon for pushing through this needed reform. The government's bill passed the Diet on June 16 and was promulgated as law on June 22 with only minor revisions.[46]

Although the opposition parties failed to enact a vigorous strict liability law, the 1972 Diet debates proved extremely influential, for they highlighted the need to establish a national compensation system. The opposition parties were quick to point out that even with the government's recognition of strict liability, the more basic issue of the victims' compensation most probably would be left unattended.[47] First, the strict liability law covered only new cases of pollution injury, thus leaving unaided the considerable number of victims seeking assistance at the time of its enactment. Second, pollution victims still bore the burdens of a court trial—the long delays, considerable expense, difficulties of proof of complex scientific and technical questions,[48] and complications of multiparty litigation. Third, even if a pollution suit were successful, victims could not assume that a liable party would be solvent (especially so in the case of small and medium enterprises). Fourth, pollution insurance was not available since existing insurance law did not treat pollution from daily factory operations as an "accident."[49] The criticisms of the opposition parties persuaded the government to include in the strict liability law a provision promising to study the feasibility of national compensation.

But compensation was also beginning to seem wise from other perspectives.[50] By this time industry and the Liberal Democratic party had concluded that the establishment of a compensation system could undercut the victims' incentive to litigate, now greatly enhanced by the successes in the itai-itai and Niigata Minamata disease trials. Moreover, an administrative system would spread the risk of liability throughout the industrial community; in any event, industry expected the government to shoulder a substantial part of the system's financial burdens. After the enactment of the strict liability law in June 1972, the government contemplated several years of "study."[51]

The Yokkaichi decision in July 1975 abruptly interrupted these plans. Although the itai-itai and Niigata courts' decisions seemed ominous to industry (and the government), they could be distinguished as exceptional cases, restricted in their impact to mining and soda manufacturing. The Yokkaichi decision, however, could not be so easily dismissed, for the court unequivocally held defendants jointly and severally liable. The door seemed now open to thousands of others exposed to the emissions of petrochemical refineries. Because many industries released SO_x in their daily burning of fuel, a spectre arose before industry leaders of a significant part of Japan's industrial sector being held liable for pollution.[52] Indeed, shortly after the Yokkaichi decision, their worst fears were confirmed by the news that pollution victims in Kawasaki, Amagasaki, and Chiba were preparing litigation.

In September 1972, after the Yokkaichi decision, Osanori Koyama, the director general of the Environment Agency, visited Yokkaichi to assess the air pollution situation. In Yokkaichi, Koyama announced that the government would present a bill establishing a compensation system for pollution-related health injury to the next Diet session.[53] He further promised that the compensation system would be financed by polluters, and that victims would receive medical expenses and be reimbursed for lost earnings. He did not detail the exact method by which the proposed system would be financed.

After Koyama's return to Tokyo, the government began an intensive study and preparations for the new program. Between September and December, the Central Council's expert committee labored, meeting twice a month; and the Environment Agency itself established a new special office. On December 22, the expert committee published an interim report and circulated it to concerned groups, including victims, the industrial community, and local governments. Thereafter, the Central Council quickly gathered a broad range of opinion in a series of formal and informal hearings.[54] By March 1973, the expert committee had collated all comments and prepared a final draft report which it submitted to the council. The council approved the draft, presented it to the director general of the Environment Agency, and the Environment Agency itself set to work in drafting the new law. By May 18, a draft was approved by the Cabinet and presented to the Diet. Although many groups in the Diet furiously opposed the draft, it ultimately passed all hurdles. On October 10, 1973, the bill was enacted into law with a curious coalition of the Liberal Democrats, Communists, and Democratic-Socialists in support and the Socialists and Kōmeitō opposed. Necessary implementing regulations were all completed uneventfully during 1974.[55]

Section 2   The 1973 Law for the Compensation of Pollution-Related Health Injury

2.1   The Compensation System in a Nutshell

The 1973 Act establishes an administrative structure to oversee compensation payments. Pollution victims of designated diseases[56] arising in officially identified pollution areas[57] are examined by a special Health Damage Certification Council[58] of medical, legal, and other experts. Upon certification victims are eligible for reimbursement for medical expenses and lost earnings. Assistance is also provided to survivors for funeral and other expenses.[59]

Under the 1973 Act, polluters pay the entire costs of victim assistance. For example, revenues for compensation of air pollution-induced disease attributable to multiple sources associated with "Class 1 areas"[60] are derived principally from a graduated emission change,[61] and secondarily, from a tonnage tax on automobiles.[62] The emission charge is collected by a special authority and pooled at the national level; thereafter, the fund is distributed to provincial governments for payment to the victims. Compensation payments for diseases traceable to a specific substance (Class 2 area diseases) are paid directly to victim groups at the local level without prior national collection by the authority.

The 1973 Act also establishes an apparatus for review of grievances. Pollution victims may petition a prefectural governor or mayor over the disposition of compensation benefits, certification, and other actions.[63] Thereafter, the prefectural governor (or mayor) may appeal to a central grievance board under the jurisdiction of the Environment Agency.[64] Polluters and others with complaints regarding the levy can request the Director General of the Environment Agency and the Minister of International Trade and Industry to review the action.[65] It is clear, however, that in no case will administrative disposition of a case bar subsequent recourse to judicial relief.[66] Figure 1 outlines this compensation system.

2.2   Victim Compensation

Under the 1973 Act, victims are relieved of the burden of proving disease causation.[67] Yet, before a victim becomes eligible for compensation under the act, a

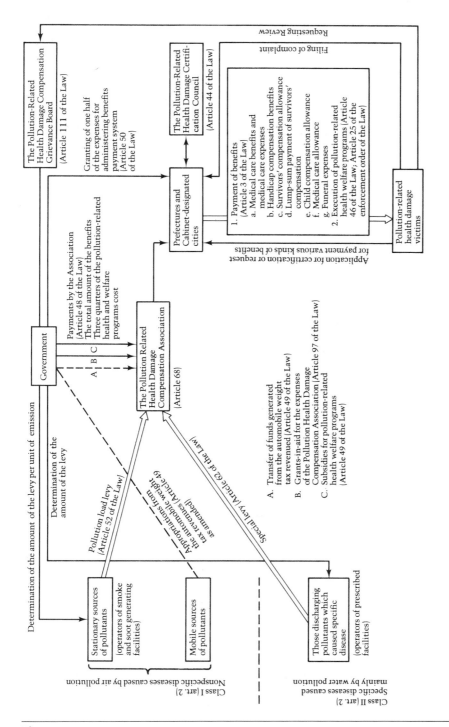

Figure 1
1973 Law for the Compensation of Pollution-Related Health Injury.

number of administrative determinations involving causation must be made. These include the official designation of specific toxic substances, pollution-induced diseases, and pollution zones. In each case administrators have to strike a compromise between scientific accuracy and administrative expediency.

2.2.1   Designation of Pollution-Related Disease

The act originally contemplated compensation for specific, officially recognized diseases, i.e., Minamata, itai-itai, and chronic arsenic poisoning as Class 2 area diseases, and four respiratory diseases, emphysema, chronic bronchitis, asthma, and asthmatic bronchitis as Class 1 disorders. These diseases were selected not simply because they were the most controversial,[68] but also because administrators possessed extensive data on their origin and effects developed during the fifteen years before the system's enactment.

Yet, the reader should not underestimate the years of prior bitter debate over the issue of causation. During the 1960s, administrators within the Ministry of Health and Welfare were called to testify before the Diet on the question of whether the diseases in Minamata, Toyama, and elsewhere were "pollution-related." The ministry had just begun to understand how epidemiological data might be effectively used along with experimental and clinical findings. But research results were not fully analyzed, the government's monitoring network was not fully developed, and companies still controlled the basic information on chemical processes. For these reasons many officials within the ministry insisted that administrative practice must defer to scientific precision.[69] Unless the best medical scientists were agreed on a disease's cause, they argued, the ministry should demur before the Diet's questions. This view dominated the government until the late 1960s.

By the early 1970s, the administration's conservative position began to change drastically. Many factors were influential. The results of earlier studies had at last been analyzed and all appeared to confirm the critical role of pollutants. The government was also now more aware of the problem. By 1968, about ten government agencies were involved in cooperative research programs on the relation of pollution to the contamination of marine organisms, soil, and food; several years earlier, only the Ministry of Health and Welfare and MITI had been concerned. All these factors influenced the official designation of Minamata and itai-itai as pollution-related diseases and recognition of the etiology of cadmium and mercury. The government made these determinations despite persistent unanswered scientific questions.[70]

The decisions in the pollution trials greatly encouraged these changes in the administration's attitude. As described in ch. 3, the trials left unanswered many questions relating to the diseases' causes. The courts' award of damages in the face of uncertainty served to justify the administrations' taking further liberties with "scientific accuracy." The courts' treatment of the technical issues gave impetus to the feeling of some high-ranking officials that scientific uncertainty could be unjustly used as a weapon against the politically weak and uninformed. As a result of the decisions, administrators became more accustomed to act under different levels of uncertainty.

Despite the obvious intention of the drafters of the 1973 Act to focus attention on diseases that occurred in the past, the system was not restricted to these maladies. Article 2 directs the prime minister to designate pollution diseases after consulting with the Central Council and the governors and mayors of the concerned prefectures and municipalities. The broad discretionary powers granted by this provision, in theory, permit certification of other diseases, like cancer, once sufficient data is compiled to demonstrate a correlation with pollution.[71]

2.2.2   Development of Criteria for Victim Certification

Although the basic procedures for victim certification are the same under the 1973 Act as in the 1969 Law, the criteria were developed only after extensive discussion within the government.[72] Class 1 areas involved principally a determination of a scientifically and administratively acceptable "nexus" to justify a patient's certification. Criteria were at last set, based on the length of the patient's residence, work, or contact with a designated area. Patients were classified by age and sex.[73]

Although the draftsmen of the criteria recognized that heavy smoking could also be a substantial factor in pulmonary disease, smoking was not given a central role in the regulations. Rather, to alleviate the burden on the individual victim as much as possible, the regulations adopted the concept of aggravation; the 1973 Act did not require that pollution be a "but for" cause. Only in the most extreme cases did the regulations permit heavy smoking to interfere with a patient's receipt of compensation benefits.[74] The drafters of the regulations concluded that differentiating between smokers and nonsmokers would be prohibitively expensive and could be unfair to many patients.

The problems of establishing criteria for Class 2 areas were different. Smoking, of course, was not in issue. The most critical problem was identifying the diseases themselves, for even after years of study, these afflictions were ill-understood. Moreover, not all patients showed the "classical" symptoms presented to the courts in the original Minamata and itai-itai trials. The drafters perceived that some cases would display a whole range of atypical symptoms, possibly reflecting different causes, individual differences, varying latency periods, and other factors. The original criteria adopted were very strict, adhering to the orthodox classification of the diseases, thus creating a bias against certification in "grey" areas. These criteria were established virtually without public hearings or other avenues for the victims or their representatives to express their views. As will be seen in sec. 3, the problems in handling grey area cases, however, have not ended, and some changes are taking place.

2.2.3   Dispensation of Compensation Benefits

The certification of victims and the dispensation of benefits is administered by the Pollution-Related Health Damage Certification Council, under the jurisdiction of the office of the provincial governor, or in the case of a municipality, the mayor's office. The council is staffed by experts from medicine, law, and other fields. Each prefecture and city has its own council that meets and screens patients. Members are forbidden to divulge confidential information obtained in the course of their service on the council.

The 1973 Act separates the administrative duties, performed by the council, from the task of writing diagnoses, assigned to the doctors. This division of administrative and technical functions stems from the period of enactment. Because many doctors at this time feared being entangled in the politics of pollution disputes, they hesitated to recommend criteria for certification and were reluctant to participate.

As noted, the act provides compensation for medical care, including medical treatment and the costs of rehabilitation, a lump sum payment to survivors, a compensation allowance for children under fifteen, a medical care allowance, funeral expenses, and compensation for loss of earnings for the handicapped. These benefits are discussed in greater detail below.

Costs of Medical Treatment

In contrast to the 1969 relief system, the 1973 Act covers the entire amount of medical bills including that portion currently paid under national health insurance.[75] Upon certification, a victim is issued a pollution-related medical care handbook that entitles the victim to a free medical examination and treatment at a designated hospital.[76] The hospital then bills the prefectural governments for the medical expenses thus incurred. Should a certified patient receive medical treatment at a hospital or clinic that has not been designated, the patient may file an application with the prefectural government for a refund.

Compensation for Disability

Compensation for disability was always among the most controversial aspects of the 1973 Act. For example, from the outset the victims had lobbied for the equivalent of the average monthly wage, but industry favored the lowest available standard for payment under various social insurance programs. The present approach is essentially a compromise between these two perspectives. Taking the Yokkaichi Court's standardized formula for lost earnings as a benchmark, the Environment Agency's standard was set between the average wage and the lower level proposed by industry.[77] The controversy was temporarily resolved[78] by fixing disability payments at 80% of the average wage, to be computed and paid monthly. The present regulations distinguish four classes of benefits differentiated by the degree of disability.[79]

Compensation to the Surviving Members of the Family of Deceased Victims

Survivors' payments are intended to compensate for damages and also to contribute to the reconstruction of family life. Regular payments are made for a period of ten years either to certain close members of the family or to those who previously bore the financial responsibility of the deceased. The amount of compensation is computed monthly on the basis of the national average worker's monthly wage by age and sex, minus an estimate of the deceased's cost of living. This amount is equal to 70% of the deceased's average wages. Where there are no eligible survivors, a lump sum is paid to other survivors in a more distant kinship relation to the deceased.

Other Compensation Benefits

A fixed children's compensation allowance is paid regularly to those having responsibility for afflicted children. The amount of benefits is computed monthly according to the degree of difficulty in living experienced by the child. A medical care allowance is paid to these victims covering various expenses such as the costs of commuting for outpatient and inpatient care. The amount of benefits is computed based on travel time or the period of hospitalization. A flat sum funeral allowance of ¥285,000 is paid to those who pay for a victim's funeral expenses.

Rehabilitation

The 1973 Act also requires the prefectural governor or the mayor of a designated city to establish rehabilitation facilities for victims of both Class 1 and Class 2 diseases. Although these facilities presumably are to be located in rural, unpolluted areas, the act leaves the details of the program to Cabinet Order.[80]

Residual Compensation for Pain and Suffering and Property Damage

The legislative history records the bitter debate surrounding two items that were not included in the compensation schedules, reimbursement for property damage and pain and suffering. During the 1973 Act's preparation, Environment Agency officials had discussed the issue of compensation for property damage, originally contemplating cases involving injury to farming and fisheries from mercury and cadmium poisoning. Some scholars had also noted that because these losses were part and parcel of the original pollution episode, it was only fair, reasonable, and practical to compensate victims suffering property losses along with those afflicted with bodily harm.

This proposal was ultimately rejected for several reasons. First, compensation for physical injury was simply regarded as more pressing. Second, the causes of property damage were in many cases less well studied. Third, the number of people suffering property damage was open-ended. It included not only fishermen but also wholesalers, retailers, sushi shop owners, restaurant owners, innkeepers, and so on. The scheme would become too complex to administer. Since it was essential that a compensation system be enacted immediately, property damage was ultimately eliminated. During the Diet debates Minister Miki reiterated that other ministries were in a better position, by virtue of their jurisdictional prerogatives, to address the problems of property damage from pollution.

The elimination of pain and suffering was due to other factors. During the Diet debates, industry and the government both took the position that pain and suffering had already been addressed in the compromise set for disability. Victims, however, insisted that this level merely constituted a floor upon which an additional amount should be paid. Industry's views carried the day.

2.2.4   Relations Between the 1973 Act, Workmen's Compensation, and National Health Insurance

Before the establishment of the compensation system, Keidanren and other industrial groups discussed integrating relief with existing health insurance programs, or somehow designing the new system from the workmen's compensation model. Neither idea was adopted. The workmen's compensation analogy was inappropriate because it involved a direct economic relationship between employee and employer, and because the proponents of compensation contemplated enterprise responsibilities extending beyond the workplace. The opposition parties, victims' groups, and some members of the government rejected the idea of integrating compensation and national health insurance because they felt this approach would sever the link between the payment of compensation and the identity of the offender, and polluters would then escape responsibility. They urged that it was unfair to defray through tax revenues costs that the courts already had determined industry should pay. Finally, designating a special class of beneficiaries entitled to a disability allowance within national health insurance (where these benefits were not otherwise available) seemed impractical. It would invite attack by victims of other diseases excluded from these benefits. For both political and practical motives it was essential to establish a new system entirely separate from national health insurance or workmen's compensation.

At present, a worker who contracts one of the designated pollution diseases may still opt to receive compensation either under workmen's compensation, national health insurance, or under the 1973 Act. In cases where a worker's individual average wage exceeds the standardized average wage, workmen's compensation still offers more attractive benefits than those to be obtained from the compensation system.

2.3   Financing Compensation

Having determined the classes of beneficiaries and categories of payments, the next problem was the determination of a budget. The drafters of the compensation system were fortunate in that they possessed, from over ten years of administrative experience, extensive data on the average costs of compensating a victim population. It was also felt that once the system was put into operation it would generate its own statistics that could assist future planning. Simply by summing all putative claims relating to the designated areas, a close approximation could be made for the first year's Class 1 budget. The most intractable problem, however, was the allocation of costs particularly for Class 1 areas. The drafters of the 1973 Act now made two critical decisions. First, they continued the practice of designating areas employed in the 1969 relief system. Second, they decided to finance compensation payments through a pollution levy.

2.3.1   Pollution Area Designation

The present designation process requires administrators again to reconcile scientific uncertainty with administrative expediency. As noted, the objective of designation is to identify an (Class 1) area where "marked air pollution has arisen . . . and where diseases resulting from the effects of such air pollution are prevalent."[81] The 1973 Act implicitly assigns the administrator the task of determining the (causal) relation between pollution and designated pulmonary diseases.[82]

The first stage of the designation process begins when the Environment Agency and other government officials survey the polluted area. These surveys usually include a demographic analysis, an assessment of various physical, meteorological and seasonal conditions, and monitoring reports on SO_x and NO_x concentrations, particulates, and other pollutants.

The next stage specifically assesses SO_x concentrations.[83] The Central Council's recommendations identify four levels of SO_x concentrations based on a yearly average monitoring sample. These are: A, below 0.04 ppm; B, 0.04–0.05 ppm; C, 0.05–0.07 ppm; and D, over 0.07 ppm. Average, monthly, and seasonal SO_x levels for three successive years are then scrutinized to examine the extent of their divergence from category A. Areas justifying in the last ten years classification as B, C, or D are recorded and the analysis proceeds to the next stage.[84]

When an area is classified as B, C, or D, the Environment Agency usually requests the local government having jurisdiction to conduct a survey of the health effects of air pollution in the area. At this point, an extensive community health survey begins and a comprehensive, thorough study is given each case.[85] The intensive population study is then supplemented by other data on pulmonary disorders among children and the elderly.

After receiving the team's report, Environment Agency officials begin their assessment of the data. Past studies of the target diseases provide a benchmark for "natural" incidence and prevalence rates. For example, the natural incidence of pulmonary emphysema for a forty-year-old man might be one in one hundred. Using this data, the administrator can begin to detect how far an area diverges from the natural incidence and prevalence rates. The regulations specify these categories: (i) two times the natural rate; (ii) two to three times the natural rate; and (iii) four times the natural rate and greater than four times. Once so classified, an area is now ready for designation.

Some administrative designations are simple. If an area has a high concentration of SO_x according to the first assessment (i.e., C or D level), and a high prevalence level with respect to the second standard (i.e., ii or iii), the area is quickly designated. If the area is given a low grade for either classification, it is excluded. More difficult problems are presented where disease prevalence is high but observed pollution is low; or conversely, where pollution is high, but disease prevalence is low.

The latter cases are the most difficult, because the administrator must compromise scientific accuracy with the objective of compensation. The polluted area must next be comprehensively reexamined. In cases where average yearly data were initially used, now monthly monitoring data are systematically scrutinized. Administrators also now pay greater attention to possible synergistic effects of NO_x, particulates, and SO_x. Some of these areas are ultimately designated. At times borderline cases are reassessed. It is, however, less usual for an area to be designated based on patient reexamination. When the review is complete, the Director General of the Environment Agency recommends to the Prime Minister's Office that an area be designated. If there is no objection, the government announces the new designation by Cabinet Order.

More frequently, however, a complicated political debate begins.[86] One important constituency is local government. Essentially, local governments must balance the benefits of qualifying for compensation funds that attend area designation against the disincentive created by designation to new firms' entering an area. Because local governments subsist mainly on tax revenues collected from local businesses, this consideration is very important. Industry usually lobbies against wide designation in order to escape the higher pollution levy imposed on factories operating within a designated zone.[87] At times the bar associations or other groups also submit opinions to the Environment Agency relating to designation.[88]

A proposal for area designation at the Cabinet level may at times create conflicts within the bureaucracy itself. For example, MITI, the Ministry of Construction, the Ministry of Transportation, or other mission-oriented agencies may take the position either that designation should not proceed or that the Cabinet Order should restrict the scope of the proposed area. After these and other issues are debated, the original proposal is often modified.

The designation process perhaps more than any other stage suggests the compromise between science and administrative expedience. As noted, the process rests upon the concept of "aggravation," that replaces scientific proof. Although "aggravation" is a less certain, more subjective benchmark, the Western reader should note that even under a more rigorous scientific standard, it is not certain that the area's pollution "caused" the diseases of those certified for compensation. Generally, the "clear" cases reflect about the same level of scientific certainty as was present in the Yokkaichi trial. And in the cases requiring reconsideration, the concept may be even more attenuated.

2.3.2   Design of the Pollution Levy

The design of a practical, fair, and efficient means of financing payments to Class 1 victims posed complex problems because of the number of stationary and mobile polluting sources and the continuing uncertainty over disease causation. The only model in operation at the time was a program established in Holland in 1972 to compensate damages caused by air pollution. Under the Dutch system, identified polluters were compelled to pay for health and property damage resulting from their activities. The system had not been deeply studied in Japan, and since it had just been established, it could serve only as a general point of reference.

The basic measure ultimately selected was a levy on emissions of SO_x. This method appeared the most simple and practical for several reasons. First, data presented at the pollution trials and the government's independent research indicated that the dioxide (SO_2) was a principal cause of the victims' maladies. Second, of all the air pollutants, the government possessed the most data on SO_2 and understood the problem of its control the best. Third, SO_x emissions could be easily calculated, because these emissions were directly proportional to the percentage of sulfur content in industrial fuel, and the rate at which this fuel was burned. Finally, since the government possessed adequate data on industrial fuel consumption, SO_x emissions could be easily monitored.

Before the law's enactment, some administrators within the Environment Agency and other ministries had suggested that emission charges might to some extent replace direct regulation. Opposition to this proposal, however, was overwhelming, because emission charges were considered to be impractical and politically infeasible. Although the levy appeared to have some regulatory benefits,[89] administrators wished to play down its attributes. Accordingly, the emission charge was designated a "pollution levy," and the entire system publicized as an emergency relief measure, not a new regulatory initiative.

Yet a levy on SO_x emissions also presented conceptual difficulties. First, it substantially extended the rationale of the Yokkaichi decision. In Yokkaichi there were only six defendant companies, but the government now proposed to levy on all polluters throughout the country. Second, in the Yokkaichi trial, the plaintiffs had demonstrated that there was a significant correlation between pollution and their maladies. But the scientific evidence on causation under the compensation system was in some cases far more obscure. In addition, some companies objected to the levy's singling out SO_x when NO_x, CO, and particulates, they alleged, had an equal, if not more important, causal role. In response to these criticisms, the administrators successfully defended the system by arguing that the designation process would be conducted with utmost fairness, and that a levy on NO_x and other substances would be assessed in the future, once a technically feasible method for imposing a charge was developed.

An equally difficult problem presented by the approach to Class 1 areas was the notion of charging present polluters to pay for past harm. To the concerned industry, this approach seemed unfair and inefficient. The government explained the apparent inequity by noting that many of the firms to be charged had operated for some time, and others already had been held liable for past harm. A second argument was that since present polluters were exposing the public to the same level of risk, a levy on SO_x was not inappropriate.

Two other aspects of the levy deserve explication. Since the levy was designed as the most simple and effective means of financing compensation payments, and not as an incentive for pollution control, virtually no thought was given to assessing a charge based on calculations of marginal cost as recommended by economic theory. For the same reason, the levy was set linearly in direct proportion to the amount of discharge, despite the fact that some evidence suggested that the costs of injury to health might actually increase more rapidly at higher levels of pollution.

Another important issue was whether to charge all emitters of SO_x, or only those sources that directly contributed to the pollution diseases within a designated area. A settlement was ultimately made whereby all would pay, but sources within a designated area would be assessed an amount nine times that imposed on other areas. This ratio was to be an approximation of the respective concentrations of SO_2 between designated and nondesignated areas.[90]

The conceptual basis for assessing industries operating outside designated areas originated in the 1969 law. Although under the earlier system payments were voluntary, the system's underlying rationale, collective social responsibility,[91] required industry as a whole to accept the financial burden of payment as a recognized social obligation. The individual allocation of the financial burden was to be determined by industry itself.

The final question was the appropriate institution to collect the pollution levy. Here again industry's primary motive was to avoid assignation of individual responsibility. Originally the Environment Agency considered establishing a special independent government organization with plenary powers. Industry, however, lobbied against this idea, proposing in its stead a private nonprofit corporation on the lines of the tax exempt public benefit corporation employed in the 1969 law.

At length a quasi-governmental body was established as a compromise. The responsibility for the determination of the budget and collection of the levy remained with the government. The association's principal officers were drawn from industry and its daily affairs managed by industry (particularly Keidanren) under guidance by the Environment Agency and MITI. As mentioned, the apparatus for collecting the levy was separated from the procedures for dispensing compensation payments. The architects of the law feared that if the entity combined both functions, those dissatisfied with some aspect of compensation (i.e., certification, the amount of benefits) would focus attack on the collection entity, and thereby undermine its operations. Separating functions, it was felt, would deflect the victims' attack from individual firms.

From the perspective of enforcement, separate collection seemed sensible. The organization of the collection body would assure industry's continuing cooperation, because industry leaders would play a dominant role in the entity's operations. In theory, the responsible industries were to police themselves.[92] Finally, having determined the principal means of financing compensation, the government refined its approach by seeking supplemental sources of revenue.

The drafters of the first interim report to the Environment Agency that discussed the new compensation system had noted that it would be unfair to place the entire cost of compensation on factories, when automobiles and motor bikes were at times significant polluters. During the subsequent debate within the Environment Agency and later in the Diet, a number of possible options were considered. Some administrators proposed a fuel tax; others advised an emission charge or tax by weight. At last it was decided that the most administratively convenient course was to supplement the system's revenues through the automobile weight tax. Since about 20% of total SO_x and NO_x emissions throughout the country derived from automobiles, the simplest, cheapest, and most easily administered means was to defray from this source 20% of the needed budget.

Financing Class 2 Payments

The problem of Class 2 payments was far more simple. The responsible companies[93] would compensate patients directly; the costs of rehabilitation and administrative expenses would be defrayed 50% by the responsible enterprises, 25% by local governments, 25% by the central authorities. Although not explicitly provided, the government itself would finance research on these diseases.[94]

During the drafting of the law, Chisso and some of the other concerned companies raised the issue of the fairness of compelling payments to victims who appeared after a factory had already terminated operations. The government, however, used the indictment of the four pollution trials to rebut this argument, urging the companies to design a more practical alternative. This they were unable to do. A final issue always underlying Class 2 payments was corporate bankruptcy. Should one or all of the companies succumb, many workers might lose their jobs, and the fund for victim compensation would be extinguished. The 1973 Act left this problem unresolved.

2.4   Grievance System

As noted, the 1973 Act also establishes an administrative board for the settlement of victim complaints relating to certification and other matters. Proceedings of the board are public and recorded, and interested persons are permitted to examine the record. The board consists of six members, appointed by the prime minister with the consent of the Diet. Board members are supposed to be individuals of integrity with experience in law, medicine, or other fields relating to compensation for health damage. The board's decision must be unanimous in order to overturn a previous ruling. Judicial review of certification or other administrative acts is permitted after the board has issued its decision. Polluters dissatisfied with any action taken by the board may seek review from the Director General of the Environment Agency and the Minister of International Trade and Industry. The act permits an administrative lawsuit to nullify any action taken by the board, after the two ministries have issued their ruling on the request.

Section 3   Assessment of Theory and Practice

It is useful to review briefly the theoretical implications of the 1973 Act in order to understand the system's inefficiencies, necessitated by its compromises with other objectives, such as equity, administrative expediency, and political necessity. The second part of this section discusses progress made in the act's implementation during the past five years.

3.1   Theoretical Note

Classical economic theory begins with the concept of "perfect competition." Perfect competition is defined as an equilibrium market condition where each firm produces at cost and no firm can influence the market price of the product. Should a firm raise its price, consumers would shift purchases, and the firm's revenues would fall. Ultimately, the firm would be forced either to cut its price to the prevailing market level, or to withdraw from the market. The classical model rests on many theoretical assumptions about economic behavior (e.g., profit maximization) discussed elsewhere that we do not repeat here.

The objective of welfare economics, the field most relevant to the present discussion, is to increase the efficiency of resource allocation in order to maximize production. The overriding goal is to achieve an equilibrium state where no further reallocation can be made, without detriment to someone (i.e., Pareto optimality). Necessarily, welfare economics emphasizes the problem of externality that is regarded as the primary cause of market failure. Externalities, simply put, are costs (for example, damage costs from pollution) unaccounted for by the firm in its decisions, hence "paid" by others (e.g., the public). Under conditions of perfect competition, externalities are assumed not to exist; each firm in theory would take account of all costs incidental to production. But, as economists readily concede, perfect competition is rarely attained. For this reason many scholars have prescribed antidotes for market imperfection.

The late A. C. Pigou argued that a corrective tax (or subsidy) could remedy the problems of market imperfection, where the tax reflected the "true costs" "caused" by an activity.[95] Imposition of these costs on the firm would ultimately contribute to more efficient resource allocation. To take the case of pollution, if a corrective pollution tax or emission charge is applied, a firm should pass its costs on to the consumers by increasing its prices; but consumers, under the classical theory, will retain ultimate control over the direction of economic activity by virtue of their options. For example, consumers can shift purchases to lower-priced goods produced by low- or non-polluting firms. These firms are then rewarded by increased consumer purchases, increased market shares, and higher profits. Polluters, however, are penalized by declining sales, lower market shares, and falling profits. Some firms may even be driven out of the market. In theory, a more productive allocation of society's resources results, for firms will endeavor to change production processes and to modify consumption of the environmental resource in order to avoid paying the tax.[96] Pigou's work has been refined by R. H. Coase, who demonstrated the inherently reciprocal aspects of any externality relationship.[97]

The theoretical literature is particularly relevant to the 1973 Act because of its use of a pollution levy for Class 1 pollution. Let us examine the implications of the design of Class 1 compensation first under conditions of perfect competition, then with price controls, and finally under the oligopolistic conditions existing in Japan.

The pollution levy is not a corrective tax in the Pigovian tradition, nor was it intended as such. As noted, its central purpose was to raise revenues for the compensation of victims, not to promote efficiency. The levy, in that it excludes pain and suffering, property losses, and other costs incidental to the polluting activity, falls short of the theoretical ideal, to impose the "true social cost."[98] So conceived, the corrective influence of the levy may be minimal. Indeed, the levy might even promote inefficiency because an increase in welfare is not necessarily achieved by reallocative adjustments. Under some conditions, such adjustments may in fact decrease welfare.[99]

Yet, the deficiencies of the levy may be offset in Japan by the fact that the levy is not the only cost-internalizing device with which a firm must contend. Polluters in Japan also bear the costs of complying with pollution control regulations, judicial decisions, and mediated private settlements, in addition to advertising and other business expenses. The levy's failure to account for all social costs should not pose a decisive theoretical problem. Indeed, the many controls a firm faces in Japan may encourage overinvestment in pollution control that would itself be inefficient. The analysis must also be qualified by how we evaluate human life. Japanese society may now be so sensitive to the risks and costs of pollution that what a Westerner would consider to be "overinvestment," many Japanese citizens might deem socially justifiable.[100]

Another problem is the act's linkage of the emission charge proceeds to the payment of compensation.[101] We deal with this first on a general level and then turn to the specific difficulty of the act's charging polluters in the present for pollution caused in the past. The principal objection of the economist to the present system is that the criteria for imposition of the system's financial burdens should be separated from considerations that must underlie the allocation of benefits. Assuming that we attribute damage costs to the individual firm, or at least to the industry (a problem to which we will return presently), economic theory indicates that the costs identified by the compensation budget should be internalized by the firm as already discussed. How we allocate the fund so generated is analytically a separate question.[102] Indeed, the use of the fund for compensation may be inefficient for several reasons. If our objective is to equate the marginal costs and benefits for each available option, an alternative use of the fund may be indicated.[103] In the present case, the availability of compensation may discourage some people from leaving a polluted area, and attract others to the area. Consequently, the air resource would continue to be overexploited, and other resources underemployed.

There is an important qualification to this argument. Planners have increasingly recognized that linking compensation payments to regulation can be useful because the transfer fund both promotes coordination of pollution controls, and mitigates "windfalls" and "wipeouts."[104] Indeed, transfer funds have been used to great effect in American land use planning.[105] The theoretical problems posed by the system's linking the emission charge to compensation may thus be offset by its practical benefits to the planner.[106]

A more troublesome problem is charging present polluters for the cost of past harm.[107] Because some polluters may not have been responsible for the compensated victims' injuries, the levy in this case imposes the "wrong" incentive. This can lead to inefficiency. Yet several factors mitigate this risk. First, there is always an interval between the discovery and assessment of damage and the imposition of damage costs on the firm. Second, the past and present risks to health of a polluting activity may be equal. In this case, the emission tax provides present polluters with the correct economic incentive. Since the rational firm will reduce emissions exactly to the point where marginal costs equal marginal revenues, it is theoretically possible that charging present polluters for past harm will not foster inefficient resource allocation.[108]

Perhaps the most theoretically troublesome aspect of the 1973 Act is the transfer payments among designated areas. This problem could develop when automobiles contribute more than 20% of the pollution of a designated area,[109] or where the number of certified victims in an area exceeds the revenues raised by the pollution levy for that area. Under these circumstances, firms in one area (A) are charged to pay for the compensation of victims in another (B). From the perspective of efficiency, firms in the former (A) area may be forced to overinvest in pollution control; conversely, polluters in the latter areas (B) may receive a windfall, since they may rely on the other firm's (A) contributions. These firms (B) may continue to underinvest in control.

The 1973 Act makes several compromises with scientific precision that also pose difficult questions. First, it establishes a 1:9 ratio to reflect designated and nondesignated areas; second, it establishes an 8:2 ratio between stationary and mobile sources; third, the pollution levy is set linearly proportional to the discharge (e.g., those firms discharging twice as much SO_x pay twice as much). The principal theoretical objection to these compromises is that they oversimplify. Damage costs may not rise on a linear basis. Indeed, there is some evidence that damage costs may rise exponentially. If so, even if the ratios of 9:1 and 8:2 correctly reflect the relative contribution of stationary and mobile sources to the nation's pollution, this does not mean the resulting damage from pollution reflects this proportion.

Two other aspects of the 1973 Act, the treatment of administrative and research costs, raise complex theoretical questions that we can only mention. As noted, polluters pay one-half the costs of the system's administration and the government covers virtually all current research costs on pollution-related disease. An efficient allocation of the economic burdens of these costs depends on our judgment of whose actions necessitated them, and who principally will benefit from these services. Although administrative costs are directly attributable to pollution, one might argue that society's benefit from the operation of the system is not great. For this reason, the act's requiring polluters to pay only one-half of administrative costs is inefficient, because polluters in theory should pay all these costs.

The problem of research costs is analytically different. Medical and other studies of pollution disease convey benefits to the society as a whole, not simply to the victims who will be compensated. For this reason, society as a whole should share research costs along with the polluters that originally necessitated the research.

The 1973 Act's exempting firms emitting discharges below an established standard presents yet a different problem. To the extent that these discharges are injurious, the responsible firms again theoretically should pay. The act's exemption, however, can be justified, since the administrative costs of collection may exceed the revenues obtained.

How do price controls and market structure affect the analysis? The steel and power industries are the principal emitters of SO_x in Japan, and they are charged the most under the system. Both industries are subject to price controls. Steel, power, and other major industries also enjoy substantial market power.

Under the classical model, price controls are generally regarded as inefficient because consumers will be unable to register their preferences and the pricing mechanism will break down. The use of the pollution levy in conjunction with price controls subjects the firm to a magnified incentive to control pollution. Since the polluter is now barred from passing the costs of the levy on to the consumer, the firm has several options. It can cut costs either by reducing output or by firing its employees; it can pass the cost on through a price increase in another uncontrolled product; it can apply for government loans; it can develop new technology. In many cases price controls discourage the entry of new firms into the industry; or they may even stimulate firms to increase prices, especially if controls are sequentially phased. This is because firms will seek to avoid the strictures of new controls. It is difficult to assess the ultimate effects of price controls in the present case, for we must evaluate whether the costs in dislocation are worth the increased incentive for prevention control.

Market structure is also relevant. Economic theory suggests that all polluting firms should pay the levy and that none should be exempted. Yet, in Japan large industry depends on the labor of many small and medium enterprises. Should these firms be forced to pay the levy, many would be driven out of business. The bankruptcy of many small firms would then disturb the production processes of the giant firms dependent on long-term contracts with these small companies. It is uncertain whether the benefits of imposing the levy on small firms would compensate for the resulting dislocations.

Recently, some commentators have suggested that a corrective tax may even lead to a net decrease in welfare when applied to a noncompetitive industry.[110] Since a monopoly controls both a product's quantity and price, some economists argue that there will be a net decrease in welfare where the total value (from society's perspective) lost by reducing quantity exceeds the costs avoided by taxing the externality. Although we are unable to assess the implications of this analysis for the 1973 Act, this theory, if correct, suggests further fundamental questions with the act's use of a pollution levy.

3.2   Assessment of the Compensation System After Six Years of Implementation

The 1973 Act soon enters its seventh year of implementation and it is therefore timely to assess its performance. This section analyzes the major functions of the act—compensation, dispute settlement, pollution control—based on a review of the current literature and interviews with victims, industrial leaders, and government officials.[111]

3.2.1   The Present State of Implementation

Table 1 records the number of certified victims. By April 1, 1977, thirty-nine Class 1 areas had been designated and 53,416 victims compensated. Children below nine and persons over sixty accounted for over 60% of these designated patients.[112] The industrial cities of Osaka, Tokyo, Amagasaki, Nagoya, and Kawasaki reported most victims of nonspecific pulmonary diseases.[113] For the same time period, there were 1,788 officially certified victims of Class 2 pollution diseases.

Table 2 suggests that the number of certified victims is increasing. As will be discussed, this point has posed significant problems for the system, because in many areas SO_x and other air pollutants are actually declining. Administrators attribute the increasing number of victims to an administrative lag in the early days of implementation and to the fact that the availability of a compensation system has encouraged people to seek compensation.

Table 3 records the approximate amounts of compensation payments expended and appropriated during FYs 1974–1976 for Class 1 areas.[114] Table 3 suggests that the costs borne by industry have increased markedly during this period. The administrative costs of the compensation system, although rising in absolute terms, have remained at 10% of total costs.[115]

Table 1

Designated areas and number of certified patients under the Law for the Compensation of Pollution-Related Health Injury (as of the end of March 1977).

Region	Diseases	Designated Regions	Number of Certified Patients
Class 1 ("nonspecific" diseases)	Chronic bronchitis, bronchial asthma, asthmatic bronchitis, pulmonary emphysema, and their complications	Southern coastal district of Chiba City, 19 wards of Metropolitan Tokyo, and Osaka City	53,416
Class 2 ("specific" diseases)	Minamata disease	Lower Agano River Basin	616
	Itai-itai disease	Lower Jintsū River Basin	57
	Minamata disease	Coastal area of Minamata Bay	910
	Chronic arsenic poisoning	Sasagaya district of Shimane Pref.	16
		Toroko district of Miyazaki Pref.	89
		Total	55,104

Source: Environment Agency.

Table 2

Number of certified victims.

	Nonspecific (Air-Pollution-Related) Disease	Specific (Water-Pollution-Related) Disease	Total
March 31, 1970	962	203	1,165
March 31, 1971	3,219	211	3,430
March 31, 1972	6,376	312	6,688
March 31, 1973	8,737	728	9,465
March 31, 1974	13,574	1,184	14,758
March 31, 1975	19,340	1,325	20,665
March 31, 1976	34,190	1,550	35,740
Dead	1,062	185	1,247

Source: OECD Report, 1976.

Statistics for Class 2 areas are difficult to acquire because companies themselves are paying the victims directly. One tentative figure for the 1977 budget for Class 2 victims is ¥58 million, which is substantially lower than the budget for Class 1. This statistic does not account for other direct payments by the companies, judicial awards, and extrajudicial mediated settlements.

Table 4 shows that the pollution levy has substantially increased during FYs 1974–1976. In 1975, the levy was imposed on approximately 7,400 "large" facilities,[116] accounting for about 90% of SO_x emissions. The proceeds of the levy exceeded ¥33,000 million and the tax was estimated to represent about 17% of the cost of fuel with a 3% sulfur content.

In April 1977, the Japanese government amended the 1973 Act's enforcement order[117] in order to rationalize the system's allocation of costs within designated areas.[118] After reviewing the discrepancy in the cost burden of various designated

Table 3
Costs and expenses under the Law for the Compensation of Pollution-Related Health Injury (in ¥ millions).

		FY 1974	FY 1975	FY 1976
Those borne by the operators of pollutants	Pollution load levy	2,342	10,950	35,631
	Appropriated from auto-mobile weight tax revenues	550	2,674	8,791
Pollution-related health welfare programs (borne by public funds)		19	34	620
Administrative expenses for benefits		277	892	2,032
Expenses for collecting assessments (borne by public funds)		307	255	435

Source: Environmental Policy of Japan (report to the 1976 OECD Conference).
Note: The figures for FY 1974 represent amounts expended; those for FY 1975 represent expected expenditures; and those for FY 1976 represent amounts appropriated.

Table 4
Pollution levy, 1974–1976.

	Per-Unit Charge (¥)	
Fiscal Year	Class 1	Class 2
1974	1.76	15.84
1975	8.59	77.31
1976	23.33	209.97

areas, the Central Council recommended that the levy be assessed on a bloc-to-bloc basis and adjusted by one quarter during the initial fiscal year (1977) to avoid dislocations. Table 5 sets out the new adjusted tariff rates for certain designated Class 1 areas.[119]

3.2.2   Reactions to the Compensation System

The Victims' Perspective

Many victims of pollution disease are currently dissatisfied with the 1973 Act. The majority of complaints strike at the compensation system's basic design, not at the quality of medical services received.[120] In 1975 an association principally of air pollution victims, assisted by the National Federation of Environmental Lawyers, prepared a highly critical position paper on the act and opened negotiations with the Environment Agency for its amendment. And in May 1975 the group presented its grievances to the Diet. Many of these victims' concerns are still unremedied.

Concerns of Class 1 Area Victims[121]

1. *The Injustice of Compensating Natural Incidence Sufferers of Pulmonary Disease Within Designated Areas and Excluding "True" Pollution Victims Living*

Table 5
Adjusted tariff rates.

Area	Tariff (¥ per m³ of SO_x discharged)	Tariff Differential
Osaka, Amagasaki	536.63	1.4
Tokyo, Kanagawa, Aichi, Mie, Kōbe	383.31	1.0
Chiba, Fuji	344.98	0.9
Okayama, Fukuoka	306.65	0.8
Other	42.59	1/9

Outside Designated Areas. Many victims believe that the system arbitrarily confers benefits based on the happenstance of a person's residence. They feel it is unfair to compensate individuals who live in designated areas and contract one of the designated nonspecific pollution diseases[122] (i.e., natural sufferers), but exclude others who, although ill from pollution, will not be compensated because they live outside a designated area. A central problem, however, is that it is extremely difficult to determine medically which victims are "true" natural sufferers and which are not.

2. *Over- and Under-Payment of Benefits.* The victims also point out that some full-time workers are compensated for disability under the system at a level far below their present earnings. This is because their salary may be *higher* than 80% of the national average wage, the act's current standard. They point out the system illogically compensates at a grossly inflated level others who do not work, or work at low-paying jobs. Since these victims under most social security programs would be entitled only to 60% of the average wage for sick leave, the system, the pollution victims assert, arbitrarily confers a windfall on these people.

3. *Abolition of Probationary Status.* Many victims want the probationary status assigned to unclear cases abolished. This classification, they feel, is demeaning and leaves victims in limbo.[123] They request that only regular categories be used.

4. *Discrimination by Sex and Age.* The victims challenge the present system's ranking of disability and other payments according to sex and age as arbitrary and insulting.[124] They feel there is no reason why the elderly and women should receive less. They ask that the compensation schedules be adjusted.[125]

5. *Compensation for Children's Delay in Education and Maturation.* The victims argue that their children have a right to unimpeded maturation and an uninterrupted education. Because this is not often possible in a family stricken with pollution disease, the victims ask that the system include compensation for children's losses attributable to the child's disease or the sickness of the parent.

6. *Compensation for Pain and Suffering.* At present the system does not compensate for pain and suffering. The victims insist that a new item be established to redress this loss.[126]

7. *Greater Weight to be Given to the Assessment of Personal Physicians.* Victims want the Pollution Certification Health Council to pay greater respect to a patient's own doctor's recommendations. They ask that the rules be amended to require, at a minimum, that a decision on classification reflect careful attention to the views of these doctors, to the victim's own feelings, and to the feelings of family members.[127]

307

8. *Expansion of Designated Areas.* The victims want more areas designated. In 1975, Toyama, Nagoya, and Kurashiki were the areas of greatest contention, and parts of these areas have since been designated.

9. *Designation of New Diseases.* The victims ask that eye, ear, nose, and throat maladies caused or aggravated by pollution, particularly photochemical smog, also be designated. The medical uncertainty of these diseases, they argue, is no greater than that of diseases currently designated.[128]

10. *Termination of Social Security Benefits.* At present victims receiving compensation must waive their rights to social security benefits. This practice, the victims allege, is unreasonable, and should be abolished.

11. *Continuation of Treatment Through Customary Medicine.* Many victims complain that the medical treatment defined under the act excludes access to various herbs, acupuncture, and other forms of traditional medicine. They request that the system compensate the costs of treatment by these and other folk remedies.

12. *New Diseases Suffered by Previously Certified Victims.* A number of victims complain that once certified it is difficult to obtain adequate compensation for a new pollution-related disease. Because it is not uncommon for individuals to suffer from more than one pollution-related disease, the victims request that administrative guidelines and other measures be established to assist them in obtaining adequate relief.

13. *The Relation of Compensation to Judicial Awards Under the Yokkaichi Decision.* Some victims complain that the benefits of the system are not made available to those who litigated the Yokkaichi case or to those who have initiated direct negotiations with the defendant companies. The victims request that this discriminatory practice be terminated.[129]

14. *Stricter Pollution Regulations Within Designated Areas.* Many victims insist that there should be more stringent regulations imposed on factories and automobiles within designated areas, and on activities originating outside designated areas but entailing adverse environmental consequences within them. The construction of a superhighway contiguous to a designated area is often cited. Some victims request a moratorium on all development within the area, while others request that the consent of victims be required as a condition to further development.[130]

15. *Retroactive Compensation.* Some victims wish to expand the benefits of the system to include compensation for injuries suffered before its establishment.

Principal Concerns of Class 2 Area Victims

The grievances of Class 2 victims, particularly Minamata disease victims from Kumamoto Prefecture, may be even more bitter than those from Class 1 areas. The long years these people suffered without official attention or redress have caused deep resentment. Their distrust naturally focuses on the dispensation of compensation benefits.

In Kumamoto the principal grievance concerns the criteria for disease certification. The charge is that the present criteria are arbitrary, unreasonably restrictive, and designed to exclude the majority of those afflicted with Minamata disease.[131] In fairness, the diagnostic problems are considerable, because there are many cases that do not exhibit the expected symptoms. Yet, the basic allegation is that those afflicted with a "Minamata-like" disease would never have become ill under natural conditions—they are the relics of Chisso's actions and they deserve compensation. For these people, certification was always a device, first manipulated by Chisso's mimaikin contract, to exclude deserving victims and to limit entitlements.[132] The compensation system, they claim, continues this perfidious practice.

The report of the Japan Federation of Bar Associations' task force on the compensation system affirms some of these complaints, and also adds further important criticisms.[133] First, the pollution committee reports that some victims are reluctant to seek assistance under the system even in Class 1 areas.[134] Second, the report criticizes the government's failure to include an adequate number of law-trained specialists in the Certification Council. The committee strongly argues that certification must involve more than a scientific determination. At times the medical and scientific questions are so obscure, while the issue of equity is so compelling that legal skills are as important as scientific or other technical expertise. Third, the report strikes at the certification *procedure*. The victims have virtually no control over the appointment of council members, the majority of whom, the report charges, are politically conservative. Finally, the report confirms the victims' repeated complaint that rehabilitation and related benefits have been grossly inadequate.[135]

Industry's Attack on the Compensation System

Since late 1975 Keidanren, the Japan Chemical Association, the Japan Federation of Steel Manufacturers, and other giant industrial groups have lobbied with the Environment Agency (principally through MITI) for modification of the compensation system.[136] Industry alleges that the scientific premises of the system are unsound, that the 1973 Act's use of the "polluter pays" principle is inappropriate, and that central and local authorities should finance a larger part of the compensation benefits. Recently, the system's assailants have also subtly begun to question whether Japan should treat pollution victims with any greater deference than other afflicted persons. The Japan Federation of Bar Associations and other various citizen associations have resolutely attempted to shield the act from these criticisms, and the battle rages as of this writing (1979).

Principal Scientific Issues

1. *Natural Incidence.* Perhaps industry's principal objection to the system's handling of scientific issues arises from industry's being forced to contribute to the compensation of natural sufferers of respiratory diseases. Industry officials note that an index of 2–3 times the natural rate of pulmonary diseases (measured by coughs, phlegm, and other symptoms) is currently used as a basis for area designation. The natural rate is presumed under the system to be 2.5–3% of the population. Recently industry has challenged this assumption, pointing out that reports from other countries indicate that natural incidence rates of some "pollution-related" diseases are as high as 20–30% of the population.[137] Industry requests reconsideration of the entire treatment of natural incidence, and suggests that if natural-incidence victims are to be assisted, these costs should be defrayed at public expense.

2. *Pollution's Role in "Pollution-Related" Disease.* Recently scientists have begun to question the assumptions behind the 1973 Act's linking pollution to pulmonary disease. For example, one writer strongly argues that about 75% of bronchitis and asthmatic bronchitis cases are not clearly correlated with air pollu-

tion.[138] Based on studies like these, industry argues that asthmatic bronchitis should be declassified as a pollution disease. This same writer suggests that the rapid increase in pollution victims, noted since the inauguration of the system,[139] may not be due to an increase in incidence rates, but rather to an "out of the woodwork" phenomenon—i.e., that the availability of benefits attracts applicants for these benefits. Other commentators point out that many factors—viruses, allergies, etc.—in addition to pollution may be responsible for pulmonary disease, and that the system must also account for the role of these factors. Finally, some industry and government officials insist that a large number of pulmonary disease victims recover,[140] and that there are many malingerers among the victims who do not recover.

3. *Compensation for Smokers.* Industry also objects to being asked to contribute to the compensation of victims who are smokers. As noted, the 1973 Act does not distinguish smokers, even heavy smokers, from other victims. There have been three general proposals to amend the act to address this problem. The first proposal denies smokers compensation benefits; the second contemplates the central and local governments' financing compensation of benefits to smokers; a third proposal imposes a general tax on smokers, the proceeds to be commingled in the compensation fund.[141] Although all three proposals are seriously being considered, no action has been taken as of this writing (1979).

4. *Criteria for Victim Certification.* Industry also charges that the criteria for victim certification are ambiguous, fail to account for important variables such as allergies, and are unsystematically, and at times subjectively, applied.[142] Government officials acknowledge the problems, but point to the difficulties in establishing adequate criteria for certification. The Environment Agency's principal worry has been the development of diagnostic criteria for Minamata disease. Although Minamata disease has been under official surveillance since the late 1950s, there are still apparently no adequate biochemical and other tests for the illness. Indeed, the usual diagnostic examination takes over three hours for a single patient. These problems are further complicated by the fact that 10–25% of the victims appearing in Kumamoto reportedly have either atypical symptoms or symptoms that do not permit easy or immediate diagnosis (the aforementioned "grey area" cases). Because of the continuing medical uncertainty it has been extremely difficult to establish meaningful criteria and standards for certification. The time required to process a single case itself taxes the present medical staff, and recent manpower shortages are a serious problem. Such problems are aggravated by the victims' deep distrust and hostility toward the government. The deterioration in the relationship has required senior officials in the Environment Agency on a number of occasions to visit Kumamoto, to discuss the problems of certification with the victims, local doctors, and the provincial authorities.[143]

Not all diseases have caused such particularly acute social conflicts and medical difficulties as Minamata disease in Kumamoto Prefecture. For example, certification of Class 1 diseases is apparently proceeding fairly well. Although much still needs to be known about the scientific bases for air pollution–induced pulmonary illnesses, a system of nationwide standards for diagnosis has been established. Diagnostic examination of these illnesses usually proceeds quickly and without interruption. After an electrocardiogram, sputum analysis, surveys of monthly coughing, and other appropriate medical data are gathered, a diagnosis is reportedly quickly made. One report indicates that in Nagoya the certifying committee usually requires about thirty seconds to review and to judge a case under the national standards.[144] Implementation of certification for itai-itai disease is apparently also progressing fairly well. The government attributes this largely to the availability of X-ray analysis and other objective testing methods.

5. *The Area Designation Process.* Industry has attacked several aspects of pollution area designations. First, Keidanren and other groups argue that the scientific bases for area designation are ambiguous. They question the reliability of the BMRC testing procedure[145] and argue that the government should place greater weight on NO_x, particulates, and other pollutants.[146] Second, the industry has pressed the government to clarify the standards for the declassification of designated areas. Declassification, industry urges, is only fair, since SO_x pollution in many designated districts has substantially declined.

Cost Allocation

Although industry does not oppose the abstract idea of victim compensation, the compensation system's present allocation of cost is being bitterly attacked. Industry's basic allegation is that the system overextends the rationale of the four pollution cases, and confuses notions of legal liability, insurance, and the already ambiguous polluter pays principle. Although the 1973 Act's compromise between these concepts may have been acceptable in the early days of implementation (when industry's financial burden was low), the system's uncertainty is now attacked as unfair because the burden of the pollution levy is substantially increasing. Industry has demanded that the standards for cost allocation be clarified and rationalized. The outcome of the present debate on allocation holds the key to the future of the 1973 Act; in the following paragraphs we discuss the principal issues in contention.

1. *The Disparity Between the Decline in SO_x Pollution and the Increase of the Levy.* Industry's most acute grievance is that although SO_x pollution has declined markedly in some designated areas, the emission charge is increasing.[147] The government has attempted to explain this apparent paradox by noting that far more victims have appeared than were previously expected and that the situation is due to initial delays in the implementation of Class 1 area designations. Industry responds that the present incidence of cost creates a disincentive for pollution control, that the levy is now unfair, and that a greater share of the system's costs should be defrayed from general tax funds.[148]

A related problem is the issue of industry's paying compensation to pollution victims who contracted illnesses before the inauguration of the system. Some firms, especially newcomers to a designated area, argue that they had no part in the victims' injuries, and that it is unfair to require them to pay for injuries caused by others. In response to the government's explanation that it is not medically certain whether many victims are "past" or whether their illnesses were of more recent origin,[149] industry insists that until the scientific issues are clarified, public tax funds should defray the costs of these ambiguous cases.

2. *Interregional Transfers.* Another major issue is the government's practice of diverting the excess proceeds of one designated pollution area to defray the deficit in the compensation budget of another. Table 6 records these disbursements and revenues for 1977.[150] Industry charges that this practice is inefficient, irrational, and unfair. Why, many companies ask, should factories in Hokkaidō pay for the compensation of victims in Kyūshū?[151] Industry demands that the standards for interregional payments be clarified and the practice either discontinued or ra-

Table 6
Disbursements and revenues in designated pollution areas, 1977.

	Disbursements (Compensation and Rehabilitation)	Revenues			Revenues
		Pollution Levy	Automobile Weight Tax	Total	Disbursements
Chiba City	740	969	146	1,115	1.5
Tokyo City	10,148	1,846	5,867	7,713	0.8
Kawasaki City	3,952	1,563	547	2,110	0.5
Yokohama City	800	458	129	587	0.7
Fuji City	777	890	192	1,082	1.4
Nagoya	3,233	593	569	1,162	0.4
Tokai	554	1,246	116	1,412	2.5
Yokkaichi	1,297	1,860	237	2,097	1.6
Kusuchō	114	34	20	54	0.5
Osaka	25,850	1,375	2,311	3,686	0.1
Sakai	3,399	2,554	367	2,921	0.9
Toyonaka	377	12	169	181	0.5
Suita	258	51	133	184	0.7
Moriguchi	701	51	—	51	0.1
Amagasaki	4,549	829	598	1,427	0.3
Kōbe	232	557	—	557	2.4
Kurashiki	1,668	4,156	45	4,201	3.4
Bizenshi-Tamanoshi	198	233	8	241	1.2
Kitakyushu	1,213	3,584	497	4,081	3.4
Omuta	835	960	178	1,138	0.4
Total	60,895	23,821	12,179	36,000	

tionalized. Although the government amended the 1973 Act in 1977 to make the incidence of costs more equitable, the practice of interregional transfers continues because some areas still require additional revenues for victim compensation.

3. *The Ratio of the Pollution Levy Between Designated and Nondesignated Areas.* Polluters have argued that the levy's 9:1 ratio for plants located in designated areas and nondesignated areas is unreasonable. If the levy is to be seen as an insurance premium to be paid by all polluting enterprises, some writers note, there is no reason why polluters in designated areas should pay nine times more than others. If the levy is to be paid by the enterprises "responsible" for pollution diseases, there is no reason for enterprises located far away from the pollution disease areas to pay anything.[152] Industry's position on these issues lends insight into its opposition to the principle of "collective responsibility."

4. *Cost Incidence of Stationary and Mobile Sources.* Keidanren, the steel industry, and several local governments have also attacked the present allocation of cost between mobile and stationary sources. They argue that the present ratio of 1:4 is unfair to firms operating in some designated urban areas, e.g., Tokyo, where there are few factories, but many cars. In these areas, they assert, automobiles cause far more than 20% of the area's air pollution.[153] In 1978 the 1973 Act was amended to extend the use of the automobile weight tax.

5. *Burdens and Exemptions.* Another frequent complaint is that the 1973 Act arbitrarily overtaxes some firms, while releasing others. As noted, art. 52 of the act exempts firms whose total emissions are below a specified standard. Industry argues that this provision unfairly grants a preference to some firms that have low total emissions, although they discharge high amounts of SO_x. Despite the harm they cause, these firms will receive an exemption. Industry has called for a complete review of this issue, and demands that damages attributable to exempted firms be defrayed by tax refunds.

Conversely, the five-fold increase in the levy, coupled with other pollution costs, now imposes a formidable burden on some firms. For example, Environment Agency officials note that Mitsui Kinsoku has been placed in difficult circumstances by having to pay about ¥500 million per year under the system.[154] MITI and the Environment Agency are closely monitoring such cases.

6. *Increased Public Financing.* Because of its various uncertainties, Keidanren and others now seek a complete reconsideration of the system's present approach to financing compensation. Keidanren argues that local governments should defray a greater part of compensation payments through local relief efforts and that the central government should expand assistance to local programs. Industry is again proposing that the entire assistance effort be integrated within national health insurance.

The Japan Federation of Bar Associations categorically opposes industry's arguments. The associations' report[155] charges that in the past industry has always resorted to such casuistic scientific arguments in order to escape social responsibility and legal liability. The report urges that industry, not the public (or public tax monies), bear the economic burdens of scientific uncertainty. The bar associations essentially are defending the "classical" rationale of Japanese environmental policies since the 1970s.

But the report goes beyond this. It urges further study of NO_x pollution, and states that new pollution areas should be designated and present ones expanded based on this research. Finally, the report emphasizes that greater official and industry attention (and investment) must be given to the restoration (genjō-kaifuku) of environmentally devastated regions.

3.2.3   Dispute Settlement

An important motive behind the establishment of the compensation system was the weakening of the victims' (and later local residents') incentive to litigate, or to challenge the government or industry in other ways. To understand how effectively the 1973 Act has served this purpose we must distinguish between the formal processing of grievances and dispute settlement.

As noted, arts. 106–135 authorize persons aggrieved by a decision relating to certification or the payment of compensation to request a review by the Pollution-Related Health Damage Compensation Grievance Board. As of March 1977, eight Class 1 grievances and forty-one Class 2 grievances had been filed with the Grievance Board,[156] and 154 victims of Minamata disease had appealed to the Director General of the Environment Agency.[157] Many of these appeals (about 20–25%) have since been withdrawn, and only a few cases have been decided on the merits.[158]

Although the availability of compensation appears generally to have undercut litigiousness, victims certified under the 1973 Act have brought a few well-publicized suits. The first action was initiated by 372 victims of Minamata disease against the governor of Kumamoto Prefecture for a declaration of "nonperformance." The victims charged that the prefecture had been grossly negligent in processing applications for certification.[159] On December 15, 1976, the Kumamoto District Court ruled in the plaintiffs' favor, finding the procrastination of the Kumamoto authorities illegal. As a result of the Court's decision, Governor Ichisei Sawada of Kumamoto Prefecture issued a public apology to the victims. At the same time the Environment Agency promised to expedite the processing of certification,[160] and publicly acknowledged the many difficulties attending the compensation program especially in Kumamoto Prefecture. These included the technical difficulties of certification, the inadequate number of medical specialists, the increasing number of victims, and the unsuitability of facilities.[161] In order not to delay relief to the victims further, the officials of the prefecture, after consulting with the Environment Agency, decided not to appeal the Court's decision.[162]

In a second suit, pending as of 1979, 1,817 Minamata disease patients, their families, and other local residents have appealed to the Kumamoto District Court to suspend the prefecture's public works proposed ten year cleanup for Minamata Bay. Defendants in the case are the central and prefectural governments, Chisso Corporation, and Tōyō Construction Company, the project contractor. The plaintiffs argue that the mercury-polluted sludge dredged in the cleanup would be eaten by plankton and then accumulate in Minamata Bay's fish. Further, the plaintiffs charge that members of the committee that inspected the safety arrangements were under a conflict of interest with the proponents of the project.[163]

Minamata victims have also continued their protests outside of court. The most dramatic confrontation has been the occupancy of the Environment Agency on February 24, 1977, to protest the government's failure to provide effective relief. The patients and their supporters urged the government: (1) to expedite certification; (2) to provide more benefits to the victims; and (3) to stop Minamata Bay's cleanup. Environment Agency officials initially were reluctant to have the police eject the victims for fear that this action would destroy all public faith in the agency.[164]

Although Class 2 victims, particularly from Minamata, have pressed their grievances most aggressively, three recent suits attest to the litigiousness of Class 1 victims and the possible legal significance of area designation in future litigation. The first two suits involve actions brought by air pollution victims in Kawasaki. In the first suit, filed in May 1975 in the Chiba District Court, 200 victims and other residents sued Kawasaki Steel Corporation for damages based on a theory of negligence. The plaintiffs charged that the company continued to operate its sixth furnace[165] with full knowledge that the district had already been designated under the 1973 Act, and that many officially certified pollution victims lived or worked there. While this first case was still pending, another group of residents, including commuters and other certified pollution victims, initiated a second suit against the same company. In addition to demanding ¥370 million in damages, the plaintiffs asked the Chiba District Court to order suspension of operations at the plant because the company had continued operations, ignoring the first court action.

A most interesting suit, from a theoretical standpoint, is the recent action by 112 certified victims and their families[166] against the central government, the Hanshin Expressway Corporation, and several companies operating in the huge industrial complex covering Nishi Yodogawa-ku and Konohana-ku in Osaka Prefecture and Amagasaki City in Hyōgo Prefecture.[167] The charge is that the defendants have violated the plaintiffs' environmental and other rights, and ¥2,052 million in damages is demanded. The plaintiffs argue that the area's designation as a pollution zone imposes a particular duty of care on the defendants. In addition to damages, the plaintiffs also request the court to order the defendants individually and jointly to control factory emissions so that the district's ambient air quality will meet the present national standard.[168]

The relatively small number of victim grievances and the even fewer suits can be attributed to the following factors. First, victims have generally felt that their views have been ignored in the decisions relating to certification, dispensation of benefits, and rehabilitation; they therefore mistrust officially authorized procedures.[169] Second, because victims usually process grievances by themselves without the aid of a lawyer, many believe reversal of a decision unlikely. Third, other channels of protest—the media, public demonstrations, sit-ins, petitions to the Diet—have, to the victims, proved more effective. Finally, the few victim-initiated lawsuits can be ascribed to many of the social inhibitions already discussed. That these victims resorted to the courts at all, despite the availability of compensation, provides further evidence of the weakening of traditional barriers.

As of this writing (1979) there has not been a single polluter-initiated lawsuit challenging the compensation system. This fact, too, can be explained by reasons offered in earlier chapters—the fear of aggravating already negative public opinion,[170] the strong support for administrative approaches, the polluters' belief that many district court judges are sympathetic to the victims, and finally, industry's experience that the government is sufficiently sympathetic to its demands so that a concerted lobbying effort will be adequate to obtain the government's assent.

3.2.4   Administrative Aspects of the Compensation System

Collection of the Levy

The Pollution-Related Health Damage Association[171] has collected the pollution levy thus far with success and efficiency quite beyond the original hopes of Environment Agency officials.[172] Approximately 98% of the assessed firms have paid on time, despite the sharp yearly increase in the charge.

The success of the collection procedures may be explained in part because lying is easily discovered. As noted, the levy is determined principally from the amount of SO_x emission, and this emission is a function of the sulfur content of fuel oil. Since MITI has detailed statistics on the fuel used by each industry, and because the government operates an accurate monitoring system, lying is easily detected.

Second, there are economic and social penalties for lying or failure to pay the charge. Formal statutory penalties—an arrearage charge of 14.5% and a fine of ¥50,000—play a part. Perhaps more important is most companies' fear that the government might someday retaliate.[173]

Third, the design of the association is important. Although it is legally a separate organization, the association's directors, as noted, are all drawn from industry. The close working relationship among some of these directors, and the Japan Chamber of Commerce's support of the association, have also contributed to the success of collection.

Fourth, collection is backed by an industry-wide consensus. It is difficult for an individual firm to resist paying, once the business community as a whole has "contracted" with the government to meet industry's financial responsibilities.[174]

Fifth, the same factors that militate against a polluter's litigating a grievance also contribute to the industry's faithful payment of the levy. Victim groups and the media surely would be highly critical of procrastination, and if so, corporate name and image would suffer.

Finally, collecting the automobile weight tax is simple and inexpensive since it is already a standard part of tax revenues. The transfer of part of these monies to the compensation system is merely an accounting procedure.

Status of Rehabilitation Programs

Implementation of the rehabilitation programs[175] is apparently not progressing well. Originally, these programs were to include weekly rehabilitation visits by patients to sanatoriums, the installation of air purification equipment in patients' homes, and home visits by nurses. In Class 2 areas, specially designated beds were to be assigned certified patients in hospital wards; and medical doctors working under the auspices of the compensation system were to visit outpatients in their homes. In practice, only about 20% of the funds budgeted have been spent; consequently, patients have yet to receive many of these services. As of this writing (1979), not a single rehabilitation center is fully in operation.[176]

One important reason for the program's failure is that local governments have been too poor to raise matching funds to meet grants from the central government. There is no money to pay specialists and others to staff the rehabilitation centers. But social and cultural factors are also important. It was originally thought that children and the elderly could be sent to treatment areas unaccompanied by their families. The government, however, miscalculated; most victims have been extremely reluctant to go alone to treatment facilities for more than short periods of time. And prefectural rivalries have frustrated the realization of plans to establish regional treatment centers.

Although the inadequacy of rehabilitation and other services has outraged the victims and provoked criticism even from industry, the program's present failures are understandable from the legislative history of the 1973 Act. Because the system was speedily drafted under circumstances of political urgency, little thought was given to the details of the program's financing or administration. Since rehabilitation was always a tertiary concern, its needs quite naturally were deferred as other problems developed.

The Compensation System as an Ancillary to Direct Regulation

The establishment of a national compensation system has strengthened other aspects of the administration's control of pollution. The availability of compensation has drawn out many victims who might otherwise have remained silent. These new cases have assisted the government's preparation of comprehensive statistics on the health effects of pollution—regional disease incidence and prevalence rates, damage costs by age and sex, and other data required by the system. The administrative task of compiling this data has generally further sensitized the government to the consequences of pollution and encouraged increased investment in medical research. Such statistics have also proved useful in the setting of ambient and emission standards. Finally, the compensation system has provided the Environment Agency with a strong political, if not legal, weapon. Often when a company seeks a construction or operating permit in a designated area, administrators will point to the severity of pollution and the number of victims, and attempt to persuade the firm to abandon its plans or to adopt voluntarily additional control measures.[177]

As noted, the 1973 Act's use of an emission levy theoretically provides an economic incentive for firms to reduce pollution. At present, there is no available econometric study of the impact of the levy.[178] Of course, one may note a general trend of increasing pollution control investments[179] that coincides with the increase in the levy, but this may be due to the multiple regulatory controls to which these firms are exposed,[180] and not to the levy alone.

Industry officials themselves note that the levy has not yet had an appreciable effect on most firms' decisions to invest in pollution control. But this is because the marginal cost of control is still higher than the levy. Steel industry officials stated, however, that when the levy rises in excess of ¥700 per cubic meter of discharge, as it has in the Osaka area in 1978, the industry will install pollution control equipment because the expense of the equipment (plus installation and other incidental costs) will be less than the levy. Industry officials further noted that the levy would most probably only affect the decision to install *existing* technology and would have no effect on innovation.[181]

The Fall and Redemption of Chisso

The Kumamoto District Court's decision and subsequent events have brought hard times to Chisso. Under private settlements with victims and victims' organizations, Chisso is estimated to be paying about ¥19 million in compensation to each officially certified victim;[182] as of June, 1978, the company paid approximately ¥36 billion in compensation to 1,348 persons.[183] Its accumulated gross deficit as of December 1976, was ¥27.6 billion;[184] its accumulated gross deficit as of December 1977, totaled over ¥36.4 billion.[185] Because the company's accumulated deficits have exceeded total assets for more than three consecutive years, and it has failed to pay dividends to stockholders for more than five consecutive years, the Tokyo Security Exchange has removed Chisso's stock from its list of trading securities.[186]

The government's first response to the company's increasing debt was to expedite victim compensation by enacting special legislation authorizing the Environment Agency itself to screen and designate victims through a specially created Temporary Minamata Disease Certification Council.[187] Second, the government formulated a virtually unprecedented[188] proposal for Chisso's financial rehabilitation. The scheme's central idea is that Kumamoto Prefecture itself would issue special bonds redeemable in thirty years, and would use the proceeds of the sale to extend "low interest" loans to Chisso. Chisso would then be "urged" to employ the loans to meet its compensation responsibilities. The Ministry of Finance has proposed that its Trust Fund Department underwrite 60% of the bonds to be issued by Kumamoto Prefecture; the remaining 40% would probably be purchased by the Industrial Bank of Japan, Chisso's main financial supporter, and by some local banks. In case of Chisso's bankruptcy, the central government would assume 60% of all liabilities, while the remaining 40% would be shared equally by the prefectural government and the participating banks.[189] The government's proposal has raised critical issues not only for the future of the compensation system, but also for all of Japan's environmental policies.

3.2.5   The Future of the 1973 Act

The critical issue for many environmentalists is whether the government's redemption of Chisso will fundamentally subvert the "polluter pays" principle. Although the government terms its subsidy program "extraordinary," victims' groups and many other activist organizations fear that the program creates a dangerous precedent: that polluters in emergencies can always rely on the government's support; or worse, that they will intentionally declare bankruptcy to avoid compensation.[190]

Despite the fact that the government's stated purpose in redeeming Chisso is to assure continuing support for the victims, many victims mistrust the government's motives. For example, some Minamata victims believe that the scheme signals further limitations on compensation benefits and a tightening of the standards for certification of Class 2 diseases. The transfer of these functions to the central authorities, they fear, may only hasten this process, because the central authorities are believed to be even less sympathetic than the local health certification boards.[191]

There is also some suggestion that the Class 1 designation apparatus will also gradually be dismantled. One issue is the designation of new diseases. Although the 1973 Act theoretically permits the designation of new pollution-related disease, the Environment Agency at present does not contemplate this action. Lung cancer affords a good example. Although research on the relationship between lung cancer and air pollution has been in progress in Japan for several years, and although some studies demonstrate a statistically significant correlation, lung cancer probably will never be designated. Part of the explanation is the continuing scientific uncertainty surrounding lung cancer, especially since many forms of cancer require long incubation periods. The longer the incubation period, the less reliable the epidemiological data base, and the less willing will the government be to compensate sufferers of this disease.

Yet scientific uncertainty is only a partial explanation. Indeed, the level of uncertainty associated with some forms of cancer appears no greater than that surrounding some diseases, especially respiratory afflictions, already designated under the 1973 Act. Clearly, many social and political factors are also in play. In the absence of a court decision (such as that in the Yokkaichi case) to spark a movement for official recognition, it is unlikely that cancer or other pollution-related diseases will be designated for compensation.[191a]

Similarly, it is unlikely that a levy will be applied against discharges of NO_x or particulates on the theory that these substances also cause respiratory and other diseases. Despite evidence of the hazards of these substances, administrative and technical considerations outweigh the possible benefits to the victims of charging dischargers of these substances. Because NO_x discharges depend on heat and other factors in the combustion process, they are more difficult to calculate and control. Particulates pose similar constraints. Many particulates are fugitive emissions, not easily recorded by monitoring equipment.

Industry and government spokesmen also indicate that after several new areas are designated in June 1978,[192] no other areas will be designated. Indeed, the government will then begin establishing standards for the declassification of areas where SO_x pollution has markedly declined.[193]

Will the compensation system ultimately be scrapped? We think not. Initially, the compensation system was viewed as only an emergency measure. When pollution controls would take effect, it was assumed, the incidence of victims would decline, and the need for the system would evaporate. Statistically, this has not proved true. It is now apparent that as industrialization progresses, victims will continue to appear. Politically it will continue to be difficult to terminate the system completely. But compensation will gradually be limited.

The compensation system will also continue to influence other relief efforts. Already, its general approach is reflected in relief efforts to assist victims of noise pollution, in recent proposals to assist victims of medical accidents, property damage from oil and other pollution, sunlight loss, and in recent discussions of regulating traffic flow in national parks by a "crowding tax" during peak hours. Perhaps the compensation system is best viewed as an experiment—an early effort in one country to cope with human injury from pollution, a problem that all industrialized and industrializing countries must inevitably face.

Section 4   Aiding Victims of Toxic-Substance Poisoning in the United States

Recently there have been in the United States several federal and state proposals to aid victims of toxic-substance poisoning.[194] In this section we apply the principal lessons of Japan's experience to the problem of formulating and implementing an assistance program in the United States.[195] A decision to assist the victims of toxic-substance poisoning raises three related issues. Who should be entitled to the benefits? Who should pay for the benefits? By what process or processes should these benefits be conferred?[196]

4.1   Criteria for Entitlement

The Japanese system compensates victims of designated diseases traceable to a specific toxic substance or significantly correlated with a polluted area. In each case, the basic criterion for entitlement is causation, although the system transfers the burden of proving causation (or of making a judgment on causation) from the victim to the administrators. We agree that causation can serve as a useful threshold criterion for allocating entitlements. Those victims clearly suffering from a designated or medically definable toxic substance–induced disease should receive assistance; those clearly not so afflicted, should be excluded.

Yet scientific causation cannot be the sole criterion, because, as Japan's experience tells, it is not always possible to prove causation in anything other than a statistical sense; in such cases a dogmatic adherence to "science" may cause inequity. The problems posed by the symptomatically atypical cases of Minamata disease best illustrate this point. Although industry has applauded the use of a strict standard because it excludes atypical cases of Minamata disease, the victims consider the certification standards to be arbitrary, subjective, and politically motivated. Indeed, as Minamata disease is better understood, denying relief to many victims now displaying atypical symptoms may even prove medically unwarranted. Because problems of atypical cases will probably recur under any relief system, the scientific standard should be supported by other criteria.

After determining causation on a prima facie basis, we believe the entitlements of atypical cases should be judged against the system's *primary* objective. So measured, certification may depend on the natural incidence of disease. For example, if there is no natural incidence (or if natural incidence is low) as in Minamata disease, we should err on the side of the victims. If the system's primary objective is relief, the original motive for granting assistance—need—militates in favor of entitlement; similarly, if the objective is deterrence, polluters should bear the costs of demonstrating that these atypical cases must be disqualified. The polluter usually can marshal information relevant to the issue of causation at a lower cost to society than the victim (or often even than the government).[197] Where the natural incidence of a disease is high, however, entitlement probably cannot be justified under either standard. Such victims are better assisted under other programs, and the imposition of this additional cost on industry may actually create economic incentives for overinvestment in pollution control.[198]

Should considerations of wealth matter in judging the entitlements of typical as well as atypical cases? If the system's primary motive is deterrence, a victim's wealth is irrelevant. Yet, wealth may be important if the central objective is relief, because the underlying rationale again rests on need. This seems particularly true for atypical cases where the marginal costs of an additional relief payment are high.[199]

The question of whether to assist victims who contribute to their own illness, like smokers, poses additional complexities. Entitlement in these cases should depend on whether a victim's action was the primary cause of injury. Cases where a victim's injury is primarily self-induced should not qualify, irrespective of the system's motivation. For imposing the costs of assisting these victims on industry is unfair and inefficient,[200] and the original rationale of relief—to assist the victims of man-made toxic chemicals—militates against aiding self-inflicted injury. Where a victim's action is a concurrent cause of injury, however, it seems unfair to bar a victim from relief under either a deterrent or welfare rationale. For in the former case, polluters should bear at least part of the costs of injuries they cause, and in the latter, the system should not totally exclude victims whose injuries are the same as those of others who obtain relief.

The problem of the extent of relief may be similarly analyzed. Victims whose own actions contribute to their injury (but whose actions are not the primary cause of injury) are best handled by the idea of comparative negligence,[201] for if it is unfair to deny relief completely, it is still inappropriate, under either standard, to aid fully. The system's objectives should dictate the extent of assistance to the ordinary victim. Under a deterrence scheme, all injuries (including pain and suffering) should, in theory, be compensated; while a welfare system must consider the level of payments made under other systems, whether victims have other means of relief, and whether the marginal benefits of incremental payments for secondary items, such as pain and suffering, justify their additional costs.

4.2   Criteria for Allocating Costs

Both theory and equity dictate that polluters bear the costs of the harms they cause. Japanese experience, however, cautions against extending this principle to its next logical step—of letting the compensation budgets determine the amount and allocation of financial burdens. Indeed, the fate of Chisso affords these lessons. First, the bankruptcy of the polluter can jeopardize compensation. Second, the mere prospect of a major company's bankruptcy will motivate governments—at times perhaps unconsciously—to manipulate the criteria for entitlement. Third, it may be wiser to socialize relief costs at the outset than to transmute a compensation system into a welfare program. Switching objectives in midstream produces uncertainty, undermines the effectiveness of other pollution controls, and promotes misallocation of capital and other resources.

Pollution controls have yet to force the bankruptcy of a major U.S. enterprise.[202] But the U.S. government's probable reaction, as in Japan, would be to rescue the firm in distress. The years of judicial equivocation despite Reserve Mining Corporation's flagrant disregard of public health standards attest to a continuing ambivalence. So does the apparent unwillingness of Congress and the courts to force the auto industry to meet the original auto emission deadlines. An American compensation program, copying in design Japan's pollution charge, would very likely encounter comparable financial dilemmas.

We do not urge, however, that the "polluter pays" principle be abandoned. Indeed, within the flexible enforcement scheme envisioned, this idea can powerfully serve policy objectives. Still, alternative or additional criteria must be identified to allocate the costs of a relief effort.

One option is to socialize the costs. Japanese industry, of course, has pressed this idea for many years, and there is some logic to it. For if we once concede that industrialization will always jeopardize some people's health, while everyone, to some extent, benefits from economic growth, then we all should consider contributing to the aid of those who suffer.

Placing primary reliance on tax revenues, of course, will not foreclose supplemental financing from other sources. For example, an environmental trust fund, generated from emission levies, administrative charges, penalties, or other economic incentives seems a sensible option.[203] Indeed, to the extent that appropriations for victim relief are economically justifiable,[204] revenues from this source might substantially reduce the relief system's dependence on the tax base.[205]

Finally, the substantial regulatory benefits of a compensation system will not be lost entirely by severing the link between entitlements and burdens. Cost and other data developed by a relief program, if carefully analyzed, can be valuable to administrators in designing ambient and emission standards and in enforcement of these standards.

4.3   The Relief Process

The difficulties Japan is now encountering in implementing the 1973 Act reinforce on a more general level insights gained from the problems of other administrative relief programs in the United States. In 1969 Congress established the black lung disease program, an administrative scheme for compensating coal miners totally disabled by black lung disease, adding by amendment in 1972 several important provisions.[206] These provisions required reexamination of the backlog of cases denied by the Social Security Administration during the first three years of implementation. They also imposed on the coal mining industry[207] the burden of providing benefits to worthy claimants who filed on or after January 1, 1974. Moreover, they required coal mine operators to finance the miners' benefits through participation in approved state workmen's compensation programs, or by self-insurance. In cases where an operator fails to provide appropriate relief, the United States is now authorized to extend benefits to claimants and to recoup the amount paid by a civil action against the operator.[208]

The Black Lung Disease Compensation Program has been severely criticized by both the industry and the miners, and its administrative expense concerns members of Congress. As in Japan, the victims' principal grievance has been the uncertainty of standards for certification[209] and the lack of realistic avenues of administrative review of denials of these benefits. Yet, unlike in Japan, the American mining industry has judicially challenged the constitutionality of the presumptions of the black lung compensation system that relate to liability, and also to the law's retroactive application.[210] The uncertainty of certification and review procedures and the coal industry's concerted attack have delayed the dispensation of benefits to many miners, and have substantially increased the administrative costs of this already expensive relief program.[211] The law is reported not to have significantly influenced the industry's attitudes or practices affecting worker health or safety.[212]

Although the specific weaknesses of the Black Lung Compensation Program might be overcome through more careful planning, this piecemeal effort seems an inefficient, ineffective means of addressing the larger, more complex, and often less localized phenomenon of toxic-substance poisoning.

Yet, other omnibus programs, such as workmen's compensation, are no better. Limited in coverage, with inadequate benefits, exclusions, and other shortcomings, these programs are now being attacked by labor and other groups.[213] Workmen's compensation systems also leave the unemployed members of the general public as charges of the state. And the states for the most part are politically unwilling, financially unable, and, at times, even legally forbidden to assist.[214]

An ambitious alternative to piecemeal programs like black lung or unsatisfactory omnibus approaches like workmen's compensation is H.R. 9616, introduced by Congressman William M. Brodhead. Under this bill, victims, upon proving a prima facie case of causation before an administrative panel, would be entitled to compensation for injuries associated with a subsequently designated toxic-substance-related disease. Victim benefits would include medical care, a disability allowance, rehabilitation expenses, a survivor's allowance, funeral expenses, and reimbursement for attorneys' fees. The costs of these items would be borne by the manufacturer and paid under an order of the compensation board. The system's administrative costs would be financed by a special levy on "risk," also collected from the manufacturers.[215] Grievances under the bill would be investigated and mediated by a special ombudsman.

Despite its promise, H.R. 9616 raises serious legal and political questions. One objection is that it creates a complex and expensive new bureaucratic apparatus that in the end may not bring victims speedier relief. Its requirement that victims make a prima facie showing of causation could impose a substantial burden, even with the aid of the bill's enumerated presumptions.[216] Indeed, the proposal eliminates the major benefit to the victim conferred by the Japanese approach, introducing in its place the many uncertainties victims already suffer under the black lung program. The levy on "risks" to some extent also oversimplifies the difficulties of ascertaining and quantifying these risks, and presumes that it is most sensible to defray the system's administrative expenses by these revenues.

The prospects for the passage of a proposal like H.R. 9616 of course would be substantially enhanced if extraordinary circumstances like those that compelled the establishment of the Japanese compensation system were to occur in the United States. We must not forget that Japan's 1973 Act was conceived in an atmosphere of political crisis following the Yokkaichi decision[217] and hastily prepared. From the start, it enjoyed industry's blessing, and an industry-wide consensus has continued to support its implementation. For its part, industry would never have acceded to the idea of "collective responsibility" had not a strict liability law already been enacted, and the courts had not decisively decided in the victims' favor. But now that this consensus is beginning to dissipate, the system itself is in danger of collapse.

None of these factors are present in the United States. Indeed, it is difficult to imagine an industry-wide agreement to support such a program. American industry has simply not demonstrated an inclination for self-policing in the manner of Keidanren and other Japanese industrial associations. And it is likely that American manufacturers, in contrast to their Japanese counterparts, would ask the courts to delay implementation of H.R. 9616 should the proposal muster sufficient political support in Congress.

On balance, despite all their shortcomings, we believe the courts should be preferred in the United States. They are familiar and tested, and they would avoid the need to establish a new, expensive layer of bureaucracy. A judicial award might provide at least as effective an incentive for control as a pollution levy.

Common law remedies, of course, are at present inadequate to protect the general public. An artful use of discovery, continuances, or other tricks of civil procedure can harass, retard, or dispirit a victim's court action.[218] Victim class actions, the best means to reduce the costs of litigation, rarely qualify under the federal rules.[219] The substantive rules of torts also frustrate by design. To prove negligence a victim must show that the consequences of a defendant's actions are foreseeable, at times a formidable barrier since a company may act on governmental instruction or comply with all emission limitations.[220] Perhaps the most redoubtable obstacle is causation. Often, merely identifying a toxic substance, tracing the pathway of its dispersal, or demonstrating with statistical accuracy its etiology, prove insurmountable. For the toxic-substance victim in the United States, unaided by a powerful political movement, before traditionally minded courts, relief may be impossible.

Although it is inappropriate here to develop specific recommendations for expanding judicial relief by statute,[221] Japan's experience suggests at least the basic elements of such a scheme. A model statute might include strict liability, prospectively imposed on toxic-substance polluters; a definition of the minimum showing required to establish prima facie causation; a presumption to shift the remaining burden of proof to the polluter; and various modifications to facilitate class actions and to reduce the costs of a suit.[222] The principal objection, of course, to such recourse to the courts is the years it wastes and its expense. In the next chapter we explore the uses of mediation, conciliation, and arbitration as one means of solving these and related problems.

7　The Uses of Conciliation, Mediation, and Arbitration in the Settlement of Environmental Disputes

As the discussion in ch. 1 of the Ashio and other early pollution cases indicates, Japanese society historically has favored extrajudicial settlement of disputes. In 1970, the traditional mediation and conciliation procedures were tailored to the needs of the environmental field by the Law for the Resolution of Pollution Disputes (hereinafter the Dispute Law).[1] After analyzing the government's motives for enacting this law, this chapter tentatively assesses how efficiently, fairly, and effectively the system has operated as of 1979.

Section 1　Development and Operation of the Dispute Law

1.1　Legislative History of the Dispute Law

The application of mediation and conciliation (and to a lesser extent arbitration) to environmental problems arises from a rich historical experience in Japan with these procedures.[2] Richard Beardsley, in his study of Niiike village life, analyzes the Japanese farmer's reliance on mediation as "a means of reinforcing the community against outsiders" and as a vehicle to contribute to "community cohesiveness and self-containedness."[3] Takeyoshi Kawashima has noted the prevalence of this institution even in the urban setting in his reference to the shady practices of the "makers of compromise" (jidan-ya);[4] and Walter Ames reports that the police also often act as intermediaries in disputes.[5] Dan Fenno Henderson has traced the historical development of conciliation as a flourishing institution from the Tokugawa period to the present.[6]

The earliest statutory application of conciliation to environmental disputes is found in the Mining Law of 1939. In the postwar era a number of statutes, including the Water Quality Conservation Law, the Soot and Smoke Regulation Law, the Noise Pollution Control Law, and the Atomic Energy Injury Compensation Law, incorporated conciliation and mediation procedures. The Dispute Law establishes a comprehensive and unified system based on these earlier experiments. Several factors were responsible for this development.

As noted in earlier chapters, public concern over pollution increased in the early and mid-1960s. For example, the number of complaints received by local governments rose from about 20,502 in 1966 to over 40,854 in 1969.[7] When unsatisfied, some of the complainants would resort to violent confrontation. Also, during this period, the press relentlessly attacked the government's indifference to the pollution victims, and warned of the dangers of permitting such conflicts to continue.[8] When the victims finally began to turn to the courts in the late 1960s, an alternative, more familiar, less socially disruptive method of solving disputes was viewed by the government as necessary.[9]

As noted in ch. 1, the government's original approach to pollution control emphasized no-fault liability, not dispute settlement. But when a draft proposal to include this principle in the Basic Law was first circulated by the Ministry of Health and Welfare, it was resolutely attacked by industry and MITI. Only after

the defeat of strict liability did the government begin to consider the less controversial options of dispute settlement and monetary relief.[10] The final version of
the Basic Law linked the two concepts in one provision.[11]

After the enactment of the Basic Law in 1967, the government began to give
more serious attention to a new system of extrajudicial pollution dispute settlement. There were several reasons why another approach was necessary and why
existing procedures in the air, water, and mining laws would not suffice. First,
these measures, the government soon discovered, had proved ineffective and were
not widely used.[12] Second, the Basic Law had virtually committed the government to pass new legislation; indeed, new legislation would also serve to demonstrate the government's concern about pollution. Third, many local governments
wanted special legislation in order to expand their powers to investigate and dispose of local conflicts. Fourth, a new, more comprehensive statute, the government hoped, might amplify the obvious benefits of existing mediatory procedures,
while avoiding their limitations.

Some scholars at the time suggested that the courts might not be the most appropriate institution to deal with pollution disputes, because these controversies
raised particularly complex and difficult issues. Polluters were often difficult to
identify, and the victims' injuries varied greatly, extended over wide geographic
areas, involved many people, and their causes were often scientifically obscure.
The courts, it was felt, were better suited to redress simpler controversies involving damages to a small number of people. Mediation, in theory, provided a
simpler, less expensive, and less time-consuming avenue of relief. Mediation
was also thought to reduce the victim's burden under the Civil Code of proving
causation, foreseeability, and duty of care.[13] Finally, the government wanted to
protect administrative prerogatives over pollution control. Indeed, many officials
were concerned that the courts and victims might one day use pollution controversies as a powerful means of checking government action.[14]

The government's preparation of the new law began in 1968 and proceeded with
little opposition or interruption in 1969. When a bill was finally presented to the
Diet, all political parties voted in its favor, except for the Communist party,
which abstained.

During the early 1970s, the principal legal issue involving the Dispute Law was
whether it should be an adjunct of the courts or be made a part of the bureaucracy.
Advocates of the latter view prevailed.[15] In 1972,[16] the Dispute Law was amended
to establish a separate administrative body, the Central Pollution Review
Committee. The committee was authorized to make findings on the issues of
causation or responsibility (quasi-arbitral determinations), a procedure described
in the next section.

1.2   Overview of the Dispute Law

1.2.1   Structure of the Dispute Settlement System

The Dispute Law establishes three independent subsystems for dispute settlement, the first two at the local level, the third at the central level in Tokyo. They
consist of a citizen's Complaint Referral Service (Kōgai Kujō Sōdanin), a Local
Pollution Review Board (Todofuken Kōgai Shinsakai), and a Central Dispute
Coordination Committee (Kōgai to Chosei Iinkai).

The citizen complaint system incorporated into statutory form the activities of
the local prefectural complaint counselors (sōdanin), which were authorized by

Table 1
Counselors at various governmental levels, 1976.

Level	Number of Governments That Have Counselors	Number of Counselors	Number of Full-Time Counselors
Prefectural	47	1,599	91
Cities of population over 100,000	152	1,136	378
Other cities	87	417	202
Towns and villages	178	353	30
Total	464	3,505	701

Source: White Paper 353 (Environment Agency, 1977).

local ordinances predating passage of the act.[17] Pollution complaint counselors operate in all prefectures and major cities (with a population of 100,000 or more).[18] Table 1 indicates the distribution of counselors as of 1976. Counselors consult with residents, investigate pollution incidents, and provide guidance and advice. They are required to notify the responsible agencies of pollution disputes, and they may undertake other measures considered necessary for their settlement.[19] Counselors are mostly lower-ranking staff in local government; the highest official is usually on the level of section chief. Counselors are employed full-time and often work in close contact with pollution control offices of local governments.

The Local Pollution Review Board is separate from the complaint system. The counselor system is designed principally to elicit citizen grievances; the board's purpose is to settle minor disputes. The board has jurisdiction over disputes between private parties and between citizens and the government. It is empowered to conduct mediation, conciliation, and arbitration.[20] Unlike the Central Dispute Coordination Committee (hereinafter the Central Committee), it is not permitted to make quasi-arbitral determinations.[21] Prefectural governments are also authorized to convene a Joint Prefectural Review Board[22] when more than one prefecture is involved in a dispute.

Review Boards consist of nine of fifteen members appointed by the governor with the consent of the assembly. Board members are mostly law professors, former judges, and practicing attorneys. They are forbidden, even after retirement, to disclose information learned during the course of their work,[23] and also must not assume an executive post in a political party or engage actively in political activity.[24]

The Central Dispute Coordination Committee is an independent administrative body composed of six members and a chairman. Because the committee is a semijudicial body, it is presumed to be neutral and independent. It holds hearings, enlists the cooperation of administrative agencies in collecting data, issues opinions, and provides technical information to interested administrative agencies. It can also request other agencies, businesses, scholars or other experts to research a particular problem; at times it conducts research itself. For example, under art. 18 of the Law for the Establishment of a Pollution Coordination Committee,[25] the committee may engage as many as thirty specialists for the study of matters requiring a particularly high degree of technical knowledge. The committee is empowered to issue its own opinions on pollution controversies, and it must present its findings to the prime minister or the head of the appropriate administrative agency.

The committee conducts mediation, conciliation, and arbitration, and is empowered to make quasi-arbitral determinations. The committee also has specific investigatory and fact finding powers. For example, it may conduct an on-site inspection, or request parties to a dispute to provide it with evidence.[26] In serious cases it is permitted to initiate dispute settlement even when not expressly requested to do so by the parties. At its discretion, the committee may permit the participation of injured individuals not originally parties to a dispute.

Article 24 of the Dispute Law limits the jurisdiction of the Dispute Coordination Committee to specific categories of disputes, and a special Cabinet Order further limits jurisdiction[27] to "cases involving air or water pollution causing death or physical impairment and the risk of the spread of such pollution; air and water pollution-induced damage to animals or plants exceeding ¥500 million; pollution cases involving more than one prefecture; and matters referred (to it) by prefectures."[28]

1.2.2   Mediation, Conciliation, and Related Procedures

The Dispute Law permits the use of various extrajudicial dispute settlement procedures distinguishable by function and in some cases by legal effect. The simplest, least formal[29] procedure is mediation (wakai no chūkai). Usually, this is conducted at the parties' initiative. Frequently, the mediator carries messages between the parties, arranges meetings, and generally expedites communication.[30] The parties may represent themselves during mediation, employ a lawyer, or with the permission of the mediation committee, appear through a representative who is not a lawyer.[31] Parties to mediation usually memorialize their agreement in a written mediation agreement (wakaisho). Because this document is deemed a contract under art. 695 of the Civil Code,[32] once the parties agree to be bound by a particular mediated settlement they lose their right to litigate their dispute. Parties, of course, always retain the right to litigate should a settlement break down.[33]

Conciliation (chōtei) is often more formal than mediation. At times, a conciliation begins when mediation has failed. And generally conciliators will participate more directly in a dispute than mediators. The Central Dispute Coordination Committee may commence mediation or conciliation without the consent of the parties where negotiators have not proceeded amicably or where a dispute could have serious social consequences.[34] The three-member[35] conciliation committee may request the parties to produce documents related to a case or, where necessary, conduct an on-site visit to a polluting factory to inspect its records.[36] The conciliation committee is authorized to formulate a draft agreement for the parties when they appear unable to reach a settlement; its proposal is then deemed accepted by the parties unless objected to within 30 days.[37] Conciliation settlements also have the legal effect of a contract, and, thus, are identical to a mediated settlement.[38]

A third procedure authorized, although not often utilized, is arbitration (chūsai).[39] As in the West, arbitration is the least flexible technique, because parties agreeing to arbitration are bound by the determination of the arbitrator(s).[40]

In addition to the above procedures, the Dispute Law also authorizes the Pollution Coordination Committee to make quasi-arbitral[41] determinations on the issues of responsibility and causation (sekinin saitei, genin saitei). Determinations on causation and responsibility are made by a subcommittee of three to five persons, who are central Dispute Coordination Committee members. The proceeding can be initiated by the application of either disputant.

In responsibility determinations, the subcommittee makes factual findings on the question of illegality, the monetary amount due for compensation, and other issues usually addressed in court.[42] Although the parties may prepare and present evidence in this proceeding, more often the subcommittee itself will investigate, compile, and interpret the evidence. In causation determinations, the subcommittee makes factual findings on the causal relation between a polluter's activity and the victim's injuries. The procedure is identical to responsibility determinations. Once the issue of causation is resolved, other problems, such as the assessment of compensation, are settled either directly by the parties themselves or by some combination of discussion, mediation, and conciliation. Even if litigation is commenced, a prior determination on causation can facilitate a speedy court decision. When appropriate, a subcommittee conducting a responsibility determination may on its own authority also make a finding on causation.

Quasi-arbitral determinations also have the legal effect of a contract. A subcommittee's decision becomes a binding contract within thirty days after it is proposed, provided that neither party has initiated a lawsuit before that deadline.[43] After thirty days, the parties lose their right to litigate issues of causation or responsibility. An agreement to conduct a quasi-arbitral determination, however, does not prevent the parties from litigating the same set of facts at any time consistent with the statute of limitations before the arbitral determination becomes final. Where the issues of responsibility or causation are litigated, the court will make a de novo determination, although it may consider the subcommittee's findings along with other evidence. The determinations of the committee are not accorded greater weight than other evidence.[44]

1.2.3   Relation of the Dispute Law to the Courts and to Pollution Regulation

Unlike the approach to mediation of some other Asian countries,[45] the Dispute Law does not require that the parties first submit to extrajudicial settlement before going to court. In practice, the parties usually prefer to preserve both options; at times they may even litigate first, and after litigation has begun, resort to extrajudicial settlement. Although disputants are usually encouraged to resolve their differences out of court, the Dispute Law's principal attraction is its speed, flexibility, and low cost.

The dispute settlement system relates to the courts and to the administration of pollution control in several important ways. For example, under art. 42.26,[46] a court may suspend its hearing of a case when a party applies for a determination on responsibility or causation. Similarly, under art. 42.32,[47] the court may itself request the Dispute Coordination Committee to make a determination on causation. In this way, a court may use the quasi-arbitral procedures to analyze evidence important for a decision.[48]

Because quasi-arbitral determinations on causation and responsibility can have important social consequences, the Dispute Law directs the Pollution Coordination Committee to notify the prefectural governor of its decision immediately, and the committee is also authorized to transmit its opinion on a case to the appropriate administrative agencies. In this way data from the dispute settlement system is made available to the administrative processes at the central and local levels.

From a theoretical perspective, settlements negotiated under the Dispute Law also constitute a separate, independent level of private control of pollution. As the cases included in sec. 2 reveal, the parties by private agreement may provide for more stringent standards than those set by regulation, establish new standards for unregulated areas, or identify remedies beyond the powers, capacities, or proclivities of a court.

Section 2   Cases and Comments

2.1   The Conduct of a Conciliation Proceeding

An administrative conciliation proceeding closely follows the format of judicial conciliation.[49] The meeting is closed to the public and conciliators may not reveal the contents of the negotiation.[50] Generally, the principals appear in person, or occasionally with a lawyer. The committee's permission is required should a party seek to appear through a surrogate who is not a lawyer.

The meeting usually opens with the chairman of the Conciliation Committee introducing the parties. The chairman emphasizes that the conciliators will be fair and unbiased, and stresses that all information will be kept in strictest confidence. Thereafter, the chairman calls upon the complainant to speak first. After the opening statement, the chairman or other committee members may question the complainant. Next, the respondent is heard and he too may be questioned. When a dispute is particularly acrimonious, the committee may meet with the parties individually and in private. After thus gathering the basic facts, the committee recesses to decide its next course of action.

At this point, the committee may visit a polluting factory, discuss the matter with local officials, or assemble a team of technical experts to gather further information. Thus enlightened, the committee is now in a position to draft a conciliation agreement for the parties. The more usual practice, however, is for the committee to assume a more passive role, to continue to meet with parties individually and together. During these subsequent sessions, the committee attempts to persuade each principal to offer a concession in the hope of identifying a mutually agreeable solution.

The conciliators' conduct throughout the proceedings is governed by custom. Conciliators must avoid becoming angry, agreeing with one side, or refusing to transmit seemingly unreasonable demands. Acceptance of drinks, other gratuities, or private visits to the parties' homes during the course of a proceeding are all forbidden. Committee members are expected to disqualify themselves in cases of bias or conflicts of interest. As in a judicial conciliation proceeding, a good conciliator in an environmental dispute must be a tactician—a sense of timing is critical, so that the parties do not reject a useful suggestion because it is pressed too enthusiastically. Flexibility and innovation are hallmarks of the trade, although at times a conciliator's deft use of precedent can also be persuasive.[51]

Once a settlement is reached, the committee's job is to see that the parties completely understand and acknowledge the agreement. Before the agreement document is presented to the parties, the committee analyzes it to root out misleading phrases and to assure that, if necessary, the settlement is judicially enforceable as a contract.

2.2   Case 1:[52] Noise and Other Pollution from a Metal Recycling Factory (Shizuoka Prefectural Pollution Dispute Conciliation Committee)[53]

2.2.1   Introduction

A limited liability corporation began operating a metal recycling factory in Shizuoka. By 1968, the factory had become a large operation. During this period, forty-one local residents began to complain that the factory had completely disrupted their lives by its noise and vibrations and obstruction of sunlight. They sought conciliation requesting that the company terminate its operations at once and move to other premises.

2.2.2   Disposition of the Case

This case has proceeded through conciliation since March 24, 1973, and the respondent has made the following claims: (1) After hearing the complaints of local residents, the company had tried to correct the causes of disturbance, but upon inspection found this to be technologically difficult. (2) The company plans relocation within five years. But during the interval, in order to reduce noise and other disturbances, the company will undertake to (a) modify the factory building; (b) improve operational procedures; (c) educate its employees. The Conciliation Council investigated these assertions and on April 28, 1974, prepared a conciliation proposal. The committee was able to obtain the agreement of thirty-three members of the forty-one original petitioners in this dispute. The committee terminated conciliation with respect to the remaining eight residents.

2.2.3   Conciliation Agreement

Article 1. *The company undertakes to terminate operations at its present premises with all deliberate speed, but no later than December 31, 1976.*

Article 2. *Within one month of terminating operations, the company will remove all sound-proof walls of its plant that are not connected to the roof.*

Article 3. *Thereafter, should the company wish to use or to dispose of these premises for any reason, the company must discuss the matter with the relevant administrative authority.*

Article 4. *Throughout the interim period, until the termination of its operations, the company will try to control the noise emanating from its plant. The company undertakes to accomplish the following measures immediately: (a) construct walls at the northern and southern ends of the premises; (b) restrict the time of operations to 8 A.M. to 6 P.M.; (c) curtail operations on Sunday; (c) refrain from incinerating evil-smelling materials that produce particulates; (e) refrain from depositing iron materials in the residential areas; (f) refrain from blocking the pathway in front of the factory; (g) refrain from creating noise or other disturbances by loading and unloading materials and by operating a magnetic crane.*

Article 5. *The company will compensate the complaining residents for inconveniences up to the date of this agreement, and during the period of its future operations, until the point of termination of operations. The company shall pay a total of ¥200,000 to these residents by June 30, 1974. The allocation of this sum will be distributed by the representatives of the residents' group.*

Article 6. *After these promises are faithfully performed, the complaining parties will refrain from litigating this matter further.*
Article 7. *The expenses of this mediation procedure will be borne by both parties.*

2.2.4   Comments[54]

The company's position is similar to that of many small or medium enterprises in Japan. Its principal difficulty was in financing and installing the necessary pollution control technology. This situation was addressed by art. 4 of the conciliation agreement that identifies an interim solution in order to alleviate the hardship of an immediate termination of operations. Article 2 requires the removal of the soundproofing wall of the plant, because this wall interrupted the free flow of sunlight to surrounding residences.

2.3   Case 2: Noise and Vibrations from a Textile Factory (Ishikawa Prefectural Conciliation Committee)

2.3.1   Introduction

The noise and vibrations of a textile factory in Ishikawa Prefecture, owned by two companies, have considerably disturbed neighboring residents, who claim they can neither relax by day nor sleep at night. These residents referred the matter to the Local Review Board during April 1973, and requested either that the plant be moved or measures undertaken to stop its tremors. No request for compensation was made.

2.3.2   Disposition of the Case

After listening to the opinions of the parties, the Conciliation Committee began its independent investigation of the case with the aid of a number of experts. Although offensive noise and vibrations were not observed at first, the committee continued its observations. During subsequent months, the parties logged the factories' operations and the degree of disturbance, and the committee initiated various studies of noise levels, taking measurements from all directions and at various distances. Thereafter, pursuant to art. 34, sec. 1 of the Dispute Law, the committee recommended acceptance of a conciliation plan. On June 5, 1974, the parties accepted the committee's proposal.

2.3.3   Conciliation Agreement

Article 1. *The companies admit their responsibility for the disturbance to the local environment caused by the operations of their textile plant.*
Article 2. *The companies undertake to limit the noise from their plant within six months to a level below 60 dB at residences. The companies will take those measures necessary to reduce vibrations (including the refurbishing of residents' houses after obtaining their approval). During this period the local authorities will continue to monitor noise from the factory.*
Article 3. *Should the companies wish to expand the facilities in the neighborhood of their plant, they must adhere to these standards of disturbance.*
Article 4. *Both parties will independently bear the costs of this mediation.*

2.3.4   Comments[55]

1. Although it is often difficult to assess the types of harm people suffer from vibrations, it is at times equally difficult for a responsible company to develop a technological solution to this kind of pollution. As stipulated in the conciliation agreement, the two companies agreed to restrict the noise level from the plant to below sixty dB. This is a general measure and no stipulation was made as to *how* this result would be achieved.

2. At the time of the conclusion of conciliation, there was no legislation that dealt directly with the noise problem. Thus, the case illustrates how conciliation can be used to establish through private bargaining a separate regulatory guideline (i.e., 60dB). Only in 1976 was special legislation for land vibrations enacted. Under the Vibration Control Law (Shindō Kisei Hō) (Law No. 64, June 10, 1976), the Environment Agency later established general standards and guidelines that are now implemented in designated areas at the local level. The 1977 standard under art. 1 of the Standards Pertaining to the Regulation of Vibrations Produced by Special Factories, etc. (Tokutei Kōjō tō ni oite Hassei Suru Shindō no Kisei ni Kansuru Kijun) is 60–65 dB. Thus, the limit set under the 1974 conciliation agreement was equal to the most stringent regulations for this activity established as of mid-1977.

3. The use of conciliation was very valuable in the present case, because it is still difficult to obtain injunctive relief from the courts. Conciliation, however, stimulated the parties to take affirmative action by joint agreement. Despite the fact that the agreement is broad and abstract, it served a useful purpose as a guideline. Moreover, performance is to be achieved in as congenial a way as possible. Should there ever be disagreement over the contents of the conciliation agreement, the new amendments to the Dispute Law permit the Conciliation Committee to continue to offer advice and otherwise to supervise implementation.

4. The case represents one of the rare uses of art. 34 of the Dispute Law, which states:

When the parties find it difficult to reach a settlement, the conciliation committee after considering all the circumstances may execute a conciliation agreement for their acceptance. If this agreement is not explicitly rejected by either of the parties within 30 days, it shall become binding.

The reader may question the theoretical basis of this provision, since it would seem that it violates the basic purpose of conciliation, to evidence the reciprocal agreement of the parties. Yet, a suggested conciliation agreement can serve several useful purposes. First, where both parties have reached a critical point in the negotiations, but for one reason or another they are unable to alter their positions, this provision provides a mechanism for reconciling differences. Although the parties might not independently have reached this particular solution if left to their own resources, the committee's proposal is often sufficiently cogent to persuade the parties to bury their differences.

2.4   Case 3: Heavy Metals Discharged into the Watarase River (Ashio Case) (Central Committee)

2.4.1   Introduction

Nine hundred and seventy-one farmers of Gumma Prefecture sought compensation for damage to agricultural crops caused by heavy metals discharged from the

Ashio factory of the Furukawa Mining Company. Discharged heavy metals had been carried by the Watarase River into rice paddies downstream, accumulating there for many years. The farmers requested ¥3,877,856,150 for damage caused during the period from 1952 to 1971.

The Conciliation Committee received their application on March 31, 1972, and after twelve sessions, both parties agreed to the following settlement proposed by the Committee on May 12, 1974.

2.4.2   Conciliation Agreement

Article 1. *The company admits having caused damage by its discharges into the Watarase River of copper and other heavy metals. It agrees to pay the farmers suffering injuries a lump sum of ¥1,550,000,000 as compensation for past and continuing damages to all farming lands in the Morita District of Ōta City, the affected area. This amount covers damages such as the reduction in value of farm products, the cost of adopting heavy metal control measures, the delay in modernizing farming equipment, and losses necessitated by having to replace contaminated soil.*

Article 2. *The company shall deposit into the farmers' bank account an initial payment of ¥785 million by June 11, 1974, and the remaining amount of ¥765 million by February 5, 1975.*

Article 3. *The company shall be responsible for the distribution of the compensation payment according to a fixed schedule, prepared by the conciliators and agreed to by the parties.*

Article 4. *The company shall be responsible for settling any future compensation claim brought by a third party for damages in the affected area during the period specified in the schedule.*

Article 5. *The company shall make every effort to improve its facilities in order to prevent future discharges of heavy metals from slag piles in the vicinity of the Ashio plant.*

Article 6. *In cooperation with responsible government agencies, both parties will strive to implement the Contaminated Agricultural Soil Treatment Plan based on the Law to Prevent Contamination of Agricultural Soil.*[56]

Article 7. *In view of the farmers' wishes, the company will try to conclude as soon as possible a pollution prevention agreement with the Gumma Prefectural Government and Ōta City in order to prevent the recurrence of pollution from the Ashio factory.*

Article 8. *Both parties hereby confirm that their dispute over damages in the affected area during the period specified in the schedule is fully settled by this conciliation. The parties will hereafter cooperate to fulfill the contents of this agreement. Both parties will settle amicably through consultation future disputes over damages arising over the collapse of slag piles or other unanticipated accidents.*

Article 9. *Each party shall share the costs of the conciliation.*

2.4.3   Comments

The Ashio conciliation is historically interesting because it records the final settlement of a pollution dispute that continued intermittently since the Meiji era. The farmers who were party to this negotiation claimed that about 500 hectares of farmland were substantially damaged by toxic chemicals that were washed onto their lands when the Watarase River flooded. (One report notes that the soil had been contaminated with 1,000–2,000 ppm of copper that caused extensive crop damage.)

The "pollution prevention agreement" that the Furukawa Mining Company promised to conclude with Gumma Prefectural Government was in fact reached in July 1976.

2.5   Case 4: Minamata Disease Victims and the Japan Chisso Corporation

2.5.1   Introduction

On March 20, 1973, the Kumamoto District Court delivered its decision in a suit brought by the first group of Minamata patients. The court awarded compensation payments of ¥18 million, ¥17 million, and ¥16 million (plus interest), depending on the seriousness of a patient's symptoms.

After the court decision, the Central Pollution Coordination Committee began conciliation with several groups of patients and the Japan Chisso Corporation. The committee recognized three different categories of compensation payments according to the degree of a patient's symptoms, as stipulated in the decision. In the conciliation agreements, however, the following amounts were added as a special adjustment allowance: ¥60,000 per month for a gravely ill patient, ¥30,000 per month for a moderately ill patient, and ¥20,000 per month for a patient still only slightly afflicted.

In 1973, the company and the certified patients agreed that the patients would be classified by the Central Pollution Coordination Committee (or a committee appointed by the governor of Kumamoto Prefecture on their application), and that the company would contribute ¥300 million to a fund established for the welfare of the patients.

By June 30, 1974, 386 patients had applied for conciliation and of these 174 had concluded conciliation agreements. The following are the articles of conciliation concluded in the case of a moderately ill patient.

2.5.2   Conciliation Agreement

Article 1. *Compensation Payments. The company assumes the responsibility of compensation for the affliction of Minamata disease and agrees to make the following payments:* [57]
(a) Solatium. ¥17 million of solatium plus annual interest of 5% from the date of execution until April 10, 1973; ¥11 million for the period of April 11 to May 31, 1973; and ¥1 million for the period from June 1 to the date when payment is to be made.

The following provisional payments that have already been made shall be considered as partial payments of the amount of solatium specified above: ¥5 million on April 10, 1973; ¥11 million on May 31, 1973.

The payment of ¥400,000 shall be deemed a partial payment of accumulated interest. The remaining amount of payment of ¥1 million and interest shall be made on July 17, 1974 at the Minamata factory of the company.
(b) Medical Expenses (To be paid after July 9, 1973). The company shall pay the victims medical expenses in an amount equal to that stipulated in the Law on Special Measures for the Relief of Pollution-Related Injury to Health (hereinafter, Special Measures).
(c) Nursing Expenses. After July 9, 1973, the company shall pay for nursing ex-

penses in the amount specified (to be amended by the Compensation Law) plus ¥10,000 per month.

(d) Special Adjustment Allowance. ¥30,000 per month for the period from April 27, 1973 to May 31, 1974; ¥35,000 per month for the period after June 1, 1974. The company shall remit a monthly payment on the 20th of each month. The provisional payment of ¥282,667 paid before June 20, 1974 shall be considered a partial payment of the sum due before June of 1974. The remaining sum shall be paid on July 17, 1974 at its Minamata factory.

(e) Funeral Expenses. Should a victim die (from Minamata disease), the company shall pay funeral expenses of ¥233,000 to the person responsible for administering such victim's funeral. The company shall tender payment immediately upon the request of said victim's representative at its Minamata factory.

Article 2. The amounts of payment stipulated under the above headings, (d) and (e), shall be adjusted to changes in the cost of living according to the comprehensive consumer price index figure of the previous year for Kumamoto City provided by the Statistics Bureau of the Prime Minister's Office on June 1, 1975. The same procedure shall be applied on the same date every two years thereafter.

Article 3. Should a victim's condition in the future require an increase in the amounts stipulated under section headings (1) and (4), said victim may apply to the Conciliation Committee to amend this agreement.

Article 4. When payments are revised according to the preceding paragraph, the company shall pay the increased amount from the time of said application.

Article 5. The company may apply to the Conciliation Committee for the determination of whether a victim's spouse, child, and parents are also entitled to obtain solatium for said victim's affliction.

Article 6. Should a victim die from Minamata disease (or from its complications), his or her successors may apply to the Conciliation Committee for a determination of whether they are entitled to obtain solatium. The amount of such solatium will also be determined at that time.

Article 7. The company shall promote the welfare of Minamata patients and their families by setting up nursing facilities for those requiring constant care, rehabilitation, and post-care training, and by finding jobs for the patients and their families.

Article 8. The company shall faithfully fulfill its responsibilities and cooperate with responsible government agencies and local governments. The company shall adopt concrete measures to prevent pollution and to restore the environment including clean-up operations in Minamata Bay. The company also promises to conclude at the earliest time possible a pollution control agreement with local government and faithfully to comply with all pollution control agreements already in existence.

Article 9. Both parties agree that the dispute between them is hereby fully resolved and that they shall cooperate fully in implementing the above articles.

Article 10. *The company shall bear the costs of this conciliation proceeding.*

2.6   Case 5: Destruction of Fisheries in Tokuyama Bay by Water Pollution[58]

2.6.1   Introduction

Two hundred and forty-four persons operated a fishing business in Tokuyama Bay in southern Japan. They belonged to two cooperative fishing unions in Tokuyama

City, and conducted their business in accordance with the rights and obligations of these unions. On November 29, 1973, they requested the Central Pollution Investigation Committee to conciliate their dispute with twelve companies that operated petrochemical factories near the bay. The fishermen alleged that these companies had permitted sludge to pile up in the bay, had caused the outbreak of red tides, and had damaged their fisheries. They made these demands:

1. The companies pay compensation in the amount of ¥1,007,000,000 for the reduction in the fishing catch from 1957 until 1961, for the death of shellfish under cultivation, and for the increasing costs of fishing boat fuel necessitated by relocation to other fishing grounds.

2. The companies remove sludge that had piled up in the marine areas.

3. The companies terminate all offensive discharges from their factories. The Conciliation Committee held twelve conciliation meetings and conducted an on-site investigation with the cooperation of an expert committee. On June 2, 1975, the following settlement was reached. In its investigation, the committee attempted to clarify the relationship of the pollution of the bay to red tides, and the companies' separate contributions to the bay's pollution.

2.6.2   Conciliation Agreement

Article 1. *The companies recognize that factory discharges are one element in the pollution of the water quality of Tokuyama Bay and that these discharges have damaged fisheries by reducing the catch of fishermen who operate a fishing business in these marine areas. The companies agree to pay compensation for injuries to fisheries up to the present in the amount of ¥150 million and to pay ¥50 million toward the establishment of a fisheries promotion fund and toward the defrayment of costs incurred by the complainants in disposing of this case.*

Article 2. *By August 1, 1975, the companies promise to pay this entire amount into an account designated by complainants' representatives.*

Article 3. *The dispensation of these monies (referred to in paragraph 1) shall be the responsibility of the four representatives of the fishermen. The use of the above mentioned fisheries promotion fund and its administration shall be the responsibility of the fishing unions to which complainants belong.*

Article 4. *If in the future a third party requests compensation for damages to fisheries noted in art. 1 of this conciliation agreement, the complainants' representatives shall be responsible for disposing of this matter and it shall not disadvantage the companies in any way.*

Article 5. *It goes without saying that the companies, henceforth, shall faithfully observe the established effluent standards for factories. In addition to complying with these standards, the companies shall endeavor to clean up their operations.*

Article 6. *If in the future a relief fund system is established, either by the state or local governments, the companies shall cooperate with this (effort).*

Article 7. *Except for separate agreements pertaining to the mercury poisoning aspects of this case, both parties agree that this conciliation agreement fully settles their dispute up to the present time. Both parties further agree that if harm should befall the fisheries due to an accident at the factory, or some other unpredictable circumstance, the parties will conduct themselves in a spirit of sincerity, and shall settle their dispute harmoniously; and that the agreement in this case shall be implemented through the Fishing Unions to which the complainants belong.*

Article 8. The costs of this conciliation proceeding shall be borne by each party.

2.6.3   Comments

This case is typical of many disputes involving red tides in the Inland Sea. In this case the number of possible pollution sources was a critical problem. For example, the companies claimed that many other sources had polluted the waters of the bay—sources like runoff from rivers, pollutants in these rivers, and discharges into the bay from the city of Tokuyama. A determination of what part the companies' factories had played in the pollution of the bay thus was critical to the disposition of the case. The companies asserted that the amount of compensation that the fishermen had demanded exceeded the amount of pollution for which they were responsible. Although the conciliation committee employed the services of a group of experts and conducted an on-site investigation, its survey was inconclusive. The committee was not able to explain the cause of the red tides, nor the extent of each company's responsibility.

The committee ultimately resolved the problem as follows. Because there was an apparent connection between the pollution of the bay and the companies' activities, the committee first calculated the monetary amount of damages suffered by the fishermen based on the data gathered. Taking this into account, the committee then noted the amount that the companies already were informally paying to reimburse the fishermen. Using this sum to reflect the extent of the companies' factories' contribution to pollution of the bay, the committee concluded that payment in the total amount of ¥200 million would be fair compensation. Thereafter, the committee sought to persuade the parties of its decision. Article 1 of the conciliation agreement sets forth their agreement.

The amount paid was deemed to be compensation for damages to fisheries suffered by all persons utilizing the bay. To clarify this point, the companies and the two fishing unions to which the fishermen belonged exchanged memoranda supplementing art. 4. The companies exchanged a similar memorandum with the Fishing Cooperative Association in Yamaguchi Prefecture.

The conciliation agreement does not require the companies to remove sludge because of the prohibitive cost involved, and because this was deemed to be a matter for the central or local authorities to consider first. The committee felt that the local authorities were in a better position to determine how the costs of this undertaking should be allocated.[59]

Although the committee considered carefully the extent to which the companies' operations had contributed to the pollution of the bay, and the illegality of the companies' conduct, the conciliation agreement does not require the companies to terminate operations. This compromise reflects the companies' agreement to respect existing pollution standards, and their pledge to clean up.

The case also raises a number of interesting legal questions. In order for the fishermen to have been legally entitled to the removal of sludge or the termination of the factories' operations, they would have had to show an infringement of property or other rights. But there was no clear proof that these rights had been infringed. Also uncertain was whether the fishermen's claim was to the infringement of their own rights, or to rights held in the name of the fishing union. If the latter, we are unsure to what extent the fishermen can claim infringement of these rights.

Another interesting aspect of this agreement is the compensation fund for the promotion of fisheries established in art. 6. Such funds are often created in Japan in cases where pollution arises from many, not easily identifiable, sources.[60] Since many people had polluted the bay for years, and since the companies' operations were relatively recent, it was unlikely that the fishermen would have been able to establish causation in a court action. In this case, conciliation provided relief where no remedy was possible at law.

2.7   Case 6: Osaka International Airport Noise Pollution (Central Committee)

2.7.1   Introduction

About 20,000 residents living in the vicinity of Osaka Airport, belonging to 721 separate groups, began a conciliation proceeding in the Central Committee with the representative of the Minister of Transportation as the opposing party. Osaka Airport had been built by the State, opening in 1964. Conciliation continued from February 15, 1973, to March 25, 1975, during which time the parties met twenty-one times. The residents asserted that their rights of quietude were grossly infringed, and made the following requests:

1. After 1981, when the Kansai International Airport was scheduled to open, Osaka Airport would be closed;

2. Until such time, the central government should introduce noise control measures to reduce the noise level to 70 WECPNL[60a] in the vicinity of residences;

3. The central government should pay the complainants ¥500,000 compensation for past pain and suffering, and ¥150,000 for future pain and suffering up until the point of implementation of the foregoing noise prevention measures.

The Central Committee divided the residents into several groups, met frequently, and conducted an on-site survey. Between October 28, 1975 and November 14, 1975, the committee was able to reconcile all parties except 700 members of the Itami and Amagasaki groups.

2.7.2   Conciliation Agreement

Article 1. *In order to reduce aircraft noise from Osaka Airport and to harmonize the use of the airport with the need for a tranquil living environment, the central government shall implement, systematically and sequentially, necessary airport-area noise-source countermeasures. The central government's objective shall be the attainment of the ambient standard for aircraft noise.* [61]

(a) With an eye to meeting the "Five-Year Improvement Goals" set forth in the ambient standards, the central government shall implement noise pollution countermeasures such as improving aircraft materials and operating procedures. The central government shall also reduce the level of noise by 5 WECPNL in areas over 85 WECPNL, determined at the time the ambient standards were established. The central government shall assist those residents of areas exposed to over 85 WECPNL by installing noise prevention devices, maintaining buffer zones, and by compensating for relocation, when indicated by the severity of the noise.

(b) With the long-range goal of attaining the "Ten-Year Improvement Goals" in the ambient standards, the central government shall try continuously to improve aircraft materials and to update its operating procedures. Moreover, the central government shall attempt to reduce the level of noise by 10 WECPNL in the areas over 85 WECPNL determined at the time the ambient standards were

339

*established. At the same time, the central government shall introduce system-
atic land use measures and shall develop methods for soundproofing residences.
And the central government shall attempt to reconstruct a desirable living area
for the residents. Finally, the central government will seek to implement the
ambient standards as quickly as possible.*

*Article 2. Since a transition to a "large, low-noise aircraft," meeting interna-
tionally recognized airport-noise standards, is contemplated by the ambient
standards for the areas surrounding the airport, to the extent possible, the central
government shall disclose (provide and explain) relevant information (documentation) to residents and take other necessary measures.*

*Further, the central government will endeavor to devise necessary measures
as quickly as possible to carry out improvements needed to accommodate a
low-noise engine.*

*Article 3. In order to contribute to the reduction of aircraft noise, the central
government will endeavor to improve operating procedures, such as rapid ascent
and delayed flap techniques. The central government shall also give proper consideration to flight safety.*

*Article 4. While considering airport services, the central government shall also
study seriously the amount of noise that might be reduced by limiting the
number of takeoffs and landings. In addition to commissioning large low-noise
aircraft, the central government shall immediately reduce the number of jet
takeoffs and landings at the airport. Also, within six months after commissioning low-noise aircraft, the central government shall try to reduce the number of
jet takeoffs and landings to approximately two hundred per day (except in the
periods during the New Year season and the Bon festival).*

*Article 5. To preserve the living environment of residents during the evening,
and to assure that no great inconvenience is caused by the operation of national
and international flights, the central government shall try to advance the scheduled time of the airport's last flight as much as possible.*

*Article 6. In servicing areas surrounding the airport, the central government
shall attempt to carry out the conversion of the areas designated as Type 3 areas
into green belts and buffer zones in accordance with the Airport Area Maintenance Plan. Further, the central government shall purchase the land in those
areas designated as Type 2 areas, and seriously consider compensating for the
transfer of homes. Moreover, the Osaka International Airport Area Maintenance
Organization shall effectively conduct the sale and distribution of substitute
land sites.*

*Article 7. In implementing construction of soundproofing in residences, the
central government shall formulate a plan for the Osaka International Airport
Area Maintenance Organization. The organization shall handle flexibly and
realistically the implementation of its policies. . . .*

Article 8. The central government shall cooperate with the cities involved. . . .

*Article 9. The central government shall manage the Syndicated Association for
the Control of Pollution from the Airport efficiently and shall pay attention to
the resident's requests that noise-proof telephones and "television noise modu-
lators" be installed.*

*Article 10. In view of the topographical conditions of areas surrounding the
airport, the central government shall encourage the development of safe operating procedures and shall strengthen safety precautions for fuel tanks located on
the airport sites and for the transport of fuel. In addition, to deal with unforeseen
contingencies, the central government shall cooperate with the relevant local*

340

governments in expanding safety facilities, such as fire control equipment, located in the areas surrounding the airport. The central government also undertakes to maintain an adequate fire rescue system.

Article 11. The central government shall implement these airport aircraft noise reduction measures in due course and in accordance with the degree of the noise level. It shall rationalize the designation of areas serving as standards for these measures, and within a suitable period of time, it shall reassess these areas in cooperation with relevant local governments.

2.7.3   Comments

The agreement reached represents only a partial conciliation. The residents' first request was considered premature in view of the importance of Osaka Airport and the fact that the Kansai Airport had not been built. Because the central government steadfastly denied responsibility, agreement on the third item (compensation) was also not immediately possible. The second item (noise reduction) was considered the most feasible area for negotiation in view of the residents' continuing injury.

The agreement attempts to mitigate some of the more injurious aspects of the use of the airport. For example, the agreement provides that the central government will accomplish various noise control measures in stages. Although large, low-noise aircraft were coming into worldwide use, local residents strongly distrusted these airplanes. Many residents did not believe that these planes were necessarily less noisy; indeed, because of these planes' heavy exhausts, some people felt that they might be more polluting. For this reason, the agreement stipulates that the government will explain these matters fully to the residents.

At the time of conciliation, there were two hundred and forty arrivals and departures; the daily number of arrivals and departures of domestic flights after 9:00 P.M. was 6; and there were 11 international flights after 9:00 P.M. per week.

In order to conclude a conciliation agreement, it was necessary to instill confidence in the local residents and to cultivate their trust. In view of the fact that the conciliation was conducted with 20,000 people, and that an agreement was ultimately reached with about 15,000, this conciliation was a considerable accomplishment.

Section 3   A Tentative Assessment of the Dispute Law

It is still too early to draw conclusions on the impact of the Dispute Law. Yet, because of its potential interest for those working in the environmental field in the United States and other countries, we include this tentative assessment. The Dispute Law is designed to achieve several related purposes. First, it seeks to anticipate controversy by defusing a grievance at the outset through the counselor system; second, it facilitates the settlement of a dispute; third, it contributes to pollution control and environmental protection. This section analyzes how effectively, efficiently, and fairly the Dispute Law has met these objectives.

3.1   Anticipation of Disputes

Table 2 gives the number of complaints reported from 1966 to 1975. The number of complaints increased until 1972, and declined thereafter. The greatest increase was in 1970. Grievance counselors received about 80% of these complaints; the rest were processed by local environmental offices and health centers. Table 3 breaks down the complaints by source for a single year (1975).

The increase in complaints reflects the citizens' growing awareness of the need for governmental action to remedy pollution. Environment Agency officials offer two explanations for why so few complaints were filed in 1973 and 1974. First, because Japan's economic growth rate declined during 1973 and 1974, many industries reduced output, and operated for shorter hours. The decline in the total number of complaints may in part be attributed to this reduction in operations and to the consequent decline in pollution. Second, pollution control measures were beginning to take effect during this period.

The types of complaints have tended to display a consistent pattern. Most complaints were against manufacturing establishments; the majority concerned noise and vibration; there were fewer complaints about offensive odors and water pollution. Infringement of psychological or aesthetic interests (66% in 1973) are most common, followed by property damages to crops; complaints of injury to health have been generally less frequent. Ordinarily, only a limited geographical area is involved, usually comprising less than ten households.

In practically all reported cases, counselors or local officials have conducted on-site investigations. Although the measures used to dispose of complaints vary, administrative guidance is the preferred procedure.[62] Even where polluters have not violated emission standards, officials have at times suggested supplementary control measures. In most cases, local officials themselves mediate a dispute; at times, at the suggestion of a local official, a polluter will offer to compensate an injured party. Most complaints are ultimately settled in one way or another. In 1974, for example, 31% of the complaints were processed within one week, 53% within one month, and 67% within three months. The average processing period was forty days.[63]

Generally the system of grievance counselors has worked effectively, as attested by its relatively wide use. We suggest these reasons for its success: First, local people are familiar with the system and its procedures are simple. Second, counselors have generally been accessible to the people. Third, as government officials, these counselors possess some power over polluters. In fact, some grievance counselors have been local pollution control officials.[64] Thus counselors can usually marshal technical and scientific expertise and launch a meaningful investigation of a dispute. At times they may also assist a polluter in obtaining a loan for the purchase or installment of pollution control equipment. The official status of grievance counselors and their capacity to be useful to both local people and polluters have greatly contributed to the system's success.

3.2   Resolution of Disputes

A principal function of the Dispute Law is to lessen the pollution victim's burden of proof on various technical legal issues, thereby facilitating the harmonious settlement of disputes. How successfully has the system served this end, and which procedures have been preferred?

Table 4 shows that a high percentage of disputes processed under the Dispute Law between 1971 and 1976 were settled by the end of 1976, with the percentage highest at the local level.

At the local level, most cases have been small in scale and have involved noise and vibration. Usually the injured parties have numbered less than ten. About

Table 2
Number of complaints, 1966–1975.

	1966	1967	1968	1969	1970	1971	1972	1973	1974	1975
Number of complaints	20,502	27,588	28,970	40,854	63,422	76,106	87,664	86,777	79,015	76,531
Increase over the previous year		7,086	1,382	11,884	22,579	12,673	11,658	−987	−7,762	−2,484
Percent increase		35	5	41	55	20	15	−1	−9	−3

Source: White Paper 354 (Environment Agency, 1977).

Table 3
Sources of complaint, 1975.

	Total	Air Pollution	Water Pollution	Soil Pollution	Noise and Vibration	Ground Subsidence	Odor	Other
Number of complaints	76,531	11,873	13,453	593	23,812	68	17,516	9,216
Percent	100.0	15.5	17.6	0.8	31.1	0.1	22.9	12.0

Source: White Paper 355 (Environment Agency, 1977).

Table 4
Disputes handled by the Central Committee and the local boards, 1971–1976.

Agency	Central Committee				Local Board				Total			
Status of Cases	a	b	c	d	a	b	c	d	a	b	c	d
Period												
70/11/1—71/12/31	6	6	1	5	22	22	12	10	28	28	13	15
72/1 /1—72/12/31	12	17	1	16	22	32	15	17	34	49	16	33
73/1 /1—73/12/31	32	48	8	40	26	43	25	18	58	91	33	58
74/1 /1—74/12/31	31	71	29	42	27	44	16	29	58	115	45	71
75/1 /1—75/12/31	39	81	16	65	23	52	30	22	62	113	46	87
76/1 /1—76/12/31	62	127	42	85	20	42	22	20	82	169	64	105
Total	182	—	97	—	40	—	120	—	322	—	217	—

Source: White Paper (Environment Agency, 1976).
a, newly accepted
b, docketed
c, settled
d, pending.

60% of the enterprises complained against have been involved in manufacturing; the remainder have been mostly construction companies or common carriers. Of these cases, fifty-seven reached agreement by March 1975, thirty-five included some form of compensation, and twenty-eight required pollution abatement or termination of operations. In eight cases, the parties agreed to some form of environmental restoration. The average time for processing a given case was 8.4 months, and 75% of the cases were settled within a year.

One possible explanation for the success of local dispute settlement is that these controversies are often less serious and more easily managed. Local officials have also played a critical role for the same reasons noted above.[65]

Table 4 also suggests that the Central Committee has played a comparatively minor part in dispute settlement. Although Central Committee officials cannot explain this phenomenon,[66] a number of pollution victims interviewed provided insight at least into their own perspective. Essentially the Central Committee is viewed as merely a part of the bureaucracy, or worse, as an appendage of the LDP.[67] For this reason, victims regard the committee as biased and its establishment to have been politically motivated. Although committee members feel that they are as "neutral as the courts," many victims still do not view the courts as "neutral." And some victims consider Central Committee members to be actually biased. The Central Committee, like the rest of the central bureaucratic apparatus, is also felt to be more remote and impersonal than local dispute settlement institutions. Finally, some victims are simply unwilling to settle their grievances out of court, for they seek an official, prescriptive recognition of the justice of their cause.[68]

Although it is too early to conclude that arbitration, quasi-judicial arbitration, or mediation are proving ineffective, it is clear from table 5 that conciliation has been by far the preferred procedure.

Historic and institutional considerations help explain the apparent preference for conciliation. First, even before the enactment of the Dispute Law, mediation of pollution disputes had a checkered record. The government attempted mediation with little success in the late 1950s in Minamata. Although mediation is generally considered flexible because it leaves the initiative of negotiations to the parties, the Ministry of Health and Welfare during this period tried to make

Table 5
Types of pollution cases and procedures employed.

Type of Case	Procedure				
	Conciliation	Mediation	Arbi-tration	Quasi-Judicial Arbi-tration	Total
Air pollution	20 (1)	5	2	1 (1)	28 (2)
Water pollution	159 (146)	6	3 (1)	1 (1)	169 (148)
Noise pollution	89 (23)	13	—	3 (3)	105 (26)
Soil contamination	7 (2)	1	—	1 (1)	9 (3)
Offensive odor	7	—	—	—	7
Ground subsidence	1	—	—	3 (3)	4 (3)
Total	283 (172)	25	5 (1)	9 (9)	322 (182)

Source: White Paper (Environment Agency, 1977).
Note: Data is given for the period from November 1, 1970, to December 31, 1976. Numbers outside parentheses give cases handled by local committees; numbers in parentheses give cases handled by the Central Committee.

the procedure more resemble arbitration. For example, before mediating the Minamata dispute, the ministry attempted to extract written assurances from the parties that they would respect the result of any decision. Although these assurances (kakuyakusho) would not have had binding legal effect, social pressure at the time on a nonconforming party was great. Chisso quickly accepted the ministry's proposal and offered written assurances, but some victims refused to participate. This group established its own faction (soshōha), which later initiated litigation. Mediation has also been attempted from time to time by local prefectural governors. Problems, however, have occurred because citizens' groups view provincial governors as politically motivated, or too involved in the dispute themselves to be effective.

Second, most citizens now greatly distrust procedures that preempt litigation. We suggest this is the principal reason for the small number of arbitration and quasi-arbitral determinations during 1970–1976. Since pollution disputes have evoked violent emotions and deep distrust, it is unlikely that the parties in most cases would confide the disposition of a controversy to a government-appointed arbitrator.

Third, conciliation seems to combine just the right mix of formality and flexibility. Conciliation is a more forceful procedure than mediation. The conciliation team can exhort the parties to adopt one of several options, and, as noted, may even draft an entire program of settlement. But, as in mediation, the parties need not meet face to face; as we have noted, the conciliators often meet with the parties separately, passing views back and forth, identifying various options, exploring with the parties the costs and benefits of various settlements. Yet, conciliation is a less summary proceeding than arbitration or quasi-arbitral determinations. The parties preserve the luxury of choice.

Fourth, the evidence suggests that conciliation has indeed proved effective. Despite their authority and official status, conciliation committees have often encouraged parties to identify mutually satisfying options, and have usually refrained from imposing preconceived settlements.[69]

The system has functioned in one important respect differently than originally anticipated. Rather than serving principally to intercept litigation, conciliation has more often provided a means to implement the courts' decisions.[70] In this respect, the Central Committee's ministrations to victims of noise pollution from Osaka Airport or of Minamata disease have been very valuable, because the committee has relieved these people (as well as the courts) of having to relitigate many painful questions.[71]

3.3   Pollution Prevention and Environmental Protection

Recent extrajudicial settlements are increasingly conferring benefits on members of local communities who often are not even parties to the negotiations, to the extent that these settlements serve as an additional legal means of protecting the environment. For example, some agreements supplement existing standards with new or stricter limitations; others require polluters to prepare an environmental impact assessment, or condition a company's continuing operations on the establishment of a pollution injury compensation fund. Many agreements stipulate that before commencing any new environmentally hazardous action, a company must explain its motives and needs to local residents and obtain their approval.

A creative exchange is now developing between the administrative, judicial, and extrajudicial processes. Although most remedies identified by extrajudicial settlements are principally inspired by the bargaining process itself, others adopt, or seek support in, newly created judicial remedies. And although most Japanese courts, like their American counterparts, have been extremely cautious in experimenting with new remedial solutions, local mediated settlements have on occasion inspired judicial action. The administration has also borrowed ideas developed in extrajudicial settlements and, as noted, administrative actions have in turn granted legitimacy to extrajudicial processes.

Although it is difficult to assess the extent to which the parties have fulfilled their promises in extrajudicial settlements, Central Committee officials suggest that compliance is of a high order. They point out that failure to respect an agreement risks the same governmental displeasure as disrespect of administrative guidance. But the parties under the Dispute Law are even more constrained, because violation of an agreement constitutes a breach of contract.

The reader may wish at this point to review the major legal controls a polluter now faces in Japan. First, polluters must observe regulations set under statute. Second, a company usually concludes a local pollution prevention agreement (kōgai bōshi kyōtei) (ch. 5). Third, an extrajudicial mediated settlement with local residents may also be necessary. Fourth, a firm will have to pay its share of the national compensation fund. Finally, a polluter still must meet the standard of care established by the courts (chs. 3 and 4). Japanese (and American) courts may establish a higher standard of care than implied by statute; in other words, even though a company meets an existing pollution regulation, the company still can be held liable. And the standard of conduct which the mediated settlement (or pollution prevention agreement) identifies may at times be more stringent than an existing regulation or judicially created standard. These legal measures, together with public pressure, constitute a formidable basis for social control.

3.4   Considerations of Fairness and Efficiency

In theory the Dispute Law redresses fundamental inequities in bargaining. Where polluters hitherto have had a monopoly over information, the Dispute Law aids the victim, who is usually in a weaker bargaining position, in securing expert assistance and technical information. Moreover, a conciliation team will make its own independent assessment of the facts. Polluters in theory may not influence a committee's decision.[72]

To some extent the system is also serving to redress arbitrary aspects of the judicial process. For example, residents often feel that a company's operations are impinging on important interests. Yet, if these residents were to litigate their claim, it is likely that doctrinal or other technical considerations would bar relief. Extrajudicial procedures avoid this. They aid the victim and at the same time gradually draw the society's attention to interests in need of protection. Although the Dispute Law, to the extent it influences low-cost bargaining, in theory contributes to efficient resource allocation,[73] an assessment of the system's efficiency is beyond the scope of the present study.

From the perspectives of the parties, recourse to the procedures in the Dispute Law is clearly cheaper than litigation. For example, parties need not pay the costs of an expert's testimony in a proceeding, nor the travel expenses of conciliators, mediators, or arbitrators. Application fees for conciliation are estimated to be about one-eighth court document filing costs,[74] and attorney's fees in a conciliation proceeding are usually low.

One aspect of Japan's pollution dispute resolution system, its duplication of function, may seem inefficient to the Western reader. Thus, pollution disputes are processed not only under the Dispute Law, but also by local officials, by Civil Liberties Commissioners, in civil conciliation, and by other institutions.[75] Although in some ways these criss-crossing institutions supplement each other, on occasion they perform identical functions. Is such duplication needlessly costly and inefficient?[76] We think not. Under the Japanese system, a party at least enjoys an array of specialized, technical, and generalized institutional options. Although this mixed arsenal may be at times functionally duplicative, it has also been useful in settling many complex controversies.

Section 4   Exploring the Uses of Mediatory[77] Institutions in the United States

Many countries are now beginning to explore the uses of extrajudicial intermediary institutions in settling environmental disputes.[78] In this section we focus on the implications of the Dispute Law for the United States,[79] although we hope our analysis is also applicable to other countries.

We must remember at the outset that mediation, conciliation, and arbitration are not unique to Japan; they have long played an important role in labor law in the United States and other countries, and various commentators have urged the application of these procedures in other areas.[80] One field of experimentation in the United States has been civil rights, where, under the Civil Rights Act of 1974,[81] a special system of Community Relations Service Officers has been established. This system reportedly has processed civil rights controversies cheaply and effectively.[82]

U.S. environmental disputes do not seem less amenable to extrajudicial settlement than Japanese conflicts. Indeed, if anything, they may be more so, because the disputes may lend themselves more easily to bargaining. For example, although at times bitter,[83] environmental disputes in the United States have not torn the very fabric of society as did the Minamata tragedy; nor are political ideologies as often intransigently in issue. Further, American environmental disputes have usually involved amenities that one would think are more easily exchanged than human health, the central issue in many of the famous mediated settlements in Japan. That many American environmental controversies involve prospective damage, rather than relief for injuries already suffered, need not undermine mediation's usefulness. The Osaka Airport conciliation (case 6, sec. 2.7) suggests that mediation is also useful in complex situations involving past and prospective harms and numerous parties.

4.1   Who Should Conduct Mediation?

Many American readers (especially lawyers) may find it hard to understand that lawyers may not be the only profession suited by training to be mediators. In Japan, the members of the Central Council currently include a judge,[84] a doctor and former director of the Medical Bureau in the Ministry of Health and Welfare, several other former high-ranking officials,[85] and a law professor and a lawyer. The investigatory staff (shinsakan) directly under the committee members also includes men of professional distinction—two former judges, two former officials of the Ministry of Agriculture, two from MITI, five from MHW, and one from the Prime Minister's Office.[86] The Japanese approach thus mobilizes a variety of technical, scientific, and legal skills and emphasizes extensive administrative, particularly governmental, experience.[87]

A recent report on the settlement of a complicated environmental dispute in the Snoqualmie-Sohomish river basin in Washington suggests that nonlawyers in the United States also can skillfully perform a mediatory function.[88] Lawyers, of course, could also serve as effective mediators, although American law schools have not traditionally emphasized training in mediatory skills.[89]

The process of settling the Snoqualmie-Sohomish conflict is also instructive in that the governor of Washington apparently played a critical role by supporting, hence legitimating, the action of the mediators.[90] In Japan also the mediator's official status has contributed at least at the local level to the system's effectiveness.[91]

4.2   Institutionalizing Extrajudicial Environmental Dispute Settlement

As the contribution of mediatory institutions becomes more generally recognized in the United States, greater attention may need to be given to their harmonization with judicial procedures. Let us briefly consider the possible benefits of mediation within the statutory scheme for compensating victims of toxic substances discussed in ch. 6. For the reasons already discussed, mediation might reduce costs, save time, further lighten the plaintiffs' evidentiary burdens, and provide the parties with a means of identifying remedies that many courts might now be reluctant to recognize. Indeed, part of the function of the Dispute Law—the fact-finding services of the investigator—are already essentially embodied in some American environmental laws authorizing the appointment of a master.[92] The actual services of mediation could be performed in conjunction with the master's services either directly by the judge through greater use of pretrial conferences,[93] or by a special staff affiliated with the courts.[94]

It is also interesting to reflect on the possibilities of the Dispute Law from the perspective of the debate in the United States over the establishment of an environmental court.[95] Supporters of an environmental court urged that it could marshal expertise critical to the settlement of environmental issues,[96] reduce the workload of ordinary courts,[97] contribute to decisional uniformity,[98] and remedy judicial and administrative delays.[99] Despite these ostensible benefits, the environmental court was ultimately rejected because of jurisdictional and other concerns.[100] We invite the Western reader to consider whether an approach similar to Japan's Dispute Law might not capture the considerable benefits envisioned by the defenders of the environmental court, but avoid in large measure the pitfalls so successfully argued by its critics.

IV **Japan's Environmental Law and Policy in International Perspective**

8   Japan's Contribution to International Environmental Protection

Earlier chapters have discussed Japan's environmental law in isolation. But Japan is no longer in anything but a geographical sense an island. And environmental issues increasingly can be found in the dominant concerns of today—security, free trade, energy, and food.

This chapter focuses on Japan's contribution to the protection of the environment beyond her borders.[1] Here Japan's performance falls far short of her impressive domestic achievement. Indeed, to some extent, Japan's domestic success has been attained at a cost to other countries and to the international environment. Section 1 of this chapter explores the evidence in support of this contention and seeks to explain the contradiction. The discussion centers on three case studies—the proposed construction of the Port Pacific industrial complex in Palau, Micronesia; Japan's position on marine pollution control at the U.N. Law of the Sea Conference; and the Japanese policy on whaling.

Section 2 addresses the question of whether Japan bears any legal or moral obligation to adopt a more affirmative policy towards the environment beyond her borders, and, if so, what response is appropriate. Should Japan control the activities of her nationals abroad by extraterritorial legislation or by resort to administrative guidance? How should guidance be developed and what form should it take? These questions are important, because Japan's increasing economic and political power in Asia and the Pacific places her in a unique position to influence the development of new international environmental institutions.

Section 1   Japan's International Environmental Record

1.1   The Pledge and the Performance[2]

At the 1973 Stockholm Conference on the Human Environment, Buichi Ōishi, head of the Japanese Delegation, made the following pledge on behalf of the delegation:

The Japanese delegation approves of the Draft Declaration on the Human Environment prepared through the efforts of the Intergovernmental Working Group and recognizes the preservation of the human environment to be an imperative goal for mankind. [The delegation] strongly supports this declaration in the interests of ourselves and our posterity. . . .

The Japanese delegation fully endorses the establishment of a U.N. Environment Committee to coordinate environmental activities and to expedite improvement of the environment, and supports the creation of a voluntary United Nations Fund for the Environment. The Japanese delegation pledges that Japan will contribute up to 10 percent of the target, if major developed nations also make substantial contributions to this fund . . . .

Now is the time that the developed countries should extend as much cooperation as possible to the developing countries for their speedy attainment of development and prosperity. At the same time, the developed countries should not fail to give consideration to prevent the developing countries from following the path of environmental destruction Japan has trodden. . . .

Japan, as a nation having a serious interest in the seas, wishes to carry out pos-

itively the comprehensive measures against the pollution of the seas and earnestly hopes that international standards for the structure and operation of vessels will be expeditiously strengthened to prevent pollution of the seas. Tankers and other vessels entering Japan are obliged to dispose of bilge and ballast water. My government intends to see that the obligation is fully performed and that pollution of Japanese waters by oil is prevented. At the same time, we believe that oil discharge should be decreased as much as possible through the provision and use of disposal facilities in the crude-oil loading harbor, or through the improvement of "load on top" systems.

In March of this year, the Japanese government concluded with the U.S. government a treaty for the protection of migratory birds. It is the first treaty in the world that has been concluded for such a purpose between two countries separated by an ocean. The migratory birds, innocently flying across national boundaries, may well be called the symbol of international cooperation for the protection of nature. Very recently our country came to a basic agreement to have a similar treaty with the Soviet Union. Japan also hopes to conclude treaties for the exchange of information concerning the protection of the habitat of migratory birds and their ecology with Canada, as well as with China, Australia, and other countries of Southeast Asia. I expect that such treaties for the protection of migratory birds will be successively concluded in other areas of the world and that a network of such treaties will ultimately cover the whole earth. . . .

This section examines how faithfully Japan has fulfilled this pledge.

Japan's remarkable domestic environmental achievements contrast with her bleak international performance.[3] The discrepancy is perhaps most vividly expressed in Japan's reluctance to ratify a single major environmental convention. For example, as of 1979 Japan had refrained from ratifying the Convention for the Prevention of Marine Pollution from Land-Based Sources,[4] the Convention Concerning the Protection of the World's Cultural and Natural Heritage,[5] the Convention on International Trade and Endangered Species of Wild Flora and Fauna,[6] and the International Convention on the Establishment of an International Fund for Compensation for Oil Pollution Damage.[7] Japan originally also opposed the Convention on the Prevention of Marine Pollution by Dumping of Wastes and Other Matters[8] because at the time Japanese industries wished to discharge cadmium and mercury.[9] Ironically, this was the very period when the itai-itai and Minamata disease controversies were at their height.

Japan has also adopted conservative antienvironmental positions in international negotiations. For example, at the 1973 International Maritime Consultative Organization (IMCO) negotiations, Japan actively opposed segregated ballast and double-bottom requirements for tankers; at the U.N. Law of the Sea negotiations, Japanese delegates insisted on flag-state standard-setting and enforcement jurisdiction.[10] Japan has also continued to resist international control of fluorocarbon aerosols within the OECD and elsewhere.[11] Finally, Japan has generally been indifferent to the environmental ramifications of deep sea mining.[12]

Ignoring the spirit of the Stockholm Conference in some international negotiations, Japan has actively violated it in others. Perhaps the most celebrated example is its intransigent opposition to a 10-year moratorium on whaling as recommended by that conference. While repeatedly asserting the economic and nutritional importance of the whaling industry, Japan has not offered any alternatives to violation of the moratorium.[13] A second example is the annual slaughter of an estimated 300,000 to 700,000 migratory sea birds by the Japanese fishing industry.[14] Despite the fact that many species involved have been listed for protection under the 1972 U.S.–Japan Migratory Bird Treaty,[15] Japan until recently did not offer any practical solution to the problem.

Japan has also been reluctant to impose legal controls on the environmentally hazardous activities of its nationals abroad, despite the fact that Japanese industry's advance in Southeast Asia and other developing areas has provoked increasingly bitter pollution-related controversies since 1970.[16] To date, none of Japan's environmental laws have been administratively interpreted to apply extraterritorially, and there is also little apparent governmental interest in using administrative guidance to protect the environment outside her borders. The Environmental Agency's International Office is understaffed, underfinanced, and includes few experts on these issues. The Ministry of International Trade and Industry (MITI), the Foreign Ministry, the Ministry of Transportation, and other agencies and branches of the government have thus far demonstrated little commitment to extraterritorial environmental matters.

1.2   Case Studies

1.2.1   Case 1: Port Pacific in Palau

On March 3, 1977, a coalition of fourteen foreign conservationist organizations[17] filed a petition[18] with the House of Representatives of the Diet to protest the construction of a crude transshipment station (CTS) at Palau, situated in the Western Caroline Islands. The coalition alleged that the complex (konbināto), would have devastating consequences on Palau's physical and cultural environment, and would undermine the Palauans' aspirations for self-determination as guaranteed by the U.N. Trusteeship Agreement. The environmentalists' petition was virtually unprecedented, and the Diet initially demurred on the grounds that the right of petition was restricted to Japanese citizens and aliens resident in Japan. Nine days later, however, the Diet reversed its twenty-eight-year policy.[19]

On March 24, 1977, the U.S. Senate Committee on Energy and Natural Resources began hearings on the superport project. At these hearings, American, Palauan, and Japanese groups submitted their views, along with testimony by the project's promoters, the U.S. Defense Department, and others. On June 15, 1976, a coalition of environmental and other groups, including the Consumers Union of Japan, requested the U.S. National Oceanic and Atmospheric Administration (NOAA) within the Commerce Department to designate Palau's coastal waters as a marine sanctuary under Title III of the Marine Protection, Research, and Sanctuaries Act of 1972 As of this writing, the merits of the Palau superport project are still being debated within the Japanese and U.S. governments, and its future remains unclear.[19a]

Summary of the Facts[20]

Palau and its Status within the United Nations

Palau, the westernmost island of the Caroline group of Micronesia, lies about five hundred miles due east of the Philippines, about 1,500 miles south of Japan, and several miles north of the equator. There has been relatively little industrial development in Palau despite about 400 years of occupation by outside powers (Spain 1589–1899; Germany 1899–1919; Japan 1919–1945; United States since 1949).

At the close of World War II Japan relinquished its control of Palau, which it

355

previously administered under a mandate from the League of Nations. In 1947 the U.N. Security Council concluded an agreement with the United States, wherein Palau and five other districts of Micronesia were designated a "strategic trusteeship" and the United States was made the area's new administrator.

The trusteeship is scheduled to terminate in 1981. One district, the Marianas, has elected to become a commonwealth of the United States, while others, including Palau, are still negotiating their future relationship with the United States. The superport idea was thus first presented when the future status of Palau and Micronesia as a whole was still extremely uncertain.

The Origin of the Superport Project

The Palau superport was the brainchild of Robert Panero, an American and a former member of the Hudson Institute. The term "superport" was understood to mean a very large concept, including a port, transshipment facilities, refineries, smelting works, oil storage, nuclear power plants, and a whole range of associated industries.

In 1974, when the idea of the superport was first discussed, a number of other alternatives to Palau were considered: Okinawa, Taiwan, Hong Kong, the Philippines, Indonesia, Singapore, Malaysia, Guam, and Australia. All were rejected either because of their physical unsuitability, geographic impracticality, or political instability. From the outset mainland Japan was regarded as unsuitable because of the stringency of its pollution controls. Thus the decision was reached that Palau would be the site.

Palau has several natural deep water harbors and a surrounding coral reef serves as a buffer against the sea. Palau is within American military control; thus the promoters considered it politically secure. Palauan environmental laws appeared to be virtually nonexistent and its fifteen thousand people or so, it was thought, would offer little resistance. The proximity of Palau to Japan also made it seem very attractive, since supertankers stopping there could avoid the hazardous Malacca Straits; they could instead traverse the Lombok and Makassar Straits. In 1974 Palau became even more appealing when the Nixon administration opened the area for the first time to investment from all countries and created the Board of Economic Development for the Palau District.

In 1974 Panero became associated with Nisshō Iwai Corporation and the Industrial Bank of Japan, and Panero's consulting firm, Panero Associates, was requested to prepare a report on the Palau superport. Soon thereafter (October 1974) a report bearing the imprint of the Mitre Corporation was privately circulated.[21] During this period, the Japanese government grew increasingly interested in the Palau superport, and initiated, through former MITI Vice-Minister Morozumi, an investigation into oil supply routes alternative to the Malacca Straits. At the behest of Nisshō Iwai and the Industrial Bank of Japan, allegedly with strong encouragement from Japanese and U.S. government officials, Panero Associates completed a "final report" on the Palau complex in May 1975. The report extolled the virtues of Palau and stated that key officials in the Japanese and U.S. governments had already given assurances that the project would be approved.[22] By the end of 1975 the Iranian government, the National Iranian Oil Company, various smaller American oil companies, and three Japanese firms, Maruzen, Daiko, and Teijin, all expressed interest in the proposed development.

Initial Palauan Response

News of the planned superport first circulated in the Palauan newspapers in January 1975. During the subsequent months there were many rumors of the involvement of Trust Territory administrators in the project. During the same period there were also reports of CIA and naval intelligence activity in the area, and of various attempts to bribe prominent local people. The High Chief of Palau was said to have been offered a bribe by a CIA agent in early 1975 in return for cooperation. This offer was reportedly rejected. A correspondent for the Pacific Daily News reported that she was offered a bribe in 1975 to gather intelligence. This offer, also, was rejected. Other reports noted that the project's promoters had begun to entertain prominent members of the Palau Legislature in Japan.

Such rumors increasingly alarmed many Palauans, particularly some traditional leaders. On May 15, 1975 fifteen local chiefs wrote to Edward Johnston, High Commissioner of the Trust Territory, noting that they "have been denied any meaningful or specific information about the CTS proposal; the effect of this veil of secrecy is to render any measure of the attitudes of the people of Palau toward the CTS proposals a farce."

On November 5, 1975, members of Ollei Hamlet petitioned the District Administrator "to express [their] objection and opposition to the way the superport proposal has been handled by foreign companies." The petition requested a moratorium on the project, public hearings, a neutral investigation, and a referendum to determine the Palauans' reaction to the superport. These requests were ignored by the project's promoters and the Trust Territory authorities.

During the ensuing months the superport was promoted more aggressively in Palau, while at the same time opposition to it increased. On January 20, 1976, the Palau Legislature voted 13 to 10 (4 absent) to approve a resolution to invite representatives of the promoters to Palau to conduct a feasibility study. The next day an antisuperport coalition introduced a resolution to petition the U.N. Trusteeship Council to undertake an objective and comprehensive feasibility study. Within two weeks of this resolution, the antiport coalition, known as the "Save Palau Committee" (later, the "Save Palau Organization") formed a local group to investigate the port project. The committee was chaired by the High Chief of Palau. The members of the organization expressed concern over the port's potentially disastrous environmental consequences and over the secrecy of the negotiations.

Gradually, the superport project came to the attention of those outside Palau. American environmental groups began to send letters to the Department of the Interior, U.S. Congressmen, and others, expressing their concern over the project's environmental risks. In August 1975, a number of scientists who had visited and had previously worked in Palau attended the thirteenth Congress of the Pacific Science Association, a body composed of marine biologists and other scientists. The superport project was discussed at this meeting. The association adopted a resolution condemning the superport project because of its environmental hazards.

Despite these indices of concern, the Trust Territory authorities continued to encourage the promoters of the superport. In early 1976 these authorities admonished the Pacific Science Association for its antisuperport resolution, and refused a team of marine biologists, dispatched by the association, permission to inspect the reefs. At the same time, an Iranian marine biologist was given free access to the reefs and use of the territorial government's boats and facilities.

The Feasibility Study Contract

In early 1976, the Japanese, American, and Iranian promoters of the superport moved to conclude a feasibility study contract with the Trust Territory authorities. At the same time public anxiety and suspicion over the activity of the foreign companies in Palau increased. In April, at the Second Regular Session of the Palau Legislature, The House of Chiefs introduced a resolution "that the governments of the United States, Japan, and Iran and all private firms involved are hereby urged not to undertake any feasibility study regarding the proposed supertanker port project, unless all prefeasibility and conceptual studies in existence are fully disclosed to and understood by the people of Palau."

On April 30, 1976, a contract was concluded between the Industrial Bank of Japan, Nisshō Iwai Company, and the Trust Territory government; this contract authorized Nisshō Iwai and the Industrial Bank of Japan to conduct a feasibility study of the superport. The study was to take twelve to eighteen months at a cost of $5 million.

In addition to permitting the contractors to engage in core drilling, sampling, seismic surveys, and a variety of other tests, the contract contained the following two provisions:

Article 4c: The report submitted by the contractor will be treated by the government as the property of the contractor, and confidential as such, for a period of five (5) years after submission of the above report or until earlier released by the contractor and may only be released by the Director, Resources and Development, Trust Territory, to authorized government personnel during this period. After the above five (5) years (or earlier release by the contractor) the information will be in the public domain.

Article 6c: If the initial reconnaissance and feasibility study conducted pursuant to this agreement indicates the feasibility of the project or projects considered and further study is required, and if the contractor has complied with the terms and conditions of this agreement, the government will recommend to the District Foreign Investment Board having cognizance of this matter that a business permit will be granted to the contractor with the exclusive right to enter into specific areas and conduct additional study and engineering activities designed to further detail the extent and scope of such project and, if commercially viable in the judgement of the contractor [emphasis supplied], to develop and execute such projects subject to such terms and conditions as the government shall prescribe.

The contract for the feasibility study was initially kept secret from members of the Palau Legislature. The Vice-Speaker himself was reportedly amazed when he was first shown a copy of the contract in July 1976.

Subsequent Developments

During 1976 the promoters continued to receive assurances from the Japanese government and Trust Territory authorities. MITI, for example, appropriated ¥24 million (approximately $83,000) for a feasibility study. In addition, a nongovernmental foundation, headed by Hideaki Yamashita, a former MITI official, established a special "Palau Committee" to study and to promote the superport proposal.

In June 1976, the Save Palau Organization, disturbed by the increasing momentum that the project was gaining in Palau, formally petitioned the U.N.

Trusteeship Council to oversee a neutral feasibility study. The Trusteeship Council responded to this petition by noting in its formal June–July Report that it had grave doubts whether Nisshō Iwai Company and the Trust Territory administration could conduct an objective assessment. The Trusteeship Council recommended that an independent "body of experts recognized to have no vested interest in the matter be called to review any assessment prepared by Nisshō Iwai and the other parties."

During the fall of 1976 environmental groups in the United States and Europe grew increasingly troubled by the Palau project. Repeatedly, the Natural Resources Defense Council (NRDC) and the Environmental Defense Fund (EDF), two U.S. environmental organizations, wrote letters to Toshio Tsuji, president of Nisshō Iwai, requesting clarification and information on the project. These letters were never answered. In November 1976 the Congress of the World Wildlife Federation, with a membership representative of over fifty governments, made the protection of Palau its overriding priority.

On December 14, 1976, EDF and NRDC notified Nisshō Iwai that the secrecy and exclusivity provisions in the contract were illegal and threatened litigation. On December 20, 1976, NRDC telegramed MITI Minister Toshio Komoto, urging reconsideration of MITI's financial support of the Nisshō Iwai–Industrial Bank of Japan feasibility study. Thereafter, on January 10, Nisshō Iwai acceded to the EDF-NRDC request by striking the secrecy provisions.

After these events, concern over the superport increased in the United States and abroad. In January, 1977, Senator Daniel K. Inouye, chairman of the Senate Select Intelligence Committee, conducted hearings on reported CIA bugging of the Micronesia status negotiations; and Senator Henry M. Jackson announced that he would conduct hearings within the Senate Interior and Insular Affairs Committee specifically on Palau. In Europe, the International Union for the Conservation of Nature stated that it intended to investigate the environmental ramifications of the project, and, if necessary, to have the endangered sites set aside as an international marine sanctuary.

The Diet Petition

While these events were under way, fourteen environmental groups from the United States, Europe, and Palau decided to petition the Diet. The principal purpose of the Palau petition was to forestall the project through the political process at an early stage. Too often, it was thought, environmental lawyers wait until a case is "judicially" ripe. But then the courts frequently deem it unfair to enjoin a project after its promoters, relying on government encouragement, have made a substantial financial investment. An important objective of the petition was to establish the risks of the project at the outset. This record might then be used at a later time by Palauans in a damage suit against the Japanese and other promoters. The petition's detailed allegations were designed with the elements of "foreseeability," "duty of care," and other specifics of Civil Code litigation in mind.[23] The outcome of such a hypothetical suit is, of course, not easily predicted. The petitioners reasoned that if the future plaintiffs could demonstrate injury and overcome procedural hurdles under the Civil Code, it would be unlikely that a Japanese court would dismiss the suit simply because of its transnational nature. In the following pages we summarize the arguments advanced in the petition.

*"The Construction of Port Pacific Will Endanger Human
Health, Devastate Palau's Environment, and Degrade
International Marine Areas"*

The construction of the superport presented distinct dangers—to the health of
the Palauan community, to the local environment, and to international marine
areas. The most dramatic hazard to human health, the petition alleged, would be
the explosions of tankers and onshore refineries. The Palauan community was
completely unprepared for accidents like the S.S. *San Sinea* explosion that had taken
eight lives and injured fifty persons just a few months earlier (January 11, 1977).

A second serious hazard to human health would be heavy metal and other discharges. The promoters' own reports had noted that various heavy metals, synthetic organic compounds, chlorine, and other chemicals would be discharged directly into Palauan waters. Since the Palauans depended for their food supply on
fish and other marine life, Japan's own experience with Minamata and itai-itai
diseases indicated that the discharge of these toxic substances would risk
poisoning large numbers of people.

The petitioners noted that the environmental impact of the superport might be
equally profound. Many scholarly and scientific articles had indicated that Palau
was a virtual wonderland of exotic, and in some cases endangered, animal and
plant species. An article appearing in the Audubon Magazine[24] some months before had noted that Palaun waters teemed with life—dugongs, crocodiles, green
turtles, hawksbill turtles, sea snakes, goat fish, parrot fish, porcupine fish, but-
terfly fish, rabbit fish, squirrel fish, unicorn fish, trumpet fish, surgeon fish, rud-
der fish, sail fish, snappers, dolphins, wrasses, grouper, shark, rays, barracuda,
moray eels, sea cucumbers, jellyfish, star fish, marine snails, octopi, feather-
worms, clams, and coral of all descriptions. Since the superport's proponents intended to construct the complex directly on and contiguous to the coral reefs, the
petitioners argued, it would be necessary to dredge, scrape, blast, reinforce, and to
landfill the reefs. Construction itself would therefore immediately destroy
thousands of marine organisms and the reefs themselves. Construction would
also produce siltation that could damage or kill contiguous reef areas. The re-
configuration of the natural shoreline would change salinity and temperature; it
might also drastically alter animal habitats in various small "sea lakes." And the
parts of Palau's marine environment that were not thus destroyed, could well
be claimed by erosion, runoff, sedimentation from onshore construction, or by the
likely catastrophe of a major oil spill.[25]

Although the most immediate adverse impact of the project would be to the
local Palauan environment, the superport also posed environmental risks of international concern. First, many plants and animals in Palau were recognized to
be of international importance either because they were unique or because they
were endangered. Some of these animals had already been officially listed on the
schedules to the Endangered Species Convention. Indeed, many scientists were
urging that the entire marine ecosystem of Palau be designated a marine sanctuary
under the World Heritage Convention.

Another problem was the impact of the development on extraterritorial marine
areas. Many tankers transiting the Malacca straits would be redirected to the
Lombok-Palau route, and the petitioners were concerned that the redirected
traffic might adversely affect (because of the likelihood of oil spills, bilge dumping, etc.) these biologically productive areas. Although the petitioners did not
cite evidence in support of this contention, they argued that the international

community at least had an interest in assessing such impacts before a decision was made on the Palau site.

> *"Since the Environmental Dangers of Port Pacific Would Be*
> *Unacceptable in Japan, Port Pacific Should Not Be*
> *Sanctioned in Palau"*

Petitioners next argued that since the project could not pass Japan's own stringent environmental controls, it should not be foisted onto the Palauans. In Japan, at the central level, the brief argued, the superindustrial complex would have to comply with the National Land Use Law[26] and the Industrial Plant Siting Law.[27] The discharge of waste from the facilities would be strictly regulated by the Water Pollution Control Law,[28] and air emissions would be controlled by the Air Pollution Control Law.[29] Tanker traffic to and from the facility would come within the control of the Marine Pollution Prevention Law,[30] that forbids the discharge of oil or waste of any kind, and noise and evil smells would be controlled by other laws.

In Japan, at the local level, the konbināto would have to conform to the stipulations of various ordinances, and to the provisions of various land use plans. No doubt the promoters would also have to conclude a pollution control agreement with local authorities and, possibly, with local citizens' groups. The pollution control agreement would make the promoters responsible for all damages and impose even more stringent standards than those prevailing under central or other local-level regulations. The project would also be tightly supervised through various recommendations, suggestions, advice, and other forms of control issued under administrative guidance.

The petitioners also pointed out that recent Japanese court decisions had recognized a developer's legal obligation to initiate wise environmental planning at the earliest stages and had also mandated full and complete public disclosure.[31] The brief argued that the Nisshō Iwai Company, the Industrial Bank of Japan, and the project's other promoters had not discussed their plans in a spirit of openness and full disclosure. Rather, they had sought at every point to confuse the Palauans and to hide their activities through technical contractual provisions, reminiscent of the notorious mimaikin contract negotiated by Chisso in Minamata.

The project, the petitioners urged, should very much concern the Diet. First, the project would ultimately demand governmental subsidies, which the appropriate budgetary committees should scrutinize. Second, and more important, the Diet was at that time already considering legislation requiring environmental impact assessment procedures, public disclosure, and public participation for major development projects. In view of the national importance of the debate on environmental assessment, the Diet itself had the right to be informed about what the Port Pacific promoters intended, the exact amount the government had already appropriated and committed in financial guarantees, and the availability of less environmentally fragile sites for the superport.

> *"The Construction of Port Pacific in Palau Will Violate the*
> *Stockholm Declaration on the Human Environment and*
> *Various Treaty Obligations Which Japan Has Assumed to*
> *Protect the International Environment"*

361

The petitioners contended that various international mandates gave the Diet not only the right to investigate the Palau project, but also imposed a responsibility of doing so. The first claim was that the principles developed at the Stockholm Conference, subsequently incorporated and adopted by the U.N. General Assembly (with Japan's concurrence), had imposed on states the "highest duty of collaboration in a spirit of openness and full disclosure toward the goal of protection of the environment."[32] Specifically, petitioners cited principles 2, 4, 7, and 21 of the Stockholm Declaration.

Principle 2 of the declaration calls for the safeguarding of the "natural resources of the earth, including the air, water, land, flora, and fauna and especially representative samples of natural ecosystems [which] must be safeguarded for the benefit of present and future generations through careful planning or management."

Principle 4 notes that "man has a special responsibility to safeguard and wisely manage the heritage of wildlife and its habitat which are now gravely imperiled by a combination of adverse factors."

Principle 7 mandates that "states shall take all possible steps to prevent pollution of the seas by substances that are liable to create hazards to human health, to harm living resources and marine life."

Finally, principle 21 holds that "states have . . . the responsibility to ensure that activities within their jurisdiction or control do not cause damage to the environment of other states or of areas beyond the limits of national jurisdiction." The petitioners argued that principle 21 also imposed a correlative responsibility on states to supply the international community with information on activities under their jurisdiction or control that would entail significant adverse risks to the environment.

Besides pointing out general obligations under the Stockholm Declaration, the petitioners urged Japan to respect obligations to protect the environment assumed under bilateral treaties. For example, under the Migratory Bird Treaty, Japan and the United States had undertaken "to cooperate in taking measures for the management, protection, and prevention of the extinction of certain birds." Specifically, art. 6 of the treaty states that:

Each contracting party shall endeavor to take appropriate measures to preserve and to enhance the environment of birds protected. . . . In particular, it shall:
(a) seek means to prevent damage to such birds and their environment including especially damages resulting from pollution of the sea . . .

The petitioners contended that this imposed on both countries not only the broad obligation of refraining from engaging in any activity that would endanger the migratory birds listed under it, but also the narrower obligation to control those acts of their nationals that would destroy the birds' habitats. The petitioners noted that many species of migratory birds visited Palau, and that it was certain that oil and other discharges of the supertankers and the industrial complex would destroy the habitats of these birds. The petitioners urged the Diet to investigate the conduct of Nisshō Iwai Corporation and the other promoters that would place Japan in violation of its obligations under this treaty.

The petitioners' final argument was that Japan had repeatedly recognized the international importance of the protection of the marine environment and the preservation of unique and endangered animals and plants. For example, the Convention of the High Seas, the International Convention for Prevention of Pollution of the Sea by Oil, and the amendments to the International Convention for the Prevention of Pollution of the Sea by Oil, all ratified by Japan, imposed at least general obligations on states of controlling oil pollution on the high seas. In addition, Japan had signed and adopted, although not yet ratified, various conventions which specifically covered many of the animals and plants that would be endangered by the Palau Project. These included the Convention for the Prevention of Marine Pollution from Land Based Sources, the Convention Concerning the Protection of the World's Cultural and Natural Heritage, and the Convention on International Trade in Endangered Species of Wild Flora and Fauna. The petitioners urged that by signing and adopting these conventions, Japan had at least created the expectation in the world community that neither she nor her nationals would engage in any activity that might seriously endanger the animals and plants listed therein.

> "Japan is Responsible to the World Community for the
> Social, Economic, and Political Consequences of Port
> Pacific in Palau"

In addition to its serious environmental repercussions, Port Pacific also promised disastrous social, economic, and political consequences in Palau. The petitioners asked the Diet to consider the possibility of such events and, if necessary, to prevent them.

One such concern was the socioeconomic impact of the superport. Studies of comparable projects in other countries[33] had all indicated that the construction of a deep water terminal was never an isolated event, but rather necessitated the construction of new refineries and other industries. The complex would demand acres of land, housing, roads, schools, and other infrastructure to support an increasing population of workers. Industrial development in one area would stimulate polluting developments in contiguous areas.

Although the project undeniably would promote Palau's economic development, there was no assurance that the Palauans would share in the profits, or that a different project might not foster Palau's economic growth at lower socioeconomic cost. One option was tourism. Although at the time only 5,000 tourists visited Palau each year, projections for 1981 had estimated an increase of about 200%. Although still in a fledgling state, tourism represented an important future souce of revenue. A major oil spill might ruin this industry overnight.

A second prospect was the offshore tuna and local mariculture industry. Although considered to have great potential, this industry was extremely fragile. An oil spill, explosion, or even daily discharges of waste within the reefs, would destroy plots under cultivation and the bait fish upon which the offshore fishing industry depended.

Expanding offshore fishing, mariculture, and tourism would at least assure the Palauans of some control over their destiny and bring them some return on their investment; the superport project would not. The petitioners feared that since few Palauans possessed the requisite managerial skills for executive posts, the best jobs would go to foreigners. Moreover, studies of other superports had indicated that these developments greatly burdened local tax systems, which had to finance public investments in schools, roads, and sewage treatment facilities. New industries such as the proposed superport are often granted tax exemptions, and the Palauans feared that they might well be forced to shoulder a principal part of the resulting tax burden.

The promoter's own studies estimated that 15,000 foreign workers would be brought in to work on the superport complex. This event would cause the

Palauans overnight to become a minority in their own islands, with catastrophic social consequences. For example, there was the problem of evictions. Construction of the harbor would require that an estimated 1,500 Palauans would have to leave their homes. The U.S. military's interest in leasing 30,000 acres on Babelthaup, the chief island of the district, would intensify the pressures on the Palauans to relocate. Since the Palauan kinship system was based on land holdings and land rights, the expected relocations would seriously undermine the basic foundations of Palauan society.

Equally serious were the political ramifications of the project. In November 1974 the Palauans had expressed their desire, by resolution of the District Legislature and in other ways, to pursue status negotiations with the United States separate from the rest of Micronesia. The Palauans were thus already questioning whether they would elect to become a U.S. commonwealth or seek independence or some other form of self-determination.

The establishment of the superindustrial complex would have adversely affected such Palauan efforts toward self-determination. The petitioners were concerned that Palau would become a one-industry town inhabited by outsiders and controlled by the U.S. Navy and Japanese industrial interests. Under those conditions it would be impossible for the Palauans to exercise any control over their political destiny.

A final concern was the "boom town" effect. The superport's construction would last from three to five years, during which time the islands would be transformed to service the foreign construction workers. Yet when construction was completed, the demand for foreign labor would dissipate and the workers would leave.

The petitioners argued that this project should concern the Diet, and indeed the entire Japanese people. First, the development was being built for the convenience of Japan, and Japan would be its principal beneficiary. Second, Nisshō Iwai, the Industrial Bank of Japan, and others among the project's promoters were Japanese nationals. They would be in charge of the physical construction of the complex; they would most immediately profit. Third, the Japanese government, through MITI, had actively promoted the project and had promised financial assistance toward its implementation. The project was being, and would be, assisted with national tax funds. Finally, Japan, as a member of the United Nations and as a former member of the Security Council, bore a fiduciary duty to the people of Palau and to the world community not to permit Palau to be injured in the manner described.

In support of these arguments the petitioners cited art. 76b of the U.N. Charter, that made the basic purpose of strategic trusteeship the promotion of "the political, economic, social, and educational advancement of the inhabitants of the trust territories, and their progressive development toward self-government or independence." The petitioners also cited art. 6 of the Trusteeship Agreement; this imposed specific obligations upon the United States, as administrator of the trusteeship:

In discharging its obligations under art. 76(b) of the charter, the administering authority shall:

1. foster the development of such political institutions as are suited to the Trust Territory and shall promote the development of the inhabitants of the Trust Territory toward self-government or independence as may be appropriate to the particular circumstances of the Trust Territory and its peoples and the freely expressed wishes of the peoples concerned; and to this end shall give to the inhabitants of the Trust Territory a progressively increasing share in the administrative services in the Territory; shall develop their participation in government; and give due recognition to the customs of the inhabitants in providing a system of law for the Territory; and shall take other appropriate measures toward these ends;

2. promote the economic advancement and self-sufficiency of the inhabitants and to this end shall regulate the use of natural resources; encourage the development of fisheries, agriculture, and industries; protect the inhabitants against the loss of their lands and resources; and improve the means of transportation and communication;

3. promote the social advancement of the inhabitants and to this end shall protect the rights and fundamental freedoms of all elements of the population without discrimination; protect the health of the inhabitants. . .

The sad fact, however, was that the United States, and the territorial government (which is under the jurisdiction of the U.S. Department of the Interior) had repeatedly ignored these obligations.[34] The petition summarized various instances of dereliction, focusing on the Trust Territory government's approval of the feasibility study. The brief alleged that the Trust Territory government was well aware that such approval:

would create expectancy and reliance interests in the promoters, build up political pressure in favor of the superport, and also engender fears and expectations among the people of Palau. . . . Prior to negotiating and signing the feasibility study contract, the Trust Territory authorities should have made full and complete disclosure to all the people of Palau of the terms of the contract, the political significance of its signing, and the effect a finding of "commercial feasibility" would have on ultimate approval of the project. Because of the potential conflict between the interests of the Japanese companies, the U.S. government and the people of Palau, special steps should have been taken at the time of signing the contract to safeguard the interests of the Palauan people. Much more than mere "commercial feasibility" was at stake. Therefore, the Trust Territory authorities should have called at the time of signing the contract for an independent study by a neutral body of experts recognized to have no vested interest in the matter, as was later recognized by the Trusteeship Council. Such a study would include in addition to commercial feasibility, a thorough assessment of the environmental, social, economic, and political impact of Port Pacific on Palau and the implication of this project for the unity of Micronesia.

From the perspective of international law, the issue presented by the Palau superport was whether the members of the United Nations generally, and Japan in particular, were absolved from their obligation to the Palauans simply because the United States, the designated administrator of the area, had failed to meet its own responsibilities. The petitioners argued that Japan was not relieved of its separate obligation to protect the Palauans; but rather, that, along with other members of the United Nations, Japan had the responsibility of calling the territorial administrators to an accounting.

What was the minimum responsbility of a nation in Japan's position? Measured by any legal definition, whether by Japanese, U.S., or international law, the duty of a fiduciary was to be informed. The Diet as the representative of the people of Japan thus had the right as well as the responsibility of ascertaining for itself how this development might injure the Palauans, and to report its assessment of the risks and benefits of this development to the world community.

There was a final reason why Japan should concern itself with the fate of Palau.

Irrespective of whether the complex was built or not, Nisshō Iwai and the other promoters must have known of the special international status of Palau, of the potentially overwhelming impact of the superport, and the need for public disclosure. Yet they had continually refused to disclose information necessary to Palau's independent assessment of the environmental, social, and political risks of the development; they had frustrated independent investigators opposed to the project.

The conduct of Nisshō Iwai, the Industrial Bank of Japan, and the other promoters, the petitioners charged, violated obligations that Japan had recently recognized to insure the good conduct of its enterprises abroad. In the wake of years of flagrant misbehavior by multinational enterprises, the international community through the OECD had decided to establish a code of conduct for multinational enterprise. To this end, Japan had voted on June 21, 1976, as a member of OECD, in favor of a Declaration on International Investment and Multinational Enterprises. Indeed MITI had just promulgated guidelines implementing the declaration.

In essence, the declaration instructed enterprises when dealing with poorer areas of the world:

(7) not (to) render—and they should not be solicited or expected to render—any bribe or other improper benefit, direct or indirect, to any public servant or holder of public office . . .

(9) to abstain from any improper involvement in local political activities.

Multinational corporations were also required to conduct their operations openly and to foster public understanding.

Item 1 of MITI's promulgated guidelines stated that "investing companies should try to develop cooperative and harmonious relations with most countries. . . ." Item 2 held that business must be based on "mutual trust." And item 5 of sec. 4 prescribed that "Japanese investors when investing overseas should strive to avoid criticisms by the local inhabitants of exporting pollution by doing the best within their ability to cooperate in the work of conserving the natural environment." The petitioners requested that all the Japanese promoters of the superport respect the international mandate that Japan had recently recognized.

Relief Requested

Based on the above arguments, the petitioners requested the following:

1. The Pollution Countermeasures Committee, the Budget Committee, and the Foreign Relations Committee (and any other appropriate committees) in the House of Councillors and the House of Representatives independently conduct a question and answer session on the Palau superport.

2. That the continued promotion of the project by Nisshō Iwai, the Industrial Bank of Japan, and others be the subject of oversight hearings by the Diet.

3. That petitioners be permitted to present evidence in their possession to the Diet relevant to the Palau superport.

4. That representatives of Nisshō Iwai, the Industrial Bank of Japan, and other promoters of the project be called to testify.

5. That an appropriate committee of the House of Councillors and the House of Representatives establish a team to visit Palau and to investigate for the Diet the superport proposal.

6. That the Diet pass a resolution for a budgetary appropriation to finance an assessment of the environmental, social, and political impacts of the superport in Palau, to be conducted by an independent team of experts.

7. That if in the judgment of the Diet it is considered useful and appropriate, parallel hearings[35] be conducted on the Palau superindustrial complex by appropriate committees in the Diet and in the U.S. Congress.

8. That, where useful, the Diet seek the assistance of the Trusteeship Council, the Environmental Programme (UNEP), the Social and Economic Affairs Council, the Organization for Economic Cooperation and Development, and other concerned organizations of the United Nations.

Procedural Problems Raised by the Diet Petition

As noted, the Diet Petition Office initially took the position that only Japanese citizens and resident aliens possessed petition rights.[36] (There was only one precedent, wherein a resident of Taiwan had petitioned the Diet; and that petition had been rejected.[37]) The response of the petitioners to this rested on three grounds.

First, the petitioners argued that, in the absence of a clear and explicit intent to discriminate based on alienage, the Japanese Constitution and laws should be interpreted broadly.

The provision at issue was art. 16 of the constitution, which states:

Every person[38] *shall have the right of peaceful petition for the redress of damage, for the removal of public officials, for the enactment, repeal, or amendment of laws, ordinances, or regulations and for other matters, nor shall any person be in any way discriminated against for sponsoring such a petition.*

The question was whether "every person" included nonresident aliens.

The Petition Law also contained several pertinent provisions. For example, art. 79 (Presentation of Petitions) states:

Those who would file a petition in the Houses of Parliament must present their petition by the introduction of a member.

Article 80 (Disposal of Petitions) states:

1. Petitions will be disposed of after consideration by a committee [in either House] of Parliament.

2. Those petitions decided by the committee as unnecessary to refer for deliberation by the House shall not be referred. However, if more than 20 representatives so request, the petition shall be deliberated upon.

Article 81 (Referral to the Cabinet) states:

1. If it is regarded as appropriate for the Cabinet to handle a petition accepted by the Diet, such petition shall be referred to the Cabinet.

2. The Cabinet must report every year to the Diet on the process of disposal of petitions in the above paragraph.

Article 82 (Petitions and the Independence of Each House of Parliament) states:
Neither House shall interfere with the separate receipt of petitions.

Finally, art. 5 states:
Petitions subject to this law shall be received and disposed of sincerely.

The petitioners argued that the Diet Petition Office's interpretation contradicted these provisions. Moreover, to deny petitioners on the pretext of residence, they maintained, would violate the mandate for sincerity (seijitsu ni) expressed in art. 5 of the Petition Law.

The petitioners' second argument was that art. IV.1 of the 1953 Treaty of Friendship, Commerce, and Navigation between Japan and the United States required the Diet Petition Office to accord them nationality treatment. Article IV.1 states:

Nationals and companies of either party shall be accorded national treatment and most-favored-nation treatment with respect to access to the courts of justice and to administrative tribunals and agencies within the territories of the other party, in all degrees of jurisdiction, both in pursuit and in defense of their rights. It is understood that companies of either party not engaged in activities within the territories of the other party shall enjoy such access therein without registration or similar requirements.

The petitioners' arguments raised several questions. First, the treaty does not explicitly mention the Diet. Is the Diet Petition Office an "administrative tribunal or an agency within the territory of the other party"? Even if it is not, does the fact that reference to the Diet is omitted express the intent of the parties to exclude it? The basic problem again was one of interpretation.

Second, what "rights" were involved in this case? Moreover, what law protected or defined these rights? If questioned on this issue (the question was never raised), the petitioners would have responded that two sets of "rights" (or at least "interests") were at issue: (1) interests covered by the various conventions cited (i.e., the Migratory Bird Treaty and other international agreements), and (2) the petitioners' environmental rights and rights to human dignity as interpreted by recent decisions of the Japanese courts.[39]

The petitioners' final argument was that, since nonresident aliens are permitted to petition the U.S. Congress, the Diet Petition Office should reciprocally recognize the petition rights of nonresident aliens in Japan. Although the Petition Office did not challenge this contention, the petitioners' argument would have been that an alien present within the United States has a legal right to frame and circulate a petition to the president of the United States and to comment freely on governmental processes and actions.[40]

It is difficult to assess the significance of the Diet's change in position, especially given the uncertainty of the petitioners' legal arguments. On the one hand, the alteration of a policy of twenty-eight years may merely have reflected the Petition Office's assessment that an antiquated rule required modification. Undoubtedly, the speed of the Diet's volte face suggests the political sensitivity of the issues raised, and perhaps the LDP's desire of keeping Palau from becoming a cause célèbre.

1.2.2   Case 2: Japan's Position on Marine Pollution Control at the Third U.N. Law of the Sea Conference[41]

Japan's position on marine pollution control was first submitted to the preliminary U.N. Seabed Committee meeting in Geneva on August 13, 1973.[41a] Since 1973, Japan has adhered virtually without modification to its original position.

The text of the Japanese position follows.

1. A coastal State party to this convention may investigate and prosecute natural or juridical persons under the jurisdiction of other contracting States when such persons have discharged or dumped any harmful substances in contravention of generally accepted international rules and standards, provided that:
(a) There is sufficient evidence required by the law of the coastal State enacted in conformity with generally accepted international rules and standards, and
(b) The discharge or dumping has occurred in areas adjacent to its territorial sea, the maximum limit of which areas shall be _____ nautical miles from its coast.
2. In taking actions referred to in par. 1, the coastal State shall ensure that the maritime activities of the natural or juridical persons referred to in the same paragraph are not unduly interfered with.

3. The coastal State shall inform the other contracting States referred to in par. 1, as well as competent international organizations, of the results of such investigation and prosecution.

The concept of a zonal approach to marine pollution control originated in the infant Environment Agency in 1972.[42] The Environment Agency in its first flush of creation argued strongly within the government that it was no longer possible or desirable, in light of the erosion of the traditional freedom of the high seas, for Japan to maintain a conservative position on marine pollution issues.

For this reason, the Environment Agency proposed a 50-mile pollution control zone. The concept of a pollution zone was based on the example of other states, particularly Canada. The 50-mile limit was based upon a prohibition in the 1954 International Convention for the Prevention of Pollution of the Sea by Oil[43] (amended in 1962, 1969, and 1971), which forbade the discharge of oil and certain other substances except in areas 50 nautical miles beyond the coastline of a state from which its territorial sea is measured. The 50-mile rule had also been incorporated in art. 4(5)(iv) of Japan's Marine Pollution Prevention Law,[44] a law which Buichi Ōishi referred to as "epoch-making" in his address to the Stockholm Conference.[45]

But the ambitious Environment Agency proposal met with strong opposition within the government, particularly from the Ministry of Transportation.[46] In principle the ministry opposed the pollution zone proposal, because it was concerned that other coastal states, particularly developing countries, might abuse their newly acquired powers to interrupt tanker traffic bound for Japan. Specifically, the ministry believed that the coastal states would manipulate their right to police pollution within the zone to set design and construction standards for tankers traversing these waters. The ministry's fear centered around the narrow, accident-prone waters of the Malacca straights. The final version of the Japanese proposal presented at Geneva reflected a compromise between the two agencies, wherein the breadth of the pollution zone, in deference to the wishes of the Ministry of Transportation, was left blank, and the proposal phrased as ambiguously as possible.

By the time of the first substantive session of the Seabed Conference, held at Caracas in 1974, the positions of many nations on marine pollution were becoming clearer. The positions fell into three general categories.[47] Those in the first group, exemplified by traditional maritime powers such as Great Britain, relied on the flag-state principle: jurisdiction over a vessel polluting the sea would rest with the state whose flag it bore. The Soviet Union supported this conservative position. The second category, represented by the position of the United States, left jurisdiction to the state whose port the ship had voluntarily entered. The third category recognized a coastal state's antipollution zone adjacent to the territorial sea. Based on factors such as the extensiveness of the zone and the exclusivity of control, this category was further divided into "radicals," such as Australia, Canada, and most of the developing coastal states, and "moderates," such as France and Japan. Australia ultimately identified its antipollution zone with its claimed 200-mile economic zone. The jurisdictional limits of Japan's position at the opening of the Caracas session, however, were unclear.

The 50-mile rule was defined during the Caracas meeting by Shigeru Oda, principal advisor to the Japanese delegation, as "reasonable breadth."[48] However, the idea of a 50-mile zone was not presented as a formal supplement to the original proposal; rather, it was offered in the form of a "suggestion." In elaborating on the virtues of the 50-mile proposal, Oda later explained that "to claim jurisdiction over a wide zone in which it would be practically impossible to exercise control would not solve the problem." While a number of delegates listened to Oda's explanation without question, others, such as the British delegate, were manifestly concerned. Noting that the United Kingdom had found it extremely difficult to apply existing conventions and agreements even within three miles, the delegate stated diplomatically that he "was, therefore, particularly interested in the comments of the Japanese delegation on the costs and difficulties involved in controlling an area which might extend as far as 50 miles."[49] Not repudiated, yet not officially embraced, Oda's 50-mile suggestion still rests in a state of limbo as the official Japanese position on marine pollution control.

Japan's zonal pollution control approach is somewhat unusual in other respects. Most striking is the vagueness of the language, and the considerable omissions in the proposal. First, it is unclear what measures a coastal state can adopt to control pollution within its jurisdiction. Uncertain is whether the coastal state is entitled merely to establish standards within the zone, or whether it can also enforce these standards. The provision only stipulates that proceedings may be initiated against a foreign polluting vessel when such action is "in contravention of generally accepted international rules and standards." Second, the proposal is silent on the structural standards and operational procedures that foreign vessels must meet, for there is no reference to the application of the IMCO "segregated ballast" and "load-on-top" rules. Third, the original Japanese proposal did not specify the breadth of the zone. Finally, the proposal focused narrowly on oil pollution from vessels. Unlike presentations by other nations, Japan's position omitted any mention of additional controls on marine pollution.

How may we explain the peculiarities of Japan's position? To a large extent it reflects Japan's vulnerability in the Law of the Sea negotiations generally, for the sea is quite literally Japan's lifeline. Japan relies predominantly on sea transport for her import of millions of tons of raw materials each year, roughly 250 million tons of which is crude oil; and for her export of roughly 6,000 million tons of finished goods. Japan's 38 million gross ton merchant marine exceeds, in real terms, that of Liberia's 66 million tons, because of the latter's inclusion of various flags of convenience.[50] Some 200 tankers—ninety of over 200,000 tons —link Japan with the Persian Gulf and other oil-producing areas. Japan's sensitivity to any impediment on the free flow of maritime trade is thus understandable.

These were the dominant concerns for the drafters of the 1973 position. Of course, the Japanese delegates also recognized that some controls were needed on the increasing number of foreign vessels plying Japanese waters. But this problem was considered to be of secondary importance. Indeed, if coastal regulations were made too strict, some members of the Japanese delegation feared that other countries might retaliate against Japan's distant fisheries. The primary objective of the drafters was to design a document ambiguous enough to preserve all options.

Ministerial rivalries also played an important part in the formulation of the Japanese position. To date, policy has almost overwhelmingly been determined by the Ministry of Agriculture, Forestry, and Fisheries and the Ministry of Transportation and their advisory bodies. The Foreign Ministry has served primarily as a conduit for maritime commercial interests, and its views have generally been subordinated to those of the other two agencies. For example, when the Ministry of Agriculture, Forestry, and Fisheries at last decided that Japan must accede to the 200-mile fisheries zone, the Foreign Ministry disagreed, urging that a final decision on the issue be deferred. The Ministry of Agriculture, Forestry, and Fisheries prevailed.

In matters of marine pollution control, the warring loyalties of the various ministries together militate against an affirmative policy. Neither the Ministry of Agriculture, Forestry, and Fisheries nor the Ministry of Transportation perceives a unified stance in this area to be in the national interest. The former has been more concerned with the impact of foreign pollution control limitations on Japan's distant fishing operations; the latter is concerned about the obstruction of Japan's merchant trade. And the Foreign Ministry views a strong stand on pollution control as an irritant to Japan's sensitive relations with developing countries, many of whom feel that the strong environmentalist positions urged by some developed countries constitute a neocolonialist subterfuge designed to frustrate their own efforts toward industrialization. Given the strength of these perceptions, and the increasingly peripheral role of the Environment Agency in the Japanese delegation, Japan's undistinguished position on the protection of the marine environment becomes more understandable.[51]

1.2.3   Case 3: Japan's Position on Whaling

The Development of Japanese Regulation of Whaling

Although Japanese literature mentions coastal whaling as early as 230 B.C., the industry remained unorganized until around 1600.[52] Thereafter, fishermen began spearing whales on a large scale in the coastal waters of present-day Mie and Wakayama prefectures in Southern Honshū, and around 1670 net whaling was introduced in Imura Bay, Kyūshū.

Japan's modern era of whaling began around 1899 with the formation in Yamaguchi Prefecture of the Pelagic Fishing Company (Nippon Enyō Gyogyō Kabushiki Kaisha), and during the first years of the twentieth century the industry grew rapidly. By 1907 there were twelve companies operating from twenty-two land stations in Japan and Korea.[53]

The destruction of coastal whale stocks, however, prompted the Japanese government to regulate the industry more closely. In 1909 the companies were organized by law into an association; catcher boats were required to obtain a license from the Ministry of Agriculture, Commerce, and Administration (Nōshōmushō); the number of catcher licenses was limited to thirty.[54] Additional legislation in 1934 and 1938 reduced the number of catcher licenses to twenty-five and imposed minimum-length limits for various whale species.

The industry continued to grow within the terms of these early restrictions, and by 1945 three companies were in control: Japan Marine Producers, Ltd., Ocean Fisheries, Ltd., and Polar Whaling, Ltd. In 1946 the SCAP authorities directed that the old law regulating the coastal industry be brought within the framework of existing international whaling conventions.

The principal legal control on whaling since 1949 has been the Fisheries Law.[55] Jurisdiction over whaling is entrusted principally to the Ministry of Agriculture, Forestry, and Fisheries. Under art. 52, "persons desirous of operating any fishery by vessel" must "obtain a license for each vessel." By Cabinet Order, whaling has been designated a "fishery" requiring a license. Other ministerial regulations identify the whale species that may be killed and those that may not,[56] impose quotas on the number of whales to be taken,[57] specify hunting seasons and time periods,[58] hunting areas on the high seas and areas deemed off limits,[59] approved hunting methods and equipment,[60] and methods of disposing of whale carcasses.[61] The rules also require the industry to report each whale killed to the

Minister of Agriculture, Forestry, and Fisheries.[62] The law requires the minister to consider the conservation of whale stocks.[63]

But this mandate is undercut in many ways. Although art. 58 requires the minister to determine that granting a permit will not diminish the effectiveness of conservation programs, neither the law nor its implementing regulations specify the conservation principles or criteria to be applied. Similarly, the results of research on the state of stocks need not be incorporated into conservation criteria. And since the Fisheries Law grants jurisdiction over whaling to the Minister of Agriculture, Forestry, and Fisheries, the Environment Agency is effectively excluded from important policy decisions. Finally, although a number of provisions provide an opportunity for an aggrieved party to seek agency review of a negative ruling on a permit application,[64] members of the public wishing to challenge a permit approval (perhaps on the grounds that the permit would endanger a whale species) are not accorded a comparable dispensation. A third-party whaling company, however, would probably have standing to challenge an approval to a competitor.[65]

In 1976 the structure of Japan's whaling industry was modified when six major whaling firms, Hokuyō Hōgei, Kyokuyō, Nihon Hōgei, Nippon Suisan, Niito Hōgei, and Taiyō Gyogyō, consolidated their operations into the Japan Joint Whaling Company (Nihon Kyōdō Hōgei Kaisha), originally capitalized at $10 million. Kyokuyō, Nippon Suisan, and Taiyō Gyogyō each took an equal 32% interest in the new company, and the three smaller corporations held the remaining equity. The first president of the new company was the director general of the Japan Whaling Association, a former director of the Fisheries Agency. The Joint Whaling Company has since purchased three mother ships previously operated by the parent companies, and has reduced the total number of catcher vessels to twenty.[66] Kyokuyō, Taiyō Gyogyō, and Nippon Suisan continue to maintain close ties with major industrial enterprises in Japan, the United States, and Europe.

Conflicts and Cooperation Between the United States and Japan Over Whaling

In December 1970, Walter Hickel, the U.S. Secretary of the Interior, added the eight largest species of whales to the U.S. endangered species list, thereby banning the import of products made from any of these animals.[67] The secretary's action compelled the shutdown of the last American whaling station.

In 1972 the Stockholm Conference adopted (with the whaling nations abstaining) Resolution No. 33:

It is recommended that governments agree to strengthen the IWC, to increase international research efforts, and as a matter of urgency to call for an international agreement under the auspices of the IWC and involving all governments concerned for a ten-year moratorium on commercial whaling.

Japanese negotiators angrily dismissed the ten-year moratorium as scientifically unjustifiable, asserting that the resolution had been railroaded through the conference.[68] But the United States, now out of commercial whaling, and badgered by its own conservationists, enthusiastically applauded the moratorium, adopting it as official policy.

With the 1973 International Whaling Commission meeting, U.S.–Japan relations on whaling began to deteriorate. One observer noted that Japanese and Russian delegates were "visibly shaken" by the IWC position and that "throughout the meeting the Soviet and Japanese representatives reiterated the threat that unless the talk of a moratorium or significant quota reductions ceased, the International Observer Scheme might be stopped, and the IWC might be wrecked."[69] Although the U.S. proposal failed to obtain the necessary three-fourths majority, the commission voted to retain a conservative quota on minke whales, to divide the sperm whale quota in the southern hemisphere into regional quotas, and to phase out Antarctic fin whaling by 1976.[70] Within ninety days after the meeting, both Japan and Russia had formally objected to these regulations. In 1973 Japan and the U.S.S.R. reportedly exceeded the minke whale quota by about 50%.[71]

The international response was furious. In 1974 the National Oceanic and Atmospheric Administration (NOAA) certified to the U.S. president that both countries had violated the Pelly Amendment. The Pelly Amendment to the Fishermen's Protective Act of 1967[72] authorized the president to ban the import of fish products from any country that conducted fishing operations in a way that might diminish the effectiveness of international fisheries programs. In its letter of certification, NOAA emphasized that although the 1946 Whaling Convention permitted member countries to file objections (thereby escaping legal obligation), it was legitimate to apply the Pelly Amendment where the IWC quota represented a consensus of the Scientific Committee of the IWC and of other member nations.[73]

Outside government, many American and a few European whale conservationist groups organized a boycott of all Japanese and Soviet imports, irrespective of their connection with whaling. The boycott was to last until these countries publicly committed themselves to whale conservation. By the spring of 1974, reportedly twenty-one American conservation and environmental groups, totaling over five million members, had pledged to support the boycott.[74] Japan and the U.S.S.R. adopted more conciliatory positions at the 1974 IWC meeting, and the president decided not to impose restrictions under the Pelly Amendment. But the boycott continued.

So also did other congressional proposals seeking to impose economic and other sanctions against Japan and the U.S.S.R. For example, Congressman John D. Dingell, the former chairman of the House Subcommittee on Fisheries and Wildlife Conservation, proposed H.R. 80, which would have allowed the president to embargo the import of any or all products from countries violating the Pelly Amendment.[75] Another proposal, by Congressman Alfonso Bell (H. J. Res. 448), urged a trade embargo on all products produced or distributed by an enterprise engaged anywhere in commercial whaling.[76]

Both these proposals were opposed by the Commerce and State departments, and the bills were ultimately defeated. The two agencies argued that the bills might provoke whaling nations to retaliate; in any event, the bills raised serious questions under the General Agreement on Tariffs and Trade (GATT). Since 1975, the executive branch has continued to favor international negotiations on the whaling issue and has sought to discourage the use of the Pelly Amendment.[77]

United States–Japan relations over whaling became even more complicated after the United States began claiming a 200 nautical mile fisheries zone.[78] Although Japan sought to protect her substantial whaling interest in this area,[79] American environmental laws now effectively bar Japanese operations.[80]

The United States has also frustrated Japanese whaling on the high seas. For example, at the request of the United States and other conservationist countries, the IWC in 1977 slashed Japan's quota.[81] In 1978 Japanese whaling industry leaders again urged the government to withdraw from the IWC should that commission adopt a Panamanian proposal to revive the ten-year moratorium.[82]

Section 2   An Interpretation

Two overriding considerations explain Japan's uncertain commitment to international environmental protection. First, the citizens' movement, the bar, and the media, the groups that sparked domestic action in the 1960s and 1970s, have for the most part shown little concern over the actions of Japanese companies abroad; there appears to be little, if any, commitment to the protection of the environment beyond Japan's borders. Second, the structure and behavior of the Japanese bureaucracy have militated against strong positions on international environmental issues.

2.1   The Uncertain Constituency for International Action

Why have the Japanese environmental movements, the bar, and the media remained to a large extent indifferent to international environmental matters? Perhaps the simplest explanation is that at least until the early 1970s, the victims' and citizens' groups perforce concentrated on the need for reform at home. The victims, for example, were simply overwhelmed throughout most of the postwar period in struggling for the domestic support; there was little time to reach out to victims in other countries. Then too, the victims' movement was always a grass roots phenomenon, a congeries of fragmented, at times ideologically divided, groups. Although by now (1979) various citizens' organizations have established national federations, the citizens' movements have tended to focus on local problems. A shortage of funds and a paucity of members with expertise in international issues have also seriously circumscribed the contribution of these groups to international environmental protection.

The media's position on international environmental issues has been complicated by its perception of Japan as ill-understood, encircled by an uncompromising, even hostile world. Since the protection of whales or the Palau superport touch paramount national concerns—food, energy, trade, security—absent a clear national consensus, the media has been reluctant to adopt an unequivocal stance on international environmental protection.

Different priorities and perceptions have also frustrated effective collaboration between Japanese and Western (particularly American) environmentalists. As in all human relations, the forging of trust has been a barrier. Consider the cases of accidents in Japan involving American-manufactured and -supplied nuclear fuel and equipment.[83] Some years ago, American lawyers suggested that in view of the careless governmental attention given nuclear plant siting in Japan, and the inadequacy of relief at the time under Japanese law, an alternate strategy—a suit brought in the United States by Japanese victims against the American manufacturer—should be considered. The Americans noted that an important policy was at issue: American producers and suppliers of nuclear equipment should bear responsibility for any consequent harm suffered by others anywhere in the world.

Although the proposal seemed acceptable enough to American environmental groups, Japanese victims and their lawyers remained unconvinced. "Who are these foreign lawyers?" "Why should we trust them?" "What is their ideological position?" "How can it be reconciled with our own?" The proposition that a remedy might lie within the jurisdiction of a foreign court was to the victims of Oita, Osaka, or Kurashiki as remote as some far-off star.

Another important part of the problem has been a different sense of priorities. American environmentalists, for example, have been nonplussed by the Japanese animal protection groups' apparent insensitivity to the slaughter of whales. But many Japanese consider the foreigner's attachment to whales to be silly and sentimental, since they see no reason to protect whales yet slaughter cattle. Radical foreign actions like the boycott, various children's campaigns, and protests to the emperor are viewed in an even more sinister light—hysterical, racist acts, promoted by American industrial interests.

Different perceptions of the suitability of "international" action have also impeded cooperation. During the spring of 1977 various American environmental groups became increasingly concerned over the environmental repercussions of U.S. approval of Japan's first nuclear spent fuel reprocessing facility at Tokaimura, Ibaraki Prefecture. In their negotiations with the U.S. State Department, the American groups pointed out that nuclear waste reprocessing had been enjoined in the United States as unacceptably dangerous,[84] and therefore the United States government should not promote its development in Japan. The American environmentalists urged the State Department at a minimum to assess the risks of the Tokaimura facility to Japanese citizens and to the international environment.

When the State Department refused, the American environmentalists prepared for litigation and began soliciting the help of various groups in Japan. No difficulty was anticipated; various Japanese groups had already begun protests against the Tokaimura plant, and litigation to terminate operations at the complex was in progress.

To the surprise of the Americans, none of the local Japanese groups or their lawyers wished to participate in the American action. The requests for assistance had perhaps been too precipitous, for as one of the participating lawyers later explained, the groups were not given ample opportunity to evaluate the chances of the contemplated suit's success. But there was also a psychological preference at play—Tokaimura was a "'Japanese problem," and Japanese, not foreign, laws should control. Litigation against the facility more appropriately should be brought in Japanese courts.

Despite such impediments, there is now some hope of change. The Japan Consumers Union and other activist groups have begun to speak out on international issues. The Japan Federation of Bar Associations, through its Environmental Committee, is also, if incrementally, becoming more internationalist. In August 1976, the federation dispatched a team of forty lawyers to the United States to study environmental impact assessment procedures, the regulation of nuclear power plants, and the general role of public interest lawyers in the United States. On their return to Japan, the team established a new group with a small budget, to take responsibility for international environmental problems. Another group, principally of lawyers and law professors, the Committee on Human Environmental Problems, has forged working relationships with international organizations. Its most ambitious project to date involved working with the U.N. Economic and Social Commission for Asia and the Pacific (ESCAP) and the U.N. Environment Programme (UNEP) on the problems of formulating environmental legislation for rapidly industrializing nations in the Asia-Pacific area.

2.2   Japanese Bureaucracy and Politics

As a general rule, the structure of the Japanese bureaucracy discourages affirmative positions on extraterritorial environmental issues. As noted, jurisdiction over extraterritorial environmental issues is distributed among traditional mission-oriented agencies, none of whom have favored strong international action. For example, international trade and environment comes within MITI's competence; jurisdiction over whales and other marine resources falls under the Ministry of Agriculture, Forestry, and Fisheries; supertankers are the Ministry of Transportation's concern; jurisdiction over environmental problems in the U.N. Trust Territories is the Foreign Ministry's problem. Since none of these agencies perceives its mandate to include concern for the protection of the environment beyond Japan's borders, and, because their constituencies are similarly unconcerned, there is no basis for action. Indeed, affirmative action could subject an agency or official to stern criticism from other ministries, politicians, the business community, and affected interest groups.

To some extent, the bureaucratic career system also impedes innovation, especially in a new, controversial area such as international environmental law. Indeed, environmental activists within the government risk serious professional frustrations. If a senior official of the Environment Agency, for example, sought to promote extraterritorial environmental controls, it is unlikely that his office would even be permitted to submit a bill to the Diet. Since the Environment Agency is closely affiliated with the Prime Minister's Office,[85] the Cabinet would most likely withhold approval on the presentation of the bill to the Diet. This would kill the bill at the outset.[86] If a MITI official aggressively promoted international environmental protection, and if such a bill were presented to the Diet (a remote possibility), the ministry would risk offending its supporters. The environmental activist within government would become increasingly isolated and ineffective; he would be criticized; eventually his reputation would be tarnished. A poor work record and enemies among a ministry's constituents would also affect retirement, because many officials look to contacts within their ministry's constituency to obtain a comfortable livelihood after leaving public service.[87]

Lack of information and inexperience also contribute to the indifference of the Japanese bureaucracy to international environmental issues. One problem is that international issues often concern the destruction of wildlife and problems of resource management, areas that even domestically have received scant attention.

The government has also been concerned that its businessmen might be placed in an unfavorable international competitive position, should Japan require additional investment in international pollution control, when other countries do not insist on comparable investment. Because many Japanese policy makers still view the nation as economically fragile, unilateral attempts to protect the extraterritorial environment are generally deemed unwise. Moreover, the absence of any real authority to enforce international environmental mandates has permitted Japan to advance her own perceived national interests without fear of effective sanction. Finally, Japan has wished to avoid accusations of "environmental imperialism" from the less developed countries.

Although political, structural, and institutional considerations all contribute to Japan's conservatism on international environmental issues, we must remember that Japan's performance is not an exception to an otherwise affirmative "internationalist" foreign policy. Indeed, Japan has consistently maintained comparably cautious positions on other international issues such as security, human rights, immigration, refugees, and foreign aid.[88]

It is difficult to speculate on the origins of Japan's generally cautious approach to foreign policy. Her inherited tradition of isolation, suspicion of foreign motives, and a self-image of separateness undoubtedly have played a part. Perhaps more critical is the continuing influence of Japan's postwar drive for economic recovery. In the late 1940s and early 1950s the government's perception of the need for economic rehabilitation compelled extremely cautious negotiating positions on controversial international issues. To act otherwise would have risked offending the United States and other world powers, a result which Japan then could not afford. But during the late 1950s and 1960s, Japan realized that this policy of "not making waves" was also economically and politically beneficial, especially when Japan's actions seemed too small to cause adverse international repercussions. Since this policy has yielded ample dividends, the government is reluctant to relinquish it, even now in the late 1970s when Japan, almost to her surprise, finds herself a major power in a radically changed world.

Section 3   The Challenge to Japanese Policy Makers

3.1   Japan's Legal and Moral Duty to Protect the International Environment

Japan's legal duty to adopt stronger positions on international environmental issues is uncertain both under current theories of international law and under Japanese interpretations of international law. Perhaps the strongest claim to Japan's international legal obligation is the Migratory Bird Treaty, ratified by the Diet in September 1972. Both international legal theory and Japanese practice[89] recognize the legal obligation imposed by a ratified treaty.

The difficulty presented by the Migratory Bird Treaty is the ambiguity of the obligation. For example, in the Palau case, the petitioners asserted that the treaty would be violated since the construction and operation of Port Pacific would destroy the habitat of many migratory birds listed in the annex to the treaty.

A number of questions and responses are raised by the petitioners' argument. First, the Japanese government could take the position that the contracting parties had agreed only *"to endeavor* [emphasis added] to take appropriate measures to preserve and enhance the environment of migratory birds," not to assume a specific legal obligation actually to preserve and enhance these habitats in all cases, or to restrain all industrial activities that might in some way disrupt these habitats. This argument, however, seems unconvincing. "Endeavor" has been construed in other contexts to impose some obligation of making a bona fide effort; Japan's failure to conduct even an inquiry into alternative sites, or to initiate planning for the protection and enhancement of the environment of migratory birds endangered by the superport seems at least to violate the spirit of the provision.

An equally unsettled question is whether the destruction of habitat constitutes a "taking" prohibited by the treaty.[90] The Japanese government could argue, for example, that "taking" requires an affirmative, purposeful activity intended to bring a creature into the custody or possession of a hunter or fisherman. Incidental killings, the petitioners' chief concern, were simply too indirect. Of course, once the inevitable consequences of the superport were known, the government might be deemed to have intended that the activity continue, and have accepted responsibility for its consequences.

Perhaps the most difficult question presented is whether Japan can be held responsible for the acts of its nationals in Palau when the treaty explicitly states that it applied "for Japan, [and] to all areas under the administration of Japan." As noted, the Trusteeship Agreement clearly designates the United States, not Japan, as the "administrator" of the Trust Territories. The petitioners' claim would rest on two premises: (1) that the treaty's limitation did not apply to activities that were conducted jointly by both parties, in which case the obligations would be jointly borne; (2) that under the circumstances, the United Nations not the United States must be deemed the "administrator" since the United States had violated its obligations to the Palauans. Since the treaty explicitly covered the Trust Territories, Japan, as the U.N. member directly involved, must also be responsible along with the United States for the habitat of Palau's migratory birds.

The principles of the Stockholm Conference (later adopted by the United Nations) and various OECD guidelines cited in the Palau petition constitute a second basis for Japan's legal obligation to the international community. Various scholars have suggested that these pronouncements together constitute a new evolving body of customary international environmental law.[91] Japan's concurrence in virtually all these recommendations and guidelines, one might contend, has at least created expectations within the international community of her adopting a more affirmative position on international environmental issues. Some observers might also construe Japan's concurrence to be evidence of her acceptance of some international legal responsibility.

Although there is arguably a growing consensus on some principles—that states should not act unilaterally when their actions (or those of their nationals) threaten the environmental interests of other states—many states and students of international law may justifiably debate the precise legal effect of these principles. For example, are states now legally bound to give timely notice in advance to the other affected states, provide them with information, engage in consultations, or minimize any expected threat?

Japanese constitutional theory further complicates the question of international legal obligation. Article 98 of ch. 10 of the Japanese Constitution states:

The Constitution shall be the supreme law of the nation and no law, ordinance, imperial rescript or other act of government, or part thereof, contrary to the provisions thereof, shall have legal force or validity.

2. The treaties concluded by Japan and the established laws of nations shall be faithfully observed.

The provision is ambiguous in several respects. First, even if the principles adopted at Stockholm and the OECD recommendations are deemed "international custom," it is unclear whether these pronouncements would be held to be "the established laws of nations" (kokusai hōki) within the meaning of the constitution. Some scholars argue, for example, that art. 98.2 contemplates only codified international custom. It is unclear whether the Stockholm and OECD pronouncements are "codified" in this sense. Second, the constitution requires only that the established laws of nations be "faithfully observed." It does not, as other constitutions, make these laws the law of Japan. Third, it is also uncertain whether international custom is directly incorporated into domestic Japanese law without implementing legislation.[91a] Finally, the case against Japan's legal obligation is further strengthened by the Japanese government's position that assent to various principles or recommendations of international bodies such as the OECD may not be deemed acceptance of any legal responsibility.

Finally, one might argue, as did the Palau petitioners, that by signing and adopting various international conventions, Japan has "at least created the expectation that neither she nor her nationals would engage in any activity which will seriously endanger the animals and plants contemplated for protection by these Conventions." This idea finds broader support in art. 18 of the Vienna Convention on the Law of Treaties, which states:

A State is obliged to refrain from acts which would defeat the object and purpose of a treaty when:

(a) it has signed the treaty or has exchanged instruments constituting the treaty subject to ratification, acceptance or approval, until it shall have made its intention clear not to become a party to the treaty; or

(b) it has expressed its consent to be bound by the treaty, pending the entry into force of the treaty and provided that such entry into force is not unduly delayed.

Japan, however, has yet to ratify the Vienna Convention. Although one might contend that the Vienna Convention merely codifies existing custom that Japan in any event would be bound to obey, this argument falls into the same legally murky area described above.

Even if Japan may not be under a strictly legal duty to act more responsibly, may she not bear a moral obligation to do so? To many Westerners, the moral issues posed by international environmental issues are clear. All countries are morally bound to act with wisdom and restraint when they exploit resources that are common, resources that may be damaged irreversibly or never replaced. There is also an international moral ethic evolving with respect to developing countries.[92] No longer may a nation freely beggar its neighbor, especially when the exploiter is economically powerful, and the burdened nation poor and weak. If this ethic is now taking hold even in the international marketplace of free trade, how much more compelling is its logic when applied to the environment, where the consequences of ill-advised action are less easily predicted, and the costs of error far greater.

May we assume that many Japanese also perceive the moral issues in this way? Certainly the moral imperative of concern for community exists and there is a strong recognition of the need for close cooperation at the village level. And during the prewar period these institutions were harnessed to a more abstract dedication to the state. As noted in earlier chapters, Japanese environmental law has a distinctly moral underpinning, more strongly and clearly articulated than comparable Western law. Indeed the decisions in the four major pollution trials, the subsequent judicial development, the continuing debate over environmental rights, and various other proposals to impose a fiduciary obligation on the Japanese government all attest to deep concern for limitations on individual or government action that jeopardize Japanese community environmental interests.

The point is that this body of thought is not yet linked to a strong perception of Japan's belonging to a defined international community. In practice, many Japanese decision makers and commentators have tended to emphasize immediate concerns—Japan's poverty in natural resources, vulnerability to fluctuations in energy supplies, or the indifference of other countries—that serve to excuse Japan's inaction. Protection of the international environment is viewed more as a problem in public relations—of keeping up with other nations only to avoid embarrassment.

3.2   Unilateral, Bilateral, and Multilateral Action

Should the Japanese government seek to protect the international environment, many options are readily available. In this concluding section we explore the costs and benefits to Japan of unilateral, bilateral, and multilateral action.

3.2.1   Prospects for Unilateral Action

Let us return to the Palau superport case and ask: How might the Japanese government have proceeded differently? How might comparable developments be better handled in the future? A first step would have been for the Japanese government and the participating companies to have prepared an assessment of the superport's impact on Palau and on the environment outside Japan. Since the policies at stake in preparing international and foreign-country impact statements differ, these subjects must be discussed separately.

In the early 1970s when the issue of the extraterritorial application of the U.S. National Environmental Policy Act (NEPA) was first raised, the U.S. State Department and other U.S. foreign affairs agencies initially resisted the idea. Having to publicize negotiating positions or fallbacks, it was argued, would place the United States at a disadvantage; the public disclosure of another government's positions or initiatives would make foreign negotiators less willing to offer concessions.[93]

The State Department's fears to some extent have proved chimerical. A great body of information pertaining to foreign affairs is already in the public domain; the generation or publication of even more information on environmental issues, most of which is not sensitive, has not been detrimental. Moreover, most requirements of speed or secrecy in international negotiations have been addressed by the government's filing a programmatic impact statement at the outset.[94] And to the extent that any issue has deserved special treatment, NEPA recognizes that public disclosure of impact statements may be restricted by invoking the exemptions available under the Freedom of Information Act.[95] Many U.S. foreign affairs agencies today, where appropriate, file international environmental impact reports. The State Department, for example, has prepared assessments of the U.S. position for the U.N. Law of the Sea Conference and the Panama Canal treaty negotiations,[96] and other agencies have also promulgated regulations and guidelines regarding the dissemination of similar reports. Environmental impact statements are open for public review and comment.

Preparing assessments of environmental impacts in foreign countries, however, has raised more difficult problems because of the issue of sovereignty. The U.S. State Department, Nuclear Regulatory Commission, and some other federal agencies, for example, still take the position that preparing an assessment of environmental impacts in another country offends foreign governments and obstructs the right of other nations to conduct their internal affairs free from interference.[97] Although some courts have come close to recognizing the U.S. government's legal obligation, NEPA's transnational application is as of this writing uncertain.[98]

Although concern for foreign sovereignty is legitimate, we believe that this concern must not be dispositive; that in some cases circumstances will demand assessment of environmental impacts despite possible foreign sensibilities. First, we must remember that the purpose of an environmental assessment is not to impose one's own higher environmental standards on a less environmentally "aware" neighbor. Rather, it is to assist a country's decision makers in formulating environmentally wise actions. Indeed, at least under U.S. law, the results of an assessment would not mandate any preconceived substantive conclusion. In some cases economic, social, or political considerations might predominate over environmental concerns, and an action would proceed despite adverse environmental repercussions.

Second, a decision to prepare an assessment could be made on a case by case basis subject to sensible guidelines. For example, the Japanese government might consider the magnitude of the foreign environmental impact and whether it entailed significant atmospheric, open-ocean, or even transfrontier pollution; the extent to which the government or its nationals would retain control over a project either by financing, participation, veto, or by other means; the foreign government's feelings on the issue; the extent to which the Japanese government had promoted the project; the existence of satisfactory bilateral or other institutions through which information on environmental risks could be generated and transmitted. Significant, adverse environmental consequences, either within the host country or internationally, and substantial Japanese government or private involvement (either by promotion or otherwise) might well favor an assessment despite the host government's position that such action was unnecessary.

As a general rule it may not be appropriate for industrialized countries to insist that all poorer countries with which they or their nationals do business adopt strict environmental standards. A developing country may justifiably wish to operate under lower environmental standards than those existing in Japan, the United States, or other industrialized countries. Yet in some cases an industrialized country's imposing its higher pollution controls, at least on its own nationals, would be appropriate.

The Palau case seems to fit this exception. As noted in the petitioners' brief, Palau's trusteeship status placed the islands in a unique position of dependency and imposed on all members of the United Nations special fiduciary obligations. Moreover, the manner by which the promoters of the superport conducted operations denied the people of Palau an adequate opportunity to evaluate the environmental risks. The gross disparity in bargaining positions made it impossible for the Palauans to reach a balanced assessment. Finally, the Palau case was complicated by the risk of open-ocean pollution. Given the present failure of international organizations to abate transnational pollution, Japan, as other nations, arguably has a special obligation to control the activities of its nationals that threaten the world environment.

The need for Japan's imposing stringent standards on industrial activities within Japan that pollute the ocean or atmosphere beyond its borders is even more compelling than the Palau case. A good example is the fluorocarbon aerosol industry. Despite mounting evidence of the serious atmospheric effects of this industry, and Japan's significant contribution to the problem, the Japanese government has consistently refused to regulate the industry more stringently.[99] Yet in this case control would involve no possibility of Japan's intruding on foreign sovereignties. The principal beneficiary of the activity is the Japanese fluorocarbon aerosol industry; there is little benefit to the international community that will offset the risks of the activity. Yet by refusing to regulate this industry Japan can extract a competitive economic advantage when other countries decide to impose effective environmental controls.[100]

In addition to assessing environmental impacts and promulgating new regulations, the Japanese government might also consider implementing OECD's Principles Concerning Transfrontier Pollution—to which Japan subscribed in May 1977.[101] One affirmative measure would be Japan's establishing a fund to assist victims of transfrontier pollution, where relief is inadequate or unavailable in the country where injury has occurred. Here again the Palauan example is relevant. In the case of a major oil spill in Palau, it would be extremely difficult for Palauans to obtain adequate relief in Palauan courts against Japanese or American promoters. Japan's willingness to compensate environmental damage claims unsatisfied in Palauan tribunals would seem a fair and necessary component of Japan's participation in the superport project.

A second measure would be to grant claimants of transfrontier pollution equal access to Japanese administrative and judicial tribunals. The Diet's acceptance of the petitioners' brief is perhaps evidence of Japan's growing recognition of this principle. But in the Palau case most of the petitioners were citizens of countries enjoying nationality treatment under existing treaties of friendship and commerce. Japan should grant equal access to victims of environmental damage whether Japan is under a treaty obligation to their countries or not.

How might these various measures be accomplished? One general option, of course, would be legislation, since in other areas Japan has not hesitated to regulate her nationals' foreign activities. For example, Japanese distant water fisheries are subject to tight statutory and administrative controls, and firms wishing to establish branch offices, withdraw currency, or invest internationally are all strictly scrutinized and regulated.[102] Perhaps a more realistic approach, at least during the next few years, would be administrative guidance. As noted in ch. 5, the Japanese bureaucracy is in a unique position to formulate and implement a new environmental policy, should a need for one be seen. Indeed, the government could immediately adopt each of the cited measures without even having to obtain prior Diet approval. As elsewhere, administrative guidance (of Japanese industries' foreign operations, or of industries whose actions within Japan entail adverse effects outside Japan) could provide a fair and flexible means of assuring compliance.

3.2.2   Strengthening Bilateral and Multilateral Collaboration

Since it is clearly unreasonable to expect Japan to shoulder the burdens of protecting the world environment without the help of other nations, this section examines the contribution Japan can make by strengthening bilateral and multilateral collaboration. Through more affirmative international action Japan can enhance the effectiveness of domestic environmental controls, and at the same time mitigate the potential economic disadvantage to Japanese industry from unilateral action.

The 1975 U.S.–Japan Agreement on Environmental Protection

On August 5, 1975, the United States and Japan concluded an Agreement on Cooperation on Environmental Protection, one of nine separate arrangements covering a broad range of scientific and technical matters of mutual concern to the two countries.[103] As in other areas of international relations, a complex of reasons motivated the agreement. For the United States, political factors were at least as important as environmental considerations. Comparable agreements had been concluded with the Soviet Union and West Germany, and the United States also enjoyed close collaboration with the Canadians; it was important to demonstrate that the United States government deemed the relationship with Japan to be of equal importance. The State Department, in particular, hoped that consultations on environmental problems might facilitate cooperation in other areas.

Japan viewed the agreement primarily symbolically, rather than as an opportunity for substantive collaborative work. Japanese officials believed that the United States was very eager for agreement, and since the arrangement appeared a convenient means of expediting exchanges, the Japanese government acquiesced. Generally, the environmental protection agencies in both countries welcomed the agreement because it helped expand budgetary appropriations.

In part the agreement was also the product of the two countries' mutual suspicion that the other would abuse pollution control regulations to obtain a competitive trade advantage. The United States, for example, was concerned that Japan would grant production-oriented subsidies to her domestic industry to offset the international competitive disadvantage suffered from domestic pollution controls.[104] For this reason, the United States urged that the "polluter pays" principle be made explicit in art. 5 of the agreement.[105] Japan, however, suspected that U.S. insistence on including this provision was designed to benefit American business.[106]

As a technology exchange accord, the agreement is reported to be working well, apparently to the satisfaction of both governments.[107] The United States has learned more about Japanese technology for identifying and controlling toxic substances; Japan has profited from U.S. officials' explanations on methods of sewage treatment and environmental impact assessment procedures. Close cooperation continues in research on the health effects of pollutants, management of bottom sediments, and in other areas.[108]

The issue for both Japan and the United States is not that the agreement has failed to open a fruitful exchange; the pity is that it could do so much more. For example, in addition to matters like sewage treatment or toxic substance control, the agreement could address larger and more controversial problems such as the hazards of the nuclear export program, deep sea mining, the Concorde, Palau, or Japanese whaling practices. Certainly the agreement is extremely broad. The preamble, for example, invokes both countries' responsibilities "for the protection and improvement of the global environment," and art. 3(b) permits cooperation in "other areas" by agreement.[109] The parties' present focus on narrow technical issues must be attributed more to institutional and political constraints than to legal limitations.

It is easy to imagine many new ways by which the agreement could be more productively employed. It could be used not only to educate the parties on technical matters, but also on broader legal and institutional developments. It could serve as an early warning system on conflicts between environmental, trade, and other policies. It could provide a forum for both countries to investigate together the environmental risks of open-ocean developments, such as deep sea mining, in which nationals of both countries are deeply involved. It might provide an alternative, less rigid means of exploring the environmental consequences of projects such as Tokaimura, when a full-dress environmental impact statement is deemed unnecessary. Through bilateral collaboration on problems like the Palau superport, both countries could establish a common record that might ultimately strengthen their control of the enterprises involved.

3.2.3   Japan's Future Role in Asia and the Pacific

Japan's present international record does not augur well that she will soon become the protector of the environment in Asia and the Pacific. Yet present priorities fade and leaders change. The opportunity for a noble contribution will continue.

One area for Japan to begin is her relationship with the Republic of Korea. Korea today is approaching an industrial turning point where, like Japan in the early and mid-1960s, even the leaders are recognizing the wisdom of pollution control and environmental planning.[110] And already representatives of the two countries have met to exchange information on environmental matters. Japan must recognize that Korea is now in a position not simply to copy, unthinkingly, Japan's innovations, but to do even better. Korea's Fourth Five-Year Industrial Growth Plan (1977–1981), which will locate many petrochemical and other industrial complexes (many of which are at least partly Japanese-owned) in ecologically fragile coastal areas, offers this chance. Korea has the opportunity of systematically screening environmentally hazardous projects, and of mitigating other adverse developments, that would far surpass Japan's helter-skelter reaction to similar problems during the 1960s and the early 1970s. With honest reflection on its own mistakes, Japan has a wealth of knowledge it could impart to Korea; and Korea's own experiments in environmental planning might help refine Japanese environmental policies. Both countries could jointly conduct administrative guidance, which would substantially enhance the effectiveness of statutory controls on Japanese enterprises operating in Korea.

Beyond Korea, the whole ESCAP region is now alive to the need for environmental protection.[111] Malaysia, Thailand, and the Philippines have established environmental protection agencies or commissions, and are formulating new environmental laws;[112] India has begun applying rudimentary environmental impact analyses to large-scale industrial developments;[113] even Papua New Guinea gives some attention to regulating environmentally destructive activities.[114]

There are many ways Japan could foster this development. Perhaps of greatest interest for these countries is Japan's apparent success in implementing environmental controls with little detriment to economic growth or employment and modest price inflation.[115] Japanese researchers could greatly aid these countries if they developed, collaboratively, a counterfactual econometric model, based on Japan's own economic growth since the 1950s, that selectively assessed the economic and other effects of introducing environmental protection measures at the present stage of these countries' economic development.[116]

Japan could also support international efforts to protect the environment of Asia and the Pacific far more than it has so far. For example, it could ratify and begin implementing many multilateral environmental protection treaties that await Japanese action. She could play a major role in establishing an environmental data retrieval system within ESCAP, UNEP, WHO, and other international organizations concerned with the region's environment. She could take the lead in creating new, needed educational programs on environmental law or administration under the auspices of these international organizations.[117] Japan could marshall her substantial influence to make international lending institutions like the Asia Development Bank more sensitive to the environmental repercussions of their lending policies.[118]

Could Japan also pioneer a new comprehensive treaty for the protection of the marine environment in East Asian waters? There is now ample precedent from other parts of the world—for example, the Convention for the Protection of the Mediterranean Sea Against Pollution, or the Nordic Convention on the Protection of the Environment.[119] Indeed, UNEP's Governing Council has designated East Asia's waters to be of highest priority, and Malaysia, Indonesia, and Singapore have already negotiated a tripartite agreement to guard the environmentally and politically sensitive Malacca and Singapore straits.[120] Obviously, there is need for broader effort. The ocean here is yearly despoiled by vessel dumping, seabed extraction, atmospheric fallout, and land based runoff. As man's technological dominion over these seas increases, pollution will become more serious;[121] and the sacrifice of Asia's yet uncultivated food resources will be prodigious.[122] Now must the countries of this region plot the course of their future collaboration, before their Torrey Canyon or their Santa Barbara. Is not this vision of a better future possible?

Appendix 1   A Summary of the Record

The following tables 1–7 give a rough indication of the program of pollution control in Japan between 1965 and 1975.

Tables 1–3 indicate significant reductions in SO_2, CO, and various pollutants in water, respectively.

The record in other areas is somewhat less certain. It is still difficult to assess trends concerning NO_x concentrations, although Japan has now established an extensive monitoring network for these substances. Table 4 suggests that yearly average concentrations of nitrogen dioxide have tended to increase slowly.

Data for oxidants that depend to a large extent on NO_x and hydrocarbon precursors and other factors such as wind and sunlight are similarly uncertain. Table 4 suggests an increase in oxidants as reported in a set of monitoring stations.

Trends in water quality are even less clear. For example, pollution by organic matter, measured by biological oxygen demand (BOD) found in rivers over recent years, apparently has not significantly changed.

Although recently the administration has attempted to control noise pollution in the vicinity of airports, highways, and the bullet train, it is still too early to assess accurately the effectiveness of these measures.

Tables 5–7 show the growth in antipollution equipment sales between 1969 and 1974. Recently these sales have tended to slacken.

Appendix 1

Table 1
Annual average SO₂ pollution, 1965–1974.

	Concentration[a] at Ground Level (ppm)	Monitoring Stations Meeting Quality Standards[b] (%)
1965	0.057	n.a.
1966	0.057	n.a.
1967	0.059	48
1968	0.055	59
1969	0.050	67
1970	0.043	72
1971	0.037	87
1972	0.031	34
1973	0.030	46
1974	0.024	69

Source: OECD Report 60 (1977).
[a]Averages over fifteen stations in operation since 1965.
[b]Quality standards for 1967–1971 are different from quality standards for 1972–1974.

Table 2
Annual mean CO pollution in Tokyo, 1965–1974 (ppm).

	First Source	Second Source
1965	3.2	
1966	2.9	
1967	3.1	
1968	3.6	6.4
1969	4.4	5.9
1970	5.7	3.4
1971	4.7	3.1
1972	4.3	2.3
1973	3.7	2.7
1974	3.5	n.a.

First Source: Quality of the Environment 74 (Environment Agency, 1975).
Second Source: Tokyo Statistical Yearbook 461 (Tokyo Metropolitan Government, 1973); average over four monitoring stations.

Table 3
Harmful substances in water,[a] 1970–1974.

	1970	1974
Cadmium	2.80	0.37
Cyanides	1.50	0.06
Organic phosphorus	0.20	0.00
Chromium (hexavalent)	0.80	0.03
Arsenics	1.00	0.27
Total mercury	1.00	0.01[b]
Alkyl mercury	0.00	0.00
Total	1.40	0.20

Source: OECD Report 166–167 (1977).
[a]Percentages of samples exceeding quality standards; data is based on more than 16,000 samples for 1970 and more than 160,000 samples for 1974.
[b]1973.

Table 4
NO_2 and oxidant pollution, 1970–1975.

	NO_2[a] (ppm)	Oxidants (O_3)[b] (ppm)	Oxidant[c] Warnings by various prefectures (days)
1970	0.032	0.029	7
1971	0.033	0.032	98
1972	0.026	0.029	176
1973	0.038	0.036	328
1974	0.040	0.037	288
1975	0.041	0.042	266

[a]Present State of Air Pollution in Japan (Environment Agency, 1976); average over six monitoring stations in operation since 1970; Quality of the Environment (Environment Agency, 1976).
[b]Report on Air Pollution Monitoring in 1975, 1976 (Tokyo Metropolitan Government); average over three Tokyo area monitoring stations.
[c]Warnings are issued when hourly oxidant concentration exceeds 0.15 ppm.

Table 5
Antipollution equipment sales, 1974.

Type of Equipment	Sales (%)	Type of Client	Sales (%)
Air	45	Private	67
Water	38	Public	32
Waste	16	Export	1
Noise	1		
Total	100	Total	100

Table 6
Growth of antipollution equipment sales, 1969–1974.

	¥ Million	Index
1969	143,000	100
1970	195,000	136
1971	302,000	211
1972	375,000	262
1973	488,000	341
1974	677,000	473

Source: Japan Industrial Machinery Association.

Table 7
Antipollution investment by private enterprise, Japan and selected OECD countries, 1974.

	Percent of Total Investment by Private Enterprises	Percent of GNP
Japan	4.0	1.0
United States[a]	3.4	0.4
Netherlands[b]	2.7	0.3
Sweden[b]	1.2	0.1
Germany[b]	2.3	0.3
Norway[b]	0.5	0.1

[a]From National Expenditure for Pollution Abatement and Control (U.S. Department of Commerce, 1972) and 2 Survey of Current Business 55 (Feb. 1975), adjusted by the OECD Secretariat.
[b]Calculated from replies to OECD questionnaire on Procedures for Notification for Financial Assistance in Relation to Pollution Control Expenditures.

Appendix 2　Documents

Many of the major environmental laws of Japan have already been translated into English and may be found in the International Environmental Reporter, 91:0001 (1978). As of 1979 the Reporter had published the following statutes:

Basic Law for Environmental Pollution Control

Air Pollution Control Law
   Air Pollution Regulations
   Air Pollution Standards
   Air Pollution Ordinances
   Emission Standards

Water Pollution Control Law
   Water Pollution Order
   Water Pollution Standards
   Effluent Standards

Marine Pollution Law
   Marine Pollution Order

Waste Management Law
   Waste Management Order

Agricultural Chemicals Law
   Agricultural Chemicals Order

Agricultural Soil Pollution Law
   Agricultural Soil Pollution Order

Noise Regulation Law
   Noise Regulation Order
   Aircraft Noise Standards

Public Works Law
   Public Works Order

Odor Control Law
   Odor Control Order

Pollution Crimes Law

Pollution-Related Health Damage Law
   Pollution-Related Health Damage Order

Law Concerning the Examination and Regulation of Manufacture, Etc., of Chemical Substances

English translations of several other important statutes appear in Environmental Laws and Regulations in Japan (Environment Agency, International Affairs Division, 1976). Further reference materials on Japanese environmental laws and regulations may be found in An Index to Japanese Law in Law in Japan, An Annual (1975). For a basic compendium in Japanese, see Kankyō Roppō, 1980.

With the exception of the Basic Law for Environmental Pollution Control, the materials in the present appendix have been included because they are not easily obtained elsewhere.

BASIC LAW FOR ENVIRONMENTAL POLLUTION CONTROL

Law No. 132 of 1967
Amended by Law No. 132 of 1970 and No. 88 of 1971.

Contents

Chapter I   General Provisions
Chapter II   Fundamental Policies for Environmental Pollution Control
Chapter III   Bearing of Costs and Financial Measures
Chapter IV   The Conference on Environmental Pollution Control and the Councils on Environmental Pollution Control

Chapter I   General Provisions

(Purpose)
Article 1
In view of the vital importance of environmental pollution control for the preservation of a healthy and civilized life for the nation, this Law is enacted for the purpose of identifying the responsibilities of the enterprise, the State and the local government bodies with regard to environmental pollution control and of determining the fundamental requirements for control measures, in order to promote comprehensive policies to combat environmental pollution thereby ensuring the protection of the people's health and the conservation of their living environment.

(Definition)
Article 2
1. The term "environmental pollution" ("kōgai" in Japanese), as used in this Law, shall mean any situation in which human health and the living environment are damaged by air pollution, water pollution (including the deterioration of the quality and other conditions of water as well as of the beds of rivers, lakes, the sea and other bodies of water. The same shall apply hereinafter, except in the case of paragraph 1, Article 9.), soil pollution, noise, vibration, ground subsidence (except for subsidence caused by drilling activities for mining. This exception shall apply hereinafter.), and offensive odors, which arise over a considerable area as a result of industrial or other human activities.

2. The term "living environment," as used in this Law, shall include property closely related to human life, and animals and plants closely related to human life and the

environment in which such animals and plants live.

(Responsibility of the enterprise)

Article 3

1. The enterprise shall be responsible for taking the measures necessary for the prevention of environmental pollution, such as the treatment or disposal of smoke and soot, polluted water, wastes, etc. resulting from its industrial activities, and for cooperating with the State and local government bodies in their efforts to prevent environmental pollution.

2. The enterprise, in manufacturing and processing activities, shall endeavor to take precautionary measures to prevent environmental pollution which might otherwise be caused by the use of the products which it manufactures or processes.

(Responsibility of the State)

Article 4

The State has the responsibility to establish fundamental and comprehensive policies for environmental pollution control and to implement them, in view of the fact that it has the duty to protect the people's health and conserve the living environment.

(Responsibility of local government bodies)

Article 5

In order to protect the health of the local population and to conserve the living environment, local government bodies shall take measures in line with the policy of the State and shall also work out and implement appropriate measures for environmental pollution control which take into account the specific natural and social conditions of the area concerned.

(Responsibility of citizens)

Article 6

Citizens shall endeavor to contribute to the prevention of environmental pollution in all appropriate ways such as cooperating with the State and with local government bodies in the implementation of control measures.

(Annual report, etc.)

Article 7

1. The Government shall present to the Diet an annual report on the situation with regard to environmental pollution and on those measures taken by the Government in order to control it.

2. The Government shall present to the Diet annually a document, outlining the measures which the Government is going to take to deal with the environmental pollu-

tion situation described in the report referred to in the preceding paragraph.

(Control of air pollution, etc. caused by radioactive substances)

Article 8

With regard to measures for the control of the pollution of air, water and soil by radioactive substances, the Atomic Energy Basic Law (Law No. 186, 1955) and other related laws shall apply.

Chapter II Fundamental Policies for Environmental Pollution Control

Section I Environmental Quality Standards

Article 9

1. With regard to environmental conditions relating to air, water and soil pollution and noise, the Government shall establish environmental quality standards, the maintenance of which is desirable for the protection of human health and the conservation of the living environment.

2. In the event that one of the standards referred to in the preceding paragraph establishes more than one category and stipulates that land areas or areas of water to which those categories are to be applied should be designated, the Government may delegate to the prefectural governors concerned the authority to designate those land areas or areas of water.

3. With regard to the standards provided for in paragraph 1, due scientific consideration shall always be given and such standards shall be revised whenever necessary.

4. The Government shall make efforts to ensure the maintenance of the above-mentioned standards, by implementing environmental pollution control measures in a comprehensive, effective and appropriate manner.

Section II Measures to be Taken by the State

(Emission control, etc.)

Article 10

1. In order to control environmental pollution, the Government shall take measures for the control of the emission of pollutants responsible for air, water and soil pollution, establishing standards to be observed by the enterprise.

2. In order to control environmental pollution, the Government shall endeavor to take measures to deal with noise, vibration,

ground subsidence and offensive odors, in a manner similar to that referred to in the preceding paragraph.

(Control of land use and installation of facilities)

Article 11

In order to control environmental pollution, the Government shall take necessary measures with regard to land use and shall, in areas where environmental pollution is serious or likely to become serious, also take measures to control the installation of facilities which cause environmental pollution.

(Promotion of establishment of facilities for the prevention of environmental pollution)

Article 12

The government shall take measures to promote necessary projects for the prevention of environmental pollution, such as the establishment of buffer zones, etc., as well as those projects to establish public facilities which will contribute to the prevention of environmental pollution, such as sewerage and public waste disposal plants.

(Establishment of surveillance and monitoring systems)

Article 13

The Government shall endeavor to establish systems for surveillance, monitoring, measurement, examination and inspection in order to ascertain what the situation with regard to environmental pollution is and to ensure adequate enforcement of measures to combat environmental pollution.

(Carrying out of surveys and investigations)

Article 14

The Government shall carry out surveys and investigations necessary for the planning of measures for environmental pollution control, such as those for predicting environmental pollution trends.

(Promotion of science and technology)

Article 15

In order to promote the development of science and technology which will contribute to the prevention of environmental pollution, the Government shall take the necessary measures such as the consolidation of survey and research systems, the promotion of research and development, the dissemination of the results of such research and development work, and the education and training of research experts.

(Dissemination of knowledge and information)

Article 16

The Government shall endeavor to disseminate knowledge and information concerning environmental pollution and also to make the nation more conscious of the need to prevent environmental pollution.

(Consideration of environmental pollution control in the planning of regional development policies, etc.)

Article 17-1

The Government shall take into consideration the need to control environmental pollution in the planning and implementation of regional development measures such as those for urban development and the construction of factories.

(Protection of the natural environment)

Article 17-2

In order to contribute to the prevention of environmental pollution, the Government shall, in conjunction with other measures prescribed in this Section, endeavor to protect the natural environment as well as to conserve green areas.

Section III   Measures to be Taken by Local Government Bodies

Article 18

The local government bodies shall, provided that the measures do not infringe laws and regulations, take measures in line with the policy of the State provided for in the preceding Section and shall also implement measures for environmental pollution control which take into account the specific natural and social conditions of the area concerned. In this case, the prefectural governments shall be responsible mainly for the implementation of measures covering wide areas and also for the coordination of measures to be taken by the municipal governments.

Section IV   Environmental Pollution Control in Specified Areas

(Formulation of Environmental Pollution Control Programs)

Article 19

1. The Prime Minister shall instruct the prefectural governors concerned to formulate programs relating to the environmental pollution control measures (hereinafter called "Environmental Pollution Control Programs") to be implemented in specific areas which fall into any one of the following categories, by showing to those governors fundamental policies for such programs:

(1) areas in which environmental pollution is serious and in which it is recognized that it will be extremely difficult to achieve effective environmental pollution control unless comprehensive control measures are taken;

(2) areas in which environmental pollution is likely to become serious on account of rapidly increasing concentrations of population, industry, etc., and in which it is recognized that it will be extremely difficult to achieve effective environmental pollution control unless comprehensive control measures are taken.

2. When the prefectural governor concerned has received the instruction referred to in the preceding paragraph, he shall draw up an Environmental Pollution Control Program in accordance with the fundamental policies referred to in the preceding paragraph and shall submit it to the Prime Minister for his approval.

3. Prior to issuing an instruction under paragraph 1 or giving the approval required under the preceding paragraph, the Prime Minister shall consult with the Conference on Environmental Pollution Control.

4. Prior to issuing an instruction under paragraph 1, the Prime Minister shall seek the opinion of the prefectural governor concerned.

(Implementation of Environmental Pollution Control Programs)

Article 20

The State and local government bodies shall endeavor to take measures necessary for the full implementation of Environmental Pollution Control Programs.

Section V Settlement of Disputes Relating to Environmental Pollution and Relief for Damage Caused Thereby

Article 21

1. The Government shall take the measures necessary to establish a system for the settlement, by such means as mediation and arbitration, of disputes which arise in connection with environmental pollution.

2. The Government shall take the measures necessary to establish a system which will make possible the efficient implementation of relief measures for damage caused by environmental pollution.

Chapter III Bearing of Costs and Financial Measures

Article 22

1. The enterprise shall bear all or part of the necessary cost of the works carried out by the State or local government bodies to control environmental pollution arising from the industrial activities of such enterprise.

2. The nature and amount of the costs which the enterprise shall bear under the preceding paragraph, the enterprises which shall bear such costs, the method of calculation of the amount to be borne by such enterprises, and other necessary matters relating to the bearing of costs shall be laid down in other laws.

[Chapter IV is omitted]

SAMPLE POLLUTION CONTROL AGREEMENTS

The following three documents are examples of pollution control agreements. The first is an agreement between a citizens' group and a company. This agreement is not legally enforceable because of vagueness. Even art. 6 fails to provide any legal measure not already provided in the Civil Code. The agreement, however, will serve the citizens' group as a device to initiate negotiations with the company from a favorable bargaining position. The agreement may also exert an important influence in encouraging the company to implement stronger pollution controls.

The second document is typical of a more detailed agreement. Recently, many prefectural governments have used these agreements to supplement regulations imposed by laws or ordinances. The agreement is frequently used to establish even more specific control measures tailored to the technical needs of an industry and to local conditions. The agreement prescribes a specific means of pollution control (arts. 6–11), monitoring systems (arts. 12 and 13), and moral obligations (arts. 19–23). It also provides for public participation (arts. 15–17), and strict liability for pollution damage (art. 18). Although the agreement does not require the company to obtain a permit for construction or modification of its facility, it does stipulate that the company must consult with prefectural and city authorities before further construction or modification (art. 4). Should the provisions of the agreement be violated, the prefectural and city authorities may terminate or curtail the company's operations (art. 5). The agreement, however, does not provide specific sanctions against companies that refuse to comply with an order.

The third agreement, a form contract signed in Ibaraki Prefecture in December, 1974, applies this same basic approach to controls on the operations of a nuclear facility.

Agreement 1: Iwaki City, Onahama District Joint Committee on Pollution Control Measures and Nihon Suiso Kōgyō Kabushiki Kaisha (The Japan Hydrogen Engineering Corporation) (June 4, 1970)

The Iwaki City Onahama District Joint Committee on Pollution Control Measures, hereafter referred to as A, and the NIHON SUISO KŌGYŌ KABUSHIKI KAISHA (Japan Hydrogen Engineering Corporation), hereafter referred to as B, do hereby agree on the following items concerning the control of pollution from B's coking plant.

Article 1: The purposes of this agreement shall be to control pollution and to preserve the living environment.

Article 2: A and B shall respect each other's rights and obligations, honor their faith, sincerely carry through each item of this agreement, and endeavor to accomplish the agreement's purposes.

Article 3: The substantive items of pollution control shall be determined in particular through discussion by A and B.

Article 4: Neither A nor B may refuse a negotiation proposal from the other party.

Article 5: At the proposal of A and as needed, B shall give, as shall be determined by mutual discussion, aid for the expenses of cooperative investigation as well as pollution prevention.

Article 6: In cases where damage has been caused, B shall provide A with compensation.

Article 7: When the solution to a dispute between A and B is not forthcoming, a cooling-off period will be set and another discussion shall take place.

Article 8: The term of validity for this agreement shall extend a full three years from the day this agreement is made. If, three months before the term's limit, neither A nor B has any amendments or termination proposal to offer, the agreement shall be automatically renewed. However, amendments to or termination of the agreement shall be determined by discussion between A and B, and neither A nor B shall unilaterally annul the agreement.

Article 9: In the event of doubts concerning this agreement, the agreement can be revised through discussion between A and B.

Supplementary Provisions

This agreement shall be implemented as of June 4, 1970.

In order to witness this agreement, two copies of the text shall be drawn up and A and B shall each retain one copy.

June 4, 1970

A: Chairman of the Iwaki City Onahama District Joint Committee on Pollution Control Measures—Kusano Keishō

B: President of NIHON SUISO KŌGYŌ

KABUSHIKI KAISHA (Japan Hydrogen Engineering Corporation)—Ishizaka Ichirō
Witness: Director of the Public Welfare Bureau, Fukushima Prefecture—Muraura Kōji
Witness: Mayor of the City of Iwaki—Ōwada Yaichi

Agreement 2: Hyōgo Prefecture and Himeji City vs. Kansai Electric Corporation (October 30, 1971) [Excerpt]

Hyōgo Perfecture (hereafter referred to as the "Prefecture"), Himeji City (hereafter referred to as the "City"), and the Kansai Electric Corporation (hereafter referred to as the "Corporation"), in conjunction with the construction of Generators No. 5 and No. 6 (hereafter referred to as the "Additional Generators") at the Corporation's Himeji Power Plant No. 2 and in order to protect the health of the area's residents and to contribute to both the preservation of the environment and the advancement of the public welfare, do hereby agree on the following items concerning the control of pollution from Himeiji Power Plants No. 1 and No. 2 (hereafter referred to as the "Power Plants").

Chapter 1   General Provisions
(The Guiding Principles of Pollution Control)
Article 1: In view of the present situation in which the increase of environmental damage from pollution has been precipitating a crisis in society, the control of pollution caused by the productive activities of enterprises shall be an important social responsibility of the enterprises. Recognizing the fact that the control of these kinds of pollution and the protection of area residents are important duties of local government, the Prefecture and the City shall, after taking into account the will of the area's residents, provide administrative guidance to the Corporation. The Corporation shall make the utmost efforts toward the prevention, control, and elimination of pollution.
(The Implementation of Anti-Pollution Measures)
Article 2: In order to prevent pollution before it occurs, the Corporation not only shall faithfully and adequately implement the various measures established by this agreement, but also shall maintain an anti-pollution administrative organization and manage with careful attention pollution-related facilities.

(The Discussion of New Plans)
Article 3: When the Corporation decides to construct in other areas steam-powered generating plants, currently being planned for expansion, the Corporation shall inform the Prefecture and the City of measures for the control of resulting pollution.
In this event, when there is a measure for the control of pollution beyond the conditions set for the Additional Generators, the Corporation shall, upon consultation with the Prefecture and the City, implement that measure.
(Discussion on the Establishment of Facilities)
Article 4: In cases where the Corporation attempts to construct or modify pollution-related facilities, the Corporation shall consult with the Prefecture and the City before acting and shall respect the Prefecture's and the City's opinions concerning the control of pollution.
(Measures for Times of Violation)
Article 5: (1) In order to ensure the implementation of the various measures established by this agreement, the Prefecture and the City, when necessary, may relay requests or advice to the Corporation. In this event, the Corporation shall comply.
(2) When the Prefecture and the City recognize that a violation against the pollution control measures established in this agreement has been committed, the Prefecture and the City may order the cancelation of expansion construction, the curtailment (or suspension) of the Power Plant's operations, or other necessary measures against the Corporation.
In this event, the Corporation, no matter what its reasons are, shall comply with these measures.
(3) In the event that violations referred to in the last paragraph should occur, the Prefectural Governor, upon consultation with the Prefecture and the City, may, when necessary, demand that the Hyōgo Prefecture Public Enterprise Administrator (hereafter referred to as the "Administrator") limit or terminate the industrial water supply to the Power Plants. When the Administrator has worked out the needed measures, the Corporation shall comply with these measures.
Chapter 2   Pollution Control Measures
(Measures for the Prevention of Air Pollution)

Article 6: omitted (Specific standards for oil use; emission standards for sulfur oxides, nitrogen oxides, and dust; regulations on the sulfur content of oil; and stack specifications)

(Measures for Times of Emergency)

Article 7: When an emergency situation as prescribed in Article 23 of the Air Pollution Control Law appears imminent or does occur, the Corporation shall adopt the measures described in the following sections.

(i) omitted (emergency emission standards for the sulfur content of oil)

(ii) (1) When an emergency situation is caused by nitrogen oxide and other polluting substances, the Corporation shall adopt all measures requested by the Prefecture and the City.

(2) In the event that the state of air pollution is not improved through the measures prescribed in paragraph (i) and there is a risk that the health of residents will be significantly harmed, when the Prefecture and the City demand that the Corporation adopt measures more stringent than those prescribed in paragraph (i) (e.g., the temporary curtailment of operations), the Corporation shall comply.

(Measures for the Control of Water Pollution)

Article 8: omitted (Specification of thermal pollution levels, specific standards for oil separation facilities)

(Measures for the Control of Noise)

Article 9: omitted (noise level standards)

(Measures for Times of Accident)

Article 10: (1) When accidents such as breakdowns or breakages occur in pollution-related facilities, the Corporation shall immediately adopt temporary measures, promptly inform the Prefecture and the City of the situation, and endeavor to clear up the accident.

(2) In the event of accidents discussed in the previous paragraph, when either the Prefecture or the City has ordered necessary measures, the Corporation shall comply with these measures.

(The Maintenance of Monitoring and Observation Machinery)

Article 11: (1) The Corporation shall install in each generator smoke stack at the Power Plants continual monitoring devices for sulfur oxides and for nitrogen oxides.

(2) The Corporation shall install in the Power Plants weather observation devices that measure wind direction and speed and that record air temperature.

(3) In connection with the establishment of the monitoring and observation devices described in the two preceding paragraphs, the Corporation shall install facilities from which data can be recorded and checked in the City's Inspection Center. The Corporation shall also be responsible for the maintenance and management of these facilities.

Chapter 3 Reports, Inspections, and Compensation

(Measurements and Reports)

Article 12: (1) The Corporation shall periodically measure the density of dust in each generator smoke stack at the Power Plants. It shall also measure the amount and nature of its thermal drainage and other types of waste water. The Corporation shall report the results of these measurements to the Prefecture and the City.

(2) The Corporation shall periodically report the consumption levels and components of the fuels it uses.

(The Maintenance of an Observation Network)

Article 13: (1) The Corporation shall actively cooperate in equipping the sulfur oxide and nitrogen oxide air pollution monitoring network established by the Prefecture and the City.

(2) In connection with the cooperative efforts to equip the observation network prescribed in the preceding paragraph, the Corporation shall establish and maintain facilities from which necessary items of data can be recorded and observed in the City's Inspection Center.

(On-the-spot Investigations)

Article 14: The Prefecture and the City may, when necessary, enter the Power Plants and conduct an inspection or investigation related to pollution control.

In this event the Corporation shall comply.

(The Principle of Public Openness)

Article 15: The Prefecture and the City may, when necessary, make public the circumstances of anti-pollution measure implementation at the Power Plants.

(The Publication of Emission Substances)

Article 16: (1) The Corporation shall take steps to publish the substances discharged from the Power Plants.

(2) With regard to the steps referred to in the preceding paragraph, the starting times, the

methods, and the places of these steps shall be determined through mutual discussion among the Prefecture, the City, and the Corporation.

(The Establishment of a Pollution Prevention Council)

Article 17: (1) In order to ensure the smooth implementation of the pollution control measures established by this agreement, the Prefecture, the City, and the Corporation, with participation from residents, shall establish the Kansai Electric Pollution Control Council (hereafter referred to as the "Council").

(2) In order to accomplish the purposes referred to in the preceding paragraph, the Council shall obtain information from the Prefecture, the City, or the Corporation, carry out the exchange of opinions, and conduct on-the-spot investigations of the Power Plants, when necessary.

(3) The organization and administration of the Council shall depend on guidelines which will be determined separately.

(The Settlement of Grievances and the Compensation for Damages)

Article 18: (1) When there is a resident complaint concerning pollution caused by the Power Plants, the Corporation shall undertake to sincerely resolve the complaint.

In this event, when damage against residents has been caused, the Corporation, regardless of the existence of intention or negligence, shall work out compensation and other appropriate measures.

(2) If the settlement of disputes referred to in the preceding paragraph becomes difficult, when a properly concerned party offers a proposal, the Prefecture and the City shall cooperate in mediation and other necessary measures.

(3) When, through accidents in the midst of construction or for other reasons, the Corporation and persons related to the construction of the Power Plants, cause damage to residents, the Corporation shall be responsible for providing compensation and other appropriate measures.

Chapter 4 Mutual Discussion and Cooperation on the Establishment of Facilities

(The Development of Technology)

Article 19: The Corporation shall endeavor to develop pollution control technology such as equipment for removing sulphur from discharged smoke. The Corporation shall also actively carry out improvements of the Power Plants facilities.

(Cooperation on Pollution Control Measures)

Article 20: Besides the measures established in Article 13, the Corporation shall actively cooperate in other pollution control enterprises carried out by the Prefecture or the City (e.g., the investigation and monitoring of pollution, the development of pollution control technology, the implementation of local pollution control programs, the creation of green belts as buffer areas, etc.).

(Pollution Control in Related Enterprises)

Article 21: The Corporation shall actively give guidance and supervision concerning pollution control to enterprises whose operations are related to the Power Plants. In the event that complaints concerning pollution related to these enterprises occur, the Corporation shall take responsibility for the resolution of these complaints.

(Cooperation with Surrounding Areas)

Article 22: The Corporation shall actively cooperate in enterprises for the advancement of the welfare of residents in areas surrounding the power plants.

(Beautification of the Environment)

Article 23: The Corporation shall willingly work toward the afforestation and environmental beautification of the locations of the Power Plants and related business sites.

(Document Outlining Details)

Article 24: The implementation of measurements, reports, and other particulars to be carried out by the Corporation shall depend on a separately determined Document Outlining Details.

(Revision of the Agreement)

Article 25: When attempting to modify this agreement, the Prefecture, the City, and the Corporation shall conduct a discussion and conclude a revised agreement.

(Other Provisions)

Article 26: When doubts arise concerning the interpretation of this agreement as well as the interpretation of its successors and other particulars not established in this agreement, the Prefecture, the City, and the Corporation shall determine these points through discussion.

Supplementary Provisions

(1) This agreement shall be in effect from the day of its conclusion.

(2) The agreement concluded by the Prefec-

ture, the City, and the Corporation on December 9, 1969, shall terminate.

In order to witness the conclusion of this agreement, six copies of the original document shall be made and one copy shall be retained by each of the witnesses and concerned parties signing this document.

October 30, 1971

The Representative for the Governor of Hyōgo Prefecture

The Lieutenant Governor of Hyōgo Prefecture—Hitotsudani Inosuke

The Mayor of the City of Himeiji—Yoshida Toyonobu

The President of the Kansai Electric Corporation—Yoshimura Seizō

Witness: Osaka Bureau Chief of the Ministry of International Trade and Industry—Tsukamoto Yasao

Witness: Chairman of the Himeji City Assembly—Inoue Yoshinobu

Witness: President of the Himeji City Association for the Promotion of Local Autonomy—Kita Enji

Agreement 3: Safety and Environmental Protection Around a Nuclear Facility in Ibaraki Prefecture

Ibaraki Prefecture (hereafter A), the town where the nuclear facility exists (hereafter B), the towns close to B (hereafter C), and (the name of the enterprise which operates the nuclear facility) (hereafter D), in recognition of the fact that D has the responsibility of maintaining the safety of the area where its nuclear facility (for the study, development, and utilization of nuclear power for practical purposes) is located, and with the purpose of protecting the people and the environment (of the area), hereby agree as follows.

(Disposal of the radioactive waste)

Article 1

D shall not discard radioactive waste from its facility, either liquid or gas, where the radioactive content of such waste exceeds the limits defined by law.

2. D shall develop techniques for decreasing radioactive waste, and must operate the facilities consistent with a numerical goal based on the results of a total evaluation, and (formulated in the spirit) of "as low as practicable."

(Storage of radioactive waste)

Article 2

D must take appropriate precautions for storing and managing radioactive waste and other radioactive material. These precautions are in excess of that which is required by law.

(Prevention of pollution and environmental preservation)

Article 3

D shall take the measures necessary to prevent pollution of the air, water, etc. which may be caused by its activities, and, in order to preserve the environment of the area, take necessary (counter) measures, such as planting grass and trees on its site.

(Agreement concerning new construction)

Article 4

When D plans to extend its nuclear facility, or plans new construction which is closely related to the nuclear facility, or changes these plans, it shall obtain the approval of both A and B, except in cases involving incidental (not closely related) facilities.

2. A shall consult with C when the above situation requires.

(Strengthening the inspection system)

Article 5

D shall work toward strengthening the system used to inspect the disposal of radioactive waste, watch and measure the waste, and record the results.

(Guidance of trustee enterprises)

Article 6

When D entrusts the operation of the nuclear facility to a trustee, it must educate and train the trustee on safety, providing full guidance, in order that the trustee does not fail to be safe in its activities.

(Accident prevention measures)

Article 7

D shall strengthen its accident prevention system, and shall cooperate in a positive way with the accident prevention plans for the area.

(Self control)

Article 8

When necessary (in an emergency situation in its operations) to prevent accidents, D shall take all required steps, including the suspension of the operation of all or part of the nuclear facility.

(Safety measures)

Article 9

A and/or B, in the interest of safety, can request that D suspend the operation of its nuclear facility or that D improve its ways of operating if one of the following situations exists:

(1) When it is recognized, from on-site in-

spection, that it is necessary to take special measures to maintain the safety of the area.

(2) When there is a recognized, emergency need to prevent any accidents caused by D's operational activities.

2. When a request is made pursuant to the above, D shall take the necessary steps in good faith, and report on the measures it took to A and B. (Addition to nuclear power plants only—"Nevertheless, in the case where the operation of part of or the whole facility is terminated, it must be done in conformity with the applicable law.")

3. If, from an on-site inspection (as defined in Article 11, Section 2) or from information received pursuant to Article 13, C determines that it is necessary, C can request that A and B take the (necessary) steps outlined in Section 1.

(Compensation for damages)

Article 10

When the operation of D's nuclear facility causes damage to the people in the area, D must provide compensation in good faith.

2. When there is damage which appears to be caused by the operation of D's nuclear facility, and when there is a dispute as to whether the responsibility is D's, D must endeavor to settle the dispute, respecting the results of the joint investigation conducted by A, B and the Cs who have been affected.

(On-site inspection)

Article 11

A or B can, within the limits of the requirements of this agreement, request that D report (as defined elsewhere), or have their staff conduct the necessary on-site inspection of D's facilities.

2. After obtaining agreement from A and B, C's staff can conduct on-site inspections within the limits of the requirements of this agreement.

(Accompanying the on-site inspection)

Article 12

When B or C orders an on-site inspection pursuant to Sections 1 and 2 of the previous article, and if a disorder in D's nuclear facility is, or is feared to be detrimental to the environment or health of the people in surrounding areas, B or C can accompany their appointee on the inspection.

(Communication)

Article 13

D must, pursuant to the conditions described elsewhere, provide A, B, and C with all relevant information, such as D's safety policy.

(Respect for the opinion of the inspection committee)

Article 14

D shall respect (give credence to) the following from the report of the Radiation Inspection Committee for the Tokai Area, Ibaragi Prefecture.

1. Radiation inspection plan

2. Evaluation of the results of the radiation inspection

3. Results of the investigation of the environmental impact of radioactive waste disposal

4. Results of the investigation of environmental surveys

(Cooperation with investigations)

Article 15

D shall cooperate with safety research or investigations conducted by A, B, and C.

(Note)

Article 16

Details necessary to effectuate this agreement will be established in memoranda, following talks among A, B, C, and D.

(Talks)

Article 17

Changes in the terms of this agreement, questions on this agreement, or provisions not covered by this agreement shall be resolved in discussion among A, B, C, and D.

In order to ratify this agreement, copies of the original document will be made, and, after A, B, C, and D have signed them, each will retain one copy.

December 16, 1974

A Governor
   Ibaraki Prefecture
B Mayor
   Town where site is located
C Mayor
   Neighboring town
D Representative
   Nuclear facility

FINAL REPORT OF SEVEN MAJOR CITIES' INVESTIGATION TEAM (October 1974) [Excerpt]

1. The Restoration of the Cities in an Automobile Society.

The destruction of the environment caused by automobiles has led to serious changes such as the aggravation of traffic accidents, the expansion of traffic congestion, and the intensification of air and noise pollution. It has become the central theme of current urban problems.

The central government has pursued a policy of a high rate of growth for the Japanese economy and has attempted to protect the development of business with the automobile industry, which embraces a great many related industries, at the center.

Under this policy the level of automobile production has continued a surprisingly sharp increase of 28% in its yearly rate since 1960 and now production has reached 7,000,000 cars or 6.6 trillion yen (in 1973), with 25,000,000 cars being privately owned.

The startling motorization is bringing about not only environmental destruction but also the dissipation of natural resources and is reaching the point where it controls the basic structure of today's society, economy, and cities. Now it [motorization] monopolizes the city streets and snatches away even a breath of fresh air or a night's sleep, one's minimum-rights as a human.

Once the automobile indicated prosperity but now it has become the new symbol of poverty.

It must be said that because of its economic policy, the central government has been lacking in measures which would give the automobile industry prior guidance as to efficient methods of preserving the environment for the citizens.

Inspired by the "Muskie Law," which took effect in January, 1970 in America, Japan too has clarified its [anti-pollution] measures such as 1975 and 1976 regulations.

We cannot help but feel that if the two major automakers which have especially advanced technology had many years ago made a total effort in combatting the problems, and if the central government had played a leadership role with a stricter attitude towards technical oversights, the situation would be far and away advanced,

and today we would be beyond arguing for and against the enforcement of these 1976 regulations.

The automobile industry, under the protection of the central government, attained today's high growth rate while blaming the vast exterior waste of economy on the citizens and the local government. However, in proceeding with the logic of capital without further reflection, they [the automobile industry] have poured out the beguilement of the pursuit of profits and neglected the effort to improve the environment.

At this point things are reaching a state of affairs where, in order to recover fundamental human rights and the right to control their cities, citizens cannot help but defend themselves against the central government and the automobile industry which produce a social structure referred to as an "automobile society" and control the cities by the logic of capital.

In a response to the situation from the standpoint of protecting the environment for city residents, local governments have devised possible limited measures, but they are impatient with a difficult state of affairs for an effective policy, namely that legal measures relating to automobiles are almost the exclusive possession of the central government.

In a word, the 1976 regulations have become a place of confrontation of the logic of capital with the logic of the citizens.

We highly value the fact that the mayors of the seven cities representing the cities of Japan have attempted to grapple with the problem, and we have great expectations for future policies.

2. An evaluation of the Technical Possibilities of 1976 Regulations.

We, the investigation group, have overcome the investigative difficulties of the concealment of technical information by the two large automakers in the automobile industry, Nissan and Toyota. We have determined that even at the present levels of technology it is possible to reach the emission standards in the regulations or something appproaching their standards during 1976.

Toyota and Nissan put the technical limits of emission levels at four times those of the regulations on the premise of not changing various kinds of past efficiencies, such as the competition for speed. We cannot accept these contentions which are backed only by commercialism and are connected to environmental destruction. Also, we cannot approve the fact that the central government, based on this sort of contention, reconsulted the Central Commission for Pollution Control about the enforcement of 1976 regulations.

If the automakers say it is impossible, then the automakers' side has the responsibility to show actively and concretely experimental data, etc. and to demonstrate its position.

3. Proposed Measures.

Along with demanding that the central government move to an operation that will not lessen the regulatory amounts of the original policy, we propose the following as measures which should be taken by the seven major cities themselves for the defense of their citizens:

(1) Measures which should be justly carried out:

1. The recognition and use of low pollution cars by the seven major cities as well as recommendation for their use and agreements with the related groups.

2. Restrictions on high pollution cars on the city streets in times of photochemical smog emergencies.

3. The strengthening of parking regulations and the rapid enlargement of bus (and taxi) lanes.

4. A reexamination and increase of automobile related taxes; the establishment of a tax system and strong encouragement for low polluting cars.

The favorable treatment measures of the central government for low pollution automobiles which were sold prior to the 1975 regulations were only the slight reduction of the commodity tax, the automobile acquisition tax and the automobile tax.

In point of fact, if one desires to expand the production of low pollution cars, the central government should reduce the commodity tax, etc. all the more for low pollution automobiles and at the same time assess a high tax rate sufficient to pay the social costs on high pollution cars.

Local governments, too, should take like measures of their own regarding an automobile tax.

(2) Matters requiring the inauguration of long term investigations:

1. Restrictions on automobile production

2. A qualitative conversion of street construction

3. Another look at the traffic system (development of non-pollution automobiles, etc.)

The period of our investigation has been limited. The materials offered by the automakers have also been limited. However, in the background of the outcry of citizens making up ¼ of the total population we intended to investigate wholeheartedly and to raise questions. It will be good if this report is widely made public by the mayors of the seven major cities and stirs up new debates on the issues of automobiles and cities.

October 1974

The Seven Major Cities Investigation Group on the Problem of Regulating Automobile Emissions

LAW FOR THE RESOLUTION OF POLLUTION DISPUTES [SELECTED PROVISIONS]

Article 1 (Purpose)

The purpose of this law is to provide prompt and appropriate solutions to disputes over pollution by establishing a system of mediation, conciliation, arbitration and quasi-judicial arbitration.

Article 2 (Definition)

The term "pollution" (Kōgai) in this law will be the same as stipulated in Article 2.1 of the *Basic Law for Environmental Pollution Control.*

Article 3 (Central Committee)

Pollution Disputes Coordination Committee (hereinafter called the Central Committee) shall undertake mediation, conciliation, arbitration and quasi-judicial arbitration with regard to disputes concerning pollution in accordance with the provisions of the law and shall provide guidance to local governments in settling grievances over pollution.

Article 13 (Local Review Board)

A local government may establish a Local Pollution Review Board (hereinafter a Review Board) by ordinance.

Article 14 (Functions of Local Board)

The Review Board shall deal with the following matters: (1) it shall undertake mediation, conciliation, and arbitration with regard to disputes concerning pollution in accordance with the provisions of the law; (2) In addition to the matters provided in the preceding paragraph, it shall undertake such matters as delegated to it by the provisions of the law.

Article 15 (Organization of Local Board)

(1) The Review Board shall consist of 9 to 15 committee members.

(2) The Review Board shall have a chairperson who is elected from among the committee members.

(3) The chairperson shall manage the affairs of the Board and represent the Board.

(4) . . .

Article 17 (Duties of Local Board Members)

(1) Board members may not disclose secrets they learn in the course of their work. This applies after they leave such work.

(2) Board members shall not take executive positions in political parties and other political organizations nor shall they actively engage in political activity.

Article 17.2 (Review Board)

(1) The chairperson shall convene the review board.

(2) If there is a majority of the members as well as the chairperson not present, the board meeting cannot be opened and resolutions cannot be made.

(3) The business of the review board shall be decided by a majority of those in attendance. When there are the same number of pros and cons, the decision of the chairperson shall be followed.

(4) In the application of Article 2, if an accident should befall the chairperson, a board member prescribed in Article 15-4 shall be deemed to be the chairperson.

Article 20 (Establishment of a Joint Review Board)

The prefectures shall cooperate with other prefectures and shall be able to establish a Joint Prefectural Pollution Review Board (hereinafter called the Joint Review Board).

Article 23.2 (Agent)

(1) A party may appoint an agent who has been approved by a lawyer, a conciliation committee, an arbitration committee, or a quasi-arbitral determination committee.

(2) An approval as prescribed in the preceding paragraph may be revoked at any time.

(3) Competence of an agent shall be certified by a written document.

(4) An agent shall be entrusted with the following (when acting with regard to any one of them).

a. Withdrawal of application.

b. Acceptance of conciliation proposal.

c. Appointment of an agent.

d. Election of a representative according to Article 47.7, Paragraph 1.

Article 23.4 (Intervention)

(1) When a procedure of arbitration or quasi-arbitral determination concerning disputes over pollution related injuries is pending, a person alleging injury from the same cause may intervene in the procedure as a party (under the condition of) obtaining permission of the arbitration committee or of the quasi-arbitral determination committee.

(2) When an arbitration committee or a quasi-arbitral determination committee gives its permission as prescribed in the preceding paragraph, the committee shall hear opinions of the parties in advance.

Article 24 (Jurisdiction)

(1) The Central Committee shall have jurisdiction over mediation, conciliation and arbitration in the following disputes: (a) disputes concerning pollution in which serious injuries to human health and the living environment (as defined in Article 2-2 of the *Basic Law for Environmental Pollution Control*) exist and where such injuries extend to a considerable number of people or where there is a risk of such extension, and which are specified by a cabinet order.

(b) In addition to those mentioned in the preceding paragraph, disputes concerning pollution which affect more than one prefecture and which require a broader response, and which shall be specified by a cabinet order.

(c) In addition to those disputes mentioned in the two preceding paragraphs, disputes concerning pollution in which polluting enterprises or human activities and resulting injuries are in a different prefecture, or either or both of the above are located in more than one prefecture.

(2) The Local Review Board shall have jurisdiction over mediation conciliation, and arbitration with regard to all other disputes of pollution except for those mentioned in paragraphs (a) through (c) above.

(3) The parties may choose by agreement either the Central Committee or the Local Board for arbitration notwithstanding paragraphs (1) and (2) above.

Article 26 (Application)

(1) One or both parties to a dispute over compensation or other civil disputes relating to pollution-related injuries may apply in writing to the Central Committee or a Local Review Board for conciliation, mediation, and arbitration. Application to the Review Board shall be made through the governor.

(2) A party who applies for arbitration must have the consent of the other party to submit the dispute to arbitration.

Article 27.2 (Commencement of Mediation by the Committee or Board)

(1) The Central Committee or a Local Review Board may commence mediation upon making an investigation and hearing the opinion of the parties, by making its own resolution where there is a civil dispute over serious and extensive pollution-related injuries, where negotiations between the parties have not proceeded amicably, where the dispute may result in serious social consequences such as economic difficulty for a large number of victims if left unattended.

(2) The above proceeding shall commence upon the request of the governor concerned.

Article 27.3 (Removal from Mediation to Conciliation)

The Central Committee or Local Review Board may proceed to conciliation on the recommendation of the mediation committee members after hearing the opinions of the parties and making a determination that it is too difficult to solve this problem by mediation and that conciliation seems more appropriate.

Article 28 (Appointment of the Mediation Members)

Mediation by the Central Committee or Local Review Board shall be conducted by one or more mediators, not to exceed three.

Article 31 (Appointment of the Conciliation Committee)

Conciliation by the Central Committee or Local Review Board shall be conducted by establishing a three member committee.

Article 32 (Requirement of Appearance)

When considered necessary for conciliation, the conciliation committee may request the appearance of parties involved and can ask them their opinions.

Article 33 (Request for Documents and other Measures)

(1) The conciliation committee shall request the parties, where necessary, to produce documents or materials related to the case when the committee conciliates a dispute as provided in Article 24 (1) (a).

(2) The conciliation committee may undertake, if necessary, to ascertain the facts which caused the dispute by entering a factory, place of business or other premises and inspecting documents or materials related to the case, when the committee conciliates a dispute as provided in Article 24 (1) (a).

(3) The conciliation committee may request specialist members to assist the committee with on-site inspections provided in the preceding paragraph. (A similar power is granted to the arbitration committee by Article 40.)

Article 34 (Conciliation Proposal)

(1) The conciliation committee shall make a conciliation proposal taking all circumstances into consideration and advise the parties to accept it within 30 days, when the committee finds that it is difficult for

the parties to reach an agreement, and when the committee deems it appropriate to do so.

(2) The committee shall formulate a conciliation proposal provided for in the preceding paragraph by the majority opinion of its members.

(3) The parties shall be deemed to reach the same agreement as the proposal made by the committee in accordance with paragraph 1 unless the parties report to the committee within the prescribed period that the proposal is not acceptable.

Article 34.2 (Announcement of Proposal)

The conciliation committee may make the proposal, made in accordance with the preceding Article, paragraph 1, public with the basis for its reasoning, when the Committee deems it appropriate to do so despite Article 37. (The conciliation proceeding is not open to the public—Article 37.)

Article 42.12 (Responsibility Arbitration)

(1) When a dispute arises over the compensation of pollution-related injuries, a party who claims compensation may apply to the Central Committee in writing in accordance with the regulation of the Central Committee for quasi-judicial arbitration on the issue of responsibility for compensation (hereafter responsibility arbitration).

(2) The Central Committee may reject the application when the committee deems it inappropriate to undertake responsibility arbitration by taking into consideration such factors as the limit, degree, and extent of injuries, the existing situation of disputes, and all other circumstances.

Article 42.15 (Open Hearing)

The quasi-judicial arbitration proceeding shall be open to the public provided that when the committee finds it necessary to do otherwise for reasons of protecting the privacy of an individual or business, proprietary information, or of protecting procedural fairness or other matters in the public interest, it can do so.

Article 42.16 (Examination of Evidence)

The quasi-judicial arbitration committee may, upon application of the parties or on its own initiative, undertake the following investigations:

(1) The committee may:

(a) order the parties and witnesses to appear before the committee to testify;

(b) order an expert to appear to give an opinion;

(c) order a person who holds documents or materials related to the case to submit them to the committee which may retain them;

(d) enter the places related to the case to inspect documents and materials.

(2) The parties may attend an evidentiary investigation held on an occasion other than on the day of a hearing.

(3) The committee shall hear the opinions of the parties about the result of such evidentiary investigations when the committee initiates such investigation.

Article 42.17 (Preservation of Evidence)

The Central Committee may take measures prior to the filing of an application for responsibility arbitration, to preserve evidence upon the request of a party who will be applying for responsibility arbitration when the committee finds it difficult to secure such evidence at a later time.

Article 42.18 (Investigation of Facts)

(1) The quasi-judicial arbitration committee may investigate, or have officials of the Central Committee's secretariat investigate, the facts when the committee deems it appropriate to do so.

(2) Committee or secretariat officials under the Committee's direction may enter a party's factory, place of business and other premises to inspect documents or materials related to a case when the Committee deems it necessary to do so in making a factual investigation in accordance with the preceding paragraph.

(3) The Committee shall listen to the opinion of the parties on the results of a factual investigation when it considers such results as material to responsibility arbitration.

(4) The Committee may accompany the specialist members to the on-site inspection provided for in paragraph 2 to assist the Committee.

Article 42.20 (Effect of Responsibility Arbitration)

Where there has been a determination by arbitration of responsibility, if the parties refrain from bringing a suit for damages or withdraw a suit within 30 days after the original text for responsibility arbitration is delivered, both parties shall be deemed to have accepted responsibility arbitration.

Article 42.21 (Exception to Administrative Procedure Act)

The parties shall not bring a suit under the Administrative Procedure Law contesting either the result of responsibility arbitration or the Committee's procedural disposition (of the case).

Article 42.24 (Official Conciliation)

(1) In addition to placing cases under conciliation, the quasi-arbitral determination committee may, when suitable, obtain the agreement of the parties involved and resolve the dispute through a review board which has jurisdiction over the dispute. Furthermore, regardless of the regulations outlined in Article 24 Paragraphs one and two and Article 31 Paragraph one, the committee may settle the dispute on its own.

(2) In the event that, by virtue of the regulations set forth in the preceding Paragraph, a case is put under conciliation, when a mutual agreement is reached among the parties involved, the application for a quasi-arbitral determination of responsibility shall be considered dismissed.

Article 42.26 (Relation to Civil Litigation)

(1) A court may suspend its hearing of a case in which a party has made an application for responsibility arbitration until an arbitration decision is made.

(2) When a court proceeding is not suspended as provided in paragraph (1), the arbitration committee may suspend its own proceeding of responsibility arbitration.

Article 42.27 (Cause Arbitration)

(1) In a dispute over compensation or in other civil disputes over pollution related injury where the cause of injury is in issue, a party may apply to the Central Committee in writing for arbitration on the cause of injury (hereinafter "cause arbitration") in accordance with the Central Committee regulations.

Article 42.29 (Authorized Cause Arbitration)

The quasi-judicial arbitration committee during responsibility arbitration may, under its authority when appropriate, also undertake cause arbitration.

Article 42.32 (Referral of Cause Arbitration by Court)

(1) In civil litigation involving pollution related injuries, when necessary, a court may ask the Central Committee upon hearing its opinion to undertake cause arbitration.

(2) A court, when necessary, may have a person named by the Central Committee explain cause arbitration where the court requests the Central Committee to undertake cause arbitration in accordance with paragraph (1).

Article 49 (Disposition of Complaints)

(1) Local government in cooperation with responsible administrative agencies shall endeavor to seek appropriate settlement of pollution-related complaints.

(2) Prefectures, and cities which are designated by cabinet order shall appoint pollution complaint counselors.

(3) Cities other than those provided for in paragraph (2) and towns may also appoint pollution complaint counselors.

(4) The pollution complaint counselor shall deal with the following matters concerning pollution-related complaints:

(a) provide counseling to residents;

(b) investigate, provide guidance and advice necessary to settle complaints; and

(c) in addition to the matters provided in paragraphs (a) and (b), shall give notice to responsible administrative agencies and undertake other matters necessary to the settlement of complaints.

Notes and References

Chapter 1

1. Kōgai Taisaku Kihon Hō (1967, Law No. 132); amended as Law No. 132, 1970; No. 88, 1971; No. 111, 1973; and No. 84, 1974 (henceforth, the Basic Law).

2. Chapter 2 analyzes in depth this process of value transformation, and its impact on legal institutions and Japanese society in general.

3. The following is a summary of F. G. Notehelfer's splendid study of the Ashio case, *Japan's First Pollution Incident*, 1, J. Japanese Studies (no. 2, Spring), 351. The reader should note, however, that Ashio was not, as Notehelfer states, Japan's "first" pollution incident; rather Ashio represents the most important of many episodes in the prewar era. For an interesting historical compilation dating from 1469 of sicknesses and other injuries concerning pollution and labor relations, see Nobuko Iijima, Kōgai Rōsai Shokugyōbyō Nempyō (A Chronology of Pollution, Labor, and Employment Illnesses), Kōgai Taisaku Gijitsudōyūkai (Pollution Counter Measures Technical Friendship Association), 1977. The increasing national concern in Japan over pollution has also stimulated other serious studies in both English and Japanese on the history of Japanese environmental problems. These include: Kanson Arahata, Yanaka Mura Metsubōshi (History of the Destruction of Yanaka Village) (1970); Nobuko Iijima, *Nihon Kōgaishi Kenkyū Nōto* (*Research Notes on Japanese Pollution History*), 3 Kōgai Kenkyū (Pollution Research) No. 3, pp. 56–63 (1974); No. 4, pp. 53–60 (1974); 4 Kōgai Kankyō No. 1, pp. 60–68 (1974); No. 3, pp. 40–48 (1975); No. 4, pp. 61–69 (1975). Taku Ōshika, Watarasegawa (Watarase River) (1974); Kenneth Strong, *Tanaka Shōzō: Meiji Hero and Pioneer Against Pollution*, 67 Japanese Society Bulletin 10 (June 1972).

4. As Notehelfer notes, each pit was under the supervision of a subcontractor who hired his own labor, provided his own tools, and managed production as he saw fit. Miners policed themselves through vigilante committees and punishments were group-inflicted and very cruel. In 1890 Japan adopted a second code for mining (Kōgyō Jōrei) whose most significant feature was the establishment of a mining police system to supervise mine safety. Mining police affairs were made the responsibility of the Di-

rector of the Mine Supervisory Office, itself placed under the supervision of the Minister of Agriculture and Commerce.

4a. 1.66 kin equals one kilogram.

5. Notehelfer describes how Furukawa's efforts were substantially encouraged by a contract to supply copper to the Jardine-Matheson Company. The contract guaranteed sales for a three year period in excess of ¥6,500,000 despite a severely fluctuating market.

6. Tanaka pointed out that art. 27 of the Meiji Constitution protected the people from infringements to property.

7. Article 59 of the Mining Law of 1890 stated:
If the Director of the Mine Supervisory Office determines that there is a danger of mine pollution, he shall order the owner of the mining right to prevent such dangers or mine pollution or suspend his mining operations where necessary.
And in order to enforce the mandate of art. 59, art. 60 provided that
In cases referred to in the preceding Article, the Director of the Mine Supervisory Office, when the owner of the mining right has not immediately begun to take measures to prevent dangers or mining pollution, shall enforce the order by directing the executives and mine workers of the mine. In this case the owner of the mining right shall be under a duty to supply its supervisors and mine workers and to bear all necessary expenses.
See also n. 53 in Notehelfer, n. 3 supra.

8. Not until 1939 did the Japanese mining laws have a strong provision on no fault liability providing for compensation. For example, although art. 23, ch. 5 of the Japan Mining Law of 1873 (Nihon Kōhō) did include a kind of compensation provision, the section was too general to have any practical effect. Similarly the new Mining Law of 1890 also failed to provide an effective clause. See Zensuke Ishimura, Kōgyō Ken no Kenkyū (A Study of Mining Rights) (1960).

9. For a general discussion of the Supreme Court under the Meiji Constitution, the Great Court of Cassation (Taishin-in) see H. Tanaka, The Japanese Legal System (1976).

10. We do not believe, however, that this was the motivation of the inquiry conducted by the Tokyo Appeals Court. Even

during this early period, the court was acting as a check on the excesses of the administration. For a further discussion, see ch. 4.

11. Until very recently the spectacle of the "kemuri no miyako" (cities engulfed in smoke) was itself actually viewed as an emblem of progress. We must be careful here, however, not to read present conceptions of "pollution" into the attitudes of people during this period. For an interesting comparison see Justice J. Musmanno's discussion of the virtues of factory smoke during the midst of the depression in *Versailles Borough v. McKeesport Coal and Coke Co.*, 83 Pittsburgh Legal J. 379 (1935).

12. For a discussion of the fiscal difficulties of Meiji and Taishō governments, see John K. Fairbank, Edwin O. Reischauer, and Albert M. Craig, East Asia: The Modern Transformation (1965).

13. This point raises important questions, discussed later, on both societies' concepts of law. From the Japanese perspective administrative action is part and parcel of the law. The reader might compare ch. 5, which discusses in detail the regulatory process in postwar Japan. Similarly, legislation in Japan tends to be phrased in broad terms with specifics left to administrative regulation.

14. See sec. 1.2.4 for a discussion of this case, and sec. 1.3 for a comment on the distinctive lack of environmental litigation before World War II.

15. During the Meiji era students of Western law were only beginning to become familiar with Western concepts of rights. The legal sociologist Takeyoshi Kawashima offers this report on the Meiji government's confusion over the concept:
The lack of the concept of "right" was particularly evident on the level of court practices with respect to infringement by the government upon the interests of private persons or their associations. When the translation of the French Civil Code, in which the expression "droit civil" was translated into Japanese for the first time as "minken" (right of a subject or citizen) was under debate in a conference of the Ministry of Justice in 1870, some of the members of the conference were not familiar with the concept that citizens have any rights whatsoever, and raised the question: "What on earth does it mean to say that a citizen has

a right!" This point is expressed in exaggerated form in Kokutai no Hongi (Grand Principles of National Policy) as an authoritative statement by the government:
To conceive of relationships of dominance and obedience or rights and duties is a rationalistic view which, based on an individualistic viewpoint, considers everything to be relationships between equal persons. Our relationship between the emperor and his subjects is not a shallow flat one at all in which the emperor and his subjects confront each other, but a relationship of identification through self-denial which, starting from this new base (konpon), never loses this base (konpon). It can never be apprehended by the individualistic way of thinking (Takeyoshi Kawashima, The Japanese Mind, 1973).

16. We wonder how self-consciously these farmers acted because of the betrayal of the "social compact," as Notehelfer suggests.

17. One example is that compensation awards in the major pollution trials were received as a lump sum and then distributed among the victims. In some cases part of the awards were devoted to the financing of legal services for the victim's legal actions.

18. Jun Ui, The Singularities of Japanese Pollution, 19 Japan Quarterly 281 (no. 3) (1972). Ui suggests that these principles, "contrary to expectation," apply to pollution problems in foreign countries.

19. Ui suggests that in negotiations the polluter usually will refuse to add the sums demanded by each side and divide by two (the average solution) to identify the appropriate monetary sum for compensation; rather, the victimizer multiplies the parties' monetary proposals and then takes the square root in order to reach the "compromise solution." Ui argues that when the victim starts talking about compromise, "he will lose for sure."

20. The summary of these early cases is drawn from the special issue on pollution in 458 Jurisuto 1 (The Jurist) (1970).

21. Arbitration is generally a less favored form of dispute settlement in Japan. Because the Minister of Agriculture and Commerce himself mediated the disputes in this case, in deference to his office the parties had to pledge that they would unconditionally abide by whatever decision was reached. Compare this case with another arbitration

decision delivered in the Asano Cement case.

22. Or about $390,000. The reader should note that the indemnity payments were not distributed to individual farmers but rather were collectively used for general agricultural improvements in the area.

23. Before World War I Japan imported potash from Germany, but these imports were terminated by the war. As the price of potash rose, it became economically sensible to retrieve potash from dust.

24. Until this case, the general practice was not to compensate for mental suffering associated with loss of property, and even today the courts are reluctant to grant such relief. In addition Hitachi agreed to pay for the present discounted value of the expected profit from the small trees that were destroyed, measured by the expected life of the trees. This was a remedy far in advance of its time.

25. Reports of these court decisions may be found as follows: High Court, July 29, 1916, 1047 Hōritsu Shinbun 25; Court of Great Judicature, December 22, 1916, 8 Minroku, No. 22, 2474, (1916); Osaka High Court, December 27, 1919, 1659 Hōritsu Shinbun 11.

26. See Hideo Tanaka, The Japanese Legal System (1976) for a discussion of the Great Court of Judicature (Taishin-in).

27. This is the second example during this period (the other is the Osaka Alkali case) of an injured party turning to the courts for relief.

28. But see Edward Dalton, *Combating Disease and Pollution in the City*, in American Environmentalism, Donald Worster, ed. (1973).

29. James E. Krier, *The Pollution Problem and Legal Institutions: A Conceptual Overview*, 18 UCLA L.R. 429 (1971).

30. Jan G. Laitos, *Legal Institutions and Pollution: Some Intersections Between Law and History*, 15 Natural Resources J. (July 1975). See also S. Edelman, The Law of Air Pollution Control (1970); H. W. Kennedy and A. O. Porter, *Air Pollution: Its Control and Abatement*, 8 Vand L.R. 854 (1955).

31. Laitos, n. 30 supra, 434.

32. Laitos' article includes the following references which the reader may wish to consider in light of contemporary Japanese developments: Detroit, Michigan, Rev. Ordinances, ch. 37 ("Any owner . . . who shall cause smoke to be emitted from such structure . . . shall be liable to a fine of not less than $10 or more than $100."); New York Sanitary Code, sec. 181 (cited in *People v. New York Edison*, 159 App. Div. 786, 144 NYS 707 (1913)); Chicago Ordinances 10 1903 (cited in *Glucose Refining Co. v. City of Chicago*, 138 Fed. 209 C.C.N.D., Ill. (1905)).

33. See sec. 134 of New York City Sanitary Code (1899). The ordinance reads in part: *Owners, lessees, . . . of every blacksmith or other shop . . . shall cause all ashes, cinders, rubbish, dirt, and refuse to be removed to some proper place . . . Nor shall any owner . . . allow any smoke . . . to escape . . . from any such building . . . and every furnace employed in the working of engines . . . or used for the purposes of trade or manufacturing, shall be so constructed as to consume or burn smoke arising therefrom.*

34. St. Louis, Missouri, Ordinances 41804, sec. 5340 (cited in *Ballentine v. Nester*, 350 Mo. 58, 164 S.W.2d 378 (1942)).

35. See *Ex Parte Junqua*, 10 Cal. App. 602, 103 P. 159 (1909), in which the court upheld a Sacramento ordinance prohibiting the escape of soot from smokestacks using distillate or crude oil. Ordinances were also upheld in *People v. Lewis*, 86 Mich. 273, 49 N.W. 140 (1891); *Dept. of Health of N.Y. City v. Ebling Brewing*, 78 N.Y.S. 11 (Mun. Ct. 1902). Enabling statutes (granting legislative permission to enact smoke ordinances such as the above) were upheld in *City of Cincinatti v. Miller*, 29 W.L. Bull. 364 (1893); *City of Brooklyn v. Nassau Electric R. Co.*, 44 App. Div. 462, 61 N.Y.S. 33 (1899) and the Supreme Court's opinion in *Northwestern Laundry v. Des Moines*, 239 U.S. 486 60 L. ed. 396, 39 S. Ct. 206 (1916).

36. For example, see *Sullivan v. Jones and Laughlin Steel Co.*, 208 Pa. 540, 57 A. 1065 (1904). In *Sullivan*, the court granted an injunction that stopped a polluting industry from expanding its operations.

37. See the famous dispute between Georgia and Tennessee over the pollution in Georgia from the operation of the Tennessee Copper Company. *Georgia v. Tennessee Copper Company*, 36 S. Ct. 465, 60 L. ed. 846 (1916). The court granted an injunction prohibiting discharges from the Tennessee plant.

38. See *Price v. Philip Carey Manufacturing*

Company, 165 A. 849, 42 Am. Rep. 534 (1933).

39. See *Judson v. Los Angeles Suburban Gas Co.*, 157 Cal. 168, 106 P. 581 (1910). In *Judson* a California appellate court upheld the granting of an injunction and judgment for damages against defendant's gas works.

40. Laitos, n. 30 supra, 432.

41. Id.

42. See these references cited by Laitos: *McMorran v. Cleveland-Cliffs Iron Co.*, 253 Mich. 65, 234 N.W. 163 (1931) (refusal to restrain vessels from emitting smoke at the dock); *Pettit v. New York Cent. and H.R.R. Co.*, 29 N.Y.S. 1137 (1914) (refusal to grant damages to the plaintiff from smoke emitted by railroad pumping station).

43. The Mining Law, passed in 1873, was intended to provide a general framework for administrative control of industry and was not specifically designed to deal with the problem of pollution. The reader should note the 1877 Osaka prefectural ordinance that controlled municipal factories by means of a permit system and authorized termination of a factory's operations that injured public health. See, however, the National Parks Law enacted in 1931, which was the first legislation designed to protect the natural environment in Japan.

44. We have not been able to obtain any historical data to clarify what motivated the plaintiffs to turn to the courts in this case.

45. But see John D. Haley, *The Myth of the Reluctant Litigant*, 4 J. Japanese Studies 359 (1978), a recent study suggesting the existence of considerable litigation in other areas. For a discussion of major changes in the society's perception of the courts, see ch. 2 of that work.

46. Ui argues that (1) low investment in research and development of pollution control technology, (2) low wages, (3) protection against foreign competition, and (4) little official attention to the crucial costs of industrial activity comprised the foundation of Japan's economic development.

47. See, for example, the Asano Cement case. For some insights into social perceptions on the role of science in Japan, see Science and Society in Modern Japan, Shigeru Nakayama, et al. (1974).

48. Kōjō Kōgai Bōshi Jōrei (Tokyo Ordinance No. 72, August 13, 1949).

49. Article 18 was typical: Where the community was harmed by dust, offensive odors, poisonous gases, steams, liquid wastes, noise, or vibrations emanating from a factory, the metropolitan government was empowered to order the factory to repair its facilities, to stop or limit its operations, or otherwise to abate the offensive condition. Of course, even today in the United States and the rest of the world, local ordinances tend to be vaguely drafted.

50. Kanagawa-Ken Ordinance No. 28, December 28, 1951.

51. Osaku-Fu Ordinance No. 12, April 13, 1954.

52. Fukuoka-Ken Ordinance No. 6, April 1, 1955.

53. Pollution itself was in part the product of the extraordinary rapid growth that Japan enjoyed during this period, and this growth was largely the result of U.S. military spending during the Korean War.

54. Many local governments wanted companies to establish factories as a means of bolstering the tax base. See Ministry of International Trade and Industry (MITI) and Ministry of Health and Welfare (MHW), Taiki Osen No Shōmondai (Problems of Air Pollution) 49–51 (1956).

55. In August 1955 MHW drafted an outline of the law. MITI and the Ministry of Transportation opposed the draft. MITI was critical of MHW's preparing a law of this kind because MITI believed it had sole jurisdiction over the regulation of industry. MITI was also more sympathetic to industry. The argument that pollution control would stop economic growth soon became the chief excuse for inaction.

56. The Tokyo government initially was reluctant to act because the fishermen were from another province. Even after it asked the plant to control the effluent, the company refused.

57. Kōkyōyō Suiiki no Suishitsu no Hozen ni Kansuru Hōritsu (1958, Law No. 181).

58. Kōjō Haisui tō no Kīsei ni Kansuru Hōritsu (1958, Law No. 182). For the legislative history and details of these laws, see e.g., Nobuo Hirai and Nobuhiko Yamada, *Mizu Osen No Kisei (Regulation of Water Pollution)*, in Kōgai Hō No Seisei To Tenkai (Outgrowth and Development of Pollution Laws) 99–145 (Katō ed., 1968). Even before

these two laws were enacted, there had been some individual laws that prohibited water pollution: the River Law (Kasen Hō) (1896, Law No. 71), the Port Law (Kōsoku Hō) (1948, Law No. 174), the Fishing Port Law (Gyōkō Hō) (1950, Law No. 137), the Water Supply Law (Suidō Hō) (1957, Law No. 177), the Cleaning Law (Seisō Hō) (1954, Law No. 72), the Sewage Law (Gesuidō Hō) (1958, Law No. 75), the Water Resources Protection Law (Suishigen Hogo Hō) (1951, Law No. 313), and others. However, except for drinking water, these laws failed to provide standards for water quality and were ultimately ineffective.

59. The Director General of the Economic Planning Agency, after consultation with the Water Quality Commission, would designate an area when "industry (fishing, etc.) in the area has been substantially impaired or public health has been impermissibly affected, or some risk of these exists" (art. 5). The twenty members of the Water Quality Commission included representatives from related ministries.

60. Article 1 stated: ". . . should contribute to the natural harmonization of industry and to the promotion of public health."

61. Standards were to be set by the Director General of the Economic Planning Agency after consultation with the Water Quality Commission.

62. All these laws were criticized as serving merely as stopgap measures. As noted below, the Extraordinary Diet Session of 1970 for pollution control modified and consolidated these separate laws into one law, the Water Pollution Prevention Law (Suishitsu Odaku Bōshi Hō) (1970, Law No. 138).

The Water Quality Conservation Law also established conciliation committees, to be appointed by the prefectural governors, which were supposed to settle disputes between the factories and fishing and farming communities. The committees were to conduct negotiations over compensation and sometimes over a factory's installation of control equipment or other prevention measures. But during the twelve years the law was in force, only thirty-four cases were filed with the committees for conciliation. Chūō Kōgai Shinsa Iinkai Jimukyoku (Secretariat of the Central Pollution Investigation Committee), Kōgai Funsō Shori Hō No Kaisetsu

(Commentaries on the Law for the Resolution of Pollution Related Disputes) 12–13 (1970). Eventually Japan adopted a comprehensive approach to extrajudicial dispute settlement in the 1970 Law for the Resolution of Pollution-Related Disputes.

63. See Tsūsanshō Kigyō Kyoku Sangyō Kōgai Ka (Industrial Pollution Section, Industry Bureau, MITI), Baien Kisei Hō no Kaisetsu (Commentaries on the Smoke and Soot Regulation Law) 1 (1953). See also Baien no Haishitsu no Kisei tō ni Kansuru Hōritsu, (1962, Law No. 146).

In 1962 the government had just promulgated its ten-year plan to double national income and also a national comprehensive development plan for the expansion of large industrial complexes.

64. During the discussion of the bill at the Committee on Commerce and Industry in the Upper House, a government representative stated:
In the view of the Ministry of Health and Welfare, the government is obliged without any reservation to respect human dignity and public health. However, in order to realize this absolute value, it costs a lot to install reasonable facilities. Therefore, it is not reasonable for the government completely to adopt the ministry's standpoint when we see the whole picture. As stated in the provision, we believe, it is necessary to harmonize the policy with the sound development of industry.
Dai 40 Kai Kokkai Sangiin Shōkō Iinkai Kaigiroku (Proceedings of the 40th Diet Session, Committee on Commerce and Industry, Upper House, No. 28), p. 33.

65. During the Diet debates, the bill was modified to grant cities powers to enforce the law. The House of Representatives also passed a resolution requiring the government to grant financial assistance to small industries and promote the control of automobile emissions.

66. Baien no Haishitsu no Kisei tō ni Kansuru Hōritsu (1962, Law No. 146).

67. Or the mayor in the case of large cities. See arts. 8 and 19.

68. See arts. 5, 6, and 17.

69. The government was also reluctant during this period to reduce Middle East imports of oil of high sulfur content because of economic considerations.

70. See regulations pertaining to the control of land subsidence.

71. For an interesting discussion of the legislative process in the environmental field, see Margaret A. McKean, *Pollution and Policymaking* in Policy Making in Contemporary Japan, T. J. Pempel, ed. (1977). Another useful study in Japanese is Michio Hashimoto, Kōgai o Kangaeru (Thoughts on Pollution) (1970).

72. The uncertainty over the preemption issue prompted local officials to request a special legislative amendment. In 1963 art. 32-2 was added to the Smoke and Soot Regulation Law. This provision permitted local governments to regulate all pollution facilities not otherwise regulated by law.

73. Hashimoto in Kōgai o Kangaeru (n. 71 supra) reports that the Mishima-Numazu controversy may have been influenced by an earlier dispute with a petroleum refinery in Nishinomiya City, Hyōgo Prefecture. People there had been afraid of air pollution caused by the petrochemical industry. The serious air pollution in Yokkaichi was well known throughout the country and people were aware that the government had failed to solve the Yokkaichi problem. The local *sake* brewers in Nishinomiya opposed the construction plan on the grounds that the air pollution and depletion of ground water caused by the petrochemical industry might destroy their famous local industry. As a result the plan for the petroleum refinery was eventually suspended. This incident demonstrated an increase in public concern over pollution problems.

There are many articles on the Mishima-Numazu residents' movement. See, for example, N. Huddle and M. Reich, Island of Dreams (1975) pp. 256–263. Akio Nishioka and Tōru Yoshizawa, *Shimizu-Mishima-Numazu Sekiyu Konbināto Hantai Undō (Anti-petrochemical Industry Complex Movement in Shimizu-Mishima-Numazu Areas)* in Nihon Gyōsei Gakkai (Japanese Association of Public Administration) Kōgai Kyōsei, ed. Pollution Administration, 217–241.

74. Special Standing Committees for Industrial Pollution Control (Sangyō Kōgai Taisaku Iinkai) were established in both houses of the Diet during the 48th session (1965).

75. Our description of the drafting process relies on Iwata, ed., Kōgai Taisaku Kihon Hō No Kaisetsu (Commentaries on the Basic Law for Pollution Control), 15–106. The reader should note, however, that even before the Mishima-Numazu incident, the government had considered the idea of a basic organic statute for pollution control. As early as 1963 the Council on Population Problems, an advisory organ to the Ministry of Health and Welfare, released an opinion underlining the need for a comprehensive policy, especially one that fixed responsibility on industry. See Jinkō Mondai Shingikai (Council for Population Problems), Chiiki Kaihatsu Ni Kanshi, Jinkō Mondai No Kenchi Kara Tokuni Ryōi Su Beki Jikō Ni Tsuite No Iken (An Opinion on Points Related to Area Development, Which Should Be Given Special Attention from the Viewpoint of Population Problems) (August 17, 1963). After Takeharu Kobayashi, then Minister of Health and Welfare, visited Yokkaichi in the same year, the importance of enacting a basic pollution policy law became even more apparent. In 1964 on the request of the Ministry of Health and Welfare, the Cabinet established a coordinating body, the Liaison Council for Pollution Control Promotion (Kōgai Taisaku Suishin Renraku Kaigi) which it charged with the duty of "remaining in intimate contact with the related administrative bodies in order to promote comprehensive and effective pollution control measures." The council consisted of a number of high-ranking (vice-ministerial) officials from different ministries and agencies.

76. The Environmental Pollution Commission was composed of forty members, most of whom were university professors, although a number were former government officials and representatives of industry, and a few were journalists. The commission was divided into four subsections that immediately initiated deliberations of various problems.

77. Specifically, the standing committee in the House of Representatives resolved that the government should promote a study of techniques for the desulfurization of gas and crude oil and also formulate plans for the relocation of factories. The House of Councillors' Special Standing Committee emphasized that (1) the administration of pollution control should be consolidated within a special environment agency, and

418

an advisory body of experts should be assembled; (2) the classes of pollution subject to regulation should be adopted; and (3) the allocation of responsibility for pollution control between the central and local governments and industries should be clarified.

78. See Kōgai Seisaku no Kihonteki Mondaiten Ni Tsuite No Iken (An Opinion on Fundamental Problems in Pollution Control Policy), Federation of Economic Organizations, Oct. 5, 1966.

79. As noted, the earlier air and water laws included similar "harmony" provisions.

80. The ministry suggested that prefectures be empowered to control air and water pollution, and municipalities, noise and fumes. It also urged that governors be authorized to designate areas for pollution control programs, that each prefectural government should have a committee for settlement of disputes over pollution, and that a feasibility study for the establishment of a relief system for pollution-induced illnesses be undertaken.

81. The final report consisted of a preface, a conclusion, and fifteen sections. The introductory section, which sets forth the basic principles of a pollution control policy, stressed that

1. comprehensive measures should be taken to maintain ambient standards;
2. preventive measures, such as land use planning, should be emphasized more;
3. the central and local governments should share the responsibility for prevention of pollution, particularly for the construction of public facilities such as sewage systems;
4. pollution control measures for polluted areas should be implemented in stages with continuous consultation with regulated industries;
5. strict controls should also apply to unpolluted areas; and
6. long-term policies should properly be distinguished from short-term emergency measures.

The report limited itself to the most desperately needed controls on pollution that affected human health or property.

82. It is almost certain that consensus had already been established before the meeting began.

83. In Japan, as in many other countries, the government (i.e., the bureaucracy) usually drafts the law. Upon the draft's completion and approval by the ruling party, it is thereafter submitted to the Diet for debate.

83a. The Ministry of Agriculture and Fisheries was renamed the Ministry of Agriculture, Forestry, and Fisheries on July 10, 1978.

84. The Liaison Council was composed of fourteen (later fifteen) ministries or agencies, all of which had some connection with environmental protection.

85. Shortly after the publication of the Environmental Pollution Commission's report, a number of other agencies released reports which officials in the Ministry of Health and Welfare considered. For example, the Ministry of Construction (Kensetsushō) announced a pollution control policy that emphasized land use planning and public works projects (such as the construction of green buffer zones). See Kensetsushō no Kōgai Taisaku (Pollution Control Measures of the Ministry of Construction), November 15, 1966.

On November 28, 1966, the Industrial Pollution Subcommittee (Sangyō Kōgai Bukai) of the Commission on Industrial Structure (Sangyō Kōzō Shingi Kai), an expert advisory organization attached to MITI, published its report, On the Measures for Industrial Pollution Control (Sangyō Kōgai Taisaku No Arikata Ni Tsuite). Although some members of the commission had served on the Environmental Pollution Commission of the Ministry of Health and Welfare, the ideas in this report closely resembled those outlined in the Keidanren opinion. The report primarily blamed Japan's unusually rapid post–World War II urbanization and industrialization and the low level of government investment for the nation's pollution problems. It recommended that the central and local governments promote the redevelopment of existing industrial areas, develop land use plans for new areas, and provide industry with financial and technical assistance. Finally, the report emphasized that an ideal pollution control policy should strike a balance between the preservation of public health and the sound development of the economy.

It is interesting that on November 15 of the same year the National Living Council (Kokumin Seikatsu Shingi Kai), another advisory council of the Economic Planning Agency, submitted a report on general policy, in which the council suggested that polluters bear the responsibility for pollution. Nonetheless, the report required the central and local governments to share responsibility for pollution control with industry. [Shorai Ni Okeru Nozomashii Seikatsu No Naiyō To Sono Jitsugen No Tameno Kihonteki Seisaku Ni Kansuru Tōshin (Report on the Basic Policies Concerning the Substance of Desirable Future Life and Its Realization), Nov. 15, 1966.]

86. We will discuss the significance of this phrase shortly. See The Basic Law for Environmental Pollution Control of 1967, art. 1 (2).

87. The Liaison Counsel's draft outline provides:

Chapter 2. Establishment of Ambient Standards

1. The government should establish standards for desirable environmental conditions, with respect to air and water pollution and noise, to be maintained in order to preserve the human health and living environment.

2. In the setting of standards prescribed in 1, mutual harmonization between industries should be taken into consideration.

The issue of the extent to which collateral economic factors should be considered in the standard setting process was politically as well as legally important and would continue to be debated seriously in the ensuing years.

88. As a result, two separate organizations were proposed: the Environmental Pollution Control Council of Ministers (Kōgai Taisaku Kaigi), charged with establishing basic pollution policies, and the Advisory Commission for Environmental Pollution Control (Kōgai Taisaku Shingi Kai), a body of experts. See ch. 4 of the 1967 Basic Law. With regard to these points, the Liaison Council's draft was also more restrained than the welfare ministry's draft.

89. These questions were among those presented:

1. Should pollution by radioactive substances be regulated separately?
2. Should ambient standards for noise be set despite the technological difficulty of setting such standards?
3. Should the allocation of pollution control costs be made clear by the law?
4. What should be the respective roles of the central government and the local governments in pollution control?

5. How should the Environmental Pollution Control Council be managed?
6. How should the discussion of various pollution control measures in the existing councils be coordinated with the operations of the Advisory Commission for Environmental Pollution Control?

90. Naikaku Hōsei Kyoku: All bills prepared by government agencies have to be cleared by this bureau.

91. The government's bill was presented on May 17; the opposition parties' proposals on May 30. The Socialist party had submitted the same bill twice before, only to have it tabled on each occasion.

92. In order to save time, the Special Standing Committee for Industrial Pollution and Traffic Control in the House of Councillors initiated preliminary deliberations before the bill was passed in final form by the House of Representatives. (The House of Councillors usually has a preliminary debate on a bill before the House of Representatives passes it.)

93. Article 1.

94. Article 29.

95. Article 24 (2).

96. The Liberal Democratic party, the Socialists, the Democratic Socialists, and the Kōmeitō.

97. Kōgai Taisaku Kihon Hō (1967 Law No. 132).

98. See Shigeto Tsuru, ed., *Environmental Pollution Control in Japan* in Towards a New Political Economy 286–287 (1976), Tsuru, former president of Hitotsubashi University, is one of the foremost scholars of the economics of pollution control in Japan. The article referred to was originally written for an international symposium on pollution held in Tokyo in 1970.

99. The Basic Law for Environmental Pollution Control reflected a general post–World War II pattern of legislation also evident in other areas. See, for example, the Basic Law for Education (Kyōiku Kihon Hō) (1947, Law No. 25), the Basic Law for Nuclear Energy (Genshiryoku Kihon Hō) (1955, Law No. 186). The general purpose of all such legislation was to establish a long-range policy goal to which subsequent legislation or administrative action should be directed. The characteristics of these basic laws varied with the substantive area. For example, some basic laws, such as the Basic Law for

Nuclear Energy, leave the details of implementation (apart from the establishment of a nuclear energy power plants) to other laws, while the Basic Law for Disaster [relief] Measures includes specific provisions for relief in emergencies. As a whole, basic laws do not require special or extraordinary procedures for their enactment.

100. See, e.g., Law for the Resolution of Pollution Disputes (ch. 7), The Air Pollution Control Law (ch. 5), and The Law for Special Measures for the Relief of Pollution-Related Disease (ch. 6).

101. For example, art. 4 affirmed the government's responsibility to adopt comprehensive pollution measures and its duty to protect public health and conserve the living environment. Article 21 unequivocally recognized the sovereign's obligation to establish procedures for extrajudicial dispute settlement and the relief of pollution victims, and arts. 9 and 10 exhorted the state to establish environmental quality and emission standards. Articles 13–16 obligated the government to establish pollution monitoring, measurement, examination, and inspection systems, conduct health and other surveys, and disseminate findings widely "in order to make the nation more conscious of the need to prevent environmental pollution."

102. Another provision was art. 24 (2), in which the state acknowledged an obligation to give particular consideration to small and medium enterprises. This provision was important because these enterprises were economically too fragile to be able to finance the costs of pollution control measures themselves. Many of these enterprises were chronic violators, who, taken together, presented a major control problem. The Basic Law's sensitivity to the particular problems of small and medium enterprises is impressive. See discussion on small and medium enterprises in ch. 5.

103. These were the Central and Local Advisory Commissions for Environmental Pollution Control (Kōgai Taisaku Shingikai) (art. 27–30).

104. It must be noted, however, that other councils attached to the ministries or agencies, such as MITI's Commission on Industrial Structure, continued to play an important role in the policy decisions of these bodies. See ch. 5.

105. See art. 3.

106. See art. 22.

107. It is significant that these provisions were included at all, for Japan did not recognize the OECD promulgated "polluter pays" principle until 1972. The inclusion of these provisions several years before the court's decisions in the four pollution trials also gives insight into the ongoing national debate on the subject: Japan was beginning to come to grips with the notion that the polluter must pay, but the society simply had not reached any meaningful consensus on questions such as the extent to which enterprises should be obligated or the economic and social conditions of payment.

108. In the 1967 version of the Basic Law the notion of "harmonizing" pollution control with "sound economic development" appears twice, first in the purposes clause of art. 1 and again in art. 9.

Article 1 states:

1. This Act is enacted for the purpose of clarifying the responsibilities for environmental pollution control on the part of enterprises, the state, and local public bodies and of specifying in a comprehensive manner countermeasures against environmental pollution, thus protecting the health of nationals and conserving their living environment.

2. In conservation of the living environment provided for in the preceding paragraph, harmony with sound economic development should be considered. [*emphasis added*]

Article 9 states:

1. The National Government shall prescribe, with regard to air pollution, water pollution and noise, the ambient quality standards that are deemed desirable for the protection of the people's health and the conservation of their living environment.

2. Among the standards provided for in the preceding paragraph, the one concerning the living environment shall be prescribed with due consideration to the harmony with sound development of the economy. [*emphasis added*]

3. The standards provided for in paragraph 1 should be reviewed regularly in the light of relevant scientific progress and revised whenever deemed necessary.

4. The national government shall endeavor to secure observance of the stan-

dards provided for in paragraph 1, through the execution of environmental pollution control policies and measures in a comprehensive and effective manner.

109. Indeed, sec. 4331 of the U.S. National Environmental Policy Act of 1969 (42 U.S.C. sec. 4331, et seq., 83 Stat. 852, P.L. 91-190 (1970)) also employs the phrase "production and enjoyable harmony between man and nature"; and sec. 4331 uses the phrase: "to create in productive harmony, and fulfill the social, economic, and other requirements of present and future generations."

110. The government repeatedly stated in the Diet debates in 1967 that human health should be protected without any reservation. When the government, however, was questioned by the Diet members as to why it was necessary to include the harmonization clause, the Minister of Health and Welfare, Hideo Bō, replied:

"Health is an absolute value. Since health cannot be replaced by anything, it has an absolute priority. However, when we further desire to improve our living environment and enjoy a comfortable life, we cannot ignore our life, our wealth, and industry. We have to develop industry. In this sense, the government, while discussing the Ministry of Health and Welfare's draft, concluded that the notion of harmony with industry was quite reasonable."

Dai 55 Kai Kokkai Shūgiin Kaigiroku Sangyō Kōgai Taisaku Tokubetsu Iinkaigijiroku (The 55th Diet, Proceedings of the Special Standing Committee for Industrial Pollution, House of Representatives, no. 14), p. 2.

111. For example, the establishment of SO_x standards was delayed until 1969, and as late as 1970 the central government had still not approved pollution control programs for even the most seriously polluted areas, such as Yokkaichi, Chiba, or Mizushima.

112. For example, the Resolution on Pollution Control Measures (Kōgai Taisaku Ni Kansuru Ken) of the Special Standing Committee of the House of Representatives, May 8, 1970, reads:

Pollution control measures have been taken by the government in accordance with the provisions of the Basic Law for Environmental Pollution Control. In view of the gravity of pollution control, however, the government should promote further pollu-

tion control measures taking into account the following:

1. In order to comprehensively and strongly promote pollution control, the government should strengthen the function of the Conference on Environmental Pollution Control and further employ its coordinative functions.

2. In order to promote the installation of pollution prevention facilities, the government should forcefully take measures such as assistance and taxation necessary to finance investment.

113. The headquarters consisted of a prime minister, secretary general to the prime minister's office, and governmental officials from other ministries. For a good discussion of the establishment of this headquarters and the process of revision of the Basic Law, see Iwata, ed., Commentaries on the Basic Law for Pollution Control (Kōgai Taisaku Kihon Hō No Kaisetsu), 112–132 (1971).

114. One proposal was: "While preserving the harmony between the activities of man and nature, the government's policy goals pertaining to pollution shall be to prevent pollution and to guarantee a minimum standard for a healthful and cultured living within the spirit of Article 25 of the Japanese Constitution, and to contribute to the growth and development of the national economy."

115. Two other recommendations were (1) that the phrase "the promotion of the establishment of public treatment facilities for industrial and public waste" be added to the provision (art. 12) calling for the establishment of pollution prevention facilities; (2) the administration of the general affairs of the Conference on Environmental Pollution Control and the Central Council on Pollution Control should be transferred to the secretariat of the prime minister.

116. During the last Diet debates on the Basic Law, the opposition parties jointly submitted a bill entitled the "Basic Law for Environmental Protection," which advanced a comprehensive approach (including wilderness areas and areas of cultural importance). The opposition also proposed to enhance the powers of local governments, adopting a provision for strict liability, and establishing a new Environmental Protection Ministry.

117. After the Basic Law weakened the cen-

tral government's monopoly over pollution control, the air pollution control law, the water pollution control law, and the noise regulation law, all permitted local governments to set stricter emission standards.

118. Article 17-2 states: "In order to contribute to the prevention of environmental pollution, the government shall, in conjunction with other measures prescribed in this Section, endeavor to protect the natural environment as well as to conserve green areas."

119. During the 64th Diet session, the opposition parties voiced great concern about the protection of the natural environment. During the Diet debates the opposition pointed out the loss of fish in the Seto Inland Sea, the deterioration of national park areas from overuse by tourists, and the general international concern over the protection of the natural environment.

For example, the Natural Parks Law (Shizen Kōen Hō) (1957, Law No. 161) was amended in 1971 and 1972; also the Natural Conservation Law (Shizen Kankyō Hozen Hō), (1972, Law No. 85) was passed to expand designated nature protection areas and strengthen regulations. Nevertheless, protection of the natural environment was greatly subordinated to measures for pollution.

120. Furthermore, comprehensive land use plans developed during this period also failed to include effective environmental preservation components.

121. From this perspective the remark of Shintarō Ishihara, Director General of the Environment Agency, that the fundamental policies of the Basic Law should be reevaluated, because the 1970 amendments were the result of a "witch hunt against industrialists," should evoke less surprise than many people have accorded them. See *Ishihara Draws Public Criticisms for Anti-Environmental Views*, Japan Times, Oct. 23, 1977.

122. For a description of the initial stages of the pollution control administration, see Hashimoto, Kōgai o Kangaeru (Thoughts on Pollution) 16–34 (1970). Much of our description of the development of the pollution control administration is based on this book. Before the Environment Agency was established, eleven ministries and nine ad-

visory councils were related to pollution control administration in one way or other.

123. The council was composed of the vice-ministers from the Prime Minister's Office, the Economic Planning Agency, the Ministry of Justice, the Ministry of Finance, the Ministries of Health and Welfare, Agriculture, and Fisheries, International Trade and Industry, Transportation, Labor, Construction, Local Autonomy, and the Director General of Police.

124. The council proposed instead that a Conference on Environmental Control serve as a coordinating body within the government.

125. Environment agencies were established in Sweden in 1967, and in the United States and Great Britain in 1970.

126. The Environment Agency Organization Law (Kankyō chō Setchi Hō) (1971, Law No. 88). For a description of the bases for the reorganization that created the EPA in the United States, see Message of the President Relative to Reorganization Plans Nos. 3 and 4 (1970), H.R. doc. no. 366, 91st Cong., 2nd sess. (1970). For insight into the process of establishing comparable agencies in other countries, see Robert E. Lutz, *The Laws of Environmental Management: A Comparative Study*, 24 Am. J. Comp. L. 447 (1976).

127. For a critical examination of these laws, see Tokushū: Kōgai Rippō no Kentō to Hihan (Special Issue: An Examination and Critique of Pollution Legislation), 471 Jurisuto 1 (1971).

128. For example, the 65th Regular Diet Session passed the Noxious Odors Regulation Law (Akushū Bōshi Hō) (1971, Law No. 91), the Law Concerning Special Financial Measures of the State for Pollution Control Facilities (Kōgai No Bōshi Ni Kansuru Jigyō ni Kakaru Kuni no Zaiseijō no Tokubetsu Sochi ni Kansuru Hōritsu) (1971, Law No. 70); and the Law for the Establishment of a Pollution Prevention Organization at Special Factories (Tokutei Kōjō ni Okeru Kōgai Bōshi Soshiki no Seibi ni Kansuru Hōritsu) (1971, Law No. 107). The most important laws passed in the 1972 and 1973 sessions were the Chemical Substance Control Law, the Law for the Compensation of Pollution-Related Health Damage, the Seto Inland Sea Environment Preservation Law,

the Harbor Law, the Public Works Reclamation Law, the Natural Parks Conservation Law, and the Nature Conservation Law, which are all discussed or referred to in later chapters.

129. See, for example, Colin MacAndrews, et al. Developing Economies and the Environment: The Southeast Asian Experience, (1979). Also, Julian Gresser, Balancing Environmental Protection with Industrial Development in the Republic of South Korea, A Report of the Environmental Mission to the World Bank (I.B.R.D.), December 22, 1977.

Chapter 2

1. The reader may be interested in comparing the Japanese perspective, as recounted in this chapter, with the recent writings of Western legal scholars. These scholars challenge the prevailing jurisprudential notion that environmental quality is a barren economic commodity to be deployed beside other scarce resources (education, hospitals) in that proportion best suited to human wants. They argue that no means other than a fundamental reconstruction of values can halt the seemingly ineluctable process of environmental decline. Accordingly they seek to fashion a new environmental ethic that remands man to a more modest place in the world, that explores the legal bases for safeguarding the interests (and even the rights) of natural things, and delineates the responsibilities of the living to the unborn.

For example, see the first chapter of Richard B. Stewart and James E. Krier, Environmental Law and Policy, 2d ed. (1978); also, Laurence Tribe, *Ways Not to Think About Plastic Trees: New Foundations for Environmental Law*, 83 Yale L. J. 1315 (1974), and Christopher Stone, *Should Trees have Standing—Toward Legal Rights for Natural Objects*, 45 Southern California L.R. 450 (1972).

2. The Kumamoto Minamata victims finally filed suit on June 14, 1969. However, in contrast to the ¥522,670,000 demanded by the Niigata victims, Kumamoto's victims claimed ¥1,588,250,000 in damages. For a translation of the courts' decisions and an analysis of these cases, see ch. 3 and 6.

3. ¥200,580,000 was sought by the Yok-

kaichi victims. For a translation and analysis of this decision, see ch. 3.

4. ¥61,990,000 in damages was claimed by the Toyama victims.

5. See, for example, W. E. Smith and A. M. Smith, Minamata (1975). Also, a film, "Polluted Japan," narrated and produced by Jun Ui.

6. Michiko Ishimura, *Pure Land, Poisoned Sea*, 18 The Japan Quarterly 299 (1971).

7. It is interesting to compare the discrimination suffered by the victims of pollution disease in Japan with the experience of black children before and after the case of *Brown v. Board of Education*, 347 U.S. 483 (1954). Victims of pollution in Japan were never classified and segregated by law as racial minorities have been in the United States. However, the following parallels do exist: (1) The victims of pollution belonged to a socially and economically disadvantaged class; (2) they were few in number; (3) they were segregated from the rest of the community; (4) the dominant groups in society arguably benefited from suppressing the victims; (5) the victims were estranged from the political process; (6) the victims grew indignant over the treatment they received; (7) they finally resorted to the courts for the vindication of their claims; (8) even after the violation of their rights was clearly established in the courts, they constantly had to force the unwilling parties (industry and government) to comply with and implement legally declared mandates. See D. Bell, Race, Racism, and American Law (1973).

As early as the 1960s, the suffering of the pollution victims in Japan was viewed as a problem of basic human rights. Thus during the mid-1960s, many victims filed numerous complaints with the Civil Liberties Bureau of the Ministry of Justice. See The Governmental and Non-Governmental Machinery for the Protection of Human Rights and Legal Aid in Japan, Ministry of Justice (1970).

8. Ishimura, n. 6 supra.

9. A proposal was once made in Minamata City Council that the name of the disease be changed because it was giving the city an evil reputation. Also the labor unions were seriously concerned that a bad image might jeopardize employment opportunities. For a detailed account of these and other examples of discrimination suffered by the

Minamata victims, see the plaintiffs' final brief reproduced in 44 Hōritsujihō (no. 5), 130 (1972).

10. In rural areas parents in Japan are often blamed for their children's birth defects for similar reasons.

11. Noriaki Tsuchimoto, director, "Minamata: The Victims and Their World" (Tokyo: Higashi Productions, 1972). This quotation, as several others following, is excerpted from Frank K. Upham, *Litigation and Moral Consciousness in Japan—An Interpretative Analysis of Four Japanese Pollution Suits*, 10 L. and Society R. 579 (1976).

12. Originally Minamata was a very poor, isolated area. Chisso built up Minamata, and gradually became central to its political and economic life. Minamata came to be considered a "castle town" (jōka machi) with Chisso as its dominant manor.

13. Smith and Smith, n. 5 supra.

14. This was a Mitsui official's statement at a negotiation session with a group of victims. It is quoted by Tadataka Kondō in *Itai-Itai Byō no Gaikan* (*An Overview of Itai-Itai Disease*), 486 Jurisuto 42 (1971).

15. Id. 41.

16. During the years from the first reports of Minamata disease to the final court showdown, Chisso continued to conceal and even to suppress information critical to the elucidation of the disease's cause. This is best illustrated in Chisso's destruction of data relating to experiments with cat no. 400 in the company's laboratory, which experiments had strongly suggested that mercury in the plant's effluent was the cause of Minamata disease. (See also n. 22 infra.)

17. This was the infamous *mimaikin* (sympathy money) settlement agreement signed on December 30, 1959. The agreement carefully avoided recognition of Chisso's responsibility. The victims accepted an annuity payment of $300 for an adult patient and $100 for a child. In exchange they agreed that Chisso could terminate payments in the event that it ever was determined that Chisso was not the source of pollution. The victims surrendered all future claims, even if Chisso were later determined to be the source of their injuries.

18. In 1965 Yokkaichi City at last established a small "relief fund" to assist "certified" victims of pollution-related diseases. Thereafter other areas like Niigata rendered similar assistance covering medical expenses, a living allowance, and reimbursement of travel expenses related to treatment.

19. Thus a driver who recklessly kills a pedestrian would be deemed to have committed an immoral act. And most universities and associations (such as the bar) also include in their regulations a provision requiring the suspension and expulsion of those convicted of similar serious offenses.

20. See Takenori Gotō, *Hitokabu Undō ni Tsuite* (*On the One Share Movement*) discussed in Jun Ui, *Kōgai Genron* (Discussion of Pollution), vol. I, p. 171 (1971). In 1971 Chisso issued 156.44 million shares. The one-share shareholders called themselves "ppm" shareholders.

Upham's reference (n. 11 supra) to the statements of those victims participating in the one-share movement is instructive. Tokiyoshi Onoue, who received $5.50 from Chisso for the loss of his wife, said:

. . . I don't have much to say, but if it's this small a sum, if a company executive is going to be this stingy, maybe the executive should talk about it and decide on one or two to be sacrificed, see? And drink the mercury. Then their eyes would be opened . . . So we should put water in a big keg, then put some mercury in it, take it up to the stage and say, "At this stockholders' meeting, if some of you important people at the top, some of you from the Chisso Company, sacrifice yourselves and become victims, then maybe we'll be ready to reconsider."

Fumiyo Hamamoto, who lost both parents, commented:

. . . The Chisso Company is so hateful. When I go to Osaka, I'm going to say to President Fugashiro [of Chisso], "I'm buying your life with four million yen." I'm going to say something to him; I'm really going to go raving mad when I go to the meeting.

. . . You're a parent, too! Do you understand? Do you really understand my feelings? (Dead silence fills the auditorium.) What did you say, what did you say to me then . . . , have you forgotten how you bowed your head three times? (President: So I went to pray to your deceased. . . .)

Coming to the Buddhist altar is not enough! . . . (The people around them are silent and unmoving. She clings to the president's lapels.) How much do you think I suffered! The suffering, it's so much, I can't put it into words! You can't buy lives with money! My brother's a cripple! My parents . . . both of them! My brother's a cripple and people laugh! My parents. . . . [Upham n. 11 supra.]

21. The anthropologist Katsuko Tsurumi has suggested that Japan's modern environmentalism may be traced partly to village folk beliefs and practices. From this perspective an important influence was the village festival (matsuri), including both the first and more religious part (shinji), and the second recreational aspect (naorai). The victims' rhythmic ringing of small bells during the stockholders' meeting, she suggests, was designed to call the spirits of the dead to assist the victims in their struggle and created a solemn, almost religious feeling. See Katsuko Tsurumi, Social Price of Pollution in Japan and the Role of Folk Beliefs, Institute of International Relations, Research Paper Series A-30, Sophia University (1977).

22. See discussion of cat no. 400 in the Minamata court's opinion in ch. 3. It took over eight years until cadmium was officially recognized in 1968 as the source of itai-itai disease, and over twelve years for mercury to be recognized as the cause of Minamata disease. The role of sulfur oxides in Yokkaichi asthma (as well as cadmium in itai-itai disease) is still being debated. The victims were deeply resentful of the ways in which polluters and the government manipulated inconclusive scientific data to deny them relief.

23. Although the victims often discussed compensation, money was really not the issue for most victims. The victims (and later local residents and others) have insisted throughout that human health cannot be purchased or otherwise bargained for. This position was later extended to the living environment. Yet as Tsurumi (n. 21 supra) points out, the victims' refusal to assign monetary values to their losses produced paradoxical reactions, for "those things they really wished to regain could be counted only in terms of money." Tsurumi quoted a woman who lost one daughter and has another with congenital Minamata disease:

I am so sad about the death of my first daughter that my tears dried up. My sorrow cannot be bought with money, however, disabled patients like us cannot survive. That is the dilemma. However, I wish to return my check, and I wish to return it to you. I shall give back the money to you, so you should give back my healthy body to me. That is a fair deal!

A male patient said:

It is not money that we want. Since we are living on the coast, we most ardently long to have the sea and the mountain returned to us as they were before the pollution. Money is a nuisance, a troublemaker in the family and in the village. After all, life as in the good old days is the best—poor as we all were, we had the sea and the mountains—from the Shiranui Sea our life cannot be cut asunder. The other world in which we used to live should be brought back to us here and now. Our hope, a very slight hope, is to bring the sea back to the old days, and an even slighter hope is to have our healthy bodies of bygone days returned to us.

Tsurumi pointed out (in a private interview) that money indeed had a degenerative effect on the victims. Many of the victims' quarrels were over money, and only after they received money did the most ferocious disputes within the movement begin.

24. Tsurumi (n. 21 supra) also stresses the importance of ancestor worship as a motivating force and as a source of solidarity for the victims' movement. She reminds us that "the patients at the negotiation table refer" to all of us, "which to their minds includes their deceased relatives." She concludes, "It is the powerfully felt solidarity with the dead that sustained the survivors for the long and difficult struggle."

25. For example, the assembly of Fuchū, a town where many of the victims resided, passed a resolution allocating ¥1 million from the town's budget to finance the costs of litigation. The prefectural governor, however, opposed this resolution on the ground that it would violate the Local Self-Government Act. Despite the governor's opposition, the township continued its support of the litigation.

26. N. Huddle and M. Reich, in Island of Dreams (1975), point out that it may be erroneous to refer to these ad hoc campaigns as a single movement. Although victims'

groups began to exchange information and sometimes collaborated, they remained independent in their actions. Huddle et al. also distinguish the victims' movement(s) from the citizens' or residents' campaigns, a distinction that we also adopt.

27. See Donald R. Thurston, *Aftermath in Minamata*, 7 The Japan Interpreter 25 (1974).

28. This sample is largely adopted from Yoshirō Sawai, *Soshō no Ugoki (A Chronology of Litigation)*, 514 Jurisuto 176 (1972).

29. But the position of the unions on pollution has never been consistent. In Yokkaichi, although the Teachers' Union and the City Office Workers' Union helped the victims, the Chemical Industry Unions were far more cautious. A further complicating factor in Yokkaichi was the machinations of the Communist and Socialist parties, which sought to use the victims for political purposes. In Niigata City the unions directly assisted the victims; the lack of union support in Minamata was partly attributable to Chisso's domination of Minamata City. The support of many of the unions for the victims' cause in Yokkaichi was facilitated by the fact that many union lawyers joined the ranks of the victims' supporters.

Jun Ui has observed (in a private interview) that the smaller the union the more likely it is to be sympathetic to victims and other antipollution causes. One reason may be that activists in these small groups exert a greater influence than in larger, more bureaucratic organizations. Moreover, in many larger organizations, union leaders are often regarded by many workers and outsiders as simply an appendage of management.

There have been several cases involving labor and environmental issues. In one (the Minamata factory dispute) the court ordered the reinstatement, with back pay, of several workers who were fired for distributing pamphlets to the public. The pamphlets alleged that harmful gas and waste water from the factory had damaged rice crops in the surrounding area. (Nippon Keisanki, Mineyama Factory case, decided on March 10, 1971; see Hanrei Times (no. 263) 329). In another case, the General Oil Company in Kawasaki and Sakai penalized its workers for cooperating with the local victims' movement. After eight years the dispute

was finally mediated. Reportedly at the behest of Exxon, the management has since modified its antienvironmental position.

The reader interested in comparing Japan's experience with the resistance of American labor to environmental causes, as well as recent examples of cooperation between these two groups, should refer to Mark W. Beliczky, *Selected References for Jobs, Energy, and the Environment*, Urban Environment Conference (Sept. 1977); *The Environmentalists Try to Win Labor Over*, Business Week, October 3, 1977; and Leon Lynch, "An Overview of the Relationship of Unemployment to the Urban Environment," an address before the EPA Workshop on Jobs and the Environment (Nov. 1977).

30. See Upham n. 11 supra.

31. Thus, for example, in the Yokkaichi case, the victims' lawyers decided to sue only the companies and not the local government because they hoped to obtain the cooperation of the authorities. In fact, the statistical and other information provided by the authorities proved very useful during the trial. After litigation commenced, a group of civil service workers formed a victims' support group.

32. *Yokkaichi Kōgai ni Okeru Jūmin Katsudō (Symposium: The Residents' Activities in the Yokkaichi Pollution Case)*, 514 Jurisuto 134; *Yokkaichi Kōgai Soshō ni Okeru Soshō Katsudō: Zadankai (Symposium: Legal Activities in the Yokkaichi Pollution Case)*, 514 Jurisuto 102 (1972).

33. Because of the great pressure placed on local lawyers not to assist the victims, it was at times essential that outside lawyers come to the victims' aid. For example, in Toyama one local lawyer succumbed to this pressure and withdrew from the case. The chief of the lawyers' team, Shōriki, was reportedly approached by Mitsui officials with an introduction from an influential political figure in Tokyo. The officials asked him to serve as corporate counsel to Mitsui. When he refused, they urged him at least to withdraw as counsel to the victims.

See Kinosuke Shōriki, *Itai-Itai Byō Soshō ni Sankashite (On Participating in the Itai-Itai Disease Litigation)* in 4 Itai-Itai Byō Saiban (Itai-Itai Disease Trial) 335 (Itai-Itai Byō Soshō Bengodan, ed. 1973).

34. The cooperation of individual lawyers with activist associations within the bar is

Notes and References

also instructive. Kondō, on moving to Toyama, soon perceived the need for further support. He quickly got in touch with the Association of Young Lawyers that had also been active in the Niigata area. The association dispatched twenty lawyers to Toyama who offered their help in the plans for litigation.

35. Upham, n. 11 supra.

36. Id.

37. Id.

38. Id.

39. It is interesting that although environmental suits have raised various conflicts of interest and other difficulties for their lawyers, neither they nor other members of the profession take upon themselves the job of delineating what is the appropriate ethical role for the "new" public interest lawyer. An interesting statement is the essay by Gary Bellow and Jeanne Kettleson, "The Mirror of Public Interest Ethics: Problems and Paradoxes" (unpublished) (Feb. 8, 1977). One wonders whether, as many Japanese lawyers suggest, concern in the United States with the ethical dimensions of public-interest legal activity is the product of the substantial subsidization of this activity. For a brief review of recent bar association activities in Japan see Environmental Protection and the Role of Lawyers: Activities of the Japan Federation of Bar Associations Against Environmental Pollution in Japan (Nov. 1975), published by the Environmental Law Committee of the Association. General background information is given in Richard W. Rabinowitz, *The Historical Development of the Japanese Bar*, 70 Harv. L.R. 61 (1956). An important international comparison is made by Mauro Cappelletti, *Governmental and Private Advocates for the Public Interest in Civil Litigation: A Comparative Study*, 73 Michigan L.R. 794 (1975).

40. See n. 19 of ch. 1.

41. The ministry requested the victims to sign the following pledge:
Upon petitioning the Ministry of Health and Welfare to settle and dispose of the dispute concerning Minamata disease, we leave completely to the ministry the selection of committee members who agree to serve, and we hereby firmly pledge that we will accept without objection the conclusion reached by the commit-

tee, upon full deliberation, after hearing thoroughly from both parties about their situations and mediating differences of opinions between the parties in the process of settlement.

Only later was it discovered during a Diet investigation that Chisso had prepared two statements in advance for the Ministry of Health and Welfare. See Sangiin Shakai Rōdōiinkain Kaigi Giroku (no. 7) Mar. 18, 1969.

42. For a critical comment on this agreement, see Takehisa Awaji, *Minamata Byō Hoshō Mondai no Ichi Shiten (A View on the Compensation for Minamata Disease)*, 453 Jurisuto 73 (1970).

43. Upham, n. 11 supra.

44. For a discussion of this law, see Yoshio Kanagawa, *A System of Relief for Pollution-Related Injury*, 6 L. in Japan 65 (1973).

45. See the translation of the court's decision in the Minamata case, ch. 3.

46. Upham, n. 11 supra.

The idea of the litigation as a collective effort was also reflected in the plaintiffs' request that compensation be paid as a lump sum. The strategy was based on the belief that the suffering of all was essentially indivisible. While avoiding the necessity of proving individual damages, this approach also served to maintain the victims' groups' solidarity.

47. In addition, various social pressures were placed on the victims by the surrounding community not to litigate their grievances. In Toyama their neighbors told the victims that the lawyers would only exploit them, the lawyers were revolutionaries, and the courts were simply the pawns of the ruling political parties. Some admonitions were even accompanied by threats. The victims were warned that no one would be willing to buy rice from their villages or marry any member of their families. Under such pressure only thirty-three victims in Toyama remained determined to file suit.

48. Upham, n. 11 supra.

49. Id.

50. During the preparation of their briefs, the plaintiffs' lawyers endeavored to detail the destruction of the victims' lives by including protracted statements made by each plaintiff victim on his or her daily suffering.

Often statements provided by lawyers actively involved in other cases would also be incorporated to emphasize the commonality of the victims' suffering. This approach to pleading was designed to convey the victims' moral outrage and despair.

51. Tadataka Kondō, *Itai-Itai Byō Gaikan* (*An Overview of Itai-Itai Disease*), 486 Jurisuto 38 (1971).

52. Upham, n. 11 supra.

53. Upham, n. 11 supra.

54. Upham, n. 11 supra.

55. The initial awards to the victims were: Niigata, ¥270,420,000; Yokkaichi, ¥88,120,000; Toyama, ¥56,990,000; Kumamoto Minamata, ¥937,300,000.

56. In Japan the sovereign is still viewed by many with deference, if not awe and respect. The term "okami," literally god(s), is often used in conversation when referring to government officials. Also, these officials are considered "to descend from the heavens" (ama kudari) when seeking employment in a company that in many cases they have dealt with in their earlier governmental careers.

57. The dogeza or complete bow to the earth was performed in the Tokugawa era by commoners to lords or high-ranking samurai. It expressed complete submission and respect.

58. The ritualistic expression of penitence and the victims' resort to mediation reinforce Tsurumi's analysis of the influence of the rural festival on the victims' actions. She wrote:

The fact that even after the patients had won their case they still wished to negotiate directly with the company head is a clear indication they are following, perhaps unconsciously, the pattern of the village festival in which the affairs of the deities, the formal side, is always followed by the informal gathering of humans in naorai. Litigation is devoid of emotion and set in the formal hierarchical structure of the courtroom, whereas face-to-face negotiation is effective and the forum of communication in which the patients behaved as equals to the president and the managerial staff of the company. Tsurumi, n. 21 supra.

59. The preamble of the agreement states that "Shōwa Denkō shall compensate all damages in the past, present, and future to the health and lives of the victims for the rest of their lives; that the company shall exercise utmost care in checking and controlling industrial waste materials, . . . and it shall cease operations if some risk is foreseen.

60. This sum represented the difference between the amount awarded by the court and the amount set in the agreement. An annuity of ¥500,000 was also provided to all victims for the duration of their lives.

61. All the defendants with the exception of Mitsui waived their right to appeal.

62. In both Toyama and Minamata defendants had stopped production even before the plaintiffs filed suit. Since the Kumamoto Minamata court's decision, Chisso has reported severe financial difficulties resulting from the disbursement of compensation payments.

63. See case studies in ch. 5 illustrating operation of "administrative guidance."

64. But see ch. 8 for a discussion of contradictions with the domestic record.

65. As discussed in later chapters, the administration has been able through "guidance" to make good use of corporate sensitivities to pollution control.

66. See ch. 5 for a discussion of these changes.

67. See Law 84 (1972) amending the Water Pollution Control Law, and Law 84 (1972) amending the Air Pollution Control Law.

68. See Kōgai Kenkō Higai Hoshōhō (1973, Law No. 111); see ch. 6.

69. Kōjō Ritchi Hō (1952, Law No. 24).

70. The best examples are the current stationary source emission standards for NO_x and SO_x emission standards for automobiles. See ch. 5 for a discussion of these changes.

71. Upham (n. 11 supra) cautions against overemphasizing the influence of public pressure on judicial decision-making.

72. Hiroshi Itoh, *How Judges Think in Japan*, 18 American J. of Comparative L. 775 (1970).

73. See decisions extending automobile liability insurance to intrafamily torts reported in 667 Hanrei Jihō 3 (1972); another area is Japan's judicial recognition of a right to privacy.

74. The citizens' and victims' movements are essentially distinct phenomena, although in a number of instances the two have combined forces. See, for example, the Mishima-Numazu uprising, the Osaka Airport case, and the Sakata New Port controversy. For a review of the citizens' movement, see Jirō Matsubara and Kamon Nitagai, Jūmin Undōno Ronri (Rationale for the Residents' Movement) (1976). For a list of some publications relating to the residents' movements, see Jun Ui, Jūmin Undō no Tsukutta Shiryō: Bunken Kaidai Sono 6 (An Annotated Bibliography 6: Materials Compiled on the Citizens' Movements), 2 Kōgai Kenkyū (Research on Environmental Disruption) No. 4, 74 (1973), 4 Kōgai Kenkyū No. 1, 72 (1974).

75. An interesting historic parallel is the peasant uprising (hyakushō ikki) of the Tokugawa era. One commentator notes that the two movements display four points in common. First, the ikki and the citizens' movements are conceived as desperate struggles by local peoples in defense of their basic right to live. Both give priority to local community interest over national or public interest. In this sense, Tokugawa peasants opposed the imposition of "unjust" taxes or the corvee. Second, both movements are essentially nonideological. None of the ikki leaders were ideological like Thomas Münzer of the German peasants' war, or Hung Hsui-chuan, leader of the Taiping Rebellion. Third, both movements were imbued with a profound moral message. The ikki leaders, Meisuke Miura and Hachirō Kanno, propounded a popular work ethic that celebrated the values of diligence, honesty, and frugality. These leaders in turn demanded strict moral discipline from their followers. Finally, both movements escalated their tactics from peaceful petitions to direct, forceful action. Thus beginning in the late eighteenth century, the ikki grew increasingly more violent. See the editorial comment Peasant Uprisings and Citizen Revolts, 8 The Japan Interpreter, no. 3 (1973).

76. Ui argues that the citizens' movement has been strongly conservationist from its earliest days. See Jun Ui, The Role of the Citizens' Movement, in Jishu Kōza, The Newsletter from Polluted Japan, Special Issue, 13–30 (1975).

Ui's argument is also supported by Tsurumi's analysis (n. 21 supra). Tsurumi notes that a fierce battle developed in 1906 in response to a decree of the Minister of Internal Affairs merging the nation's shrines. The decree doomed many native shrines and the forests surrounding them to destruction. Among those devastated before the war were the shrines in Mie and Wakayama Prefectures. Kumakusu Minakata, a microbiologist and folklorist, at that time printed and distributed a series of protests backed by the people from the affected provinces in which he developed an essentially ecological argument. He described how swallows and other birds used the shrines as nests, and how when the shrines were destroyed, insects would increase and damage crops. Minakata also noted that destroying the forests would remove a major source of shade for coastal fish and that if the fish would no longer come to rest in the shallows, fishermen would be forced to go far offshore.

77. A curious movement reflecting this spirit of smallness, localism, and antimaterialism is one local community's effort to reject the "expert" advice of lawyers and bring a lawsuit pro se (honin soshō). The case involves litigation over the siting of a power plant near Lake Buzen, reported in 13 Kōgai Genron (Discussions on Pollution), discussion by Ryūichi Matsushita et al. (July 2, 1973).

78. An interesting comparison is offered in Richard B. Stewart's article, Pyramids of Sacrifice? Problems of Federalism in Mandating State Implementation of National Environmental Policy, 86 Yale Law Journal 1196 (1977).

79. See Akio Nishioka, Mishima-Numazu-Shimizu Sekiyu Konbināto Hantai Undō (The Movement Against Petrochemical Konbināto in the Mishima Area), 458 Jurisuto 118 (1970).

80. For a detailed account of this case, see the report of the Japanese Federation of Bar Associations' Committee on Pollution, Tomakomai Tōbu Daikibo Kōgyō Kichi Kaihatsu Keikaku Jittai Chōsa Hōkokusho (A Report of the Investigation of a Large Scale Industrial Development Plan in Hokkaidō) (1974).

81. Another famous example is the 1969 Annaka Pollution episode. In this case vic-

tims of the pollution from the Tōhō Zinc Smelting factory formed a countermeasures council and filed a petition with MITI for review of the factory's operations. The victims alleged that the company had constructed and operated new facilities without a proper permit and that MITI had tacitly allowed such illegal activity to continue. When MITI officials came to inspect the factory, the residents demanded that the inspection be open to the public and that all information thereafter be made public. In 1970 MITI capitulated and ordered Tōhō to clean up its operations. One MITI bureau official involved in the case committed suicide. See Tadashi Ōtsuka, *Annaka ni Okeru Kōgai to Hankōgai Undō* (Pollution at Annaka and the Anti-Pollution Movements), 458 Jurisuto 110 (1970).

82. The first example is the agreement signed by the Japan Hydrogen Industrial Company and 1,000 residents of Iwaki City in Fukushima Prefecture in May 1970. The agreement provided for compensation for pollution injury and reimbursement for the residents' expenses of installing pollution control facilities. It contained clauses assuring the residents' right to inspect the company's factory, and required the company to negotiate with residents in good faith. See Takehisa Awaji et al., *Kōgai Taisaku no Tenkai to Mondai Ten* (Evolving Pollution Control and Problematic Points) in Atsushi Satō, ed., *Chiiki Kaihatsu-Kōgai* (1974).

83. For a list of the fifty-seven major cases see Japan Economic News (Nippon Keizai Shinbun), Nov. 4, 1976.

84. It is instructive to compare these perceptions of the role of resident-initiated environmental litigation in Japan with one commentator's observation of the contribution of "public interest" litigation in the West. See John Denvir, *Towards a Political Theory of Public Interest Litigation*, 54 North Carolina L.R. 1133 (1976).

85. The social background of lawyers and clients in these cases is also markedly different from the first generation of victim cases. Generally the more recent suits involve urban and more economically well-off plaintiffs and many less ideologically motivated lawyers.

86. Defendants were prosecuted under art. 234 of the criminal code, which states: "A person, who by use of force, interferes with the business of another shall be dealt with in the same way as provided for in the preceding Article." Article 233 deals with damage to credit and obstruction of businesses, and punishes these offenders by penal servitude for not more than three years, or by a fine.

87. Article 338.4 states:
(Judgment of dismissal of public prosecution)
Article 338. The public prosecution shall be dismissed by a judgment in the following cases: . . .
(4) In case the procedure for the institution of public prosecution is void due to the violation of the provisions thereof.

88. The Tokyo Public Prosecutor's Office has since appealed the High Court's decision to the Supreme Court. The Prosecutor's Office has argued that the Terao decision was unconstitutional because it violates the concept of separation of powers, exceeds the bounds of judicial authority, misinterprets art. 14.1 of the Constitution, and is contrary to the precedents of the Supreme Court, the high courts, and other courts relating to the proper use of the prosecutorial authority. Kawamoto's laywer, Takanori Gotō, however, is arguing that the state's prosecution denied Kawamoto equal protection of the law under art. 14 of the Japanese Constitution. On Kawamoto's behalf, Gotō is arguing that analogous American precedents should be considered and applied. He principally relies on *United States v. Steele*, 461 F.2d 1148 (1972) (judgment was reversed as the government could give no explanation for its selection of defendant other than prosecutorial discretion and made no other effort to justify prosecution), and *United States v. Falk*, 479 F.2d 616 (1973) (government has the burden of proving nondiscriminatory enforcement of law; defendant is entitled to be heard on the claim that prosecution infringed First Amendment rights). See Document 3 of Appellee's Brief before the Supreme Court of Japan. In 1978 the Kumamoto District Prosecutor's Office announced that it would demand a three-year prison term for two retired Chisso executives charged with involuntary manslaughter.

89. The most recent, provocative test of the self-help doctrine will be the outcome of the

criminal prosecution of environmentalists and others arrested in demonstrations against the opening of the new Narita International Airport. The Narita case in many ways repeats and extends the basic theories this chapter has described. First is the theme of aggregation of interests. We have in Narita not simply the protests of local victims of noise pollution, but the uprising of many concerned groups—farmers upset over the loss of their land, city councilmen protesting the drastic new changes in life-style and the dislocations caused by development, local government officials protesting general inequities in labor conditions, and radicals protesting everything. (More recently the objective of many protests has shifted from the preservation of environmental interests to more generalized political confrontation.)

The Narita case also emphasizes the concerns of earlier victims' and citizens' groups: their insistence on a right to a hearing, and fair, open procedures; their desire to preserve an earlier life-style against mindless growth; their refusal to sacrifice any more so that "rich industrialists and others can further profit"; the linking of protection of health with their assault on materialism; their unwillingness to monetize, bargain, or otherwise exchange these concerns. All this, too, was stated in a highly ritualistic and symbolic style. What disturbs many about Narita, however, is the increasing polarization it suggests. As the protests become more shrill, the government's reaction becomes more severe. "Preventive detention" has recently been disinterred in the enactment of new criminal legislation to assure Narita's security. As of this writing we are unable to predict how Japanese law and institutions will reconcile the increasingly bitter conflicts between industrial development and the concerns of the citizens' movements. For a summary of the history of the Narita dispute, see *Government Rushes to Open New Airport*, 24 Japan Quarterly 268 (1977).

Chapter 3

1. The term "generation" is used more in a classificatory, conceptual sense than in a chronological one. As ch. 4 describes, a few other environmental decisions predated the courts' judgments in the four major trials, although the latter were politically and in many ways doctrinally more significant.

2. The procedural distinctions between civil and administrative litigation are discussed in ch. 4, sec. 1.1.

3. The reader unfamiliar with the Japanese court structure might note these basic facts. The Court Organization Law (Saibanshohō) (1947, Law No. 59) created five courts. The Supreme Court is the highest court and generally exercises only appellate jurisdiction (except in cases of impeachment and a few other instances). The Supreme Court includes a Chief Justice and fourteen Associate Justices. Next are the High Courts. These too are appellate courts (except for election disputes, mandamus proceedings, and a few other original matters), and they have jurisdiction over appeals from judgments of the Interior District and Family Courts. There are eight High Courts in Japan (six branches) that include eight presidents and 275 judges. Below the High Courts are the fifty District Courts (with 242 branches) and the fifty Family Courts (also with 242 branches). The District Courts are courts of general original jurisdiction. (District Courts also have appellate jurisdiction over appeals in civil cases filed against Summary Court judgments.) They are the courts in which environmental disputes most often appear first. District Court cases are usually handled by a single judge, but depending on the nature and importance of the case, by a three-judge panel. The District Courts include 870 judges and 442 assistant justices. The Family Courts, which are on the same organizational level as the District Courts, have 196 judges and 152 assistant judges. Under the District Courts and at the bottom of the judicial pyramid are the 575 Summary Courts that include 791 judges. For a general discussion of the Japanese court system, see Hideo Tanaka, The Japanese Legal System (1976).

4. As of July 15, 1968, the total number of those registered was 213 (male 28; female 185); 92 (1 male, 91 females) were verified patients and 121 (27 males and 94 females) were suspected patients.

5. They were followed by six separate groups of victims. In total, there were 514 plaintiffs. At this point the reader should be aware of the complexities and burdens of

the Japanese Code of Civil Procedure (C.C.P.). Japan does not have a special procedure for multiple party actions (class actions). Where the rights or liabilities of several persons are based on the same factual ground, they may sue or be sued as colitigants (art. 59, C.C.P.). Such suits are consolidated in one procedure; parties are joined merely as a matter of convenience. A court's decision has no binding force on persons other than those parties before it.

To accelerate proceedings with multiple parties, a court often picks some of the plaintiffs and decides their case first as in the itai-itai case. The general questions of the relationship between a polluting activity and disease, or the issue of responsibility, are common issues. The issues of causation of a single victim's disease, the extent of injuries, or the amount of damages may differ with each plaintiff. Once a court decides sample cases, then the remaining cases can make use of the factual determinations in the first decision. From a legal viewpoint, facts established in a previous decision have no binding force. In practice subsequent proceedings with the same factual grounds are usually bound by a previous decision. In the itai-itai and other pollution trials decisions for some of the plaintiffs have been followed by direct out-of-court negotiations between the defendant company and all other victims. In cases where several suits have been filed simultaneously in several courts, one of the courts may serve as a model and conduct its proceedings, while suits in other courts are temporarily suspended.

6. Article 109 of the Mining Law appears in the court opinion section IV.

7. The reader should have some sense of the evidentiary burdens that the plaintiffs face in pollution cases. There is no legal procedure to facilitate pretrial discovery in Japan. Consequently, parties are thrown back on their own resources to gather evidence. This is one important difference in the procedure in Japan and that in the United States. Its effect is to make the job of gathering evidence, especially for impecunious Japanese victim groups, a far more arduous one.

Under art. 249 of the C.C.P., a court may hold a pretrial hearing to narrow the legal issues in a case. With some exceptions, issues at a trial are restricted to those identified at this hearing. With a few exceptions, new issues and new evidence may not be addressed by the parties once limited by a pretrial hearing (art. 255).

In many pollution cases pretrial hearings are either not used or used only in a limited capacity. The pretrial hearings are limited procedurally. No witnesses are examined; no evidence presented or made a part of the record. Frequently pollution cases are so complicated that the parties refrain from narrowing the issues at the outset. They often wish to preserve their options by waiting until witnesses have been examined and other evidence submitted. This in turn will alert parties to the need for new evidence or additional witnesses.

8. Article 37 of the Code of Civil Procedure states:

In case there is or are such circumstances on the side of a judge as may prejudice the impartiality of a decision, any party may challenge him.

A party having proceeded orally or having made a statement in the preliminary proceedings before a judge may not challenge such judge; provided, that this shall not apply to cases where the cause of refusal has arisen thereafter or where its existence was unknown to the party.

9. Subsequently plaintiffs made a similar request that was also denied. In the thalidomide case, plaintiffs also unsuccessfully challenged the judge's refusal to allow expert testimony. See Masaru Ino, *Diary of a Plaintiffs' Attorneys' Team in the Thalidomide Litigation,* 9 L. in Japan 136 (1975).

10. In some cases the court ruled that ¥8 million was to be paid to a deceased person's family.

11. There are two institutionalized ways of financing legal financial aid in Japan. The first is aid in litigation, a system established by the C.C.P. The second is legal aid carried out by the Legal Aid Association. In addition to such formal assistance, labor unions and other sympathetic groups contributed to or made loans to the plaintiffs in some of the cases under discussion. Also at times, scientists and lawyers have contributed their services gratis to the victims' causes.

Aid in Litigation

Financial aid in litigation is provided for by C.C.P. art. 218, which states:

A court may, on application, grant aid in litigation to a party who does not have means to defray the costs of the suit; provided, however, that the same shall apply only when there are some prospects of winning the case.

When aid is granted, a party is allowed to postpone payments of court fees for a certain period, usually until after the suit is concluded. The aid granted covers fees for initiating a suit and expenses for witnesses. It does not cover attorneys' fees, however. Furthermore, since aid is granted only after a suit has been brought, access to the courts is difficult for the poor. Since the provisions for financial aid have been interpreted narrowly, in practice individuals not qualifying for social welfare have hardly ever received assistance. The number of recipients of financial assistance increased fairly steadily, from 79 in 1956 to 508 in 1970. This increase may have resulted from a corresponding rise in automobile accident suits dating from this period.

The financial aid system has also been used in the pollution field. In all cases discussed here, plaintiffs were granted this assistance. Although most plaintiffs in these suits were poor, a few had some financial means. In the Itai-Itai and Osaka airport cases, the courts changed their prior interpretation of art. 118, stating:

In view of the size and nature of the pollution case before us, it seems that plaintiffs have to spend an enormous amount of money for fees for expert witnesses, for the examination of evidence, for research and studies, postage, transportation, and for the preparation and reproduction of documents. Considering the large amount of these expenditures, the applicants can be said to 'not have means' in the sense used by Article 118 of the Code of Civil Procedure, although some of these individuals have attained a moderate income level.

The Legal Aid Association

The Legal Aid Association is a functional body within the Japanese Federation of Bar Associations. The headquarters are in Tokyo, and there are branch offices in Tokyo, Hokkaidō, and all other prefectures. The association is financed by government subsidies and contributions from various bar associations.

The association provides aid in civil (including conciliatory) and administrative cases. No aid is provided for criminal cases since there is a separate system that provides defense council in criminal cases.

Under the association's criteria the following persons may qualify for assistance: (1) Individuals needing relief as provided by the Daily Life Protection Law (1950, Law No. 144); (2) persons having difficulty earning a living, and who are unable to afford the cost of suit; and (3) persons under circumstances similar to (1) or (2). Assistance by the association is thus given not only to poor people, but also to persons of some means who are judged in need of help in connection with the costs of suit. Such persons can be assisted to the extent the budget of the association permits.

Within the association, a special Legal Aid Examination Committee has been organized to oversee assistance. When the Legal Aid Examination Committee decides to grant aid to an applicant, the association selects a suitable lawyer for the applicant, who then handles the case. The association advances the lawyer's costs such as costs of suit, provides a retainer, and covers bonding and other expenses. At the conclusion of the case, the individual receiving aid must return the monies advanced by the association. In a suit for compensation, such monies are frequently set off against a successful damage award. Under some circumstances, the return of whole or part of such expenses may be postponed or waived. Exemptions for the return of monies to the association may be granted by the Legal Aid Examination Committee with the approval of the Minister of Justice.

Reliable statistics on the number of applications receiving legal aid by the association are not available at present. In the period 1963–1968 some financial aid was at least theoretically available to victims of pollution. In 1968, 1,951 cases out of 6,029 applications received some form of financial assistance. During the period 1963–1968 the greatest amount of aid was provided in civil damage suits. Legal aid was not provided in a single administrative suit.

12. A translation of this agreement appears in sec. 1.4.

13. See *Komatsu v. Mitsui Kinzoku Kōgyō K. K.*, 22 Kakyū Minshū (Nos. 5 and 6) 1; 635 Hanrei Jihō 17; 264 Hanrei Taimuzu 103; Hōritsu Jihō (July 1971) 336; Hanrei Kōgaihō

1163 (Toyama Dis. Court, June 30, 1971), affirmed on Kōso appeal, 674 Hanrei Jihō 25; 280 Hanrei Taimuzu 182; Hanrei Kōgaihō 1364 (Nagoya High Ct., Kanazawa Branch, August 9, 1972); the Itai-Itai Disease case.

In this and other excerpts from subsequent decisions brackets [] separate the parts of the court's decisions that are paraphrased from the portion that is a direct translation. The numbers in parentheses refer to the pages in the courts' opinions. As much as possible we have endeavored to preserve the feeling of the original opinion, even though these opinions may seem involuted or needlessly complex to the Western reader.

14. The one important prewar exception, of course, was the Osaka Alkali case (see ch. 1). The Osaka Alkali case, however, involved crop damages, not gross physical injuries, as were presented by the Niigata plaintiffs.

15. See *Caustic Soda Firms Told to Stop Using Mercury*, Mainichi Daily News 6, December 11, 1975.

16. *Ōno v. Shōwa Denkō K.K.*, 22 Kakyū Minshū (September) 1; Hanrei Jihō (No. 642) 96; Hanrei Kōgaihō 1379 (Niigata Dist. Court, September 29, 1971); the Agano River-Niigata Minamata Disease case.

17. In November 1964 a patient with the disease was found by Niigata University.

18. In January of 1965 the plant was closed.

19. The plaintiffs claimed a total of ¥522,674,000 plus an attorney's fee of ¥68,174,200.

20. *Watanabe v. Chisso K.K.*, 696 Hanrei Jihō 15; Hanrei Kōgaihō 1641 (Kumamoto Dist. Court, March 20, 1973); the Kumamoto Minamata Disease case.

21. For a detailed description of the clinical symptoms of Minamata disease, see M. Harada, Minamata: A Medical Report.

22. Ibid.

23. Ibid.

24. Approximately $688,600 at that time.

25. There had, however, been cases of serious health injuries from high concentrations of sulfur oxides.

26. An interesting annotated chronology of the Yokkaichi case was prepared by Yoshirō Sawai, *Soshō no Ugoki*, 514 Jurisuto 176 (1972).

27. *Shiono v. Shōwa Yokkaichi Sekiyu*, 672 Hanrei Jihō 30; 280 Hanrei Taimuzu 100; Hanrei Kōgaihō, 491 (Tsu Dist. Court, Yokkaichi Branch, July 24, 1972); the Yokkaichi asthma case.

28. See the discussion of classification of payments under national compensation system in ch. 6.

29. See Tomohei Taniguchi, *A Commentary on the Legal Theory of the Four Major Pollution Cases*, 9 L. in Japan 35 (1976). In this book we do not examine parallel developments in the food and drug area. The reader should refer to the Morinaga milk poisoning case and the current controversies over SMON (subacute myelo-optico neuropathy) and polychlorinated biphenyl (PCB) poisoning. See Akio Morishima, *Hokuriku SMON Hanketsu ni no Mondai Ten (Problem Areas in the Northern SMON Decision)*, 663 Jurisuto 15 (May 1, 1978); also Takehisa Awaji, *Kanemi Soshō to Shokuhin Seizō Kanren Kigyō no Sekinin (The Kanemi Law Suit and the Responsibility of Companies Involved in the Manufacture of Food Products)*, 656 Jurisuto 44 (January 15, 1978).

30. There is no explicit principle of stare decisis in Japanese law. The most relevant provisions are art. 4 of the Court Organization Law that states: "A conclusion in a decision of a superior court shall bind courts below respecting the case concerned." Article 10.3 notes that the Supreme Court will convene en banc in hearing cases involving a decision of the inferior courts pertaining to interpretation of the Constitution or other law, contrary to a decision previously rendered by the Supreme Court. Although there is no principle of stare decisis, court precedents particularly those backed by a social consensus as were the four pollution trials have "strong persuasive force." As a general rule, given Japan's career system, few judges have been independent enough to deviate markedly from Supreme Court policy.

31. The Toyama decision also was based on art. 109 of the Mining Law that provides for strict liability.

32. See Richard B. Stewart and James E. Krier, Environmental Law and Policy 2nd ed. (1978); also Richard A. Posner, Economic Analysis of Law (1972).

33. Professor Taniguchi suggests that the judges may have been concerned that finding defendants strictly liable might

needlessly have restricted judicial interpretation of art. 709. This is because the courts probably would have had to interpret the cases according to the narrow and limited revised strict liability provisions of the air and water pollution control laws.

34. See Taniguchi, n. 29 supra. Also, Makoto Shimizu, *Niigata Minamata Byō Hanketsu ni Okeru Koi-Kashitsu no Mondai ni Tsuite (On the Problem of Willfulness and Negligence in the Niigata Minamata Disease Decision)* 642 Hanrei Jihō 89 (1971).

35. Tsunoru Ushiyama, *Koi-kashitsu o Meguru Mondai ni Tsuite (On Determinations of Willfulness and Negligence)* 493 Jurisuto 659 (1971). We must note, however, that a separate prosecution was later brought against Chisso officials in 1976. On March 22, 1979, the Kumamoto District Court convicted the former president of Chisso Corporation and Chisso's plant manager of criminal negligence. The court ruled that the in utero fetus is a proper subject of law because the fetus has a function "similar" to that of a human being. The court also dismissed Chisso's claim that the statute of limitations had run because the prosecution filed its case less than three years after the death of one of the victims. The sentence was suspended because of the age of the two executives. The court's decision has been appealed. See Int. Environmental Reporter 609, April 11, 1979.

36. For example, the studies did not evaluate in any detail environmental or social impacts, irreversible effects, or other options.

37. The use of the term "available" placed the holding several years ahead of the comparable United States standard of "feasibility" established by the federal Water Pollution Control Act of 1972, as amended by P.L. 92-500, 33 U.S.C.A. sec. 1151, et seq. For an analysis of the act's provisions concerning technology, see p. 110 of Stewart and Krier, n. 32 supra.

37a. Of course, economic analysis does not predetermine any specific result. One might argue that the court simply reached a different assessment of the society's balance between the benefits of health vs. the costs of a shutdown.

38. In some sense the courts in the four pollution trials were simply expanding the philosophy embodied in the 1970 amendments of the Basic Law that struck the provisions for "harmonizing" environmental policies with economic concerns. In 1976 the OECD was to call this philosophy "the noneconomic approach" to environmental decision making. See Organization for Economic Cooperation and Development (OECD), Environmental Policies in Japan (1976).

39. Taniguchi analyzes the issue of allocating the costs of the damage awards among the defendants (i.e., indemnification) as follows: "I believe the range of apportionment of the indemnity and its approval or denial should be decided in terms of the policy of pollution prevention. Let us suppose two industries were under joint tort liability: One abides by emission standards and is taking prevention measures to the full, and the other violates the law and is indifferent to pollution prevention. Then, a finding of indemnification would better be determined, I believe, on the basis of negligence and the degree of illegality involved." See T. Taniguchi, n. 29 supra.

40. Of course an open issue is why recent air pollution victims in Yokkaichi, not parties to the original suit and thus not bound by *res judicata*, do not file a new injunctive action. See the discussion in ch. 6 on the recent Kawasaki and new Osaka air pollution cases, which suggests that gradually, social inhibitions against seeking such relief are declining.

41. In contrast, by the time of the itai-itai trial methyl mercury had been well established as the cause of Minamata disease.

42. Defendants also argued that cadmium could not have caused the disease because no cases of the disease had been reported in other parts of the delta. The court rejected this argument because there was no proof that the amount of cadmium, the period and length of exposure, or other factors were the same.

43. Note that Japan does not currently have a jury system. See Law in Japan, Arthur Von Mehren, ed. (1963) for a discussion of an early experiment. See also *The Science Court Experiment: An Interim Report*, Science (Aug. 20, 1976) for an interesting comparison.

44. See Guido Calabresi, The Costs of Accidents (1970) and Calabresi, *Transaction Costs, Resource Allocation, and Liability Rules, A Comment*, 11 J. L. and Econ. 67

(1968) for an analysis of policy factors favoring imposing liability on the cheapest cost avoided, in this case Shōwa Denkō.

45. See T. Taniguchi, n. 29 supra.

46. Two other developments should be noted. In one case where the immediate cause of death was cancer of the liver unrelated to Minamata disease, damage claims for injury to health of the victim were denied, but awarded to the wife and children of the victim for their suffering, to the same extent as if the patient had died of Minamata disease. In a second case, a patient's death from myocardial hardening was attributed to Minamata disease and an award of damages sustained on the ground that the death of the diseased, a foster brother, inflicted as much grief on the surviving brother as if the real father had died. See T. Taniguchi, n. 29 supra.

47. By the time of the four pollution cases, the award of attorney's fees had become an established practice in traffic accident suits.

48. In practice, however, awards for mental suffering have not noticeably differed according to the social status of the plaintiffs.

49. Professor Taniguchi also notes the dangers of the subjective standard for "mental suffering comparable to damage to life"; he suggests that the more established interpretative techniques developed under art. 709 be used to assess damages for the compensation of relatives.

50. In fact the actual amounts paid in mimaikin were substantially lower than those tendered in the Tōya Maru shipwreck case, or provided under the Daily Life Security Law (Seikatsu Hogo Hō) 1950, Law. No. 144, or the Worker's Accident Law (Rōdōsha Saigi Hoshō Hoken Hō) 1947, Law. No. 50.

51. Japan now has compulsory automobile accident liability insurance that substantially accelerates relief to automobile accident victims.

52. The court avoided giving a windfall to the victims by subtracting mimaikin from the amount awarded to plaintiffs for mental suffering.

53. See T. Taniguchi, n. 29 supra.

Chapter 4

1. See Kiminobu Hashimoto, *The Rule of Law: Some Aspects of Judicial Review of* *Administrative Action*, Law in Japan, Arthur Von Mehren, ed. (1963).

2. Gyōsei Jiken Soshō Hō (1962, Law No. 139).

3. For a comparison see Eckhard Rehbinder, German Law on Standing to Sue, International Union for Conservation of Nature and Natural Resources (1972). See also Eckhard Rehbinder, *Controlling the Environmental Enforcement Deficit in West Germany*, 24 The Am. J. of Comparative Law 373 (1976).

4. See Ichirō Ogawa, *Judicial Review of Administrative Actions in Japan*, 43 Wash. L.R. 1075 (1968).

5. Perhaps the best known case outside the environmental field is the Supreme Court's decision, *Sakamoto v. Japan*, 16 Minshū 57 (Jan. 19, 1962), which involved a prefectural ordinance passed pursuant to the Public Bathhouse Act. The ordinance forbade the granting of a license to operate a bathhouse if the proposed site was less than 250 meters from the closest existing bathhouse, except in special circumstances justifying exception. The plaintiff, an operator of bathhouses, challenged the grant of a license to a competitor on the ground that the new bathhouse would be within 250 meters of the plaintiff's own establishment. The District Court dismissed for lack of standing noting that the Public Bathhouse Act was intended to protect only public health not economic interests. The Supreme Court reversed, stressing that business interests were not de facto or reflex interests but rather were legally cognizable. The court ruled that the purpose of the act was not simply protection of public health but also the prevention of unnecessary and damaging competition among public bathhouse proprietors.

6. I. Ogawa, n. 4 supra.

7. For an interesting discussion of ripeness in the environmental law field see Frank Upham, *After Minamata: Current Prospects and Problems in Japanese Environmental Litigation*, 8 Ecology L. Quarterly 213 (1979).

7a. See I. Ogawa, n. 4 supra. Administrative actions may also be challenged collaterally by an ordinary civil suit (tōjisha soshō). For example, art. 4 states:
(Party Litigation)

"Party litigation" in this Law shall mean litigation in relation to a disposition or decision to affirm or create a legal relationship between parties, which makes the other party (or parties) to the legal relationship a defendant in accordance with the provisions of laws or orders. . . .

(Note: These translations are slightly adapted versions of those appearing in Eibun Hōreisho.)

In addition there are two further types of actions. The first is a "peoples action" (Minshū Soshō) covered by art. 5, which states:

(Public Litigation)

"Public Litigation" in this Law shall mean litigation seeking the correction of acts of an agency of the State or a public entity, which violate laws or ordinances, and which are instituted by any person who is a voter and who does not have a legal interest in the action.

This action, permitted only where explicitly provided by statute, is brought to vindicate the "public interest," rather than the particular interest of an individual; accordingly, plaintiffs need not demonstrate a strong personal legal interest. A taxpayer suit under art. 242.1.2 of the Local Autonomy Law (Chihō Jichihō) (1947, Law No. 67) is an example of such an action. The second type of action is an "organ suit" (kikan soshō) which is invoked to resolve the relative jurisdiction of public bodies.

8. Naohiko Harada, *Preventive Suits and Duty-Imposing Suits in Administrative Litigation,* 9 L. in Japan 63 (1976). Intimately related with such questions is the issue of administrative discretion. Ogawa writes that the courts have acted to limit administrative discretion in two kinds of cases. The first is an external limitation where the official acts ultra vires, or oversteps his discretionary authority provided by statute. The second is an internal limitation, abuse of discretion. Here the official acts contrary to the intent of the statute. Thus, even if the act is within the literal scope of the statute, it may be an abuse of discretion.

9. Article 3 (2) states:

(Kōkoku) (Appeal Litigation)

"A suit for revocation of disposition" in this Law shall mean litigation seeking revocation of a disposition of the administrative agencies, and an act (excluding such deci-

sion, ruling or any other act as prescribed in the following paragraph; hereinafter simply referred to as "disposition") corresponding to the exercise of public power.

10. Article 3 (3) states:

"A suit for revocation of decision" in this Law shall mean litigation seeking the revocation of a decision, ruling, or any other act (hereinafter simply referred to as "decision") of the administrative agencies following a demand for investigation, objection, or any other appeal (hereinafter simply referred to as "demand for investigation").

11. Article 14 states:

(Period for limitation of bringing action)

A suit for revocation shall be filed within three months from the day on which it became known that the disposition or decision was made. . . .

2. *The period under the preceding paragraph shall be a preemptory term.*

3. *A revocation suit may not be filed after one year has passed from the day when the disposition or decision was made: provided, that this shall not apply in cases where there are justifying reasons [for the delay].*

4. *In the event that a demand for investigation may be made with regard to the disposition or decision, or an administrative agency has by mistake instructed [the litigant] that a demand for investigation may be made, and there has been a demand for an investigation, the period under paragraph 1 and the preceding paragraph shall be deemed to run from the day on which the person who has made such demand for an investigation, learned of the decision, or the day of the decision.*

12. Article 3 (4) states:

"A suit for affirmation of nullity" in this Law shall mean litigation seeking the affirmation of the existence or non-existence of a disposition or decision, or of the effectiveness or ineffectiveness thereof.

13. Ogawa, n. 4 supra.

14. Article 3 (5) states:

"A suit for affirmation of illegality of forebearance" in this Law shall mean litigation seeking to affirm the illegality of an administrative agency's inaction where it has a duty to make a certain disposition or decision with regard to an application submitted in accordance with existing laws or ordinances; and it has failed to do so in

spite of the passing of a considerable period of time.

14a. See Harada, n. 8 supra.

15. Yasuhei Taniguchi, a civil procedure expert, explains, "It is generally accepted that the plaintiff must have a grievance of his own. One cannot sue if only some other person is injured or going to be injured. This basic principle probably will not (sic) change. But the availability of private action may be enhanced should the concept of a grievance, (sic) or of being injured, change. If a new 'substantive right' is recognized, it will add to the catalogue of grievances. And this process may result in the courts' granting relief to a new kind of grievance." Yasuhei Taniguchi, A Memorandum on the Procedural Problems in Environmental (or Kōgai) Litigation in Japan, prepared for the 1973 Bonn Conference on Public Interest Law, described in Julian Gresser, *A Japan Center for Human Environmental Problems, The Beginning of International Public Interest Cooperation,* 3 Ecology L.Q. 759 (1973).

16. Articles 197–202 state:

Article 197 (Possessory Actions). A possessor may bring possessory actions in accordance with the provisions of the following five Articles. The same shall apply to a person who possesses on behalf of another person.

Article 199 (Action for Maintenance of Possession). If a possessor is disturbed in his possession, he may by an action for maintenance of possession demand discontinuance of the disturbance as well as compensation for damages.

Article 200 (Action for Recovery of Possession). If a possessor has been deprived of his possession, he may by an action for recovery of possession demand the return of the thing as well as compensation for damages.

2. No action for recovery of possession may be brought against a limited successor in title of a dispossessor, unless such successor in title was aware of the fact of dispossession.

Article 201 (Period for Bringing Possessory Actions). An action for maintenance of possession shall be brought during the period the disturbance continues or within one year after it has ceased; however, in cases where the thing possessed has been damaged by any structural works, the action

may not be brought after one year has elapsed since the commencement of such works, or after the completion thereof.

2. An action for preservation of possession may be brought so long as the danger of disturbance exists; however, if there exists an apprehension that damage may be done to the thing possessed by any structural works, the latter part of the preceding paragraph shall apply mutatis mutandis.

3. An action for recovery of possession shall be brought within one year from the time of the dispossession.

Article 202 (Relation to Actions on Title). Possessory actions and actions on title shall not preclude each other.

2. Possessory actions may not be decided upon grounds relating to the proprietary title.

17. See discussion of Minamata disease in ch. 3 for the text of art. 709. Article 710 states:

(Non-Pecuniary Damage). A person who is liable for compensation for damages in accordance with the provisions of the preceding Article shall make compensation therefore even for non-pecuniary damage, irrespective of whether such injury was to the person, liberty, or reputation of another or to his property rights.

Article 723 states:

(Reputation Injured). If a person has injured the reputation of another, the court may, on the application of the latter, make an order requiring the former to take suitable measures for the restoration of the latter's reputation either in lieu of or together with compensation for damages.

18. Personal rights are held individually and cannot be assigned.

19. Article 13 states:

All of the people shall be respected as individuals. Their right to life, liberty, and the pursuit of happiness shall, to the extent that it does not interfere with public welfare, be the supreme consideration in legislation and in other governmental affairs.

Article 25 states:

All persons shall have a right to maintain minimum standards of wholesome and cultured living.

2. In all spheres of life, the State shall use its endeavors for the promotion and extension of social welfare and security, and of public health.

20. But note that Japan does not have a separate tradition of equity courts.

21. Article 757.1.2 states:

(Jurisdictional Court). Order of provisional disposition shall be under the jurisdiction of the court having jurisdiction in the principal case.

2. The foregoing decision shall be rendered, in urgent cases, without hearing oral arguments.

22. For example, during the Tokugawa period (approximately 1600 to 1868) property owners were required to make "shade payment" (kage-shiro) should they wish to build a house that might obstruct sunlight at a neighbor's residence. See J. Wigmore, 5 Law and Justice in Tokugawa Japan 23 (Kokusai Bunka Shinkōkai, ed., 1971). See also the comments by Gail Feingold Takagi, *Designs on Sunshine: Solar Access in the United States and Japan*, 10 Connecticut L.R. 123 (1977).

23. For example, The Citizens' Alliance Against Building Nuisances formed in 1970 now claims over 164,000 members. See Steven J. Miller, *Let the Sunshine In: A Comparison of Japanese and American Solar Rights*, 1 Harv. Environmental L.R. 578 (1977).

24. Although Japan enacted a City Planning Law as early as 1919 (Toshi Keikaku Hō, 1919, Law No. 36, current version 1968, Law No. 100), it was completely inadequate to check disordered land use patterns or confused urban development. See K. Steiner, Local Government in Japan 239 (1965).

25. See N. Kawazoe, Contemporary Japanese Architecture 3 (1968).

26. Decision of Third Petty Bench, June 27, 1972, 26 Minshū (No. 5) 1067 (1972).

27. See the Supreme Court's decision, 669 Hanrei Jihō 26; see 433 Hanrei Jihō 18, for the Tokyo District Court decision, and 497 Hanrei Jihō 25 for the Tokyo High Court decision. The authors are indebted to Arthur Mitchell for his translation of the *Mitamura* decision.

28. See Kazuaki Sono and Yasuhiro Fujioka, *The Role of the Abuse of Right Doctrine in Japan*, 35 Louisiana L.R. 1037 (1975).

29. See Yasuo Kusumoto, *Comment on the Mitamura Case*, 506 Jurisuto 93 (1972); also, Katsutoshi Suzuki, *The Trend of Decisions*

Involving Injunctions for Sunlight Cases, 297 Hanrei Taimuzu 56 (1973).

30. For example, in 1926 the Anotsu District Court awarded damages because the defendant's malicious erection of a storage shed on his adjacent property allegedly robbed the plaintiff of sunlight. See Judgment of August 10, 1926, in 2648 Hōritsu Shimbun 11 (Annotsu District Court).

31. See the Osaka Alkali decision, sec. 1.2.5 of ch. 1, for a discussion of an early prewar decision.

32. Decision of the Tokyo District Court, September 27, 1972, 684 Hanrei Jihō 66 (1972).

33. Five factors are generally considered in deciding whether an injury in a sunlight case exceeds the limits of human endurance. The first involves the extent of sunlight obstruction. This is measured by the number of hours of direct sunlight reaching the plaintiff's windows at the winter solstice. Brightness and ventilation are also considered. In some cases, "claustrophobic stress" (appakuken) caused by gigantic structures is also an important factor in determining whether there has been a violation and how large the damage award ought to be.

A second factor is locality (chiikisei). The degree of allowable infringement depends on subjective community norms. The court also examines objective factors like population density, and the character of the area.

A third factor is the purpose of the building. The greater the value of the building to society, the greater the infringement the plaintiff is expected to endure.

Finally, the courts will ask whether either party could have avoided the harm. When a tenant moves into an already shadowed area, few courts have allowed a new tenant to claim damages. The court will also consider whether the developer can or could have reduced the damages by a different architectural plan and what efforts he has made to resolve the grievances of his neighbors.

34. Tokyo District Court, February 28, 1972; 660 Hanrei Jihō 32 (1972).

35. For example, between 1967 and 1974 Takagi reports that almost 80% of sunlight petitions were settled without a judicial decision.

Table 1

Zone	1st Class Exclusively Residential			2nd Class Exclusively Residential			3rd Class Mixed Residential and Commercial (Including Semi-Industrial)	
Height of proposed building	Over 7 m or 3 stories			Over 10 m			Over 10 m	
Height of imaginary plane above surface	1.5 m			4 m			4 m	
Alternative	1	2	3	1	2	3	1	2
Hours of shade cast on site between 5 and 10 m from property line	3	4	5	3	4	5	4	5
Hours of shade cast on site beyond 10 m from property line	2	2.5	3	2	2.5	3	2.5	3

36. Between 1968 and mid-1974 almost 80% of sunlight petitions were settled without a judicial decision. For a compendium of sunlight cases, see Tanaka, Nisshō no Hoshō (Compensation for Sunlight (Loss)), published by the Chūō Hōki Shuppan Kabushiki Kaisha (1977).

37. Kenchiku Kijin Hō (1950, Law No. 201).

38. Residents may file objections with the local administrative unit or the developer. The following table sets forth the maximum number of hours of shade that may be cast upon an imaginary plane above the neighboring property between 8:00 A.M. and 4:00 P.M. during the winter solstice. The figures differ for each zone. The local government chooses the appropriate clarifications according to regional characteristics and needs (table 1, from Takagi, n. 22 supra).

39. Before the amendments, many local governments employed sunshine development guidelines (Nisshō Shidō Yōkō) to regulate building standards in relation to sunlight access. Although some guidelines were fairly specific, most were vague, abstract, and often subject to misinterpretation and abuse. The *Musashino* decision has caused many local governments to rely more on local agreements as a means of regulating land use.

40. The Tokyo municipal government, for example, has established a Sunshine Problem Consultation Service and also a Committee for Adjustments of Sunshine Disputes. See Takagi, n. 22 supra.

41. Miller notes (n. 23 supra), e.g., that in Tokyo over 60% of the houses in most districts already occupy lots of less than 100 square meters. Thus, if each family were given equal space for a home and each house were guaranteed four hours of sun, no house in Tokyo could be larger than 6 square meters.

42. See Guido Calabresi, The Cost of Accidents (1970); also, *Transaction Costs, Resource Allocation, and Liability Rules*, 11 J. of L. and Economics 67 (1968).

43. For example, in 1975 a neighborhood group requested a provisional disposition to halt construction of a high-rise apartment building above the seventh story after learning that the developer had obtained a permit for an eleven-story building. A subsequent settlement allowed construction to proceed as planned on the condition that each plaintiff would receive compensation. The court noted that it was only natural for the residents to protect their environment and stated that the payments to the neighbors could be used to seek other appropriate means of relief, including lawsuits for damages. Kōbe District Court, June 3, 1975; see Takagi, n. 22 supra.

44. See Hajime Nito, A *Legal Right to Environmental Quality* (Kankyō Ken), a paper presented at the 1976 International Conference on Environmental Protection in Kyoto, Japan.

45. The proposal was drafted by Tsunetoshi Yamamura, Chairman of the Committee for Pollution Control of the Japan Federation of Bar Associations.

46. Some of the other parts of this draft law relate to the following subjects:
Article 8: Establishment of Environmental Protection Standards
Article 10: Emission Standards
Article 11: Petitions to Revise Standards
Chapter 3: National Comprehensive Environmental Protection Plan; Promotion of the Installation of Facilities for Environmental Protection
Article 18: Institution of Measures for Inspection and Forecasting
Article 19: Inspection
Article 20: The Encouragement of Technology
Article 22: Public Participation
Section 4: Disposition of Disputes and Relief of Pollution Victims
Article 24: The Establishment of a System of Strict Liability
Article 25: Allocation of Expenses
Article 26: Financial Measures of Local Governments
Article 27: Assistance to Industry
Article 37: Lawsuit Requesting Regulatory Measures
Article 39: Suits Requesting the Use of Public Funds
Article 40: Order for Request of Documentation

47. Hiroshima High Court, 693 Hanrei Jihō 27, February 14, 1973.

48. This "public corporation" was established in accordance with art. 284(1) of the Local Government Law (Chihō Jichi Hō) 1947, Law No. 67, and headed by the mayor of Yoshida town.

49. 631 Hanrei Jihō 24 (1971).

50. Haikibutsu no Shori Oyobi Seisei ni Kansuru Hōritsu (1970, Law No. 137).

51. Shortly after the Hiroshima decision, the Osaka District Court enjoined the use of a baseball field for night games by the Kintetsu Buffaloes, a professional baseball team. The decision is significant because it extended legal protection to psychological as well as physical interests. (See *Kintetsu Fujiidera Baseball Ground* Case, Osaka District Court, October 13, 1973, 717 Hanrei Jihō 23 (1973).

52. Kōbe District Court, Amagasaki Branch, May 11, 1973, 702 Hanrei Jihō 18 (1973).

53. The Hanshin Public Highway Corporation is a form of public corporation, owned and controlled by the government, often used to construct roads, harbors, and railways in Japan. See Kiyohiko Yoshitake, *An Introduction to Public Enterprise in Japan* (1973).

54. See Matsuyama District Court, February 2, 1974, 728 Hanrei Jihō 27 (sec. 4.1.6, this chapter) for an administrative law decision relating to this controversy.

55. (Genshi Ryoku Kihon Hō) 1955, Law No. 186.

56. (Genshi Ryoku Iinkai Oyobi Genshi Ryoku Anzeniinkai Setchi Hō) 1955, Law No. 188.

57. See Rex Coleman et al., "The Legal Aspects Under Japanese Law of an Accident Involving a Nuclear Installation in Japan" (mimeographed) (1963).

58. (Denki Jigyō Hō) 1964, Law No. 170.

59. (Dengen Kaihatsu Sokushin Hō) 1952, Law No. 283.

60. (Genshi Ryoku Songai no Baishō ni Kansuru Hōritsu) 1961, Law No. 147.

61. Kōchi District Court Decision, May 23, 1974, 742 Hanrei Jihō 30 (1974).

62. Haikibutsu no Shori Oyobi Seisō ni Kansuru Hōritsu (Law No. 137, 1970), art. 10 (1): *The enterprisers shall be required to manage the industrial wastes by themselves; (2) The municipalities, individually or cooperatively, may (shall) manage industrial wastes, which can be managed together with domestic wastes, and which the municipalities admit to be necessary to manage as a part of their work; and (3) The prefectural governments (as a part of their work) are entitled to treat industrial wastes when they recognize the necessity of such extensive management.*

63. Kokka Baishō Hō (1947, Law No. 125). Article 1(1) states:
The State and public corporations (including local governments) shall be responsible for the damage unlawfully inflicted upon third parties by the intentional or negligent acts of responsible public officials during the exercise of their public duties.

64. Articles 2(1) and 3(1) of the State Compensation Law state:

The State and public corporations shall be responsible for damages to third parties resulting from dereliction in their establishment or administration of highways, rivers, or other public facilities.

If the State or a public corporation is held responsible under the preceding two articles, those officials responsible for selecting or supervising the (above) public officials, or responsible for establishing or administering public facilities, and those officials responsible for paying such officials' salary, compensation, or other costs, or those bearing the cost of establishing or administering public facilities shall be liable. In cases where the person administering the facility and the person who is to bear the financial cost of such facility are different, the latter also shall be liable.

65. In March 1972 frustrated fishermen decided to resort to self-help and set large wire nettings in front of the sluice gates in both canals. The nettings caught a large amount of used vinyl sheets and virtually blocked the water flow of the canals.

66. Shizuoka District Court, May 30, 1974, 745 Hanrei Jihō 19 (1974).

67. Chihō Jichi Hō (1947, Law No. 67). Article 242 of the Local Autommy Law provides that when a local government official is suspected of illegal or unreasonable expenditures, acquisition, maintenance, etc., of public money or property any resident may request the Auditing Commission to investigate the matter in order to prevent the illegal action and/or have the official compensate the losses incurred by the local government. Article 242.2 provides that if the resident who has made such a request to the Auditing Commission is not satisfied with the result, he may file a suit to enjoin the official's action to have him refund the money to the local government, etc.

The area was designated under the Water Pollution Control Law in October of 1970, and water quality standards applicable to it became effective only after July 1971.

68. Kumamoto District Court, February 27, 1975, 772 Hanrei Jihō 22 (1975.

69. Biological Oxygen Demand, a pollution measure.

70. The Hanshin Highway, Kōchi Vinyl, and Tagonoura cases all involved local governments or public corporations.

71. See 729 Hanrei Jihō 4.

72. Osaka District Court's decision, February 27, 1974, 729 Hanrei Jihō 3, Osaka High Court's decision November 27, 1975, 797 Hanrei Jihō 36 (1975).

73. Article 9(1) of the Basic Law for Environmental Pollution Control (Kōgai Taisaku Kihon Hō) 1967, Law No. 132. With regard to environmental conditions relating to air, water, and soil pollution and noise, the government shall establish environmental quality standards, the maintenance of which is desirable for the protection of human health and the conservation of the living environment.

74. 1. Article 709 (Civil Code)
A person who violates intentionally and negligently the right of another is bound to make compensation for damages arising therefrom.

2. Article 2(1) (State Compensation Act)
If any damage has been caused to another person by reason of any defect in the construction or maintenance of a road, river, or other public facility, the state or other public body shall be liable for compensation therefore.

3. Article 1(1) (State Compensation Act)
The state and other public bodies shall be liable for compensation for damages caused by a government official who violates intentionally or negligently the right of another while in the course of exercising his public authority.

75. As noted, Japan does not have a mandatory mandamus procedure, but see N. Harada, n. 8 supra.

76. Although art. 136 of the Code of Civil Procedure (Minji Soshō Hō) 1890, Law No. 29, clearly authorizes a judge to recommend settlement at any time during a judicial proceeding, neither the district nor high court judges in the case had made this suggestion. Note that under art. 203 of the Code of Civil Procedure, a court may even draft a settlement and suggest it to the parties and, if both parties consent, the court-approved settlement has the same legal effect as a court's final judgment, at least with respect to the parties before the court. It would, of course, not bind those not party to the settlement.

77. The cost of these measures was ¥935.5 million in FY 1976 (¥718.5 million related to Osaka International Airport), and ¥409.8

million (¥258.9 million for Osaka) in FY 1977. See Quality of the Environment in Japan 194 (1978). See Int. Environmental Reporter 452, Dec. 10, 1978, and Int. Environmental Reporter 496, Jan. 10, 1979.

78. At times Japan's noise relief measures have provoked international controversies. On August 11, 1975, the Ministry of Transportation announced that a "special landing charge" would be levied on all aircraft (domestic and international) equipped with turbo prop equipment as part of the use charges for public airports. The proceeds of the charge were earmarked principally for various relief measures for victims of noise pollution in the Osaka International Airport area.

Shortly after, twenty-eight major foreign airlines servicing Japan launched a protest campaign against the charge plan. The foreign carriers included Pan American, Lufthansa, Air France, Alitalia, and Northwest Orient Airlines. On November 1, at the behest of the industry, the United States reportedly delivered two protest notes to the transportation ministry through the U.S. embassy in Tokyo. In essence, the United States argued that (1) it was illogical to collect a special landing charge from aircraft landing at Tokyo's Haneda Airport for the improvement of areas around other airports; (2) it was unreasonable to charge heavier (but quieter) aircraft, like airbuses and jumbo jets, more than other, noisier planes; (3) environmental problems created by aircraft noise in airport areas were not peculiar to Japan; (4) implementation of the special landing charge should be postponed until the International Civil Aviation Organization reached a decision on the issue. It was the first time a foreign government had challenged Japan on its landing charge system.

During early planning of the noise charge, the International Air Transport Association (IATA) reportedly expressed disapproval to the Ministry of Transportation about the measure. Thereafter, the ministry and IATA held three rounds of negotiations but no agreement was reached. On December 18, 1975, the twenty-eight foreign companies filed suit; at the same time they refused to pay the charge and demanded its revocation. As of mid-1977, the Japanese government was still holding firm and had filed a counterclaim demanding the uncollected pay-

ments. Both suits were pending at the trial level as of 1979.

79. Similar law suits have also been filed at Atsugi Air Base and at the Fukuoka International Airport. For a translation of the pleadings in the Yokota case, see J. Gresser, K. Fujikura, and A. Morishima, *Japan's Environmental Law in Comparative and International Perspective*, in Teaching Materials used at Harvard Law School, 1976–1977.

80. See "Protection and Enhancement of Environmental Quality," USFJPL 85-2, December 8, 1975, for the position of the U.S. Defense Department as of 1978.

81. The government's argument rests primarily on its construction of art. 3 and 44 of the Administrative Procedure Law that traditionally have been interpreted to narrow the availability of injunctive relief against government or quasigovernment bodies.

82. Before the Osaka High Court decision, however, at least one court had declined to adopt the distinction. In the Izumi City Crematory case, local residents sought to enjoin the construction and operation of a crematory planned by the defendant city pursuant to the City Planning Law. The court analyzed the doctrine as follows:

The entire process in this case from the decision (to build the crematory) based on articles 21(2) and 19(1) of the City Planning Law through its construction and operation might be reasonably interpreted as the 'exercise of public power' as provided in article 3 of the Administrative Litigation Law. However, the process of establishing a public facility has always been divided into the decision to establish it and the acts taken to construct it. The latter has not been interpreted as 'action constituting an exercise of public power.' Furthermore, interpreting (an act) as an 'exercise of public power' is based on a legal doctrine which is intended to expand judicial relief by recognizing an additional administrative cause of action. To use it to dismiss the present case, therefore, would be to turn the doctrine inside out.

See the decision of the Kishiwada Branch of the Osaka District Court, April 1, 1972; 663 Hanrei Jihō, No. 80 (1972).

83. See also the case of drug licensing where government discretion is even broader than

in the operation of airports, but the courts have granted relief when such action has resulted in serious injury. "SMON Decision," Kanazawa District Court, 879 Hanrei Jihō 26 (March 1, 1978).

84. See, for example, *Molitor v. Kaneland Community Unit Dist. No. 302*, 163 N.E.2d 89 (1959); cert. denied, 362 U.S. 968.

85. One exception is the right of commons (iriaiken) that traditionally protected the rights of villagers to forage on private land. Although a 1973 Supreme Court decision recognized the commons right of local residents to a piece of woodland in the town of Kizukuri, Aomori Prefecture, a decision that contradicted an early 1915 precedent that commons rights are extinguished on alienation of the common land (see Mainichi News, March 15, 1973), advocates of environmental rights have considered the extremely limited iriai concept to be of little value in strengthening their doctrinal arguments for the recognition of an environmental right.

The current debate over the recognition of the public's right to environmental quality might usefully be compared to the course of the development of other rights. In his study of the right to privacy in Japan, for example, Professor Masami Itoh describes how the abuses of power by the media created a need for judicial intervention because traditional ethical prohibitions were unable to safeguard individual privacy from intrusion. Proceeding cautiously, the courts succeeded in interpreting the Code of Civil Procedure to extend protection to individuals in those circumstances. (See Masami Itoh, The Right of Privacy (Puraibashi no Kenri) (1963)). This process is also evident in the judicial development of a right to sunshine.

86. See H. Nito, n. 44 supra.

87. It is interesting that similar arguments have been used to challenge claims of environmental rights in the United States. See William D. Kirchick, *The Continuing Search for a Constitutionally Protected Environment*, 4 Environmental Affairs 515 (1975); and Frederick R. Anderson, *Environmental Rights and Environmental Quality*, a paper delivered at the International Congress of Scientists on the Human Environment, November 17–26, 1975, Kyōto, Japan.

88. Other court decisions have also rejected the environmental rights theory. See Chiba District Court, 836 Hanrei Jihō 17 (Aug. 31, 1976); Kagoshima District Court, 675 Hanrei Jihō 26 (May 19, 1972); Ōita District Court, 925 Hanrei Jihō 3 (March 5, 19, 1979). But see a recent Yokohama District Court decision protecting individual rights to enjoyment of view when obstruction was intentional, 917 Hanrei Jihō 23 (February 26, 1979).

89. See Yoshiro Nomura, *The Development of an Environmental Impact Assessment Duty in Environmental Litigation in Japan*, a paper presented to the 1976 International Symposium in Kyōto.

90. See *Comment, Equity and the Ecosystem: Can Injunctions Clear the Air*, 68 Mich. L. Rev. 1254; also Keeton and Morris, *Notes on "Balancing the Equities,"* 18 Tex. L.R. 412 (1940).

91. To the extent that the marginal cost of noise abatement measures is lower than the marginal cost of damages (past and future), the government, like any profit-maximizing firm, should select the lower cost option.

92. See Ronald Coase, *The Problem of Social Cost*, 3 J. of L. and Economics 1 (1960).

93. See *Boomer v. Atlantic Cement Co.*, 26 N.Y.2d, 257 N.E.2d 870, 309 N.Y.S. 2d 312 (1970). For a criticism of this decision see Richard B. Stewart and James E. Krier, Environmental Law and Policy, 2nd ed. (1978).

94. The reader should note, however, that many residents oppose moving out under any circumstances, claiming that no amount of money can compensate for the disruption of their lives. This institutional fact would seem to some extent to undercut the foregoing economic analysis.

95. See 88 Hanrei Jihō 71 (May 31, 1978).

96. See The Daily Yomiuri, September 6, 1977; 862 Hanrei Jihō 3 (Sept. 5, 1977).

97. Matsuyama District Court, June 23, 1968, 548 Hanrei Jihō 63.

98. By FY 1973, the Environment Agency reports, over 40% of Japan's coastline was covered by concrete.

99. There are about 2,500 fisheries cooperatives in Japan composed of about 600,000 fishermen. Cooperatives have also been organized by type of fisheries, such as tuna fisheries cooperatives or trawlers' cooperatives. These cooperatives have their re-

spective prefectural federations which are affiliated with the National Federation of Fisheries Cooperatives.

Fisheries cooperatives by and large contribute directly to the interest of member fishermen by marketing their catches, and purchasing gear and other fishing equipment in bulk at low prices. These cooperatives maintain order in fishing activities and conserve fisheries resources by supervising fishing rights in coastal waters. The cooperatives contribute to the development of production and to the economic stabilization of members. Fisheries products processors have also organized their own cooperatives by type of products. These cooperatives are affiliated with prefectural federations and a national organization. The Japan Fisheries Association (Dai-Nippon Suisan Kai) is the sole organization representing the entire Japanese fisheries industry.

The principal objectives of the Japan Fisheries Association are to coordinate the views and opinions of the member fishermen and workers in related enterprises on questions of general importance to the industry, to conduct studies on domestic and international fisheries problems, and to take such actions on behalf of the aforementioned members as to submit petitions or statements to the Diet or competent government authorities (see Japan Fisheries Association, 1975). For a more detailed discussion of the legal structure of the Fisheries Association, the reader should refer to the *Usuki Cement* and *Date Electric Power* cases.

100. Article 31 states:
No person shall be deprived of life or liberty, nor shall any other criminal penalty be imposed, except according to procedure established by law.

101. Even apart from the deletion of the word "property" and the substitution of the phrase "except according to procedure established by law" for "without due process of law," comparative interpretation of art. 31 and the American due process clause is complicated by the interpolated clause "nor shall any other criminal penalty be imposed." There has been a wealth of Japanese academic analysis and considerable judicial exploration of the question whether art. 31 (or any other provision of the Japanese Con-

stitution) protects the individual against administrative imposition of burdens, or denial of benefits, without a fair hearing. Yet the question of whether there is indeed a constitutional right to fair hearing under Japanese administrative law remains open.

102. Koyūsuimen Umetate Hō (1921, Law No. 57).

103. Article 4(3) in part states:
If there are persons who have rights in the public surface waters located within the reclamation construction area, the governor of the prefecture may not issue a permit for reclamation unless at least one of the items mentioned below is fulfilled (in addition to the conditions of paragraph 1 of this article).
Item 2. The benefit produced by the reclamation greatly outweighs the cost.

104. Tochi Shōyō Hō (1951, Law No. 219).

105. Kasen Hō (1964, Law No. 167).

106. Cabinet Order No. 14, 1965.

107. Article 39(1) of the Fisheries Law (Gyogyō Hō), 1949, Law No. 267, states:
(Alteration, Revocation or Suspension of the Exercise of Fishery Rights for Necessary Public Interests) In case it is deemed necessary for fishery adjustments, navigation, anchorage or mooring of ships, laying of submarine cables, or for other public interests, the mayor of the metropolis or the governor of the competent prefecture may alter or revoke any fishery right or suspend the exercise thereof.

108. Article 38 of the Riparian Law (Kasen Hō) 1964, Law No. 167 states:
When there is a request for a permit for water use prescribed in art. 23 or 26, the River Administrator shall inform [except in cases when the request should be dismissed], as provided by the order of the Ministry of Construction, those persons who have previously received permits pursuant to art. 23 through 29 and those persons who have rights in the river pursuant to cabinet order (thereafter referred to as 'users of the river concerned'), of the name of the applicant, the purposes of the proposed water use and the other items prescribed by the order of the Ministry of Construction. However, the River Administrator need not inform those who, clearly, will not suffer loss from the proposed water use nor those who have given their assent to it.

Article 39 of the Riparian Law states:

When the 'users of the river concerned' receive notice pursuant to the previous article, they may express, according to the provisions of the order of the Ministry of Construction, their opinion on the proposed water use to the River Administrator, and demonstrate how they will suffer the losses from the water use.

109. Article 39 of the Fisheries Law requires a hearing for revocation of fishery rights.

Article 39(4) states:

When the Sea-Area Fishery Adjustment Commission intends to make an application pursuant to the preceding paragraph, it shall give the rights holders prior notice in writing of the reasons for such restriction or condition, and shall give them or their proxies the opportunity of explaining themselves at a public hearing and producing evidence.

110. Ōita District Court, July 20, 1971, 638 Hanrei Jihō 36.

111. Suisan Gyō Kyōdō Kumiai Hō (1948, Law No. 242).

112. See George A. Cooke, *Noise Regulation: Recent Japanese and American Standards Differ Significantly*, 1 Harvard Env. L.R. 553 (1977).

113. For a discussion of related problems of regulating railroad noise in Japan see George A. Cooke, id.

114. Tokyo District Court, May 31, 1973, 704 Hanrei Jihō 31.

115. Kōtsū Anzen Shisetsu tō Seibi Jigyō ni Kansuru Kinkyū Sochi Hō (1966, Law No. 45).

116. Article 9 states:

Eligibility of Plaintiff. Suits for revocation of disposition or decision (hereinafter referred to as 'revocation litigation') may be filed only by persons (including persons whose legal interest would be vindicated by the revocation of such disposition or decision, even after the disposition or decision has lost its effect due to a lapse in time or for any other reason) who have a legal interest in seeking the revocation of said disposition or decision.

117. Sapporo District Court, January 14, 1974, 727 Hanrei Jihō 3.

118. Article 25(2) states:

Suspension of Execution. . . . If in the event that a suit for revocation of a disposition is filed, and in order to avoid irreparable damage, there is an urgent need [to act], the court may suspend the effect of such disposition, the execution of such disposition or the continuance of a procedure, or may suspend in whole or in part the effect of the disposition, the execution of the disposition, or the continuance of the procedure (hereinafter referred to as 'suspension of execution'); provided that when suspension of the execution of the disposition or of the continuance of procedure will achieve the same purpose as would the suspension of the effect of the disposition, such suspension (of effect) shall not be made.

119. Article 10 states:

Grant of Fishery Rights . . . Any person who wishes to acquire a fishery right shall file an application for the grant thereof with the mayor of the metropolis or the governor of the competent prefecture.

120. Article 23(1) states:

Legal Character of Fishery Rights. A fishery right shall be deemed to be a real right and the provisions governing [the ownership, use and possession of land] shall apply with necessary modifications thereto.

121. Articles 8(1), 8(2) state:

Right of Members to Operate Fisheries . . . Persons who are fishery operators or fishery employees, who are qualified under the Regulations for the Exercise of Fishery Rights or the Regulations for the Exercise of Common Fishing Rights and who are, furthermore, members of a fishermen's cooperative which was established by the Fishermen's Cooperative Association or the Federation of Fishermen's Cooperative Associations of which such Fishermen's Cooperative Association is a member association, shall have the right to operate a fishery within the scope of the fishery rights and common fishery rights specified and demarcated as belonging to such Cooperative Association.

2. The Regulations for the Exercise of Fishing Rights and the Regulations for the Exercise of Common Fishing Rights mentioned in the preceding paragraph (hereinafter referred to as 'Regulations for the Exercise of Fishing Rights) shall set qualifications describing the persons who may own the right to operate such fisheries pursuant to the same paragraph, determine the area and period in which such fisheries are to be op-

erated and the fishing methods to be employed, and other matters relating to the operation of such fisheries.

122. Suishitsu Odaku Bōshi Hō (1970, Law No. 138; amended by Law. No. 88, 1971, and Law No. 84, 1972).

123. Denki Jigyōhō (1964, Law No. 170).

124. As the reader might surmise, one purpose behind the enactment of the Electric Enterprise Law was to maintain MITI's jurisdiction over this important industry against the possible encroachment of the Environment Agency.

125. The Tokyo High Court in the Nikkō Tarō cedar tree case (summary) July 13, 1973, 710 Hanrei Jihō 23 (1973).

126. Article 20(3) states:

The minister or the governor of the prefecture may authorize the business if such business fulfills all of the items mentioned below....

Item 3. The business plan contributes to an appropriate and reasonable use of the land.

127. The area in question was designated under the Natural Park Law (Shizenkōenhō) (1957, Law No. 161). Article 10(1) of this law states:

Director General of the Environment Agency may designate a special preserve area . . . when there is a special need in order to preserve the scenic beauty of the national park.

128. The reader should note that the Tarō cedar and other trees in issue were not themselves designated nor within the protected zone. See Cultural Properties Protection Law (Bunkazai Hogo Hō) (1950, Law No. 214). Article 69 states:

Designation....

2. The Minister of Education may designate especially important ones among historical sites, scenic spots or natural monuments designated pursuant to the previous article as special historical sites, special scenic spots or special natural monuments.

129. Takamatsu High Court, July 17, 1975, 325 Hanrei Taimuzu 160 (1975). See Takehisa Awaji, Ikata Genpatsu Hanketsu no Mondaiten (Problems with the Akata Nuclear Power Plant Case), 8 Kōgai Kenkyū 1 (July 1978).

130. Article 312(3) states:

Obligation to Produce a Document. A holder of a document shall not refuse the production thereof in the following cases: . . .

3. Where the document was created for the benefit of the person who has the burden of proof or in connection with a legal relation between him and the holder thereof.

131. Articles 272–273 state:

Examination of a Government Official. In the event that the court is to examine a government official or a person who was a government official as witness regarding official secrets, it shall obtain approval from the competent supervising government agency.

2. The provisions of the preceding paragraph shall apply mutatis mutandis to other public officials.

Examination of Ministers of State. In the event that the court is to examine the Prime Minister or ministers of state or persons who once occupied such office as witness regarding official secrets, it shall obtain approval from the Cabinet.

132. Tokyo District Court, Hachiōji Branch, December 8, 1975, 803 Hanrei Jihō 18 (1975). The Musashino decision is the only decision we have found where a developer or business in Japan successfully challenged the administrator's authority to set environmental pollution regulations or other controls. But the controversy continues. The Mayor of Musashino has reportedly been indicted for refusing to supply water to an apartment which was in violation of the city's guidelines. See The Daily Yomiuri, December 7, 1978, p. 2.

133. Article 44 of the Administrative Litigation Law (Gyōsei Jiken Soshō Hō, 1962, Law No. 139) provides that: "With regard to a disposition of an administrative agency or an act corresponding to the exercise of public power, temporary injunctions prescribed in the Code of Civil Procedure shall not be made."

134. Thus despite some cautious judicial extensions, a plaintiff's standing in environmental litigation still remains hamstrung by traditional doctrine. It would still be difficult for a nature protection organization, or even its members, to establish standing in an action designed to protect an endangered forest preserve. This would be true even if all members actually used the forest for recreational purposes. A fortiori,

despite Japan's historic reverence for nature, neither trees nor other natural objects have acquired standing and there is little hope of this occurring in the future. For an interesting comparison see Christopher Stone, *Should Trees Have Standing? Toward Legal Rights for Natural Objects*, 45 S. Cal. L. Rev. 450 (1972).

135. Genshi Ryoku Kihon Hō (1955, Law No. 196).

136. See Japan Constitution, Art. 13 and 25.

137. But see p. 221 for a discussion of the implications of the court's decision on the merits.

138. See 880 Hanrei Jihō 2.

139. The decisions seem to inhibit actions brought by members of the general public whose sole basis for standing is that they, like others, will be affected by the agency's action.

140. For a discussion of these developments in English, see Nathaniel L. Nathanson and Yasuhiro Fujita, *The Right to Fair Hearing in Japanese Administrative Law*, 45 Washington L.R. 273 (1970).

141. Amended by Law No. 84 (1973).

142. Note that the court apparently felt that the site's natural value could not be replaced by the administration's proposed artificial restoration plan. See Laurence Tribe, *Ways Not to Think About Plastic Trees: New Foundations for Environmental Law*, 83 Yale L.J. 1315 (1974). Compare also *Sierra Club v. Froehlke*, 359 F. Supp. 1289 (S.D. Tax. 1973) (3 Environmental L. Reporter 20248, 1973).

143. See commentaries in The Daily Yomiuri, April 26, 1978, p. 2; Mainichi Daily News, April 26, 1978, p. 12; Yomiuri Shinbun Yūkan, April 25, 1978.

144. See Mainichi Daily News, June 6, 1978.

145. See, for example, the recent decision of the Matsuyama District Court rejecting local residents' attempt to block a fishing port because they had no common right of access. Mainichi Daily News, May 31, 1978, p. 12. For a discussion of the new beach access claims and related problems see *Kankyō Hogo no Hōriron to Hōteki Shudan (Legal Theory of Environmental Protection and Legal Procedure)*, 607 Jurisuto 57 (March 1, 1976). The Tokyo High Court has also reversed the Utsunomiya District Court's ruling protecting the environmental interests of a group of residents in Tochigi Province who demanded cancellation of the province's designation of their area as a quasi-industrial zone. See The Yomiuri Daily News 2, Apr. 12, 1978.

146. Article 102(c) states:

Sec. 102. The Congress authorizes and directs that, to the fullest extent possible: (1) the policies, regulations, and public laws of the United States shall be interpreted and administered in accordance with the policies set forth in this Act, and (2) all agencies of the Federal Government shall . . . (c) Include in every recommendation or report on proposals for legislation and other major Federal actions significantly affecting the quality of the human environment, a detailed statement by the responsible official on—

(i) The environmental impact of the proposed action,

(ii) Any adverse environmental effects which cannot be avoided should the proposal be implemented,

(iii) Alternatives to the proposed action,

(iv) The relationship between local short-term uses of man's environment and the maintenance and enhancement of long-term productivity, and

(v) Any irreversible and irretrievable commitments of resources which would be involved in the proposed action should it be implemented.

Prior to making any detailed statement, the responsible Federal official shall consult with and obtain the comments of any Federal agency which has jurisdiction by law or special expertise with respect to any environmental impact involved. Copies of such statement and the comments and views of the appropriate Federal, State, and local agencies, which are authorized to develop and enforce environmental standards, shall be made available to the President, the Council on Environmental Quality and to the public as provided by Section 552 of title 5, United States Code, and shall accompany the proposal through the existing agency review processes. . . .

We have no evidence that NEPA has had a direct influence on the Japanese courts' decisions. Although NEPA has now been intensively studied in Japan, it was not well known at the time of many of the decisions under discussion. A more likely explanation for the similarity is that NEPA's mandate and that evolved by the Japanese courts may represent such basic common sense as would naturally occur to an "environmentally concerned" court.

147. See Grant P. Thompson, *The Role of the Courts*, in Federal Environmental Law, ed. Edric L. Dolgin and Thomas G. Guilbert (1974); also Harold Leventhal, *Environmental Decisionmaking and the Role of the Courts*, 122 U. of Penn. L. Rev. 509 (1974).

148. See, e.g., the opinion of Justice Bazelon in *Ethyl Corp. v. EPA*, 541 F.2d 1 (D.C. Cir. 1976).

149. See *TVA v. Hill*, 437 U.S. 153, 98S. Ct. 2279, 57L. Ed. 2d 117 (1978).

150. Along with *Ethyl*, n. 148 supra, see *International Harvester v. Ruckelshaus*, 478 F.2d 615 (D.C. Cir. 1973).

151. *Michie v. Great Lakes Steel Div., National Steel Corp.*, 495 F.2d 213 (6th Cir. 1974) 6 ERC 1444 (6th Cir. 1974).

152. The court primarily relied on automobile accident cases to support their arguments. See *Maddux v. Donaldson*, 362 Mich. 425 (1961) 108 N.W.2nd 33, *Watts v. Smith*, 375 Mich 120 134 N.W.2nd 194 (1965). See also a pollution case, *Landers v. East Texas Salt Water Disposal Company*, 151 Tex. 251, 248 S.W.2d 731 (1952).

153. But see discussion on joint enterprise in William L. Prosser, The Law of Torts (1971).

154. 67 N.J. 291, 338 A.2d 1, 423 U.S. 929 (1975 cert. den'd). See also *Ybarra v. Spangard*, 25 Cal. 2d 486, 154 P.2d 687 (1945). *Anderson* seems a substantial extension of the *Ybarra* reasoning.

155. For an important American comment, see Marcia Gelpe and A. Dan Tarlock, *The Uses of Scientific Information in Environmental Decision Making*, 48 So. Cal. L.R. 371 (1974).

156. See L. Tribe, *Trial by Mathematics: Precision and Ritual in the Legal Process*, 84 Harv. L.R. 1329 (1971); for an unsuccessful use of statistical proof in a criminal case see *People v. Collins*, 68 Cal.2d 319, 438 P.2d 33 (1968).

157. For example some diseases like asthma have a "natural incidence" with no apparent relation to pollution.

158. See *Diamond v. General Motors Corp.*, 20 Cal. App. 3d 374 (1971).

159. In affirming the trial court's refusal to permit the plaintiff to amend his complaint, the court stated: "It is improbable that we could have alleged in good faith that the conduct of every defendant named in the class action complaint injured him as an individual."

160. See Miller, n. 23 supra.

161. See Solar Energy, Research, Development and Demonstration Act of 1974, 88 Stat. 1431, 42 U.S.C. 5551-5566 (Supp. V 1975).

162. See, for example, *Fontainebleau Hotel Corp. v. Forty-Five Twenty-Five, Inc.*, 114 So. 2d 357 (Fla. Ct. App. 1959) and other cases cited by Miller.

163. See 2nd Restatement of Torts 828 (Tent. Draft November 17, 1971).

164. See n. 93 supra.

Chapter 5

1. This discussion is based on the OECD report, Environmental Policies in Japan, A Report of the Organization for Economic Cooperation and Development (1977).

2. Perhaps the clearest index of environmental improvement is the dramatic reduction of SO_2 (sulfur dioxide) concentrations since 1967. For example, in 1974, average concentrations were 50% lower than those in 1967, and as a result, the percentage of monitoring stations reporting compliance with ambient quality standards significantly increased. Concentrations of CO (carbon monoxide) have also diminished substantially. For example, all monitoring sites reporting in 1975 registered compliance with 1975 ambient air quality standards. Finally, significant progress had also been made in the area of toxic substances control. For a more detailed comparative analysis see app. 1.

3. More precisely, GNP was "reduced" by the deflationary price effects associated with nonproductive investment, but "increased" by the expansionary income effect resulting from these antipollution investments. During the first few years of implementation, the positive effects were significantly greater than negative effects, although there is a trend toward offsetting these positive effects. See Shuntarō Shishido and Akira Oshizaka, *Econometric*

Analysis of the Impacts of Pollution Control in Japan, a paper presented at the International Conference for Environmental Protection, organized by the Nippon Keizai Shinbun, May 26–28, 1976; also Yasuke Murakami and Jinkuchi Tsukui, *Economic Costs of Prevention of Pollution: A Dynamic Analysis of Industrial Structure,* from the same conference. A third model was employed by the Economic Planning Agency. The OECD report (n. 1 supra) notes that all three models "suffer from the lack of reliable data on the structure and magnitude of pollution abatement costs. Their results must therefore be interpreted with great care." According to the Shishido and Oshizaka model, antipollution policies have stimulated modest price inflation. It is estimated that during a six-year period there has been a 1.9% overall rise (about 0.3% per year) in the case of a "softer" policy. Although substantive increases have appeared in certain sectors, such as automobiles (5.8%), electricity (6.2%), primary iron (7.6%) and paper pulp (7.7%), the overall inflationary effect of pollution control policies is considered to be insignificant.

The simulation model also suggests that the share of output and employment in some industries, such as primary and fabricated metals, general electric machinery, etc., has actually increased. Apparently antipollution policies have actually benefited some polluting industries. This is attributed to the fact that some polluting industries, such as the steel industry, are the main beneficiaries of antipollution investments. These industries enjoy a positive income effect that exceeds the adverse effects of price increases.

4. See generally, Richard H. Minear, Japanese Tradition and Western Law (1970).

5. For a study of the Administrative Court see Hideo Wada, *The Administrative Court Under the Meiji Constitution,* 10 L. in Japan 1 (1977).

6. See Albert M. Craig, *Aspects of Government Bureaucracy,* in Modern Japanese Organization and Decision Making, Ezra F. Vogel, editor (1975).

7. The purge excluded or removed from public life a total of 210,288 persons; of these 1,809 (or 0.9%) were former or incumbent civil servants. The Ministry of Home Affairs suffered the greatest loss. For an evaluation of the purges see Hans H. Baerwald, The Purge of Japanese Leaders Under the Occupation (1959); John D. Montgomery, Forced to be Free: The Artificial Revolution in Germany and Japan (1957).

8. For example, decorations were abolished for a time and bureaucrats' pay declined relative to that in other sectors of the society.

9. See, for example, the use of this term in art. 15.

10. See generally, Dan Fanno Henderson, The Constitution of Japan (1968).

11. See Naohiko Harada, *Preventative Suits and Duty Imposing Suits in Administrative Litigation,* 9 L. in Japan 63 (1976).

12. For a discussion of the various commissions established to consider the reform of the bureaucracy, see Chalmers Johnson, *Japan: Who Governs? An Essay on Official Bureaucracy,* 2 J. of Japanese Studies, 1–28 (1975). Johnson notes that postwar officials work harder, are paid less, and must retire earlier. He also points out that it was unthinkable for politicians to interfere in the budgets of bureaucrats.

13. See Nathaniel L. Nathanson and Yasuhiro Fujita, *The Right to a Fair Hearing in Japanese Administrative Law,* 45 Wash. L.R. 273 (1970).

14. See Akira Kubota, Higher Civil Servants in Postwar Japan (1969).

15. See Johnson (n. 12 supra) for a discussion of the decision by SCAP authorities to retain most of the prewar bureaucracy.

16. See T. J. Pempel, *The Bureaucratization of Policy-Making in Post-War Japan,* 18 American J. of Political Science 647 (Nov. 1974).

17. But as noted in ch. 1, the model in the environmental field needs to be modified in two respects, i.e., with regard to the somewhat more active role of the Diet, at least as a "sounding board," and the slightly greater importance, particularly in the mid-1960s, of the advisory committees, because of their scientific expertise.

18. There are many terms in Japanese for this concept. "Nawabari ishiki"—"territorialism"; "kakkyoshugi"—"sectional independence"; "takotsubo-shiki gyōsei"—"octopus-pot type of administration"; or simply "sekushonarizumu"—"sectionalism." Craig (n. 6 supra) notes that even in the Edo period the bakufu exchequer

(kanjō bugyō) made explicit the specific functions and limits of competence of its various suboffices. He notes that so scrupulous was the concern of the period over bureaucratic territory that jurisdiction was held as "sacred and inviolable" (shinsei ni shite okasu bekarazu) as the person of the emperor. Territorial struggles between two offices have at times also interfered with interministerial cooperation (kengen arasoi). It has been suggested that jurisdictional rivalries may constitute a beneficial restraint on the undemocratic tendencies of the bureaucracy, for if the agencies became highly coordinated, their powers could be applied with overwhelming force. See Leon Hollerman, Japan's Dependency on the World Economy (1967).

19. This is not to say that collaboration between the Environment Agency, MITI, and other agencies does not take place. One example is the cooperation between MITI and the agency in regard to the modification of the Pollution-Related Health Damage Compensation Law (see ch. 6). Another is the current joint discussion concerning the control of fluorocarbon aerosols.

Such collaboration often occurs where (as in the former case) there is a joint statutory basis for jurisdiction or where (as in the latter case) the matter is not considered by either agency to be of great importance.

20. For an English comment see Hajime Nakamura, *Basic Features of the Legal, Political, and Economic Thought of Japan*, in The Japanese Mind, Charles A. Moore, ed. (1973).

21. The consensus system is one important reason for the few public records of arguments or even debates within the Cabinet. By the time that a decision is publicized, a consensus has already been reached.

22. See Craig, n. 6 supra.

23. Id.

24. For a discussion of the various dimensions of this practice in government as well as the private sector, see Vogel, n. 6 supra.

25. An important related issue is the role of cliques. Each ministry has its school cliques (gakubatsu), particularly involving graduates of Tokyo University. Craig (n. 6 supra) points out that while posing some threat to the solidarity of the bureaucratic structure, cliques also serve a positive func-

tion. By calling on a member of the clique, an official may facilitate consensus when a controversy has reached an impasse. Although cliques are ordinarily confined to a particular ministry, the close contacts officials have enjoyed by their common backgrounds also facilitate interministerial collaboration, when such cooperation might otherwise be blocked by jurisdictional rivalries.

26. Craig, n. 6 supra. It is important to note that the Japanese ministry does not suffer a disruptive administrative reshuffle every time there is a cabinet change, as experienced by American agencies during a change in administration. Indeed, there is a strong institutional presumption favoring continuity and stability despite political shifts. See Chalmers Johnson's discussion of the Tanaka cabinet, n. 12 supra.

27. Craig, n. 6 supra.

28. Id.

29. Id.

30. Many environmentalists consider Takeo Miki, along with Ōishi, to have been sensitive to environmental problems.

31. See sec. 3.3.

32. Johnson, n. 12 supra.

33. Under the postwar reforms there are, in theory, a number of ways in which the public is permitted to participate in governmental decision making. One mechanism is by representation in advisory councils. In practice, however, representatives of various "new" special interest pressure groups (such as an environmental or consumer organization) are rarely appointed to the councils, whose membership mainly consists of representatives of industry, labor unions, or established housewives' associations, and a few scholars and journalists.

Another means of public participation is through public hearings. Nathanson and Fujita (n. 13 supra) report that there are now 481 articles in 227 statutes requiring public hearings in 764 different types of administrative dispositions, covering practically all adjudicatory administrative actions. The authors also describe some of the defects of the public hearing system recognized at the time of their article's publication.

34. In a number of important respects, environmental controversies have actually compounded the difficulties of Japan's pub-

lic hearing system described by Nathanson and Fujita (n. 13 supra). First, there exist few statutory provisions requiring public hearings on administrative decisions entailing significant adverse environmental effects. Absence of a clear statutory basis has presented particularly acute problems for the courts in defining an appropriate standard. Second, even if a hearing is deemed desirable, it is still not clear who should have a right to be heard. Persons whose health or living environment is directly violated have generally been conceded to have some interest. But it is still unclear whether members of the general public, not directly related to a given action, can seek a hearing on constitutional or any other grounds. Finally, although local governments have at times sought to establish a public hearing requirement by local ordinance, it is not clear to what extent the public has availed itself of these provisions, or in what specific ways comments channeled through the public hearings have had any significant effect on decisions of administrative bodies.

35. For an interesting comparison with U.S. practice, see W. Pederson, Jr., *Formal Records and Informal Rule Making*, 85 Yale L.J. 38 (1975).

36. Generally, governmental processes are still less open to the public in Japan than in the United States and a few other countries. See Stanley V. Anderson, *Public Access to Government Files in Sweden*, 21 American J. Comparative L. 419 (1973). It is, of course, not altogether impossible for the public to obtain information on various matters from government officials. Indeed, each government agency has its own general in-house rules relating to the release of governmental documents. For example, under the MITI regulations, "non-secret" documents and other data may be released to the public. The rules specify various categories of "secret." "Top secret" documents are those documents whose leakage could jeopardize the national security or interest and which have been so designated by bureau chiefs. "Secret" documents are less critical and they are also designated by section chiefs. The MITI rules note that the designation "secret" is to be minimized to the extent possible. The criteria, however, are very vague and the administrator is accorded, as in other areas, broad discretion. In practice,

determinations of secrecy are not easily predictable. One official might release a document, while another official might be unwilling to disclose information of comparable sensitivity.

In 1972 the people's "right to know" became a controversial issue via the Nishiyama case. Takichi Nishiyama was a political reporter working for the newspaper Mainichi Shinbun; he had prevailed upon Kikuko Hasumi, a former secretary to the deputy vice-foreign minister, Takeshi Yasukawa, to deliver to him secret documents pertaining to the reversion of Okinawa. Both Nishiyama and Hasumi were arrested and charged with violating the National Public Servants Law (Kokka Kōmuin Hō, Law No. 120, Oct. 3, 1947; latest amendment, Law No. 117, 1971) that prohibits public servants from leaking secret government matters.

The Tokyo District Court acquitted Nishiyama but ruled that Yasumi had violated arts. 100 and 109 of the law. The Court found Nishiyama's action "justifiable" because it was aimed at news gathering, an activity protected by art. 21 of the Constitution, and also because Nishiyama's reporting had conferred a tangible public benefit greater than the harm resulting from his action. The court emphasized that all governmental affairs in principle must be open to the public in a democratic society, and that secrecy was warranted only during the formulation and implementation of policy, under circumstances where the results could be reviewed by the public. These principles, the court held, applied equally to the conduct of diplomatic negotiations.

The Tokyo High Court, however, reversed the lower court's decision and ruled that Nishiyama had violated art. 111 of the Public Servants Law that forbids the solicitation of civil servants. Although concurring with general principles articulated by the District Court, the High Court ruled that Nishiyama had deviated from "the right track of news gathering," thereby removing himself from under the Constitutional mantle. Nishiyama thereafter appealed to the Supreme Court. See Ronald G. Brown, *Government Secrecy and the "People's Right to Know" in Japan: Implications of the Nishiyama Case*, 10 L. in Japan 81 (1977).

Although the Nishiyama case served to draw public attention to the right to know, and although it may have relaxed the restrictive attitude of some government officials, it may also have produced a governmental backlash. After the District Court's decision the government actually sought to tighten control over information by its proposed amendment to the Criminal Code. The proposed amendment would impose stringent criminal penalties for the disclosure by public employees of governmental information and penalize even a private company's employee's divulgence of corporate secrets. Many commentators now fear that these amendments signal a trend toward an even less generous official policy on the people's right to know.

37. Pempel, n. 16 supra.

38. For an interesting discussion, see Richard R. Lury, *Japanese Administrative Practice: The Discretionary Role of the Japanese Government Official*, 31 Business Lawyer 2109 (July 1976).

39. At times a statute may limit discretion to a determination of applicable law or the means of ensuring compliance; however, on other occasions, a statute may authorize the bureaucracy to act in accordance with what it perceives to be the overall statutory purpose or the national interest. Such free discretionary acts are termed "jiyū sairyō kōi."

40. See Yoriaki Narita, *Administrative Guidance*, 2 L. in Japan (1968). Professor Narita begins his study of this subject by pointing out that the term "gyōsei shidō" is not a regular legal term, and is not to be found in statutes, cabinet orders, or scholarly analyses. Rather, he notes, "it may be called a usage of government offices and the mass news media."

41. See John K. Fairbank, Edwin O. Reischauer, and Albert M. Craig, East Asia: The Modern Transformation (1964) for a discussion of this period.

42. Fairbank, Reischauer, and Craig discuss the financial reforms undertaken by then Minister of Finance Masayoshi Matsukata.

43. The term "shinkinkan" that combines the characters for "parent" and "closeness" captures a sense of the close feeling of this period.

44. *See* Narita, n. 40 supra.

45. A recent interesting example of the potential for arbitrariness is the "request" made by MITI to Dow Chemical Japan, Ltd., the wholly-owned subsidiary of Dow Chemical Company. The ministry asked Dow to withdraw its application under the Law Concerning Foreign Investment (Gaishi ni Kansuru Hōritsu) (May 10, 1950, Law No. 163) to manufacture caustic soda and chlorine in Japan. Although the manufacturing of caustic soda and chlorine had been "completely liberalized," domestic competitors urged the ministry to request Dow to delete this activity from the description of business purposes in its articles of incorporation. Unless Dow acquiesced, it was made quite clear, the ministry would not issue an approval. Finally, Dow acceded to the ministry's request.

46. Malcolm D. H. Smith, *Prices and Petroleum in Japan, 1973–1974: A Study of Administrative Guidance*, 10 L. in Japan 81 (1977).

47. Kenji Sanekata, *Administrative Guidance and the Anti-Monopoly Law*, 10 L. in Japan 65 (1977).

48. Such power is "apparent" because of the constraints the prime minister must operate within, under the consensus system.

49. In fact the prime minister is always elected by the Lower House (Shūgiin).

50. See generally, Edwin O. Reischauer, The Japanese (1977).

51. This party with only one exception has been the present ruling Liberal Democratic party. In 1947 the Socialists obtained a majority in the Diet and elected Tetsu Katayama as prime minister.

52. Reischauer, n. 50 supra.

53. One example is the chief cabinet secretary or kambōchōkan, who serves as a sort of chief of staff for the prime minister. Others are the directors of the Prime Minister's Office (sōrifu), the Economic Planning Agency, the Defense Agency, and the Science and Technology Agency (see Reischauer, n. 50 supra).

54. The Secretariat is in charge of personnel, administration, accounting, and matters relating to international cooperation with the United States under the U.S.–Japan Bilateral Agreement for Cooperation in the Environmental Field (see ch. 8).

55. This bureau is responsible for planning, drafting, and promotion of basic policies relating to environmental protection, and the

overall coordination of environmental protection services performed by other agencies.

The bureau also enforces the Pollution Control Public Works Cost Allocation Law. The bureau's Pollution Control Program Division was reorganized into the Environmental Management Division which now has the following new responsibilities: (1) basic policy planning and implementation of environmental impact assessment; (2) overall coordination of work by related administrative agencies.

56. The Nature Conservation Bureau subsumed all the services that were previously performed by the National Parks Department of the Minister's Secretariat in MHW (such as enforcement of the Natural Parks Law and the Hot Spring Law, etc.) and the services performed by the Guidance Division of the Forestry Agency relating to the protection and hunting of birds and animals. This bureau also has responsibility for planning, drafting, and promoting basic policies relating to nature conservation and coordinates similar services of other agencies. These duties include the designation of national parks, quasi-national parks, special protection areas, and marine parks.

57. This bureau is responsible for establishing environmental quality standards and the enforcement of the various pollution control laws. The bureau's Air Pollution Control Division is charged with the administration of emission standards, and the use of fuel to control air pollution. The bureau also has jurisdiction over the control of noise, vibrations, and offensive odors. The bureau has also established an Automotive Pollution Control Division to address specifically the pollution problems caused by automobiles. Michio Hashimoto, who played a key role in the Japanese automobile industry's meeting Japan's stringent standards, was chief of this bureau (see sec. 3.2).

58. This bureau has jurisdiction over water pollution, ground subsidence, and soil contamination. The bureau is also responsible for establishing principles for the registration and control of the use of residue-prone agricultural chemicals. In addition, its Planning Division establishes the standards for the final disposal of wastes, and related matters.

59. The Environmental Health Department was created to oversee the implementation of the Pollution-Related Health Damage Compensation Law (see ch. 6). It is also responsible for coordinating efforts to promote scientific study of the causes of health damage induced by pollution and to take appropriate measures based on the findings of this research.

For a more detailed account of the duties of the departments of the Environment Agency, see Environment Agency 2-9, published by the Environment Agency (undated).

60. For example, MITI's Pollution and Safety Bureau, known before its reorganization in 1970 as the Mining Safety Bureau, lost its authority to establish emission standards for stationary sources (in 1973, the Bureau was again reorganized as a part of MITI's overall overhaul and subsumed in the Factory Siting and Pollution Bureau (Ritchi Kōgai Kysku)), and the Ground Subsidence and Soil Contamination councils that had formerly belonged to other ministries were integrated within the Central Council for Environmental Pollution Control.

61. For example, the National Parks Department of the Prime Minister's Secretariat, the Environmental Pollution Control Department of the Environmental Sanitation Bureau in MHW, MITI's Environmental Protection Department of the Environmental Protection and Safety Bureau, and the Water Quality Monitoring and Control Divisions of the Social Welfare Bureau of the Economic Planning Agency were all abolished.

In 1973, the Environmental Protection and Safety Bureau was revitalized by the establishment of the Factory Siting and Pollution Bureau (Ritchi Kōgai Kyoku).

62. See Shibano, Environment Agency (Kanyōchō) (1975).

63. Article 4.6 of the Environment Agency Establishment Law (Kankyōchō Setchi Hō, Law No. 88, 1971), grants the agency this power.

64. Important exceptions are arts. 14 and 15 of the Nature Conservation Law (Shizen Kankyō Hozen Hō, Law No. 85, 1972; amended by Law No. 73, 1973) that grant the director general of the Environment Agency designation and planning powers over wilderness areas, and arts. 16–21 that

grant the director general permit powers over development and other activities in these areas. In addition to these powers, art. 21 of the Natural Parks Law (Shizen Kōen Hō, Law No. 161, 1967; amended by Law Nos. 140 and 161, 1962; Nos. 13, 61, and 140, 1970; No. 88, 1971; Nos. 52 and 85, 1972; and No. 73, 1973) empowers the director general of the Environment Agency, "when he deems it necessary for the protection of the respective parks concerned," to issue an order under some conditions to restore a damaged site to its original state.

In addition to these regulatory powers, the agency also exerts some influence through its power to allocate grants (see, e.g., sec. 3.2) for environment-related research. Because of its access to a wealth of technical data and expertise, the agency has also come to serve as an informational clearinghouse for environmental problems.

65. For example, consider the activities of MITI's Industrial Location and Environmental Protection Bureau. This bureau now consists of a General Affairs Division, an Industrial Relocation Division, an Industrial Location Guidance Division, an Industrial Water Division, Environmental Protection Policy and Guidance divisions, a Safety Division, a Mine Safety Division, and a Coal Mine Safety Division. As their names suggest, these divisions are in charge of many of the major aspects of industrial accounts. For example, the Environmental Protection Guidance Division conducts administrative guidance of polluters; the Industrial Location Guidance Division is in charge of industrial location and relocation and is authorized to float bonds to raise funds to construct industrial parks.

66. For example, between 1977 and 1978 approximate appropriations for pollution control and environmental conservation for the Environment Agency were ¥35,555 million and ¥38,676 million respectively, whereas approximate appropriations for these same items during the same time period for MHW were ¥36,788 and ¥50,469 million, for the Ministry of Transportation, ¥59,628 and ¥64,795 million, for the Ministry of Construction, ¥361,529 and ¥545,729 million, and for the Defense Agency, ¥43,660 and ¥55,955 million, respectively. The estimated Environment Agency budget for FY 1979 totals ¥27,747 million, a 22.1% in-

crease over the ¥22,724 million for FY 1978. See 6 Japan Environment Summary 3 (no. 9, Sept. 10, 1978).

67. Buichi Ōishi was a medical doctor by profession. He joined the LDP and was elected from Miyagi, Tōhoku, to serve in the House of Representatives before his appointment as director general of the Environment Agency.

68. Quoted by Yoshio Ono in *Pollution Control in Japan—Firm Will and Unified Administration Needed*, 2 Industria 25–28 (no. 2, Dec. 1972).

69. In this case, Ōishi acted with legal authority because of his permit powers under the National Parks Law.

70. The council's original proposal was for a curfew between 10 P.M. and 7 A.M. at Osaka Airport and between 10:30 P.M. and 6 A.M. at Tokyo Airport. Because of the strong opposition of other ministries, the council's final report suggested that only jet flights be curtailed, and that there should be a curfew after 11 P.M. at Tokyo airport. Despite this recommendation, Ōishi strongly recommended that flights after 10 P.M. be stopped and that the number of flights at other times be reduced. Since the council is only an advisory body, the director general can ignore a council suggestion. See art. 6 of the Environment Agency Establishment Law (Kankyōchō Setchi Hō, Law No. 88, May 3, 1971.) For a further discussion of the council, see sec. 1.2.2.

71. But see sec. 1.2.2 for a discussion of the agency's control over council deliberations.

72. The rigid standards for the certification of Minamata disease patients had often staved off patients' claims (see ch. 6). In this case, nine patients whose certification had been rejected by the provincial governors appealed to the Minister of Health and Welfare for reconsideration of their decisions. (See art. 5 of the Administrative Review Act Gyōsei Fufuku Shinsa Hō, 1962 Law No. 160.) The newly appointed Director General of the Environment Agency, the successor in jurisdiction to the Minister of Health and Welfare, rescinded the governors' decisions and ordered them to reconsider the applications.

73. Environment Agency officials have continued to look to their authority to coordinate environment policies (chōsei ken) as

an independent legal basis for regulatory power. Although there is some question how this coordinating power may be used to expand jurisdiction into new areas (such as the control of fluorocarbons), its statutory authority as a coordinating body permits the Environment Agency to withhold consent from environmentally hazardous policies.

74. During his term in office Ōishi continued to make bold statements, such as "thermal power plants produce extensive pollution, so one should take time in convincing the persons concerned before building them"; or "the new overall National Development Plan will cut the whole country like mincemeat and bury it in pollution." In January 1972, Ōishi declined an invitation to attend the annual New Year duck hunting party held by the Imperial Household. Eisaku Satō, then prime minister, reportedly sarcastically remarked, "Mr. Ōishi is making a fuss about ducks, but does he not eat poultry?" Ōishi in the end was able to have the duck hunting cancelled.

Ōishi was also criticized by other powerful officials for his affirmative defense of the environment. On September 28, 1971, Kakuei Tanaka, then Minister of International Trade and Industry, openly attacked the Environment Agency's strong stand, declaring: "The agency should not depend upon idealism alone. It should make statements and take measures that are more realistic." Tanaka's statement was supported by other officials, who noted of Tanaka: "He is indeed an influential minister. We are grateful for his unequivocal statements. By contrast what is Director General Ōishi thinking about Japanese industries? Grandstand plays with trial balloons will lead us nowhere." (Quoted by Ono, n. 68 supra.)

75. An original Environment Agency draft law had contained a provision holding polluters liable under circumstances where causation would be presumed by statute, rather than scientifically verified.

76. Environment Agency officials reportedly later attempted to defend their capitulation by arguing that "the provision was the tail of a lizard", designed to divert the attack on the main substance of the strict liability bill. MITI officials declared in response: "We ordinarily give priority to environmental preservation, but demonstrate our strength where it really matters." (Quoted by Ono, n. 68 supra.)

77. See Ono, n. 68 supra. Environment Agency victories, however, have been infrequent, as is illustrated by the agency's third failure in May 1978, to push through a bill establishing a comprehensive program for environmental impact assessment.

78. Some observers assert that a basic structural problem of the Japanese administrative system is that high ranking Cabinet and other appointments are never subjected to confirmation hearings in the Diet, as is required in the United States. Rather, these positions are simply regarded as political prizes to be awarded according to the vicissitudes of interfactional dealings within the ruling Liberal Democratic party. Consequently, little attention is paid to factors such as experience, judgment, or sympathy with a given agency's tasks. This is most unfortunate, as the position of state minister, as demonstrated by the case of Ōishi, can command considerable influence and power even in a weak agency. So engrained is the practice of "political" appointments, however, that even the opposition parties have not militated for reform.

79. As noted a public hearing would not be required for a decision of this sort.

80. For example, Ishihara is reported to have said: "A collective bargaining type meeting (with victims and/or their representatives) is not only meaningless but also counterproductive, eating into my schedule for legitimate line duties. . . . I will never meet those who are unwilling even to identify themselves. I will meet only those who agree to rules prior to the meeting. I will draw a clearcut line of distinction between a collective bargaining type of meeting and that which is not. . . . I am supposed to meet with victims of pollution, say Minamata disease, but instead end up hearing impassioned pleas, not by the victims themselves, but by their sympathizers. I don't like this. . . ." (Japan Times, March 3, 1977).

81. See Doubts About the Environment Agency (Kankyō Chō e no Utagai), Asahi News 8, June 5, 1978.

82. This assertion is based on interviews with agency officials. Support for this statement can also be found in the lack of enthusiasm currently expressed by politicans for the directorship of the agency. This

contrasts with political attitudes at the time of its establishment.

83. The reader may wish to compare our brief assessment of the Environment Agency with the study Decision Making in the United States Environmental Protection Agency, undertaken by the Committee on· Environmental Decision Making under the auspices of the National Research Council.

84. See art. 27 of the Basic Law.

85. The council still acts as an advisor to the prime minister and some other ministers.

86. There have been few analytic surveys of the professional backgrounds of council members and their qualifications for service. One survey conducted by Ui, however, reports that: (1) the number of individuals with industrial connections substantially exceeds the number affiliated in one way or another with citizens' groups; (2) of the former, few individuals have been entrusted with substantial administrative responsibilities or have had experience working in the bureaucracy, and even fewer possess technical expertise; (3) only four individuals have been generally recognized to be sympathetic to the citizens' movement; (4) the majority of council members do not have substantial professional or technical expertise in the areas of their responsibility on the council; (5) and of those possessing such expertise, most individuals had either expressed no prior sympathy toward the citizens' movement or had actively taken position against it. See Jun Ui, Kōkai Jishu Kōza 65 (no. 9, Feb. 1975, Tokyo University Engineering Department Assistants' Association). Although this report is based on the subjective evaluations of the investigators, it does offer some useful insights into the existence of possible conflicts of interest within the council.

87. Originally, council membership was limited to twenty persons. After the establishment of the Environment Agency, it was increased first to eighty and then to ninety due to the increased workload associated with the establishment of an administrative compensation system.

88. See Pempel, n. 16 supra, 656–663.

89. Procedurally, upon receiving a request from the Director General of the Environment Agency, the chairman of the council refers the request to a subcommittee for deliberation. In most cases, the matter is discussed in an expert subcommittee affiliated with the subcommittee. An expert subcommittee is composed of a few council members, one of whom presides at the meeting, and usually ten experts. This group meets once or twice a month.

90. Pempel (n. 16 supra) provides examples of bureaucracies in other ministries and agencies similarly dominating other councils. See, for example, the 1963 report of the Central Education Council, the 1967 direction to the Science and Technology Council, the 1968 instructions to the Subcommittee on University Disturbances, and the 1970 directions concerning hijacking made to the Advisory Committee on the Legal System.

One important exception to the bureaucracy's domination of Council affairs was Keidanren's proposal to the council that it consider the rationalization of the administrative compensation system established in 1973.

91. Council deliberations are closed to the public, although on occasion, government officials from other agencies having an interest in the subject of research are permitted to observe the meeting. An expert subcommittee will at times call a person from the outside to obtain information on a technical matter such as the feasibility of industry's compliance with an emission standard. Pempel (n. 16 supra) notes that "even highly independent minded committees with neutrally inclined bureaucratic staffs find it almost impossible not to be heavily reflective of official thinking within the bureaucracy."

92. See art. 4(6) of Chūō Kōgai Taisaku Shingikai Rei (Cabinet Order for the Central Council for Environmental Pollution, Cabinet Order No. 350, 1970), as amended. The order provides that a decision of a subcommittee may be substituted for that of the council because the number of people on the council is so large.

93. For an example of this practice, see the discussion in sec. 3.2 on the Ad Hoc Study Group on Motor Vehicle NO_x Emissions Control Technology assembled by the Chief of the Air Pollution Division.

94. Environment Agency, Quality of the Environment in Japan 208 (1977).

95. It is interesting that the entire emphasis of this program is on technical matters.

There is no discussion in the curriculum of what in the United States is termed "policy analysis." Indeed, this subject is neither addressed in law school curricula nor anywhere in the educational structure.

96. Although at the time prefectures, cities, and towns already had their own elected assemblies, the Meiji Constitution of 1889 did not contain any provision recognizing local autonomy. Rather, the major role of local governments was to implement the tasks delegated to them by the central government.

97. Article 92 of the postwar Constitution provides (in part) that "regulations concerning organization and operations of local public entities shall be fixed by law in accordance with the principle of local autonomy." Although at the outset the phrase "principle of local autonomy" was rather broadly interpreted, later interpretations have been more restrictive. See Kurt Steiner, Local Government in Japan (1965).

98. Reischauer, n. 50 supra.

99. After the abolition of the Ministry of Home Affairs in 1947, its functions were transferred to various administrative bodies such as the Construction Agency, the National Election Management Committee, and the Local Finance Committee. In 1949 the Local Autonomy Agency was established and absorbed some of the functions of these groups. The Local Autonomy Agency was in turn reorganized in 1952 and in 1960 designated as the Ministry of Local Autonomy. By this process, the central government strengthened its control over local governments.

100. The insufficiency of local taxes is often expressed as "thirty percent local autonomy" because on an average only thirty percent of local budgets is defrayed from sources that local governments can raise with their own authority. For a discussion of the financial dependence of local governments and subsidies and grants to them, see Steiner, n. 97 supra, 283–291 and 294–299.

101. Reischauer, n. 50 supra. Articles 245 and 246 of the Local Autonomy Law provide for the supervisory power of the central government over the local governments (and the supervisory power of the prefectural governors over the inferior local governments).

102. That is, the Socialist party, the Liberal-Socialist party, the Kōmeitō (Clean Government party), and the Communist party. The New Liberal Club (Shinjiyū Kurabu) had not yet been established.

103. The reader should compare how many states in the United States have resisted taking affirmative action to protect the environment (particularly in cases of air and water pollution control) with the aggressive role of local governments in Japan. For a good study of the issues underlying the dilatory U.S. state attitudes, see Richard B. Stewart, Pyramids of Sacrifice? Problems of Federalism in Mandating State Implementation of National Environmental Policy, 86 Yale L. J. 1196 (1977).

104. Actually the first postwar pollution control ordinance was the 1949 Tokyo Metropolitan Ordinance for the Prevention of Public Hazards of Factories, followed by the Tokyo Metropolitan Noise Abatement Ordinance of 1954, the Osaka Industrial Enterprise Pollution Control Ordinance of 1954, and the Tokyo Metropolitan Soot and Smoke Control Ordinance of 1963.

By the time of the passage of the Soot and Smoke Regulation Law of 1963, ten prefectures and nineteen cities reportedly had passed such ordinances. See Kiyoshi Mori, Kōgai Bōshi Jōrei no Enkaku to Genjō (Development and Present Situation of Pollution Control Ordinances), 466 Jurisuto 24 (1970). The Tokyo Ordinance played an important role in the politics of the city during the late 1960s. As pollution worsened around 1967, Governor Ryōkichi Minobe, backed by the progressive opposition parties, promised the electorate decisive action. As his first gesture, Minobe appointed a prominent legal sociologist and progressive thinker, Michitaka Kaino, to be the first director of the Tokyo Metropolitan Research Institute for Environmental Protection. In the following year, Minobe, with Kaino's assistance, proposed the unprecedented ordinance to the Tokyo Metropolitan Assembly, which passed it forthwith.

105. See Air Pollution Control Law (Taiki Osen Bōshi Hō), 1968, Law No. 97, art. 6.

106. Id., art. 3.

107. Id., art. 18.

108. Id., art. 35. Some scholars have argued that the last three provisions may not be enforceable, because the Ordinance does not

prescribe an explicit sanction for their violation.

109. See Michitaka Kaino, National Laws Governing Environmental Protection and the Tokyo Metropolitan Environmental Pollution Control Ordinance (Tokyo Metropolitan Research Institute for Environmental Protection, July 1, 1971).

110. N. Harada, *Kōgai Bōshi Jōrei no Genkai to Sono Shimei (Role and Limit of Pollution Control Ordinances)*, 466 Jurisuto 35 (1970). When one of the authors of the present work (Akio Morishima) served in 1971 as an expert advisor to the Subcommittee on Pollution Control Ordinances in the Aichi Prefecture Pollution Control Council, the draft provisions for ordinances were sent to and reviewed by the Ministry of Local Autonomy. Although this may not have been a formally required process, at the time local officials of conservative prefectures usually did not want to act against the ministry's assessment since the ministry enjoyed powers over the allocation of financial assistance and other matters.

111. Kaino urged that the local regulations were also authorized by an "unwritten common law."

112. See, e.g., K. Mori, n. 104 supra.

113. Article 94 of the Constitution states: "Local public entities shall have the right to manage their property, affairs, and administration and to enact their own regulations within law." For an interpretation of this article, see Steiner (n. 97 supra) 127–130.

114. Article 14 of the Local Autonomy Law (Chihō Jichi Hō) (1947, Law No. 67, as amended) states: "Ordinary local public entities shall have the right to enact their own regulations concerning the subjects provided in art. 2(2), provided they are not inconsistent with the statutes and orders." The article provides that local ordinances may not conflict with statutes or orders. Article 94 of the Constitution on the other hand, only refers to the law. The concept of "law" in art. 94 of the Constitution, however, is considered to include both statutes and orders of the national authorities.

115. These pollution laws applied only to specific areas that the government designated as polluted areas. When these laws were amended in 1970, this limitation was eliminated and regulations came to be applied nationwide. Mori, n. 104 supra. The old water pollution laws did not outline the relationship between national laws and local ordinances. The Soot and Smoke Regulation Law was amended in 1963 to allow local governments to regulate by their own ordinances facilities other than those covered by the Soot and Smoke Regulation Law. On October 26, 1968, the Legislative Bureau of the Cabinet issued its formal opinion on water pollution laws and declared that local governments could regulate factories outside designated areas. They were not permitted, however, to regulate pollutants other than those specified for each designated area because national regulations had already taken into account the specific conditions of each area, and because discrepant local regulations arguably violated the harmony of industries clause of art. 1 of the Water Quality Preservation Law. 12 Naikaku Hōsei Kyoku Iken Nempō (Annual Report of the Cabinet Legislative Bureau), 1968.

116. For example, Tsutomu Muroi, *Kōgai Gyōsei ni Okeru Hōritsu to Jōrei (Laws and Ordinances in Pollution Control Administration)*, 177 Hōkagu Seminā 64, 1970. It is not clear whether this fundamental-rights argument insisted that even "inconsistent" local legislation was constitutionally permissible.

117. Indeed, these scholars argued that the appropriate role of the central authorities was to assist local governments under these circumstances.

118. See art. 18.

119. See art. 4 of the Air Pollution Control Law, and art. 3(3) of the Water Pollution Control Law.

120. See art. 2(3)(7) adopted in 1974.

121. Note should also be made of the willingness at times of the central authorities to permit local governments to impose pollution control taxes. For example, one report notes that the Ministry of Home Affairs has given tentative approval to Fukushima Prefecture's plan to introduce a special local "nuclear fuel tax." The tax revenues would be used to cover special administrative expenses for "public relations activities" necessary to inform local populations about nuclear energy developments, to monitor environmental radio-

activity, and to implement various pollution control measures (Mainichi Daily News 5, Sept. 15, 1977).

122. See sec. 3.1.2.

123. See the discussion of the Mishima-Numazu incident in ch. 2.

124. Public opinion strengthened the action of local governments and also legitimized these actions. For example, if an enterprise refused to comply with a "suggestion," local authorities could make its recalcitrance publicly known. Few companies would wish to face local residents under such circumstances. This, of course, raises the issue of the extent to which such agreements were truly "voluntary."

125. See generally Jōji Watanaki, How the Environmental Pollution Problems are Handled by the Court, the Public Authority, and the Citizens' Group of Environmentalists in Japan (Tokyo Education University Publication No. 94, 1973). The reader might compare Japan's pollution control agreements with a similar institution in East Germany. Sand reports that these local authorities, as part of their supervisory and coordinating functions, have concluded contractual arrangements with enterprises and cooperatives for joint planning and financing of environmental protection measures. Under a 1967 decision of the Neugrandenburg tribunal, polluting enterprises may be compelled to enter into such contracts on the basis of the 1965 Contract Act. (Peter Sand, *The Socialist Response: Environmental Protection Law in the German Democratic Republic*, 3 Ecology L. Quarterly 451, 1973.)

126. A number of local governments have concluded pollution control agreements for nuclear reactor facilities. See, e.g., the pact between the Hokkaidō Electric Power Company and five cities and towns east of Tomakomai, requiring the utility to take steps to prevent pollution at a nuclear power plant under construction. (Mainichi Daily News, Sept. 23, 1977.)

127. Although prefectures are allowed to establish stricter standards for dust and noxious substances, they are not allowed to do so with regard to SO_x (see art. 4 of the Air Pollution Control Law). Local governments below the prefectural level are not supposed to establish any stricter standards.

128. These agreements may also involve more than one local government. At times, several governments will cooperate in a "compact" to deal with pollution involving several jurisdictions. An example is the pact of four local governments, including Hiroshima Prefecture, to cover all forms of pollution from the Bingo Industrial Complex, covering Fukuyama in Hiroshima Prefecture and Kasaoka in Okayama Prefecture.

129. Because of the expense involved, it has often been difficult for citizen groups to monitor these agreements, a factor that at times has limited the effectiveness of the contract device.

130. See P. Sand (n. 125 supra) for a discussion of comparable legal issues raised by the institution in Germany.

131. See N. Harada, *Kōgai Bōshi Kyōtei to Sono Hōritsujō no Mondaiten (Pollution Control Agreement and Its Legal Problems)*, 458 Jurisuto 274 (1970).

132. Although not legally binding, the agreement serves as a form of administrative guidance. Consequently, companies are reluctant to challenge the agreements for all the reasons already given for their reluctance in other situations. In the present case, the company's "face" is also at stake because the agreement represents a solemn promise.

133. For example, Yoshihiro Nomura, *Kōgai Bōshi Kyōtei No Minji Hōteki Sokumen (Civil Law Aspects of Pollution Control Agreements)*, 248 Hanrei Times 10.

134. Under the Civil Code and the Code of Civil Procedure three forms of enforcement are available. (1) Execution for Delivery of Movables, Immovables, and Objects Held by Their Persons (CCP 730–732). (2) Execution by Proxy (CCP 733). (3) Indirect Compulsion (CCP 734).

135. Article 7 of the Old Land Development for Housing Law (Jūtakuchi Zōsei Jigyō ni Kansuru Hōritsu) (1964, Law No. 160) required a land developer to obtain administrative approval prior to beginning land development. A similar provision is contained in art. 32 of the New City Planning Law (Toshi Keikaku Hō) (1968, Law No. 100).

136. Three hundred guidelines of this nature are reported to have been promulgated by

1976. Hanrei Times Special Issue 290 (no. 2, Aug. 1976).

137. The reader will recall that in this case the Tokyo District Court held Musashino City's termination of the municipal water supply to a developer illegal on the grounds that the city's "enabling" guidelines provided the city with no legal authority for such action. This decision shocked many municipalities.

138. A similarly radical policy was later adopted for PCBs (polychlorinated biphenyls). The manufacture and import of PCBs was forbidden, products containing PCBs recalled, and strict controls placed on their disposal or storage. Yet there were also exceptions. See The Law Concerning the Examination and Regulation of Manufacture, Etc. of Chemical Substances (Kagaku Busshitsu no Shinsa Oyobi Seizō tō no Kisei ni Kansuru Hōritsu) (1973, Law No. 117).

139. In fairness, though, it should be noted that this field only began to be given serious attention in the early 1970s.

140. The principal central planning agency in the Japanese government is the Economic Planning Agency (Keizai Kikaku Chō), although of course each major ministry or agency conducts planning. Planning by various ministries is coordinated by the Economic Planning Agency and thereafter scrutinized by the Cabinet.

141. See Kokudō Riyō Keikaku Hō (1974, Law No. 92). This law is subdivided into land use plans and provisions regulating the land transactions. Its basic idea is "to secure a wholesome and culturally satisfactory living environment" and to assure the balanced development of the country while giving priority to the public welfare and trying to preserve the natural environment (art. 2). In comparison with the National Comprehensive Development Law of 1950, the law emphasizes environmental preservation. Specific environmental protection provisions on land use transaction it leaves to other laws such as the City Planning Law, the Agricultural Land Law, and the Forestry Law. Consequently, despite its declaration of environmental concern, the law's implementation depends on other regulations.

For a more detailed discussion of this law see, Akio Morishima, "Land Use Regulations and Land Development Planning," a paper presented to the ABA Conference, Seattle, Washington, April 25, 1977.

142. See Long-Term Environmental Conservation Program (Environment Agency, June 1977). The reader should note how environmental policies have gradually been incorporated into Japan's overall industrial planning process and how as a result the economic burdens of pollution control have been mitigated. See The Industrial Policy of Japan (OECD, 1972); Japan's Industrial Structure—A Long-Range Vision (MITI, 1978), Ira C. Magaziner and Thomas M. Hout, Japanese Industrial Policy, Policy Studies Institute No. 585, Jan. 1980. Thomas K. Corwin, The Economics of Pollution Control in Japan, 14 Environmental Science and Technology, No. 2, Feb. 1980.

143. Although the document uses the word "program," in fact, it also represents a master plan as this term is defined in the current planning literature.

144. See An Econometric Model for a Longer-Term Environmental Preservation Plan, A Report by the Expert Committee on Quantitative Analysis of the Environment (Planning Committee, Central Council for Control of Environmental Pollution, Oct. 1976). For a discussion of the Environment Agency's efforts to initiate "amenity planning" as a means of protecting historical sites, attractive urban areas, and other environments, see 6 Japan Environment Summary (no. 12, Dec. 10, 1978).

145. This information is based on interviews with government officials. At first, MITI and some of the other ministries opposed the Environment Agency's proposal to prepare the prospectus, but finally, consensus was reached. This provided the authority for the agency's action. The first case study in this chapter illustrates how the agencies imply planning powers from their jurisdictional grants of authority.

146. See the OECD report (n. 1 supra), 24. The Basic Law defines ambient standards as goals "the maintenance of which is desirable for the protection of human health and the conservation of the living environment." In the United States, ambient standards are classified in two categories: primary and secondary air quality standards. Primary standards are standards "the attainment and maintenance of which in the

judgment of the administrator, based on such criteria and allowing an adequate margin of safety, are requisite to protect the public health" (Clean Air Act, as amended, sec. 109(b)(1)). Secondary standards are those required "to protect the public welfare" (sec. 109(b)(2)). The term "public welfare" is given a wide definition that includes, but is not limited to, the prevention of "effects on soils, water, crops, climate, damage to and deterioration of property, and hazards to transportation, as well as effects on economic values and on personal comfort and well-being" (sec. 302). Secondary standards have been established at levels more stringent than those for primary ones. In Japan standards are distinguished not so much by the degree of stringency, as by the subject of regulation. For example, water quality standards for protection of human health are established in connection with cadmium, cyanide, organic phosphorus, lead, chromium, arsenic, total mercury, alkyl mercury, and PCBs, while those for the conservation of the living environment are determined in terms of pH (hydrogen ion concentration), BOD (biochemical oxygen demand), SS (suspended solids), DO (dissolved oxygen) and a number of coliform groups.

147. Article 2.2 of the Basic law defines "the living environment" to include "property closely related to human life, and animals and plants closely related to human life and the environment in which such animals and plants live."

148. But see art. 17(1)(2) of the Basic Law, mandating the government "to protect the natural environment as well as to conserve green areas." As of June 1977, the government had established air quality standards for SO_2, CO, particulates, NO_2, and photochemical oxidants.

149. As a rule, quality standards are established at the central level and with few exceptions local governments are not permitted to change quality standards.

150. Standards for pH, BOD, SS, DO, and coliforms (n. 146 supra) may vary within the same stretch of river. However, standards for cadmium, arsenic, cyanides, organic phosphorous, lead, chromium, arsenic, mercury, and PCBs do not.

151. For example, as of 1977, SO_x, soot and dust and five other toxic substances and their compounds were regulated by the Air

Pollution Control Law (art. 3); compare also art. 3 of the 1970 Water Pollution Control Law to the Cabinet Orders of 1971 and 1974 establishing standards.

152. Article 4 of the Air Pollution Law empowers local governments to establish stiffer standards than the national limitations for soot and dust and some related toxic substances. It should also be noted that many local pollution control agreements also specify standards for each plant. By the end of 1977, several prefectures in addition to the Tokyo municipality had established emission standards by local ordinance.

153. SO_x emission standards for each SO_x-emitting facility are calculated under a formula involving the coefficient k, and q (hourly volume of sulfur omitted), H (actual height of stack), Q (volume of gas emitted), V (speed of the exhaust gas), and T (absolute temperature of the exhaust gas).

The NO_x emission standards apply to a complex magnitude called "converted concentration" S, involving a coefficient N, and the quantities C (actual concentration of nitrogen oxides), and O (oxygen content of the exhaust gas).

The definition of S is such that it is not possible to meet the standard and also to increase quantities of NO_x discharged by adding fresh air. This would increase the oxygen content and, therefore, the standard.

154. See the OECD report (n. 1 supra) 28. Japan has yet to recognize a nondegradation policy. Indeed, from a theoretical perspective, the designation of various categories of pollution areas may make sense if this results in a net productive (in a resource allocative sense) gain for the nation. For a discussion of the nondegradation doctrine in the United States see N. William Hines, *A Decade of Non-Degradation Policy in Congress and the Courts: The Erratic Pursuit of Clean Air and Clean Water*, 62 Iowa L.R. 643 (1977).

155. Prefectural governors are explicitly empowered to set stricter standards for new plants.

156. Although some emission standards are a function of the type, and in some cases of the size of a polluting facility, others are not. Standards for NO_x soot and dust, cadmium, lead, and most air pollutants are defined by type of facility (Japanese report, in OECD report (n. 1 supra) 93–94, 105–

113). For instance, concentrations in grams per cubic meter per hour of soot and dust are: 0.40 g for a coal-fuel boiler, 0.10 g for a large-scale boiler, etc. SO_2 standards and water effluent standards, however, are set irrespective of the type of facility. For all these reasons, it is extremely difficult to compare Japanese emission standards to those of other nations. The OECD report (n. 1, supra) 31, notes that the fluoride emission standards in the aluminum industry are not very strict. However, this data does not account for the additional requirements a plant might face at the local level.

157. OECD report (n. 1 supra) 24–36.

158. The mass emission approach entails two theoretical advantages. First, a firm may allocate pollution discharges within its factories to minimize cost. Second, the measure insures that a given level of pollution can be maintained in the face of industrial growth and price inflation ostensibly without necessitating further governmental action. Japan's mass emission approach predates the EPA's "bubble" concept by several years.

159. See Dales, Pollution, Property and Prices (1968); Stewart and Krier, Environmental Law and Policy, 2d ed. (1978).

160. See generally, A. Morishima and J. Pugash, *Foreign Developments: II. The Standard Setting Process in Japan*, 1 Harv. Envir. L. Rev. 545 (1976).

161. Basic Law for Environmental Pollution Control (1967, Law No. 132), art. 9; Law for the Establishment of the Environment Agency (1971, Law No. 88), art. 4(6).

162. The Air Pollution Subcommittee, which is similar to others, contains seven businessmen and bankers, six government officials, five professors and members of university research institutes, and one union representative.

163. The overwhelming majority of the committee members are university professors.

164. The expert committee usually does not conduct empirical research for itself. Instead the data are supplied to the experts by specialists from the Environment Agency. In theory the experts are only supposed to consider scientific evidence on the environmental effects of a standard and the technical feasibility of its implementation. But the term "technical feasibility" allows a certain

latitude for considering economic factors as well. Although a goal might be feasible if the most sophisticated technology were used, the cost of adopting that technology on a national scale may be prohibitive. In such a case, the experts may decide that the goal is not technically feasible, when in fact it is merely too expensive.

The SO_x standards proposed by the expert committee in 1968 were relaxed by subsequent subcommittee action the same year. See Taiki Osen Bōshi Hō no Kaisetsu (Annotations on the Air Pollution Control Law) 345–373 (Environment Agency, Air Quality Bureau, 1972). Since 1968, the expert committee's standards have rarely been diluted by the subcommittee. Delay is much more common.

165. The Central Council does not review routine actions of its subcommittees. But when proposed standards are particularly important or controversial, the council will convene to discuss a subcommittee report.

166. If the Central Council fails to review the subcommittee's report, the report will have the same authority as a recommendation of the entire council. See Cabinet Order No. 350 of 1967, art. 4(6), as amended.

167. Before the council completes its deliberations, the report of the subcommittee will generally be made available to the public.

168. The Environment Agency's authority to set emission standards is granted in specific legislation. See Air Pollution Control Law (1968, Law No. 97), art. 3; Law for the Establishment of the Environment Agency (1971, Law No. 88), art. 4(16).

169. The advisory group of experts is composed almost entirely of university professors.

170. See, e.g., Air Polllution Control Law, arts. 33–37. In practice criminal penalties for failure to comply with emission standards are rarely imposed.

171. At this point, the reader might wish to compare the Japanese standard setting process with the prevailing practice in the United States and Germany. For example, let us consider the following for SO_x:

(a) Use of criteria documents. All three countries employ criteria documents. In Japan criteria documents are drawn up by an independent expert committee, whereas in the United States the federal government is responsible for their preparation. The Japanese document is confined to the effects on man, while the German and United States documents refer to general environmental effect as well.

(b) Stages leading to the adoption of standards. In the United States, the EPA has the responsibility for the different phases of standard setting. The EPA is required to publish proposals for standards at the *same* time that the criteria document is issued. In Germany and Japan, the government does not play a significant role in the process until an independent expert committee or body has submitted an initial proposal.

In the United States, the primary standards are to be met as expeditiously as possible, but no later than three years from the date of approval of the implementation plan. It is possible in exceptional circumstances to extend this maximum period for a further two years. In Japan, this period is generally longer, i.e., not less than five years, while in Germany the standards come into force at the time of promulgation. See Report on the Use of Dose and Effects Data in Setting Air Quality Standards to Provide a Basis for the Control of Sulfur Oxides: Experience in Germany, Japan, and the United States (OECD, Paris, 1975).

172. See Takeo Suzuki and Tsuneo Tsukatani, *Taiki Osen no Kankyō Kijun (Ambient Standards for Air Pollution)*, 5 Kōgai Kenkyū 55 (no. 4, 1976); Taiki Osen Bōshi Hō no Kaisetsu (Annotations on the Air Pollution Control Law) 348 (Environment Agency, Air Quality Bureau, 1972). In setting the standard the expert committees apply four criteria: (i) no aggravation of patients' symptoms based on epidemiological data; (ii) no increase in the death rate; (iii) no increase in the incidence of chronic respiratory disease; (iv) no increase in adverse effects on, or impairment of, respiratory function in children.

173. Industry also feared that stringent ambient standards would block plans for the construction of large-scale petrochemical complexes throughout the country. See Taiki Osen Bōshi Hō no Kaisetsu (n. 172 supra) 361.

174. The new limitation was based on the type of area to which the standard applied. Moreover, the minimum compliance rate was reduced from 93% to 88% with respect to the hourly standards while the daily standard was reduced from 80% to 70%. In other words, not more than 12% of the total number of hours in a year (1,150 hours) was permitted to exceed the hourly standard, and not more than 30% of the days in a year (100 days) would be permitted to exceed the daily standard. See Nishihara and Satō, editors, Kōgai Taisaku I (Pollution Control I) 92 (1969); Taiki Osen Bōshi Hō no Kaisetsu (n. 172 supra) 344 et seq.; Case History from Japan on Use of Criteria Documents in Setting Standards for the Control of Sulfur Oxides 5 (OECD, Environmental Directorate, 1975).

175. 0.25–0.30 ppm.

176. The proposed ambient standard for SO_x was as follows: The daily average of hourly concentrations should not exceed 0.04 ppm while the hourly concentration should never exceed 0.1 ppm. In determining compliance with the standard, the highest 2% of all measurements was eliminated as extraordinary. In other words, exceeding the daily standard for seven days a year, or the hourly standard for 175 hours a year, was considered to be within the allowable limit. These standards were expected to be achieved in less than five years. See Suzuki et al. (n. 172 supra). The council at the same time established ambient standards for NO_x and photochemical oxidants.

177. Suzuki et al., n. 172 supra. For an interesting and important U.S. decision upsetting EPA's attempt to establish a margin of safety for SO_x see *Kennecott Copper Corp. v. Environmental Protection Agency*, 462 F.2d 846, 3 ERC 1682 (D.C. Cir. 1972).

178. The history of the establishment of ambient standards for CO and suspended particulate matter was much less controversial than either that of SO_x or NO_x. In comparison to SO_x, there was substantial pathological and other data on the adverse effects of CO on human health and the mechanism of its effect on blood was well known. Establishment of an appropriate standard was thus not very difficult. On December 22, 1969, after eleven months of deliberation, the Council on the Living Environment of MHW made the following recommendations to the ministry: (1) The average hourly values for eight consecutive hours should not exceed 20 ppm, and (2) the average hourly values in 24 consecutive hours should not exceed 10 ppm.

In this case the safety factor was not considered because synergistic effects were not foreseen (Suzuki et al., n. 172 supra). On February 20, 1970, the Cabinet established the ambient standards for CO as recommended by the council (Taiki Osen Bōshi Hō no Kaisetsu 383) (n. 172 supra).

The case of the establishment of ambient standards for suspended particulates was not greatly different. Because the effects of these substances had been well studied in the United States and Europe as well as Japan, the issue did not provoke much controversy. Moreover, the technology for dust collection had already been developed. On December 22, 1971, the Air Quality Subcommittee of the Central Council proposed ambient standards for particulates. These were cleared by the director general and approved by the Cabinet on January 13, 1977, in the following amounts: (1) For 24 consecutive hours, the average hourly value should be less than 0.1 mg/m³ and (2) in any event the hourly value should be less than 0.20 mg/m³ (Taiki Osen Bōshi Hō no Kaisetsu, n. 172 supra).

179. One reason was that high atmospheric concentrations of NO_x were not observed until 1965.

However, despite the inconclusiveness of the evidence, epidemiological studies conducted in six large cities during the years 1970–1971 seemed to show that the rate of incidence of respiratory disease increased when the hourly average of NO_2 concentrations exceeded 0.042 ppm. The results of animal experiments showed that when rats were exposed to NO_2 at concentrations of 0.5 ppm for a few hours each day, permanent organic deterioration of epithelial cells would occur. And other experimenters reported that rats experienced similar effects from SO_2 exposure in concentrations of 0.5 ppm for a few hours each day, although SO_2 did not cause organic changes.

This description of the actions of the expert committee is based on an interview with Michio Hashimoto on November 15, 1976.

180. On June 20, 1972, the committee determined that the daily average hourly value for NO_2 should not exceed 0.02 ppm. Emission standards for NO_2, however, were not established at this time because the information needed to set the standard was not available. Similarly, there was no available data pertaining to the toxicity of photo-

chemical oxidants, although the toxicity of these substances, it was thought, was similar to, or greater than, that of NO_2. Because of the paucity of data, ambient standards for photochemical oxidants was established on an interim basis in order to prevent the acute effects of these substances. This standard limited the hourly concentration to 0.06 ppm.

181. As before industry challenged the standard as scientific nonsense. Industry also argued that because the government had failed to establish emission standards for NO_x, it possessed no basis for ascertaining whether the ambient standard was even technologically feasible. This had not been so in the case of SO_x. Industry also pointed out that the prevailing U.S. standard was substantially lower, and that there was no scientific or other rational basis for setting the Japanese standard higher than the U.S. standard.

182. At the same time, it was agreed that large-scale industry should meet an intermediate standard (i.e., 60% of the days of a year should meet the daily average of 0.02 ppm) within five years.

183. For example an American expert, Carl Shy of the University of North Carolina, attacked the Japanese standard at a meeting of the Industrial Air Pollution Control Association of Japan (International Lecture Conference, Nov. 6, 1973).

184. Only about 9% of the monitoring stations reported compliance with the standard and 84.3% (or 556) of such stations reported the NO_x situation unchanged from the preceding year. (Mainichi Daily News, Dec. 19, 1977.)

185. These ranged from automobiles, aircraft, and ships to factories and households.

186. On March 20 the Expert Committee on Criteria for NO_x of the Central Council also issued a report on permissible levels of NO_x in the human body. The report concluded that the permissible hourly level should be between 0.1 and 0.2 ppm, and the permissible yearly average set at 0.02–0.03 ppm. These limits were based on criteria presented at the WHO meeting, Tokyo, 1976.

The WHO report indicates maximum short term exposure to NO_x should be set at 0.5 ppm and that exposure exceeding 0.1–0.17 ppm extending over one hour should not occur more than once a month. The permissible levels set by the Environment Agency's committee were determined by multiplying WHO data by a safety factor of 3–5 converted to a daily and yearly average. See 1 Int. Environmental Reporter (no. 6, June 10, 1978).

187. For example, the Tokyo Metropolitan Government has repeatedly and strongly urged the Environment Agency not to relax the standard and has even threatened to establish its own ambient standard, which agency officials insist it is not legally permitted to do.

188. See 6 Japan Environment Summary (No. 8, 1978). The Environment Agency's action has since been challenged by fifteen Tokyo residents who seek to enjoin the government's action. The plaintiffs allege that the Environment Agency did not follow proper procedures under the Basic Law. The government asserts that the plaintiffs are not competent to challenge the scientific basis for the new standards. One of the plaintiffs has testified that if the standard is upheld some pollution victims would not qualify for compensation under the Pollution Related Health Damage Compensation Law (see ch. 6). The case is the first judicial challenge to the Environment Agency's decision-making process. See Int. Environmental Reporter 470 (Jan. 10, 1979).

189. 1972, Law No. 73.

190. The Minister of International Trade and Industry is supposed to draw up an industrial relocation plan (art. 3), although as of this writing only an outline of this plan has thus far been published. At present, Tokyo, Osaka, and Nagoya are designated as "removal promotion areas," while depopulated areas such as Tōhoku and Hokuriku are designated as inducement areas.

191. Both the municipality and the factory may receive up to ¥100,000 each on the basis of ¥5,000 per square meter of floor space.

192. Since the Industrial Relocation Promotion Law (Kōgyō Saihaichi Sokushin Hō), 1972, Law No. 73, has been in force, the central government has spent nearly ¥100 billion per year for such measures.

193. See art. 2 of the Factory Location Law (Kōjō Ritchi Hō) (1959, Law No. 24), which mandates the Minister of International Trade and Industry to conduct these studies.

194. See arts. 4 and 6.

195. See art. 11.

196. Article 9 permits the minister to recommend "necessary" pollution prevention measures when a factory's emissions in combination with others pose difficulties for the preservation of air or water quality; art. 10 allows the minister to order a factory's modification if an enterprise fails to heed the government's recommendations during the 90-day period.

197. See also the similar provision in the Air Pollution Control Law.

198. See the Law Concerning the Establishment of Pollution Prevention Systems in Specified Factories (Tokutei Kōjō ni Okeru Kōgai Bōshi Soshiki no Seibi ni Kansuru Hōritsu) (1971, Law No. 107).

199. As of 1975, local governments maintained 1,216 monitoring stations and 74 computerized systems for processing environmental data.

200. In recent years there has been a marked increase in the number of facilities subject to pollution regulations. For statistics on this subject see Kankyō Tōkei Yōran (Gist of Environment Statistics) 34–35 (Environment Agency, Bureau of Planning and Coordination, 1977).

201. See amendments to the Local Autonomy Law that empower local governments to control pollution. Recently the number of officials assigned to pollution control has greatly increased, although at the same time this has imposed a substantial financial burden on local governments. For example in the fiscal year 1975, local governments spent ¥1,425,800 million ($5,280 million) on pollution control (¥533,300 million from prefectures and ¥892,500 million from municipalities). This sum was ¥228,900 million higher than the amount spent in 1974 and ¥472,100 million higher than the amount spent in 1973. The breakdown of total expenditures on pollution control in 1975 was as follows: ¥1,277,200 million (89.6%) was spent on construction of facilities such as sewage treatment plants (¥850,000 million) and solid waste treatment plants (¥249,800 million); ¥42,300 million (3.0%) for the payment of personnel; ¥4,298 million

(.03%) on the purchase of monitoring devices and other equipment; ¥21,600 million (1.5%) on the maintenance of monitoring systems; ¥15,800 million (1.1%) on the administration of the pollution victims' relief system, and ¥64,700 million (4.5%) on miscellaneous items. Due to recent tightening of local government budgets, the burden of expenditures for pollution control has become quite substantial, and it is feared that future increases in pollution control budgets may not be as great as before.

Local governments also grant financial assistance to small factories for the installation of pollution control facilities. In the 1975 fiscal year, a total of ¥38,283 million was granted to 4,653 factories. See Kankyō Hakusho (Environment White Paper) 392 (Environment Agency, 1977). Article 245(4) of the Local Autonomy Law authorizes the appropriate minister to request local governments to supply information concerning the areas within their delegated powers. Article 246(1)(2) authorizes the prime minister to request local governments to carry out their responsibilities where the prime minister deems that a local government's administration is illegal or improper and the "public interest" is threatened. The Local Autonomy Law also provides that local governments shall protect the safety, health and welfare of the residents (art. 2(3)(i)).

202. See Art. 14 of the Air Pollution Control Law.

203. Article 33 stipulates imprisonment with enforced labor up to a year and fines up to ¥200,000.

204. We must, however, be careful about what we term "coercive." Although enterprises are "free" to negotiate pollution control contracts, generally they must sign contracts before beginning operations that greatly limit a company's freedom of negotiation.

205. There were 41 in 1973, 30 in 1974, and 20 in 1975.

206. There were 31 in 1973, 31 in 1974, and 12 in 1975.

207. (Hito no Kenkō ni Kakaru Kōgai Hanzai no Shobatsu ni Kansuru Hōritsu) (1970, Law No. 142). Henceforth, "Environmental Pollution Crime Law."

208. For example, art. 3 states:
1. A person who, through failure to exercise necessary care in the operation of his business, endangers the lives or health of the public by discharging those substances which adversely affect the health of persons, in the conduct of activities of industrial plants or places of business, shall be punished by imprisonment with or without prison labor for not more than two years or a fine not exceeding ¥2 million.
2. A person who, as the result of the offense mentioned in the preceding paragraph, causes the death or bodily injury of another shall be punished by imprisonment with or without prison labor for not more than five years or a fine not exceeding ¥3 million.

Although this provision is certainly an innovative first effort to apply economic sanctions or penalties (in addition to other sanctions) to a polluter's activity that jeopardizes public health, the provision has been interpreted restrictively. (The law's original draft by the Ministry of Justice included the phrase, "a person who created a situation that may possibly cause a risk of harm." However, this idea was excluded during Diet discussion, because it was considered to be too broad a definition of a crime.) According to one interpretation, the foreseeability of damage of a similar kind is required and, therefore, punishment could not be imposed at a stage where there exists a mere possibility of some kind of injury. As a result, it is likely that cases will not be prosecuted until actual injury to human health is imminent.

209. Thus, art. 5 states:
If, in cases where a person has discharged those substances which adversely affect the health of persons, in the conduct of activities of industrial plants or places of business, to such extent that the lives or health of the public are already being endangered and the danger may have been caused by the discharge of those substances, it shall be presumed to have been caused by the substances discharged by that person.

However, this provision is not very important, because the presumption is interpreted to apply only where a party has discharged an amount of pollutants so great that this discharge *alone* could cause injury. In a pollution case, the most difficult issue often is to show what amount of some pollutant is necessary to cause injury to the public. Consequently, if the prosecution can prove that the amount discharged by the defendant was sufficient to cause injury, it is not difficult thereafter to connect the pollutant discharged to the injury involved, unless other parties also discharged substantial amounts of the same pollutant. This provision cannot be applied to cases where each of several polluters discharged relatively small amounts of pollutant, which caused damage in the aggregate. Even though every year police refer a small number of cases (four in 1976) to prosecutor's offices, after five years of enforcing the law, only a few cases have been prosecuted under it. None have yet been decided.

210. Article 4 states:

In case the representative of a corporation, or the proxy, employee, or worker of a corporation or of an individual, commits any of the offenses mentioned in the preceding two Articles, in connection with the business of the corporation or individual, not only the violator but, also the corporation or individual, shall be punished by fines prescribed in the Articles concerned.

211. There is, however, some trend toward an increase of arrests and prosecutions, although the absolute number of these cases still remains low. For example, the annual number of arrests for pollution crimes rose from 482 in 1971 to about 4,597 in 1976 (Kankyō Hakusho (Environmental White Paper) 385, 1977). The number of pollution crimes handled by the Prosecutor's Office has also risen: In 1972, of 2,177 cases received, there were 1,492 indictments (a prosecution rate of 69%); in 1977, of 6,236 cases handled, there were 4,540 indictments (73%) (Kankyō Hakusho 359).

In 1979 the Japanese courts reached two landmark decisions on a company executive's criminal negligence for causing pollution-related injury. On March 22, the Kumamoto District Court convicted two Chisso executives. In another case Nippon Aerosol Company and four of its employees were found guilty under the 1970 Law for the Punishment of Crimes Relating to Environmental Pollution. The company was ordered to pay a $10,000 fine and the employees given four-month suspended sentences (Int. Environmental Reporter, Apr. 11, 1979).

212. The Mizushima oil spill is a good example. On December 18, 1974, an oil tank in the yard of the Mizushima Refinery of Mitsubishi Petroleum Company burst and 16 million liters of heavy oil spilled into the Seto Inland Sea. The damage to fisheries and the shore was devastating (it is said that the amount of damages exceed ¥50 billion including cleaning costs). Although the police and prosecutors office investigated the case, only the contractors of the tank were indicted. No one in the Mitsubishi Company was indicted, because the company had neither dumped oil into a public sea nor violated statutory regulations.

213. Article 19 provides:

1. The Prime Minister shall instruct the prefectural governors concerned to formulate programs relating to the environmental pollution control measures (hereinafter called "Environmental Pollution Control Programs") to be implemented in specific areas which fall into any one of the following categories:

(1) areas in which environmental pollution is serious and in which it is recognized that it will be extremely difficult to achieve effective environmental pollution control unless comprehensive control measures are taken;

(2) areas in which environmental pollution is likely to become serious on account of rapidly increasing concentrations of population, industry, etc., and in which it is recognized that it will be extremely difficult to achieve effective environmental pollution control unless comprehensive control measures are taken.

2. When the prefectural governor concerned has received the instruction referred to in the preceding paragraph, he shall draw up an Environmental Pollution Control Program in accordance with the fundamental policies referred to in the preceding paragraph and shall submit it to the Prime Minister for his approval.

3. Prior to issuing an instruction under paragraph 1 or giving the approval required under the preceding paragraph, the Prime Minister shall consult with the Conference on Environmental Pollution Control.

4. Prior to issuing an instruction under paragraph 1, the Prime Minister shall seek the opinion of the prefectural governor concerned.

214. See Kankyō Hakusho (n. 201 supra) for

a comparison of the present ambient standards for the designated pollution control areas and the ambient standards for the rest of the country.

215. As of January 1977, seven phases of a pollution control program, covering fifty areas, were approved by the prime minister. In the first phase, the Yokkaichi, Mizushima, and Chiba-Ichihara areas were designated in 1969 and programs for these areas were approved in 1970. Since then, every year, a new phase has been announced by the government. All large urban and industrial areas have been covered. See The Quality of the Environment in Japan 96 (Environment Agency, 1977). For a discussion of the substantial annual growth (with the exception of 1976, when there was a slight decline) in public investment in pollution control public works, see Kankyō Hakusho (n. 201 supra) 97.

216. See Kōgai Bōshi Keikaku no Genjō to Kadai (The Present Situation and Problems of Pollution Control Programs), Kokumin Seikatsu Sentā (Center for Economic Welfare), 1972.

217. At present, the prime minister issues instructions to prefectural governors on the establishment of regional programs in accordance with the fundamental policies established by the Cabinet. The prime minister's approval is essential for the formulation and implementation of the program.

218. Law Concerning Special Government Financial Measures for Pollution Control Projects (Kōgai no Bōshi ni Kansuru Jigyō ni Kakaru Kini no Zaiseijō no Tokubetsu Sochi ni Kansuru Hōritsu) (1971, Law No. 70). This law also covers the redemption of local bonds for public works. Also, the central government usually subsidizes one quarter of the local government's costs of constructing waste water treatment plants. Under the law, the central government pays half of the construction costs and subsidizes half of the costs of redeeming local bonds for construction in designated areas. In addition to the government's investment in public works, each industry must invest large amounts of its own money for the reduction of pollution. See Pollution Control Public Works Cost Allocation Law (Kōgai Bōshi Jigyōhi Jigyōha Futan Hō) (December 25, 1970, Law No. 133).

219. This, however, may be due to other factors as discussed in sec. 3.1. In 1973, the ambient standard for SO$_x$ was met at 165 stations (32%) out of a total of 514, and in 1975, 442 stations (73%) out of 602 met these standards. In 1973 although the ambient standard for BOD in rivers was met at 112 (33%) points out of a total of 339, in 1973 183 points (50%) out of 369 achieved the standard. However, the ratio of achievement of NO$_x$ standards has been very low in all areas (on average, only 2%). In some areas, the BOD level has even become worse.

220. See Kōgai Bōshi Keikaku no Genjō to Kadai (n. 216 supra) 8–17.

221. For example, when calculating pollution levels, programs assume unchanged rates of economic growth and industrial development; sewage treatment plant construction is implemented without considering measures to limit population growth.

222. As noted there is no legal prohibition against degradation.

223. The corporation is under the supervision of the Environment Agency and established and regulated under the Environmental Pollution Control Service Corporation Law (Kōgai Bōshi Jigyōdan Hō) (1968, Law No. 9; 1971, Law No. 88). See The Environmental Pollution Control Service Corporation, 1976 (published by the corporation).

224. Title to the complex is usually transferred to the participating companies after a period of 10–20 years. The terms of the contracts with enterprises vary. Usually small and medium enterprises receive a special dispensation. For example, the initial down payment for small and medium enterprises for the construction of a pollution control facility would be 20%, but for a larger enterprise 30%. It would be 5% versus 10% in the case of the construction of a housing complex or factory relocation. The payment period and interest rates also vary. For pollution control facilities, small and medium enterprises usually pay 4.5% during the first 3 years and 5.0% thereafter, in contrast to 6.05% and 6.25% (in 1979) for larger enterprises. For other construction or relocation the rates are 5.75% for small and medium enterprises versus 6.25% for larger firms.

225. Larger companies also may receive

these benefits, although small and medium enterprises are the principal recipients.

226. For example, according to the corporation's 1977 statistics, 27 companies are now participating in a factory relocation project in Amagasaki City; 13 companies in Kawachi-Nagano City have jointly collaborated in the construction of a noise-proofed worker-housing project.

227. See special depreciation under arts. 43.1 and 44.1 of the Special Measures Income Tax Law (Sozei Tokubetsu Sōchi Hō) 1957, Law No. 26, accelerated depreciation under Ministerial Ordinance 15, art. 2 (1965) of the Finance Ministry. See also exemption from fixed property tax under local tax law (Chihōzeihō), Supplementary Regulations, arts. 14.1–14.5. See also arts. 73.14.5, 73.27.5, 568.2, and 701.34.8 of the same law.

228. The importance of the pollution problems caused by small and medium enterprises is reported in the papers submitted by many developing countries invited to an APO Top Management Symposium, Tokyo, May 26, 1976. The participating countries included Thailand, Papua New Guinea, Republic of Korea, Bangladesh, Republic of China, Hong Kong, India, Indonesia, Iran, Nepal, Pakistan, Philippines, and Sri Lanka.

229. In Japan the burden of pollution controls on smaller enterprises is often increased by local requirements.

230. This discussion relies principally on Yugeta, "Small Industries and Environmental Protection," a paper prepared for the APO Top Management Symposium on the Environment, Tokyo, May 26, 1976.

231. In 1975, there were forty-seven such persons and the program cost ¥43 million. Expenses are shared by the central and local governments.

232. Yugeta (n. 230 supra) notes that in 1975, forty-seven chambers of commerce participated at a cost of approximately ¥108 million. It is important to note the contribution made by local social institutions that promote collective action. Often counseling occurs not only on an individual basis, but also collectively. In this connection various local industrial associations have proved most useful not only as a source of funds, but also as a forum for collective action. The Yugeta study gives the example of the

affirmative action taken by the Yaizu Marine Products Cooperative Union that, with local assistance, mobilized its members to participate in a collective effort to construct a fish waste processing center and to introduce other affirmative environmental protection measures. Because of cost constraints, these measures would have been very difficult to implement had there not been close cooperation between local government and the union. Such collective efforts are common and contribute importantly to the successes to date in the government's assistance of small enterprises.

233. In 1956 21.9% of total Japanese energy demands were supplied by petroleum; by 1960 this percentage had increased to 37.7%; in 1970, petroleum was used for 70.8% of all Japanese energy needs. Comprehensive Statistics on Energy (Resource and Energy Agency (Sōgō Enerugī Tōkei)).

234. Baien no Haishutsu no Kisei tō ni Kansuru Hōritsu (1962, Law No. 146). There was no idea of nondegradation. Indeed, undesignated areas were left unregulated.

235. The *Smoke and Soot Regulation Law* was weak in other ways. Electric and gas generating businesses were exempted from local control (MITI retained direct control of these industries), and no provision was made for the regulation of mobile sources of pollution.

236. During the five years of the law's existence, only twenty areas were designated.

237. For example, in the Yokkaichi area the emission standard for SO_x was set at a level ranging from 0.18 to 0.22% of maximum stack output, while actual maximum discharge of SO_x from all stacks of large factories was only 0.17%.

238. See Eiji Ono, Yokkaichi Kōgai Jū Nen no Kiroku (Ten-Year Record of Yokkaichi Pollution) 214 (1971).

239. This oil was imported primarily from Arabian countries and consequently contained a high percentage of sulfur. See Taiki Osen Bōshi Hō no Kaisetsu (n. 172 supra) 3.

240. Basic Law, art. 10.

241. Id., art. 11.

242. Id., art. 19.

243. Air Pollution Control Law (Taiki Osen Bōshi Hō) (1968, Law No. 97). The passage

of this law was, as often, the product of considerable interministerial struggle and compromise. In response to the mandate of the Basic Law, MHW began the preparation of the new law. The ministry's proposal centered on a permit system, applicable to the construction of polluting facilities in polluted areas; it also authorized controls on public utilities, the regulation of automobiles (left unregulated in the Soot and Smoke Regulation Law), emergency regulations imposing restrictions on the use of high sulfur fuel, and the tightening of emission controls. MITI opposed this draft, particularly the permit system, and in its place tendered a counterproposal that employed factory relocation measures and financial aid to relocating industries. MITI also resisted any outside effort to assert jurisdiction over the public utilities, which MITI guarded jealously. On April 27, 1968, the Cabinet approved the bill for the Air Pollution Control Law and submitted it to the Diet, where it was discussed extensively in the Special Standing Committee for Industrial Pollution in each house. The Diet debates focused principally on the areas where the earlier MHW draft had been weak. For example, the bill reflected a compromise reached between MITI and MHW to delete the original MHW proposal for a permit system. In its stead a notice system, identical to that in the earlier statute, was inserted. (See, Proceedings of the Special Standing Committee for Industrial Pollution Control, House of Representatives, 58th Diet, No. 8 (May 3, 1968), p. 14 (testimony of Mr. Mutō); also Proceedings of the Special Standing Committee for Industrial Pollution Control, House of Councilors, 58th Diet, No. 12 (May 22, 1968), p. 4.) A second issue involved the deletion of regulations pertaining to fuel use. Here a MITI official argued that regulation was inappropriate as the government was now conducting extensive administrative guidance in this area. (Proceedings of the House of Representatives, No. 9 (May 9, 1968), p. 14 (testimony of Mr. Yajima).) Finally, the issue of jurisdiction was debated. Opposition party members contended that for standards to be sufficiently stringent to protect health, MHW should at least maintain jurisdiction jointly with MITI and the other ministries. The government, however, argued strongly that MITI had long main-

tained jurisdiction in this area and, in the case of public utilities, had imposed controls more stringent than would be possible under pollution control laws (Proceedings of the House of Representatives, No. 8, p. 22 (testimony of Mr. Yajima)). Despite MITI's opposition, the government was able to push the bill through the Diet, although both the Special Pollution Control committee in both houses adopted resolutions urging the government to attend to the issues raised during the debate.

244. See Japan country report to 1976 OECD Conference for lengthy discussion of the k-value approach.

245. The Air Pollution Control Law also included those areas anticipated to become major industrial sites. The Soot and Smoke Regulation Law only applied to already heavily polluted areas.

246. Article 17. Formerly, prefectural governors merely had the right to recommend that factories adopt these plans.

247. Article 4.

248. The primary motivation for the amendment of the Basic Law, of course, was the sentiment that all pollution should be reduced, not only SO_x pollution.

249. The law had had a provision similar to the Basic Law's provision calling for harmonization between "preservation of the living environment and the sound development of industry." As seen in ch. 1, this clause had been criticized because it gave priority to economic growth over environmental preservation. In 1970 it was struck from the Basic Law and from the 1970 Air Pollution Control Law (see art. 1).

250. The practice of designating special control areas was abolished. Under the old law, only polluted areas (and areas that might soon become polluted) were designated. By the time the law was amended, only thirty-five areas (in twenty-six prefectures), covering 11,700 square kilometers and a population of 35 million were regulated. The amendments reflected the suggestion that pollution be prevented before it became serious. Although the idea was far from nondegradation, regulations were made applicable to unpolluted areas as well as polluted areas.

Emission standards were to be established for hazardous air pollutants ("toxic substances," to use the law's term) such as cadmium, chlorine, fluorine, lead, and NO_x (arts. 2(1)(i) and 3(2)(iii)). The government would specify other substances by cabinet order. Under the old law, only SO_x, soot, and particulates had been regulated. Toxic substances were controlled only in cases of accident. With the new amendments, the number of substances subject to regulation was expanded. Particulates created by means other than combustion (such as crushing and piling, which had not previously been regulated by the law) were also covered by the new law (arts. 2(4)(5) and 3(2)(ii)(18)). Furthermore, hydrocarbons and lead were added to CO as substances in automobile emissions (art. 2(6)) to be regulated.

251. As discussed, the power of local governments had been considered limited. The law finally allowed prefectures to establish emission standards by ordinance more stringent than those of the central government (art. 4). In other words, the law provided minimum emission standards that could be strengthened by prefectures to meet the degree of pollution in their areas. Yet the law did not allow prefectures to set more stringent regulations for SO_x. The government feared that stringent regulations for SO_x would cause greater demand for low-sulfur oil and, consequently, harm the government's energy policy. The law also allowed local governments (including municipalities) to establish emission standards for substances that were not otherwise regulated by national law (art. 32). The old law had allowed local governments to set emission standards for facilities not regulated by national law. It is unclear, however, whether local governments had the right to regulate substances other than those covered by national law.

Penalties against violators of the law were strengthened. Until the law was amended, violations of emission standards were not punishable. Only when factory owners failed to abide by orders for the modification of plans or facilities, would owners be punished. With the new amendment, a person discharging pollutants in violation of emission standards could be indicted immediately (arts. 32 and 33).

252. The government had been reluctant to implement any kind of fuel-use regu-lation because such regulations, it was feared, might harm its energy policy. The amended law provided that, in urban areas where the volume of fuel use fluctuated greatly according to season, prefectural gov-ernors could require polluters to comply with standards for the use of fuel in order to reduce air pollution in their areas (art. 15). The right to regulate fuel use itself, how-ever, was abridged.

253. Governors could also request their local public safety commissions to regulate traffic (art. 23), especially air pollution along highways (art. 21). The Minister of Interna-tional Trade and Industry could also on re-quest (by a governor) regulate pollution caused by public utilities (art. 27(4)). These provisions expanded the prefectural gov-ernor's powers over pollution control. The jurisdiction of the central government agen-cies over pollution, however, remained al-most unchanged.

254. See Field Survey of Diffusion and Rele-vant Meteorological Conditions (MITI, un-dated).

255. These areas included Kashiwa, Ōita, Mizushima, Chiba, Osaka, and Harima. These were all areas designated for future industrial development. Under Japan's First Comprehensive National Development Plan, formulated in 1962, fifteen cities were designated under the name of "New Indus-trial Cities." The plan also proposed that growth zones be designated and fostered as strategic areas of industrial development in areas outside of the highly concentrated areas. Six such "special areas for industrial growth" were designated by 1964. See the OECD report (n. 1 supra) 56. The decision to initiate administrative guidance reportedly was motivated as a gesture to propitiate the public, and because it was felt that there was not enough time to enact a law that would regulate future development. (Based on interviews of MITI officials.)

256. Implementation of this proposal was, however, deferred several years. (Based on interviews of MITI officials.)

257. See Uchida, Conditions for the Desul-furization of Petroleum (Sekiyu no Teiyōka Taisaku no Jihō).

258. The committee's final report submitted to MITI in 1969 hailed the latter methods as highly desirable from the perspectives of both efficiency and effectiveness. To

achieve the desulfurization of oil, the report also called upon the government to work out a system of compliance with each of the major refineries such that users of heavy oil would be strongly encouraged to use medium grade oil; and medium grade oil users would be encouraged to employ oil low in sulfur content.

259. But note that in later years many of the other options have also been developed.

260. Because Middle East oil was primarily heavy, high sulfur oil, Japan did not have easy access to low sulfur oil, and thus this option was not feasible.

261. Based on interviews of MITI officials.

262. This is the usual pattern. By the time of publication, a policy has already been decided and is well under way.

263. The project was part of the National Research and Development Program initiated in 1966. The principal purpose of the program was to support at government expense the development of technologies usually requiring large budgets and long-range programming and high risk. In addition to these, other criteria for selecting projects are that these projects be conducted in cooperation with universities and industry, and they entail significant economic impacts on the mining and manufacturing industries. Since the establishment of the program, fifteen projects have been selected at a total expenditure between 1966 and 1977 of about ¥90 billion. Other research and development projects have included seawater desalination, comprehensive automotive traffic control technology, urban waste resource recovery systems, high performance computers, deep sea remote control drilling equipment, and the electric automobile. See National Research and Development (MITI, Program Agency for Science and Technology, 1977).

264. Uchida, n. 257 supra.

265. For example, the Japan Petroleum Company agreed to introduce a new process at its plant by October 1973; Idemitsu introduced a new process in October, 1967, etc. For the industry as a whole a 1.7% reduction in heavy oil sulfur content was achieved in 1967; by 1968 a 1.6% reduction was achieved for existing equipment and 1.5% for new equipment. These percentages were reduced 1.5% and 1.4% respectively in

1969. See Uchida, Conditions for the Petroleum Desulfurization Policy (Sekiyu no Teiyōka Taisaku no Jōkyō).

266. Petroleum industry representatives report that about ten companies participated in this three-year cooperative research effort and that the government's total contribution was about ¥100 million.

267. The relationship between these administrative actions and the standards remained loose, with the standards serving as a basic target for administrative action.

268. Sekiyu gyōhō (1961, Law No. 128). Article 3.1 states: "The Minister of International Trade and Industry in accordance with ministerial order shall establish in each year a five-year petroleum energy supply plan."

269. Based on interviews with steel officials. The government's contribution was about ¥500 million.

270. For example, the major electric power companies (Hokkaidō, Tōhoku, Tokyo, Chūbu, Hokuriku, Kansai, Chūgoku, Chikoku, and Kyūshū) all concluded pollution control agreements, many of which required the introduction and use of desulfurization technology, research and development of more advanced technology, the purchase of special land sites, etc. Steel industry representatives have noted in interviews that their industry was particularly affected by these agreements because local activism took place precisely when the industry wished to expand. Frequently the steel industry, in addition to meeting the strictures noted above, also had to submit to the implementation of stricter emission standards than those prevailing nationally, various limitations on plant operations, and requirements pertaining to the use of special processes. In the case of steel and other industries, pollution control agreements addressed a broad range of pollution problems (dust, NO_x, various water pollution problems) and were not confined to control of SO_x. Steel industry officials have noted that the pollution control agreements provided the greatest incentive for the development and application of pollution control technology, an incentive exceeding that of the k-value emission standards. Having to meet the sundry requirements of many diverse agreements, industry officials report,

caused confusion within the industry. But chaos was avoided to some extent once the technology became widely available.

271. These compacts (Kentōshi Kyōgikai Hōshiki) have often been concluded in the land use area to facilitate coordination of decision making between central, prefectural, and municipal governments. At times the initiative will come, as in the present case, from the central level, although at other times from local governments themselves. The "agreements" often contain discussions of common problems and goals and the establishment of various guidelines and understandings designed to promote joint action.

272. If MITI's "regulatory" authority had been challenged between 1965 and 1967, the ministry could easily have justified its actions by reference to the broad scope afforded by its organic statute. See Ministry of International Trade and Industry Establishment Law (Tsūsanshō Sangyō Shō Setchi Hō), 1952, Law No. 275.

273. For example, as of 1976, the industry had the luxury of choosing from over thirty types of smoke desulfurization equipment to meet existing emission requirements.

274. Based on discussion with petroleum industry officials in Keidanren.

275. For example, on April 24, 1962, a resolution was passed by the Standing Committee on Social and Labor Affairs of the House of Representatives. Similar resolutions were also passed by the Special Standing Committees on Industrial Pollution Control of the House of Councilors and the House of Representatives on June 24, 1966, and June 25, 1966, respectively.

276. The maximum concentration for used cars was 5.5%. In 1970 and 1972, respectively, the ministry also ordered the installation of positive crankcase ventilation and fuel evaporation controls.

277. The establishment of emission standards would ordinarily not be handled by the Central Council. However, because automotive emission regulations had become a political issue, the problem was referred to the Central Council as a "basic matter related to environmental pollution control," under art. 27 of the Basic Law.

278. 42 U.S.C. sec. 1857 et seq.

279. It should be noted that all firms in the Japanese industry were in approximately the same position in 1971 and 1972 with respect to the development of emission control technology. However, industry officials noted that smaller makers may have had some advantage because they were producing fewer models.

280. The average emission levels for 1975-model cars were as follows: 2.1 g/km CO, 0.25 g/km hydrocarbons, and 1.2 g/km NO_x. For 1976-model cars the standard for NO_x would be 0.25 g/km.

The reader is warned, however, that conclusions from comparative analysis of auto emission standards must be drawn with extreme caution because of differences in testing procedures. An emission test procedure consists basically of two things: (1) instruments that measure chemicals and (2) driving cycles, a way of exercising a car's engine while chemicals are being measured. Different instruments may give different results; different driving cycles are almost certain to yield different results. A recent report by the U.S. EPA points out that whereas the instruments used to measure chemicals are basically the same in the United States and Japan, driving cycles are different. The report argues that "cold start" tests and higher speeds in the U.S. cycle test produce higher levels of emissions and are in large measure responsible for the discrepancy in U.S. 1981 Federal Standards for hydrocarbons, CO, and NO_x and 1978 Japanese emission standards. See generally, Fact Sheet: Comparison of the Japanese and United States Auto Emission Standards (EPA, FS-25).

281. In the meantime (December 1972) the Environment Agency announced new interim standards for 1973. These standards were 2.6 g/km CO, 3 g/km hydrocarbons, and 3 g/km NO_x. Thereafter, the Ministry of Transportation revised its vehicular safety standards to implement the new controls pursuant to art. 19 of the Air Pollution Control Law, which states:

(Maximum Permissible Limits)

1. The Director-General of the Environment Agency shall establish the maximum permissible limits on the amount of emissions of motor vehicles generated under certain conditions and emitted into the atmosphere.

2. In case the Minister of Transportation es-

tablishes by an order pursuant to the Road Vehicles Act necessary measures for the control of motor vehicle emissions, he shall take care to secure the maximum permissible limits referred to in the preceding paragraph for the purpose of controlling air pollution by motor vehicle emissions.

282. These "hearings" were private meetings with industry officials. They were not meetings open to the public, such as are conducted in the United States.

283. The chronology of the emasculation of the Clean Air Act automotive emission standards is as follows: April 1972, Ruckelshaus refused to suspend the 1975 CO and hydrocarbon standards. The D.C. Circuit Court of Appeals then remanded this decision to the EPA in *International Harvester v. Ruckelshaus*, 478 F.2d 615; on April 11, 1973, the EPA Administrator granted a one-year extension for implementation of the 1975 CO and hydrocarbon standards; on July 30, 1973, the EPA Administrator granted a one-year extension of the 1976 NO_x standards; in 1974, Congress amended the Clean Air Act to freeze the EPA interim standards for CO and hydrocarbons through 1976, thus postponing the levels originally set for 1975 until 1977 (in addition, Congress also postponed until 1978 the original 1976 NO_x standard of 0.4 g/km, thus continuing through 1977 the EPA interim standard of 2.0 g/km); in 1975 the EPA administrator granted a one-year extension of 1977 CO and hydrocarbon deadlines; in 1977, Congress again amended the Clean Air Act, continuing the freeze on existing limitations of CO, hydrocarbons, and NO_x through 1979. For a complete discussion of this subject see Richard B. Stewart and James Krier, Environmental Law and Policy 2nd ed. (1978).

284. Environment Agency officials noted that although the actions of the mayors of the seven cities did not greatly influence agency decision making, they did significantly affect public opinion which thereafter grew even more hostile toward the automobile industry. The mayors' report also stimulated the agency to assemble an independent team of experts to study problems of the industry's compliance with the 1976 emission standards.

285. The basic objection was that the report was scientifically unsupportable. During this period the Special Standing Committee on Industrial Pollution of each house held hearings.

286. The report urged that standards be set at 0.6 g/km for cars weighing 1,000 kg or less, and 0.85 g/km for all other cars. It is interesting to note that during this period the House of Councilors' Special Committee on Pollution and Environment held extensive hearings on the 1976 NO_x standard. At one hearing, Mr. Robert Fedor of Gould, Inc., was invited to speak, an extremely unusual event. Gould reportedly stated that a new catalyzer system could cut down nitrogen oxide emissions by 0.23 g/km. The existence of a device that might enable the industry to meet the NO_x standards reportedly greatly embarassed the government and particularly the LDP. (Daily Yomiuri #2, Dec. 7, 1974.)

287. The public, however, was not united in its opposition to postponement. According to one report many automobile industry workers supported the industry. See for example the position of the 550,000 membership of Japan Federation of Automobile Workers Union (Jidōsha Rōren). (Mainichi Daily News 4, Nov. 16, 1974.)

288. That is, 0.84 g/km or less (average 0.6 g/km) for cars weighing 1,000 kg or less and 1.2 g/km or less (average 0.85 g/km) for all other cars.

289. The four-member body was called the "Group for the Study of Motor Vehicle Nitrogen Oxide Emission Control Technology" and was composed largely of internal combustion engine experts. The group was organized by the chief of the Air Pollution Bureau of the Environment Agency and financed by the bureau's research funds. The organization of this body, as we will soon see, was a deft move; the body became a useful instrument in the agency's guidance of the industry.

290. See the Report of the Group for the Study of Motor Vehicle Nitrogen Oxide Emission Control Technology, Motor Vehicle Emissions (Environment Agency, Dec. 1975). During this period MITI apparently was promoting interindustry technical cooperation in controlling emission gases. According to one report an agreement of cooperation was concluded under MITI's auspices between Toyota, Suzuki, and Daihatsu. (Mainichi Daily News 6, Dec. 23, 1975.)

291. Nihon Keizai Shimbun (Japan Economic Journal) 8, Jan. 20, 1976.

292. Asahi Shimbun (Asahi News), Feb. 14, 1976. For an interesting insight into the managerial bases for Honda's success in controlling auto emissions, see Koichi Shimokawa, "An Innovator Succeeds: Honda's Entry Into the World-Wide Automobile Industry," a paper presented to the Symposium on Technology, Government, and the Automobile's Future, Harvard Business School, Oct. 19–20, 1978. Shimokawa stresses the company's sense of social commitment at the time of inaugurating its project to develop the CVCC engine.

293. Actually, Toyota was able to market more cars than usual during the extension. This action subjected the company to great criticism both from the public and MITI.

294. Asahi Shimbun (Asahi News), Mar. 18, 1976.

295. Environment Agency, Report of the Group for the Study of Motor Vehicle Nitrogen Oxide Emission Control Technology, Motor Vehicle Emission (Oct. 1978).

296. Nihon Keizai Shimbun (Japan Economic Journal), Aug. 3, 1976.

297. During this period, the Special Pollution Control Committee in both Houses called industry leaders and government officials to testify on the problems of compliance. The hearing apparently motivated the Director General of the Environment Agency to take a firm stand, to the extent that it focused public attention on the question of whether or not the agency would capitulate.

298. For an account of this episode, see Mainichi Daily News 12, Oct. 26, 1976. The Minister of International Trade and Industry and the Director General of the Environment Agency also met on a number of occasions with industry leaders to discuss the problem of compliance.

299. Air Pollution and Motor Vehicle Emission Control in Japan (Environment Agency, Air Quality Bureau, Automotive Pollution Control Division, Mar. 1977). Among the tax incentives employed were the following: For cars meeting the 1978 emission standard, ¥20,000 reduction per car in commodity taxes; between April 1 and November 31, ¥10,000 reduction for electric-powered passenger cars, and commodity tax halved. Acquisition taxes for cars meeting the 1978 standards were reduced by 0.25% and between April 1 and November 31 by 0.125%. Between April 1, 1977, and November 31, 1978, 2% reduction in acquisition taxes for electric-powered automobiles. See Kankyō Hakusho (Environment White Paper) (1977). Although these tax incentives were deemed beneficial by the industry, their direct influence on the development of emission control technology is generally discounted.

300. The government guaranteed at these meetings that each firm's trade secrets would be scrupulously protected.

301. Some industry officials noted that the health effects of NO_x were used as a particularly powerful sales point by their lower-polluting competitors.

302. For a discussion of the importance of "image" in foreign relations see Jōji Watanuki, Self-Images of Japan and the United States in a Changing World, in Japan, America, the Future World Order, Morton A. Kaplan and Kinhide Mushakoji, editors (1976).

303. This, of course, was a primary motive for the Environment Agency's publication in 1976 of the study group's report. We do not, however, possess evidence that the sales of "bad" companies, e.g., Toyota, tended to fall off during this period.

304. The Japanese automobile industry did not use several arguments employed by its American competitors. For example, the industry never argued that as a result of having to comply with emission standards, profits would fall, and, therefore, the industry would have to lay workers off. In Japan, in fact, neither sales nor profits ever did drop and in any event the concept of laying workers off would have been unthinkable. As one Toyota representative put it, the company and the workers viewed their plight as a common one, and when times got tough they simply would "bail out the boat together." The representative also noted that the LDP would have sternly reprimanded any company that attempted to lay off workers under such circumstances. Traditionally, layoffs are infrequent in Japan's "life" employment system. Compare International Harvester v. Ruckelshaus, 478 F.2d 615 (D.C. Cir. 1973).

305. One industry official interviewed

stressed the continuing public view of the government as "okami" (gods). Thus it was felt that the general public might seriously frown on a company's challenge to government action as a form of lèse majesté, an advantage that competitors would quickly seize.

306. The Environment Agency also rewarded progressive companies by purchasing low-pollution cars for the agency's use.

307. Hiroshima Prefecture also proposed a surcharge on high-pollution cars, but this measure was never implemented.

308. Reportedly Osaka Prefecture also decided to prohibit its public employees from commuting in their own automobiles as a further step in its drive to control photochemical smog. (Mainichi Daily News, July 18, 1975.)

309. It is interesting to compare this case to the example of the phasing out of lead in gasoline. In 1970, as a result of poisoning of the residents of Tokyo's Ushigome Yanagichō area by lead from auto exhausts, control of lead in gasoline became a major political issue. At the initiative of Seiji Katō of the Socialist party, the Pollution Control Committee(s) in the Diet held hearings, the upshot of which was a strong indictment of lax government controls. In response to pressure from the Diet, MITI's Automotive Pollution Counter-Measures Subcommittee under its advisory Industrial Structure Council's Pollution Committee launched a full-scale investigation. In September 1974, this committee filed its report outlining basic policy guidelines and various enforcement measures. MITI thereafter approved the council's basic recommendations and initiated the following actions. First, lead additives in regular gasoline would be phased out by 1975. Second, lead in premium gasoline would be reduced to 0.7 cc/gal in 1975. This administrative action was supported by guidance of the following kinds. First, the government developed a labeling system for automobiles, including labels for (a) unleaded gasoline; (b) unleaded gasoline for high-speed cruising; (c) mixtures of leaded and unleaded; (d) leaded gasoline. These labels were also displayed on agricultural, forestry, civil engineering, and construction machinery, and consumers were urged to buy unleaded gasoline. To finance the lead

phasing-out program, the government, through the Japan Development Bank, also provided loans, and MITI subsidized the research and development costs of the industry.

The case of phasing out of lead in gasoline affords another example, where an important pollution problem was effectively addressed solely through administrative guidance without the enactment of any explicit statutory provision to authorize such action. Compare *Ethyl Corp. v. EPA*, 541 F.2d 1 (D.C. Cir. 1976), for the key decision in this area upholding EPA discretion in the United States.

310. An issue always present in Japan's regulation of auto emissions was the international impact of domestic controls. Although Japanese newspapers warned as early as May 1973 that the foreign car import market might someday be threatened, serious foreign dissatisfaction with Japan only began to crystalize after April 1975, when the government substantially tightened its regulation of the industry. By May 1976, senior MITI officials were announcing that foreign car imports would fall by one-third, and that many "favorite" European cars (Austin Mini 1000, Jensen, Panther, Peugot, and Maserati) would eventually be shut out. In the spring of 1976, small Japanese dealers in cars like Lotus Europe, Porsche, and Trans Am began preparing for bankruptcy.

Between June and August 1976, the Environment Agency and other branches of the government commenced a series of hearings with representatives of the major European and American auto manufacturers and with delegates from the Japan Automobile Importers Association (JAIA). These groups requested the government to alter its policy in the following respects. First, the foreign auto manufacturers demanded that Japan postpone enforcement of its proposed 1978 standards, particularly the nitrogen oxide limitation of 0.25 gm/km. In addition to arguing that Japan's standard was the strictest in the world and scientifically unsupportable, the automobile makers' representatives alleged that Japan's regulations inevitably would increase costs to the consumer, waste fuel, and decrease vehicular performance. Second, the foreign companies charged that Japanese import controls wasted time and that certification procedures were unfair and obstructionist. Ultimately, Marumo

Shigeoka, director general of the Environment Agency, announced that foreign automobiles would be granted a two-year exemption, beginning April 1, 1978, from the new 1978 controls.

311. MITI has also surveyed NO_x since 1975 and dust since 1976.

312. Local governments also used this data to draft pollution control agreements and to establish local emission and effluent standards.

313. Kakushu Kōkyō Jigyō ni Kakaru Kankyō Hozen Taisaku ni Tsuite (The Items Agreed Upon at the Cabinet Meeting Concerning Environmental Protection Measures Affecting Public Works Projects) (June 6, 1972).

314. Kōjō Ritchi Hō (1959, Law No. 1959). This amendment made impact assessment mandatory whenever a large-scale development of factory land was planned (art. 2) and required the Minister of International Trade and Industry to formulate guidelines for factory locations (art. 4).

315. Koyūsuimen Umetate Hō (1921, Law No. 57).

316. Kōwan Hō (1950, Law No. 218).

317. In early 1974 the Director General of the Environment Agency conferred with the Central Council to obtain its recommendations on an environmental impact assessment system. In June 1974 the council's expert committee released an interim report on the technical aspects of the system. On December 22, 1975, another expert committee published a legal study outlining the U.S. experience under NEPA (National Environment Policy Act, 1969). The study concluded that an environmental impact assessment system was needed.

318. The Socialist's bill argued that the environmental criteria for a permit were too difficult to determine.

319. The Bar Association document mandated assessment of new legislation and the establishment of administrative policies.

320. The government's decision was also based on the recommendation of an expert committee under the Central Council that conducted an in-depth study of U.S. NEPA. Another stimulus for the government's decision was the Ushibuka sewage treatment plant case, decided by the Kumamoto District Court on February 27, 1978.

321. For example, the second draft would have limited public comment to those residents living in the affected area.

322. By 1972, at the time of the incident, a part of this phase had already been complete.

323. See Tsunetoshi Yamamura, *Procedural Aspects of Environmental Impact Assessment (EIA) in Japan: A Proposal of Legal Policy*, 2 Earth L.J. 255 (1976). Yamamura, an attorney in Osaka, was the principal draftsman of the Federation of Bar Associations' proposed EIS law.

324. The Bar Association report stated that the Environment Agency's final approval would not be appropriate until Hokkaidō Prefecture could demonstrate that harbor operations would not damage the environment.

325. It is important to note that the Environment Agency did not possess explicit statutory authority to issue these guidelines, but rather had to rely on its general "coordinating authority" under art. 7.2 of the Environment Agency Establishment Law. Other agencies of the Japanese government also gave some, albeit superficial, attention to the task of administrative guidance. For example, in 1973 MITI established the Office of the Environmental Impact Assessment Advisor (Kankyō Shinsa Komon), an informal body of about 15 persons, to review the environmental effects of large-scale industrial projects. MITI also claims that atomic and other power stations have prepared assessments in response to its guidance of these industries.

326. See 4 Japan Environmental Summary (No. 10, Oct. 10, 1976).

The Environment Agency's guidelines set forth the following:

(1) the underlying principles for conducting environmental assessments;

(2) a guide to the local situation;

(3) an analysis of the actual conditions of the local environment;

(4) the nature of the development project;

(5) an analysis of the project's environmental effects and the permissible scope of the activity;

(6) a plan for environmental control;

(7) a summary of findings.

In addition, the guidelines required that:

(1) draft reports be made simple, logical, and written in plain language to permit the pub-

lic to be able to understand and comment on them;

(2) draft reports be published and open to general public inspection;

(3) public hearings be held to permit the local residents to express their opinions, to allow the government and developers to explain the report;

(4) comments of the public be made a part of the final report, i.e., the report made to reflect public sentiment;

(5) the final report contain a summation of the opinions gathered from local residents and others (including expert opinions explaining what has been done in regard to the matters raised by the residents).

327. The guidelines generally followed the earlier Mutsu-Ogawara format:

(1) The Bridge Authority was required to conduct the assessment and to prepare a draft report.

(2) Thereafter, the authority after consultation with the governors of Okayama and Kagawa prefectures was mandated to hold hearings on the draft and to explain its contents to the public.

(3) The authority's final report had to give full consideration to public comments, with copies to be sent to the governors of the two prefectures, the heads of concerned cities, towns, and villages, and the director general of the Environment Agency.

(4) The Honshū Bridge procedures emphasized the need to account for social, cultural, and historic factors.

The project called for the construction of three suspension bridges (with as many truss bridges) totalling 6,678 meters. Each bridge would have a four-lane expressway and a double track railway, and total costs were estimated at ¥670,000 million. See 5 Japan Environment Summary (No. 8, Aug. 10, 1977). In November 1977, the Bridge Authority released a 450-page draft assessment that stressed the economic benefits of the project and minimized the environmental effects. Damages to scenery, the authority said, could be repaired by reforestation.

328. The ordinance applied both to new plants as well as to operating facilities that were required to file periodic reports.

329. Hyōgo and Fukuoka prefectures subsequently set up guidelines following the model of the Kawasaki ordinance. The Tokyo Metropolitan Government began preparation of new guidelines that would require environmental assessments of all industrial and other activities that could significantly affect the environment. The Tokyo guidelines would apply to natural, historic, and cultural "impacts" including air, water, soil, and noise pollution, vibrations, odors, and land subsidence. A draft of the guidelines was to be presented to the Metropolitan Assembly in June, 1978. See 1 Int. Environmental Report (No. 6, June 10, 1978).

On June 30, 1978, the Ministry of Construction announced plans to implement its own environmental impact assessment program, after the Environment Agency failed to press a more comprehensive bill on this subject through the bureaucracy. The regulations apply to large-scale projects funded by the Ministry of Construction and related public corporations, such as the Japan Housing Corporation. The regulations do not provide for public hearings and private projects are not included. See Int. Environmental Reporter 262 (Aug. 10, 1978). The significance of the Kawasaki ordinance is that it provides for public hearings. Such hearings were held on November 23 and 25 on a plan by the Tokyo Electric Power Company to build a 420,000 cubic meter storage terminal and gasification plant. For an assessment of these hearings, see Int. Environmental Reporter 414 (Dec. 10, 1978). For a discussion of the Atomic Energy Safety Commission's new procedures for public hearings, see Int. Environmental Reporter 475 (January 10, 1979).

330. See Frederick R. Anderson, NEPA in the Courts (1973) for a discussion of the history of this statute.

331. For a discussion of this phenomenon, see Chie Nakane, Japanese Society (1970).

332. Technically the supervisory authority of superior administrative organs over inferior bodies is not termed administrative guidance (gyōsei shidō) but "supervisory guidance" (shiki kantokuken no katsudō no shidō). But in many respects the process is similar.

333. Article 4 of the Basic Law states: "The State has the responsibility to establish fundamental and comprehensive policies for environmental pollution control and to im-

plement them, in view of the fact that it has the duty to protect the people's health and to conserve the living environment."

334. For example, to cite but a few of many examples, see *Kennecott Copper Corp. v. Environmental Protection Agency*, 462 F.2d 846 (D.C. Cir. 1972) (striking down EPA's attempt to establish a "margin of safety" for national secondary ambient air quality standards); *District of Columbia v. Train*, 521 F.2d 971 (D.C. Cir. 1975) (invalidating in part EPA's transportation control plans); see also *Brown v. EPA*, 521 F.2d 827 (9th Cir. 1975) vac. 431 U.S. 99 (1976) and *State of Maryland v. EPA*, 530 F.2d 215 (49th Cir. 1975), 431 U.S. 99 (1976) that reached similar conclusions. Then, of course, there is also the *Jarndyce v. Jarndyce* of environmental law suits, *Reserve Mining Co. v. Minnesota Pollution Control Agency*, 294 Minn. 300, 200 N.W.2d 142 (1972) and its progeny.

335. For a few of many critiques of the Clean Air Act of 1970, see Henry D. Jacoby and John D. Steinbruner, Cleaning the Air (1973); James E. Krier and Edmund Ursin, Pollution and Policy (1977).

Whether the American automobile industry might have developed the requisite technology in time to meet the 1975 and 1976 deadlines in the Clean Air Act is raised by a recent case dealing with General Motor's allegations of patent infringement by a Japanese automaker.

336. See *International Harvester v. Ruckelshaus* (n. 304 supra).

337. Compare the discussion in ch. 6 of how the effects of the compensation system have focused Japanese society's attention on the problem of pollution disease.

338. For a discussion of some of the following ideas, particularly the suggestion of greater governmental involvement in research and development, see Henry D. Jacoby and John D. Steinbruner, Cleaning the Air (1973). The reader unfamiliar with the problems of implementing the Clean Air Act in the United States can refer to Richard B. Stewart and James E. Krier, Environmental Law and Policy, 2d ed. (1978).

339. These standards may well turn out to be the same as those set by Congress.

340. For a theoretical analysis of the benefits of combining economic incentives and direct regulation in a mixed system, see ch. 6; also Michael Spence and Martin Weitzman, "Regulatory Strategies for Pollution Control" (unpublished manuscript) (July 1976).

341. See generally Jacoby and Steinbruner, n. 335 supra.

342. One proposal imposes a flat tax on gasoline sold in a "fresh air district" subject to subsequent periodic rebates measured by an official smog rating. This rating would be based on standard tests measuring the amount of air pollutants emitted per gallon of gasoline consumed by the taxpayer's vehicle under simulated average traffic conditions. See ch. 4 of Stewart and Krier (n. 338 supra).

343. Trade secrets, of course, are given appropriate protection.

344. One idea might be to use the procurement power. For example, as in Japan, the government might purchase or use only the low-pollution cars of environmentally "progressive" makers.

345. See Jacoby and Steinbrunner (n. 335 supra) 58.

346. It is interesting that although this system of charges and revolving funds has been widely considered and usefully employed in Japan, such measures have not been used in the automotive field. One possible reason is that although the Japanese government was of course aware of the benefits of such ideas, the approach actually adopted was proving equally effective. For a proposal similar to that of Jacoby and Steinbruner (n. 335 supra) see Julian Gresser, Balancing Industrial Development with Environmental Management in the Republic of Korea, Report of the Environmental Mission, World Bank (I.B.R.D.) (December 22, 1977).

347. In fact the U.S. government actually brought suit against a number of auto manufacturers on the grounds that their effort to collaborate in developing technology violated the antitrust laws. Under a subsequent consent decree the auto industry was prohibited from: (1) restraining in any way the individual decisions of each auto company regarding the date when it would install emission control devices; (2) agreeing not to file individual statements with governmental agencies concerned with auto emissions and safety standards, and from filing joint statements on such standards unless the governmental agency involved expressly au-

thorized them to do so; (3) continuing the 1955 cross-licensing agreement and refusing to grant royalty-free licenses on auto emission control devises under patents subject to the 1955 agreement to all who may request them; (4) agreeing to exchange their companies' confidential information relating to emission control devices or to exchange patent rights covering future inventions in this area; and (5) continuing their joint assessment of patents on auto emission control devices offered to any of them by outside parties, as well as their practice of requiring outside parties to license all of them on equal terms. For a critique of the U.S. administrator's approach in this area, see David Andrews, *Antitrust Laws Meet the Environmental Crisis: An Argument for Accommodation*, 1 Ecology L. Quarterly 840 (1971).

348. See 49 U.S.C. sec. 642; Andrews (n. 347 supra) 846; 56 Stat. 23 (1942).

349. In fact, sec. 308 of the Clean Air Act actually authorizes cross-licensing under specified conditions. Pursuant to a U.S. District Court order, patent holders may be required to license patents to others to enable them to meet federal emission standards (42 USC sec. 1857 h-6 (1970)). See also W. Schwartz, *Mandatory Patent Licensing of Air Pollution Control Technology*, 57 Virginia L.R. 719 (1971).

350. For a discussion of possible conflicts in Japan between the flexible institution of administrative guidance and the prescriptions of the Japanese antitrust laws see Sanekata, *Administrative Guidance and the Anti-Monopoly Law*, 10 L. in Japan 65 (1977).

350a. This part of the manuscript was completed before the announcement of EPA's "bubble concept."

351. In this connection further study of Japan's Pollution Control Public Works Cost Allocation Law (1970, Law No. 133) could be useful.

352. Conn. Gen. Stat. sec. 22a-6b (1979).

353. For an analysis of the Connecticut scheme, see ch. 6 of Stewart and Krier (n. 338 supra).

354. However, Korean government officials expressed great interest in the Connecticut enforcement scheme to one of the authors (Gresser) during the World Bank's environ-mental law project in Korea. See Gresser, World Bank Report (n. 346 supra).

355. As noted, East Germany has already begun to experiment with this idea.

356. The project was sponsored by the Georgetown University Center for Strategic Studies and reportedly involved more than one hundred economists, lawyers, engineers, and scientists. The accord resulted from a lawsuit brought originally by NRDC, EDF, and other environmental groups. See Mainichi Daily News 6, Feb. 11, 1978. See also, *Current Developments*, Environmental Reporter 1913 (1978); *Current Developments*, Environmental Reporter 1919 (1978).

357. It may be that in the United States the principal impetus for such extrajudicial compacts will be provided more by the courts than, as in Japan, by an administrative authority. However, this should not diminish the effectiveness of this highly effective and efficient institution. A further example is the $1 billion cleanup plan adopted by TVA as part of a court-induced settlement with the EPA. See *TVA's Environmental Flip-Flop*, Business Week (Feb. 26, 1979).

358. See, for example, the recent Supreme Court decision in *Vermont Yankee Nuclear Power Corporation v. Natural Resources Defense Council*, 435 U.S. 519, 55 L.Ed.2d 460 (1978) upholding agency discretion pertaining to the nuclear fuel cycle. *Train v. Natural Resources Defense Council*, 421 U.S. 60 (1975) and *Ethyl Corp. v. EPA*, 541 F.2d 1 (D.C. Cir. 1976). Also, many of the cases cited in the following articles analyzing the merits of administration action: Norton F. Tennille, Jr., *Federal Water Pollution Control Act Enforcement from the Discharger's Perspective: The Uses and Abuses of Discretion*, 7 Environmental L. Reporter 50091 (Dec. 1977); David L. Bazelon, *Coping with Technology Through the Legal Process*, 62 Cornell L. Rev. 817 (1977). For a general discussion of the wane of the delegation doctrine see Louis L. Jaffe and Nathaniel L. Nathanson, Administrative Law Cases and Materials (1976).

359. See Tennille (n. 358 supra) 50091.

360. Id.

361. Id. We, of course, do not suggest that the "citizen suit" provisions in various federal environmental laws be modified.

362. See a discussion of this problem in Sanekata, n. 350 supra.

363. See Turner T. Smith, "Federal Environmental Statutes: Some Dilemmas for the Regulated," remarks at Seventh Annual Conference on the Environment, Airlie House, Warrenton, Virginia, April 21, 1978; also Donald G. Hagman, Urban Planning and Land Development Control Law 557 (1975).

Chapter 6

1. See, e.g., Toxic Substance Control Act, P.L. 94, 469, 90 Stat. 2003 (1976); (the Japanese) Law Concerning the Examination and Regulation of Manufacture, Etc. of Chemical Substances (Kagaku Busshitsu no Shinsa oyobi Seizō tō no Kiseī ni Kansuru Hōristsu) (1973 Law No. 117, 1973). A general survey appears in William A. Irwin, Toxic Substances Laws and Enforcement (Rabels Zeitsschrift, Max-Planck Institute, 1976).

1a. As this book goes into print, other countries are also beginning to attend to victims of pollution.

2. Kōgai Kenkō Higai Hoshō Hō (1973, Law No. 111) (hereinafter the "1973 Act").

3. See Yoshio Kanazawa, ed., Chūshaku Kōgai Taikei (Systematic Commentaries on Environmental Laws) 153–183 (1973). Today not only pollution disease victims, but also victims of other serious maladies, are assisted at the local level. For example, Ashiya City recently announced that it will establish a relief system for sixty designated diseases (Mainichi Daily News 12, June 25, 1978).

4. Nanyō Town also budgeted assistance for various pollution-induced eye diseases.

5. Other prefectures and townships have since followed suit.

6. The central government also contributed some money to finance medical research on these diseases.

7. The reader should recall that local governments were caught in a dilemma. They desperately needed tax revenues from new industries, but still had to mollify local residents. See chs. 2 and 5.

8. The council was established on December 25, 1959, by the Ministry of Health and Welfare, only five days before Chisso concluded its mimaikin contracts. The council was subsequently reorganized and transferred down to the prefectural government.

9. Kanazawa (n. 3 supra) 155–161.

10. As pointed out in ch. 2, the residents' and victims' movements to some extent joined forces in the Mishima-Numazu uprising.

11. Kōgai Seisaku No Kihonteki Mondaiten Ni Tsuite No Iken (Opinion on the Basic Problems of Pollution Control Policy) (Keidanren, October 5, 1966).

12. The reader will recall the proposal was substantially weakened in the Council for Pollution Control (Kōgai Shingikai); see Chūkan Hōkoku (Interim Report) (Council for Pollution Control, 1966). For a discussion of the legislative history of the Basic Law, see Iwata, editor, Kōgai Taisaku Kihon Hō no Kaisetsu (Annotation of the Basic Law for Pollution Control) 13–116 (1971).

13. Kōgai Ni Kansuru Kihonteki Shisaku Ni Tsuite (Basic Measures Regarding Pollution Control) (Council for Pollution Control, Oct. 7, 1966) stated that "when human life, health, property, and other personal interests are injured beyond the limits of ordinary endurance, the person who caused the damage should be responsible for compensation." This vague statement was substantially weaker than the concept advocated by the victims and their supporters.

14. The committee's final report also urged that a relief system for pollution damages be established. For example, sec. 10 of the report stated that "the establishment of a system for compensation of damages caused by pollution and other relief measures shall be studied." However, industry even during the council's debates had approved the idea of a fund. See the Keidanren opinion (n. 11 supra), which stated that: "The idea of a fund system for relief of pollution victims has been put forward, but there are difficulties in implementing the idea, because the causes of pollution are unidentified and unclear. It is for the government to resolve these problems by defraying the costs of the fund system." See Yasuhiro Okudaira, Kōgai Taisaku Kihon Hō Rippō Katei no Hihanteki Kentō (Critical Examination of the Legislative Process of the Basic Law for Pollution Control), 458 Jurisuto 170.

15. Article 21 of the law provided that "the government shall take necessary measures to establish a system for the settlement of disputes connected with environmental pollution by mediation or arbitration" and art. 21.2 stated that other measures should be undertaken "to establish a system to make possible the efficient implementation of relief measures for damages caused by environmental pollution." During the final Diet debates in both houses, the opposition was also able to push through resolutions calling upon the government to endeavor to recognize strict liability at some future time.

16. For a discussion of the government's implementation of art. 21.1, see the discussion in ch. 7 of the establishment of a statutory environmental dispute mediation and conciliation system in 1970. See Kōgai Ni Kakaru Funsō no Shori Oyobi Higai no Kyūsai Ni Kansuru Hōritsuan Yōkō (A Draft Outline of the Law Regarding the Resolution of Conflicts Related to Pollution and the Redress of Damages) (MHW, Pollution Department, Dec. 1967). See Ichiro Katō et al., *Kenkyūkai: Kōgai No Funsō Shori To Higaisha Kyūsai* (Study Group: Conflict Resolution and Redress of Victims in Pollution), 408 Jurisuto 14 (1968).

17. For a general discussion of the development of social insurance, see Outline of Social Insurance in Japan (Social Insurance Agency, Government of Japan, 1976).

18. The reader should note that in 1955 Japan had established a no-fault automobile accident compensation system under the Automobile Accident Liability Compensation Law. The law required that every automobile owner obtain compulsory liability insurance. Underwriting was required by the law by any non-life casualty underwriter. The Japanese government reinsured 60% of each policy, while 40% was reinsured by an underwriters' pool to share bad risks. At the same time the law created a statutory presumption in favor of liability, which shifted the burden of proof of negligence to the car owner and/or driver. See generally W. G. Shimeall, *No-Fault Auto Insurance: The Japanese Experience*, 9 The Forum 771 (no. 5, 1974).

The automobile liability insurance system is inappropriate for the pollution field for several reasons. First, unlike automobile accidents, pollution cases are polycentric. For otherwise similar pollution-injury cases the relevant class of polluters or victims could be quite different. Second, the issue of causation is frequently scientifically and legally obscure, as discussed in the text. Under such circumstances the risk of pollution injury is extremely difficult for a private carrier to assess, and as a consequence, great opposition could be expected in seeking to induce the private sector to support such a scheme.

19. Japan's 1951 Mining Law had imposed no-fault liability on the owners of the mining right for damages caused by mining activities. The owner was required to post a bond as a security, principally against damage to property. In 1967, when the Japanese government was seeking to develop a relief system, the mining security deposit applicable to property damage was only indirectly relevant to the types of risks involved in the pollution field.

Similarly, the liability system established by the 1961 Law for the Compensation for Atomic Power Accidents was also of little use, since it concentrated liability on a single party, the atomic energy facility operator. Concentrated liability, however, was thought unhelpful in the pollution field, where many polluting sources could contribute at the same time to an individual's injury. See generally *The Law Concerning Compensation of Nuclear Damage*, 7 Japanese Annual of Int. L. 57–54 (1963); also Genshiryoku Songai No Baishō Ni Kansuru Hōritsu (Law Concerning Compensation for Atomic Energy Damage) (1961, Law No. 147).

20. There was no explicit legal requirement that companies contribute proportionately. Rather, the proportionality principle was based on notions of fairness. But see art. 3 of the Basic Law, dealing with the responsibility of enterprises.

21. The council's work is described in Yoshihiro Nomura, Kōgai No Funsō Shori To Higaisha Kyūsai (Conflict Resolution and Relief of Victims in Pollution) 53–63 (1970).

22. Nomura (n. 21 supra) 60.

23. Where causation could not be shown, these individuals proposed that industry contribute one-fifth of the costs of the system.

24. The subcommittee's report was based more on an institutional or cultural construct of collective responsibility, rather than on a strictly legal theory of joint liability. See the discussion of the Yokkaichi court's decision in ch. 3. Compensation for property damage was also eliminated. The need for relief for property damages, although important, was considered less important, and also analytically a separate issue.

25. Kōgai Ni Kakaru Funsō No Shori Oyobi Higai No Kyūsai No Seido Ni Tsuite No Iken (An Opinion on the Systems for Dispute Resolution and Redress of Damage in Pollution) (Central Council for Pollution Control, October 18, 1968).

26. The Diet failed to pass the two bills during the first 1969 session.

27. Kōgai Ni Kakaru Kenkō Higai no Kyūsai Ni Kansuru Tokubetsu Sochi Hō (1969, Law No. 90). See generally Yoshio Kanazawa, *A System for Relief for Pollution-Related Injury*, 6 L. in Japan 65–72 (1963). The Dispute Law was passed in 1970. See ch. 7.

28. Bronchitis, asthma, chronic bronchitis, and emphysema.

29. The eye diseases covered by Nanyō Town's plan were not designated.

30. Under art. 14(2), industry paid 1/2, the central government 1/4 (in the case of a city 1/6), prefectural governments 1/4 (in the case of a municipal government 1/6) of medical payments. The system's administrative costs were allocated as follows: the central government 1/2 (or 1/3); prefectural governments 1/2 (or 1/3); city governments 1/3.

31. Industry strongly preferred voluntary contributions through a self-governing body to compulsory collection for these reasons: (1) industry felt it could better control the collection agency; (2) there might be a loss of face in compulsory collection; (3) compulsory collection might imply legal liability; (4) voluntary contributions gave the appearance of how concerned industry was about the plight of victims.

32. There are many examples of such industry-wide collective efforts in Japan (see ch. 5). Katsuko Tsurumi of Sophia University makes the interesting suggestion that this collective behavior may have roots in similar campaigns in rural villages. For a general discussion of this institution, see Richard Beardsley, John W. Hall, and Robert E. Ward, Village Japan (1959). In the present case, large economic federations like Keidanren were instrumental in promoting industry-wide consensus and thereafter in establishing the relief fund.

33. Article 16.

34. Article 18.

35. The 1969 Law did not permit subrogation by local governments.

36. Articles 8, 9(2), and 22.

37. The 1972 U.N. Stockholm Conference for the Human Environment gave some attention to the 1969 Law.

38. For a note on how shame prompted the Japanese Foreign Ministry initially to conceal the severity of Japanese pollution from the outside world see Japan Times, Feb. 16, 1973.

39. See Guiding Principles Concerning International Economic Aspects of Environmental Policies (OECD, May 26, 1972).

40. Until the Yokkaichi decision, most industry leaders greatly preferred a free donative approach to relief; after Yokkaichi most industry leaders were reconciled to a compensation system. Japan's recognition of the "polluter pays" principle, however, indisputably had some influence on governmental attitudes.

41. Other factors, of course, were also at play. In 1970, pollution, rather than being alleviated by the enactment of the Basic Law, was actually becoming worse. In the summer of 1970, photochemical smog attacked various Tokyo wards, cases of lead poisoning were reported at a busy Tokyo intersection, and paper mill sludge paralyzed a port and fishing industry in Shizuoka Prefecture. These incidents caused the public to demand stricter pollution controls and stimulated the government to reorganize its pollution control programs.

Another factor was the continuing national debate over the allocation of pollution costs. National opinion was turning increasingly in the direction of imposing a larger share of these costs on polluting enterprises. See art. 22 of the Basic Law; also Kōgai Bōshi Jigyōhi Futan Hō (Pollution Control Public Works Cost Allocation Law) (1970, Law No. 133), that imposed various prevention and abatement costs on industry.

42. Since 1970, Satō's pledge had forced the government to decide many difficult legal, as well as highly political, issues. These included: Should strict liability apply to all seven types of pollution provided by the Basic Law? What types of injuries should be compensated? Should strict liability legislation be consolidated within existing laws, made comprehensive, or merely established by an amendment of some existing laws? Should remedies based on strict liability be extended to fields other than pollution, such as injuries covered by the food and drug law? Should a provision be included creating a presumption on the issue of causation? Should each polluter be jointly and severally liable in the case of damages caused by multiple sources? Finally, should the new legislation grant rights to the injured to seek injunctive relief against polluters and to require responsible agencies to take other affirmative measures to counter pollution? In general, industry and MITI, and other development-oriented agencies opposed many of these proposals, seeking as narrow a construction of strict liability principle as was politically possible.

At the time that the Diet passed fourteen new pollution laws in 1970, the government was not able to submit a strict liability law. Three of the opposition parties (the Socialists, Democratic Socialists, and Kōmeitō), however, joined to propose a comprehensive bill for a strict liability law at this time. This legislation, however, failed to pass. The Communist party also submitted a bill amending the Basic Law, and proposing a drastic strict liability principle (with wide coverage, a presumption of causation, joint tort liability, and secondary liability of the government). This, too, was killed in the Diet. During the discussion of these bills the public was highly critical of the government. In response, the Headquarters for Pollution Control, in April 1971, publicized a draft outline of a strict liability law. However, this law was limited in scope. It applied only to health injuries caused by air and water pollution. Moreover, SO_x pollution was also excluded due to industry's strong opposition. Curiously, the various LDP actions could not reach consensus even on this lukewarm proposal. At the same time, the opposition parties proposed a new bill to the 1971 Diet and criticized the government for breaking its pledge. Mindful of the approaching upper house elections, the government responded by repeatedly promising to submit its bill in the next Diet session.

In July 1971, the newly established Environment Agency took charge of the preparation of a strict liability bill. To this end, the agency organized a research project consisting of prominent lawyers who intensively discussed many of the critical issues listed above. Upon receiving the recommendations of these and other experts, the agency published an outline of a strict liability bill on March 2, 1972. The outline was a cautious proposal to amend the air and water pollution control laws; the previous government draft had eliminated air pollution caused by SO_x and dust because it was felt harsh to impose strict liability under conditions of multiple-source pollution. It was limited in application to health injury; polluters responsible to only a small extent for pollution injury were partially immunized from liability. The bill established a statutory presumption for causation; victims needed only to prove that their disease was caused by specific pollutants discharged by defendant(s); and strict liability was not recognized to be retroactive.

This outline attracted wide attention and public opinion was largely favorable. Industry, however, feared the consequences of this legislation and particularly opposed the statutory presumption on causation and the proposed pro rata allocation of liability in cases of joint pollution.

43. The opposition parties argued that industries should also be strictly liable for property damage and for injuries caused by substances other than those explicitly set forth in the pollution laws. In cases of multiple-source pollution, the opposition proposed that polluters be jointly and severally liable.

44. The Environment Agency had originally proposed the idea of a statutory presumption of causation, but this was subsequently deleted under pressure from MITI and the LDP. This deletion, however, outraged the bar as well as the general public, for by this time the itai-itai and Niigata Minamata court decisions had already recognized this concept. The government's action was construed as an effort to retreat from and subvert the victims' court victories. Immedi-

ately after the government's bill was filed with the Diet on March 22, the Socialist, Democratic Socialist, and Kōmeitō parties joined forces to submit a counterbill that set up a more comprehensive strict liability system. The opposition parties also proposed a system of guaranteeing disbursement of payments.

45. The opposition also proposed that residents be permitted to compel responsible agencies to take affirmative steps to deal with pollution, and that plaintiffs in pollution injury litigation be entitled to force a defendant, through discovery, to produce documents relevant to the trial.

46. Although the government's bill originally provided that the strict liability rule would not apply to damages caused by prior activities, the revision shifted the burden to the polluter of proving that damage occurring after the law's enactment was caused by polluting activity prior to enactment.

Because the LDP had an absolute majority in the Diet, there was practically no possibility that the opposition parties' bills would pass. The opposition parties could only manage to attach resolutions calling for future reconsideration of nearly all the points raised by their bills. Although legally meaningless, these resolutions gave political satisfaction to the exasperated opposition.

47. At the time of the debate on the strict liability law, the opposition also proposed legislation to establish a compensation system.

48. The reader will recall that the original provision alleviating the plaintiff's burden of demonstrating causation was deleted in the final version of the government's bill.

49. Outline of Social Insurance in Japan (n. 17 supra).

50. Actually, the issue of a compensation system had been discussed within the LDP at least as early as March 1972. At the same time that the party shelved the statutory prescription on causation, the LDP's Committee on the Environment resolved to establish a subcommittee to consider the advisability of setting up an administrative compensation system. Its purpose would be "to realize relief for the pollution victim and fair assumption of costs by industry."

51. See Masamichi Funago, Chikujō gaibaishō Sekinin Hō (Law for Strict Liability for Pollution Damage) 68 (1972). However, in view of the committee's attitude toward the strict liability law, one may speculate that the party's real reason for establishing the subcommittee was to avoid the opposition parties' (and the public's) accusation that the LDP was seeking to obstruct the victims' rights to recovery. A second motive undoubtedly was to shift the industry's financial burden imposed by the strict liability law to the general tax base.

At the same time that the LDP began to study the issue of compensation carefully, the Environment Agency also directed its advisory organ, the Central Council for Pollution Control, to consider the allocation of pollution costs. On April 17, 1972, the agency instructed the council to prepare its answer to the questions, How should the costs of pollution in Japan be allocated? Who should bear the costs of compensation for damages arising from environmental pollution? The first question concerned the general issue of how to implement the "polluter pays" principle in Japan; the second question asked specifically what type of compensation system was appropriate in the pollution field. Subsequently, two expert committees were formed in the subcommittee, one to study the allocation of costs, and the other, the formulation of a compensation system.

52. An interesting Western parallel is the history of workmen's compensation legislation. When the courts began to weaken the fellow-servant rule and other powerful common-law employer defenses to a workman's damage suit, the rising costs of litigation as well as of insurance premiums soon prompted employers to seek alternative ways of coping with the problem of worker injury. As a result, industry began collectively to establish a common compensation pool as a means of spreading the risk and of undercutting the incentive to litigate. See Lawrence M. Friedman and Jack Ladinsky, *Social Change and the Law of Industrial Accidents*, 67 Columbia L. Rev. 50 (1970). For a proposal in the United States to collapse some parallel barriers in the public nuisance fields, see John Bryson and Angus Macbeth, *Public Nuisance, the Restatement (Second) of Torts and Environmental Law*, 2 Ecology L. Quarterly 241 (1972).

53. At the same time, Koyama promised that the government would also initiate comprehensive planning measures aimed at the attainment of the ambient standard.

54. Hearings of this nature at the council level were rather unusual in Japanese administrative practice, but the issue was so urgent and the conflict of views was so great that the council needed some form of public participation to reach a conclusion.

55. The 1973 compensation system subsumed many of the local relief programs, although some local governments have continued to supplement the compensation system with local relief.

56. Diseases designated at present are bronchitis, pulmonary emphysema, chronic bronchitis, Minamata (mercury poisoning) disease, itai-itai (cadmium poisoning) disease, and chronic arsenic poisoning.

57. See sec. 2.3.1 for a discussion of the procedures for area designation.

58. Articles 44 and 45 provide that the council be comprised of no more than 15 persons appointed by the prefectural governor. Council members are forbidden to divulge confidential information relating to the cases. Government officials may serve on the council in an expert capacity.

59. Articles 19–43 cover medical care benefits, compensation to the handicapped and survivors (including a lump sum payment), a child compensation allowance, medical care allowance, and funeral expenses.

60. These classes are defined by the nature of the diseases found in them. In Class 1 areas the diseases are those for which no substance regarded as the "but for" cause of disease can be identified. Class 2 area diseases are traceable to a single substance.

61. An "emission charge" refers to a levy on discharges of polluters for their use of an ambient air shed; an "effluent charge" is a corresponding levy on water pollution sources. Under the 1973 Act all dischargers pay the emission charge, irrespective of whether they meet emission *standards* or not. Most charges of this sort are calculated based on marginal control costs. As will be described, the 1973 Act uses damage cost estimates as a basis for calculation of the levy. See Effluent Charges on Air and Water Pollution (Environmental Law Institute, 1973); Anderson, Kneese, Stevenson, and Taylor, Environmental Improvement Through Economic Incentives (1977). For an early study of this law see Julian Gresser, *The 1973 Japanese Law for the Compensation of Pollution-Related Health Damage: An Introductory Assessment*, 5 ELR 50229. For a translation of the 1973 Act, see 91 Int. Environmental Reporter 1.

62. The procedure developed by special amendment to the 1973 Act (Law No. 85, June 1, 1974) transfers a portion of the regular tonnage tax to the Class 1 area fund to be used for compensation purposes.

63. Articles 106–108 provide inter alia that complaints should be filed with the prefectural governor under whose jurisdiction the action was taken.

64. Articles 111–135 stipulate that the Pollution-Related Health Damage Compensation Grievance Board consist of six members who serve for three years, and are appointed by the prime minister with the concurrence of both houses of the Diet. Rules are also provided for meetings, removal from office, disposition of complaints, and the compensation of board members.

65. Articles 109 and 110.

66. Theoretically, a victim is permitted judicial recourse for uncompensated injury, even for the same injury, if payments received are inadequate. This raises several yet unanswered questions. Will a firm's payment of the pollution levy alleviate the burden of showing causation or responsibility in court? In what ways is the burden of proof of issues affected? What is the probative value in litigation of epidemiological analysis used in the administration of the 1973 Act? May a victim use administrative determinations of any of these issues as a basis for injunctive or other equitable relief? Answers to these questions differ. Administrators in the Environment Agency note that the act may alleviate proof problems of causation in a civil action with respect to the toxicity of a substance. They question, however, whether an administrative determination will eliminate a plaintiff's burden in court of proving the pathway or source of a substance. Some scholars assert that the law will be helpful here too. Unlike many U.S. workmen's compensation programs, Japanese workmen's compensation laws do not bar a victim's recourse to the courts.

67. This is not to suggest that the 1973 Act ignores this crucial question. Rather, the administrative practice under the act has refined the basic approach to causation developed in the pollution trials into a general methodology for disease, area, and victim designation.

Causation is handled in part as a statistical question. When a "significant" correlation between disease and pollution is identified, causation is inferred. Epidemiological analysis plays a crucial role in these determinations. Clinical and experimental data, as well as mortality and morbidity statistics are also carefully considered. Administrative determinations are also guided by the use of presumptions, and the doctrines of joint liability and foreseeability of harm as developed in the pollution trials. At times, however, since administrative agencies have wide discretionary powers in these areas, proof of causation is regarded as sufficient under circumstances that would probably not have been adequate to support judicial findings.

A basic general tool in probability analysis is the concept of "order of risk." Order of risk refers to a probability of occurrence of disease when an individual is exposed to certain environmental conditions. It is derived from prevalence and incidence rates of pollution disease, and clinical and experimental studies of pollution. For example, a recent study describes the order of risk of exposed populations to Kanemi oil poisoning (polychlorinated biphenyls or PBCs) or itai-itai disease as very high, i.e., one person in ten or one in a hundred, respectively. The order of risk of these diseases for the entire Japanese population is described as relatively low, i.e., one person in a million. See M. Hashimoto, "An Approach to the Problems of Risk and Damage for the Formulation of Pollution Policy" (unpublished, Environment Agency, Aug. 1972). An approach similar to this is discussed in Hamilton, *Assessment or Risk of Damage to the Environment*, in Environmental Damage Costs (OECD, 1974).

68. Chronic arsenic poisoning may be an exception. This disease was not really the subject of as extensive a national debate as were mercury or cadmium poisoning. However, the designation of the other diseases influenced the designation of this disease.

69. This description is based on discussions with senior officials in the Environment Agency.

70. The 1969 relief system first adopted on the national level the idea of designated diseases; earlier local programs had already used the concept.

71. The provision, of course, does not reveal the kinds of pressures that would be focused on the government before any such administrative determination.

72. As is usual the regulations were based on an interim report on medical experts in the Central Council. After the report's completion, the Environment Agency conducted a series of field investigations, discussing draft criteria with local governors and mayors. Thereafter, the agency reexamined the report of its advisory body, and after consultation, wrote its final criteria.

At last a set of criteria was established for both Class 1 and Class 2 areas. The criteria for both classes were similar in that they identified specific disease symptoms which victims had to display in order to receive certification. Under the criteria, symptoms were classified according to various technical medical classifications, subjected to various standardized testing procedures, and differentiated by severity. Although the establishment of criteria for the clear-cut case was easy, difficult problems were presented by less clear-cut cases for both classes of diseases.

73. Before setting the criteria, the Environment Agency conducted an intensive study of patients in the Yokkaichi area.

74. Even if it is quite clear that a victim is a heavy smoker, he cannot be barred from certification for this reason alone. Article 43 does permit the government, on the advice of the examination committee, to reduce compensation benefits when an extraneous factor, such as smoking, materially contributes to the disease.

75. The 1969 system covered only those expenses unsupported by national health insurance.

76. Special hospitals and clinics are designated as institutions treating designated pollution-related cases.

77. The Central Council had originally recommended that the level be placed between the average wage (100%) and the workmen's compensation level (60%).

78. See sec. 3.2.2 for a discussion of the continuing controversy on this subject.

79. For example, special and first class patients receive 80% of the average wage, differentiated by age and sex (special class victims receive additional benefits covering nursing care); second class patients receive 40%; and third class patients about 24% of the average wage. The categories are distinguished by the severity of impairment, although it is very difficult to decide to which class some patients belong. The classification scheme was derived from the Yokkaichi decision.

80. By Cabinet Order, governors are to undertake rehabilitation by taking patients to unpolluted sites to breathe clean air, supplying special equipment, and providing home medical counseling.

81. Article 2.

82. The basic methodology of designation is based on administrative practice developed since the early 1960s (principally using Yokkaichi data) and refined by the courts' analysis in the four pollution cases. See ch. 3.

83. Sulfur oxides are used because of the wealth of data on their health effects and because they are thought to be a principal factor in the four designated pulmonary diseases.

84. See the recommendations of the Central Council pertaining to area designation, November 25, 1974.

85. A useful tool in conducting such interviews is the British Medical Research Council (BMRC) questionnaire. BMRC data includes not only disease prevalence but also other factors like age, sex, smoking habits, etc. The technique is particularly helpful for people between the ages of 40 and 50.

86. Until this point the assessment has been primarily conducted within the Environment Agency and neither the general public nor even the concerned industries have yet substantially influenced the process. Most of the relevant data has been generated by the research of the central government in cooperation with the local authorities.

87. Although the 1973 Act does not prohibit industry from litigating the issue of designation, industry's case would be legally weak in light of the strong presumption supporting administrative discretion; it would also be politically unwise.

88. Generally, the role of the public is peripheral. The process is usually conducted without public hearings or official explanation.

89. Polluters under the 1973 Act can avoid the levy by using desulfurization technology or by switching to less polluting fuels. A special provision also exempts small or medium enterprises.

90. As already noted the emission charge is currently correlated with SO_x. The present ratio of SO_x pollution within and outside designated areas is 9:1. After comparing this ratio with the required budget, a basic pollution charge was derived of ¥1.76/cm³ of SO_x for nondesignated areas and ¥15.84/cm³ for designated areas. The charge was subsequently raised to ¥77.31/cm³ for designated areas and ¥8.59/cm³ for nondesignated, corresponding to the new ratio.

91. Industry in those days was willing to accept collective responsibility, but not liability.

92. Articles 18–105 of the 1973 Act reflect these objectives. The officers of the association are all drawn from industry or government and are appointed by the Minister of International Trade and Industry and the Director General of the Environment Agency. The association's officers conduct the daily management of the affairs of the association. The association also has a board of up to twenty trustees. The board is required to deliberate on important matters relating to the conduct of the association's affairs, upon request of the president of the association. According to the law, trustees are selected from industry officials in the large national economic federations. Each year the association prepares a budget and a business and finance plan that it submits to the Director General of the Environment Agency and the Minister of International Trade and Industry. The association is permitted to borrow money and to hold title in its own name to various securities and financial obligations.

93. Chisso and Shōwa Denkō for Minamata disease, Mitsui Metal Mining Company for itai-itai disease, and Sumitomo Kinzoku for chronic arsenic poisoning.

94. Major postwar research on pollution-

induced diseases began with the epidemiological and clinical studies of Minamata disease at Kumamoto University and with experimental work on animals; postmortem analyses commenced in 1957. A similar data base has been developed in some of the other pollution cases over the past 19 years. The Environment Agency and the recently formed National Institute for Pollution Research have also begun studies on diagnostic problems of itai-itai disease and cadmium, diagnostic problems of atypical Minamata disease, pathological-histological research of the human lung, epidemiological studies of Minamata disease, health surveys of residents in the vicinity of inactive or abandoned mines, and clinical studies of serious health damage from air pollution. In FY1974, ¥73 million (about $240,000) were appropriated for additional studies of the effects of cumulative substances on the human body, research on the effect of heavy metals on human health, and research into the accumulation of chemical substances.

95. A. C. Pigou, Economics of Welfare (4th ed., 1962). Of course, economic theory relies on a concept of causation that greatly simplifies the underlying scientific issues. Because scientists still do not fully understand the etiology of pollution-induced diseases, efficient resource allocation predicted in theory may not be feasible to the extent that theory depends on a precise scientific determination.

96. Economists debate the extent to which the result would be the same under oligopolistic or monopolistic market conditions. In the former case, since the so-called price leader can undercut any other firm that might seek to reduce its price, no firm may have a real incentive to reduce costs even where cost-reducing alternatives, such as new technologies or less polluting fuels, are present. Yet, these arguments are weakened by the following: (a) a corrective tax may still create an incentive because the oligopolist or monopolist can increase profits by avoiding paying the tax; (b) the whole industry may become inefficient and simply be replaced by a new industry, as canals were by railroads. See Schreiber, The Ohio Canal Era (1965).

97. Ronald H. Coase, The Problem of Social Cost, 3 J. L. and Economics 1 (1960).

98. Polinsky points out that government's possession of information and the attitudes of the parties toward "bargaining" may influence the extent to which compensation schemes contribute to efficient resource allocation. See A. Mitchell Polinsky, The Efficiency of Paying Compensation in the Pigovian Solution to Externality Problems (Harvard Institute of Economic Research, Discussion Paper 614, April 1978).

99. This effect is predicted by the "theory of the second best." For a general discussion of this idea, see R. G. Libsey and K. Lancaster, The General Theory of the Second Best, 24 R. Economic Studies 11 (1956).

100. See ch. 2 for a discussion of the change in values in Japan. Changed values affect consumer preferences that in turn influence the allocation of society's resources. For this reason, the allocation of resources in Japan may differ from a Westerner's notions of efficiency. See the discussion of Japan's "noneconomic" approach in the 1977 OECD report on Japan's environmental policies.

101. For a theoretical discussion of this and related problems, see A. Mitchel Polinsky, Controlling Externalities and Protecting Entitlements: Property Right, Liability Rule, and Tax-Subsidy Approaches, 8 J. of Legal Studies 1 (1979).

102. See William J. Baumol, On Taxation and the Control of Externalities, 62 American Economic Rev. 307 (1972).

103. For if a given option's marginal costs exceed its marginal benefits, we should reallocate our resources to more productive uses.

104. Donald Hagman and Dean Misczynski, editors, Windfalls for Wipeouts: Land Value Capture and Compensation (1978).

105. John J. Costonis, Development Rights Transfer: An Exploratory Essay, 83 Yale L. J. 75 (1973).

106. This seems particularly true in the Japanese approach to environmental planning.

107. Medically, however, it is not certain whether the harm occurred in the past or not.

108. Another approach is to assess a lump sum tax for past harm and to impose an emission charge to control present discharges. In theory, a lump sum tax does not affect a firm's marginal cost schedule;

hence, the risk of resource misallocation is avoided. Some economists believe that lump sum taxes, because they are unexpected, may be inequitable. See Talbot Page, *Failure of Bribes and Standards for Pollution Abatement*, 13 Natural Resources J. 677 (no. 4, Oct. 1973). It is also possible that the exaction of a separate lump sum tax would have been so bitterly opposed in Japan that its allocative efficiency would have been offset by the high costs of assuring its acceptance.

109. The reader will recall that 20% of compensation payments are financed from the automobile weight tax and that automobiles are thought to contribute 20% of SO_x pollution.

110. See James N. Buchanan and Gordon Tullock, *Polluter's Profits and Political Response: Direct Controls Versus Taxes*, 65 American Economic Rev. 139 (1975); also Dwight R. Lee, *Efficiency of Pollution Taxation and Market Structure*, 2 J. Environmental Econ. Management 69 (1975).

111. For the most part the present discussion relies on statistical information supplied by the Environmental Agency and on interviews with the senior government officials who throughout have been responsible for the implementation of the 1973 Act. To a lesser extent we have looked outside official sources, supplementing personal experiences and observations by interviews with different affected groups, and by secondary sources, such as scholarly writings and position papers prepared by industry, the bar associations, and others.

Although we have been fortunate in the full cooperation of the Japanese government, our primary reliance on governmental sources for information poses its own problems. Most empirical legal research in the environmental field in Japan is conducted within the government, dictated by governmental needs, and rarely published. This situation presents formidable difficulties for the researcher. For example, we were unable to obtain from the government data on how much each individual firm is currently paying under the levy. Similarly, the economic impact of the charge on industry, a question of considerable interest to Western scholars, apparently had not been studied either by government or by industry as of this writing (1979). Another issue is per-

spective. As much as possible, we seek in this chapter to appraise the compensation system in its own setting as neutrally as possible. We are concerned with how fairly, efficiently, and effectively the system has achieved its purposes and how it has been received by those it most intimately affects, the victims and industry.

112. Quality of the Environment in Japan 184 (Environment Agency, 1977).

113. For example, Osaka City (designated December 27, 1969) reported 11,092; the Eastern and Southern areas of Amagasaki City (designated December 1, 1970) reported 5,171; Tokyo (wards of Chiyoda, Chūō, Murato, Shinjuku, Bunkyō, Taitō, Shinagawa, Ohta, Meguro, Shibuya, Toshime, Kita, Itabashi, Sumida, Kōtō, Arakawa, Adachi, Katsushika, and Edogawa) reported 8,336; and Kawasaki City (designated December 27, 1969) reported 2,801. On February 3, 1977, the Environment Agency released the findings of an intensive study of some polluted areas (Chiba, Osaka, and Fukuoka prefectures) entitled A Population-Based Survey of the Influence of Compound Air Pollution on Health. This report noted that in some cases there was a significant correlation between respiratory ailments and high concentrations of atmospheric pollutants, particularly nitrogen oxides, dust, and other particulates. See Mainichi Daily News 12, February 4, 1977; 5 Japan Environment Study (no. 3, Mar. 10, 1977).

114. In 1977 approximately ¥63 million and in 1978 approximately ¥70 million were paid to victims under the compensation system. To date (1979) approximately ¥200 billion (or about $1 billion) has been collected to compensate Class 1 victims. This figure is unadjusted for inflation and changes in the exchange rate.

115. Governmental expenditures for environmental health programs have increased since 1975. For example, the government's budget for the compensation system in 1975 was ¥5,749 million; in 1976, ¥10,703 million; and in 1977, ¥14,259 million. Also, in 1977, ¥442,752 million was budgeted to establish a special Minamata Disease Research Center; ¥216,615 million was budgeted to continue research on cadmium and other toxic substances; and ¥95,724 million was budgeted for

the enforcement of preventative programs against toxic substances. In contrast to compensation for physical injury, government expenditures for compensation of property damages appear to be decreasing. Thus, subsidies to dealers of marine products suffering pollution losses dropped during 1975–1976 from ¥715 million to ¥519 million and relief measures to fishermen for oil pollution damage dropped from ¥194 million to ¥148 million. See 4 Japan Environment Summary (no. 5, May 10, 1976); 5 Japan Environment Summary (no. 6, June 10, 1977).

116. Defined by regulation as sources emitting more than 5,000 m³/h in designated areas, and 10,000 m³/h in nondesignated areas. See Environmental Policies in Japan, a Report of the Organization for Economic Cooperation and Development (OECD) 47 (1977).

117. Enforcement Order of the Pollution-Related Health Damage Compensation Law, Cabinet Order No. 295, August 20, 1974.

118. The total levy paid by the Osaka bloc (i.e., wards) was previously small in comparison with benefits paid to Osaka certified victims; whereas polluters in Kurashiki (Okayama) and Kita Kyūshū and Ōmuta (Fukuoka) cities appeared to be overcharged.

119. 5 Japan Environment Summary (no. 4, April 10, 1977).

120. This is understandable, since the general standard of medical services delivered through the health care system is reported to be high.

121. Much of this discussion is based on our interviews of victims and government officials.

122. For example, some people become sick with asthma irrespective of their exposure to pollution.

123. Class 2 area victims express a similar concern.

124. For information on the present schedules for disability and family allowance, see Japan Country Report to the OECD Conference in 1976 referred to in this chapter.

125. This grievance is shared by Class 2 area victims.

126. Some industry and government officials, however, insist that the payment of 80% of the average wage represents a compromise that includes partial compensation for pain and suffering.

127. Again, this concern is also shared by Class 2 area victims. Class 2 area victims are also particularly aggrieved by the system's failure to compensate for pain and suffering. There are also reports from some areas that doctors have been reluctant to treat pollution victims, especially Minamata, because they do not want to become entangled in the politics of these cases.

128. A typical example is children's nosebleeding allegedly induced by air pollution. See Kids' Nosebleeds Caused by Autos (a report of air pollution in Amagasaki, Kobe, Ashiya, and Nishiomiya), Mainichi Daily News 12, January 22, 1977.

129. Of course victims would not be entitled to compensation for the *same* injury. The charge, however, is that the victims, once having instituted negotiations, are discouraged from changing course and looking to the compensation system.

130. See discussion of the Kawasaki litigations in sec. 3.2.3.

131. Mr. Teruo Kawamoto, one of the leaders of the Kumamoto Minamata victims, noted in an interview that there may be as many as 200,000 victims, but that only a little over 1,000 have received compensation. So many victims distrust the system that only about 7,000 have applied for certification as of 1979.

132. The mimaikin contract referred to victims certified by the council (i.e., "Kyōgikai no nintei shita mono").

133. See Kōgai Higaisha Kyūsai wa Kore de Yoika? (How Good is the Relief of Pollution Victims?) (Japan Federation of Bar Associations, Pollution Measures Committee, May 1977).

134. As noted, many Kumamoto Minamata victims distrust the system.

135. See "Status of Rehabilitation Programs" in sec. 3.2.4 for a discussion of this issue.

136. The federation has presented the government with several position papers. See, for example, Keidanren position papers (Feb. 8, 1977, Aug. 22, 1977, and Mar. 20, 1978); Japan Industrial Association memorandum to MITI (Jan. 4, 1976); Japan Steel Federation memorandum (Dec. 16, 1975);

All-Japan Chemical Manufacturers Association position paper (Oct. 31, 1977).

137. The Japan Industrial Association memorandum to MITI (Jan. 14, 1976) states that natural incidence among pulmonary disease patients may be as high as one-half of the population.

138. See Kyūtoku, *Kōgai Kenkōhigai Hōshō Seido no Moten wa Koko Da! (Here Is the Blind Spot of the Pollution-Related Health Damage Compensation System)*, Sangyō to Kankyō (Industry and the Environment) (Apr. 1978).

139. Kyūtoku notes the rapid increase in victims since 1974. For Class 1 areas, he gives: 1974, 19,281; 1975, 34,185; 1976, 53,414; 1977, 60,090. For Tokyo: 1975, 6,386; 1976, 15,933; 1977, 18,980.

140. One figure frequently cited by government officials is 80%.

141. This idea was also suggested within government circles. One of the principal writers on the tobacco compensation problem is Umeda. See his *Kōgai Hoshō to Tabako (Pollution Compensation and Tobacco)*, published in Shinnai Kogaku Taikei (April 14, 1976). An interesting comparison is New York's nicotine tax, although proponents of the New York system do not seriously discuss using the proceeds of the tax to compensate air pollution victims. See William Drayton, Jr., *The Tar and Nicotine Tax: Pursuing Public Health Through Tax Incentives*, 81 Yale L. J. 1487 (1972).

142. Related complaints are that there are only infrequent assessments of how many victims are recovering, and that there are few official efforts to accelerate the victims' recoveries.

143. Based on interview with Michio Hashimoto, former chief of the Air Pollution Division.

144. Government officials, however, also note that some criteria for Class 1 area disease certification are also ambiguous. For example, there is no unified medical definition for chronic bronchitis. One test, the release of sputum for three months, is determined from patient interviews, and is, therefore, felt to be highly subjective.

145. See n. 85 supra.

146. Although industry initially urged that a levy also be imposed on discharges of NO_x, a recent Keidanren position paper seems to suggest that in light of the uncertainty surrounding NO_x pollution (referring to the debate over the NO_x ambient standard), a levy on NO_x emissions would be inappropriate.

147. For a hard-hitting critique, see Ijiri, *Kōgai Kenkō Higai Hōshō Seido o Kangaeru (Reflections on Compensation for Health Related Pollution Injury)*, Sangyō to Kankyō (Industry and the Environment) 63 (Jan. 1978).

148. Industry officials also point out that art. 2 of the 1973 Act defines Class 1 regions as areas where "marked air pollution has arisen over a considerable area. . . ." If SO_x pollution has markedly declined, they argue, these areas should no longer be considered Class 1; hence, they should be declassified.

149. Another government argument is that pollution (including that generated by newcomers) aggravated the victim's health problems.

150. From Ijiri (n. 147 supra).

151. In effect, industry is challenging the basic notion of "collective responsibility" upon which the system relies. Of course, one response is that even though Hokkaidō polluters may not cause injury to Kyūshū victims, Hokkaidō polluters expose Hokkaidō residents to the same risk of injury from pollution; therefore, it is appropriate to charge them. There is, however, still the issue of whether the levy should be the same as that paid by polluters in Kyūshū. Others ask should not a different criterion be used in Hokkaidō (for example, marginal costs) if the intent of the levy is regulation? If it is simply compensation, then the differential should theoretically be paid by Kyūshū polluters. They assert that the government cannot assess the levy indiscriminately, simply because it needs the money. A similar debate has recently arisen in the case of a noise pollution levy on aircraft used principally to compensate victims of noise pollution in Osaka. In this case, 27 foreign carriers have sued the Ministry of Transport, alleging that the levy is irrational.

152. See 1977 OECD Report (n. 116 supra).

153. Keidanren has suggested that costs might be allocated more equitably under the formula

ratio of mobile sources $= \dfrac{a_i/(a_i + b_i) \times N_i}{N_i}$

where N_i = number of victims in the i-th designated area; a_i = quantity of pollution discharged by mobile sources (or more precisely, the level of contribution to monitored environmental pollution) in the i-th area; and b_i = the quantity of pollution derived from stationary sources (or similarly, the level of contribution to monitored environmental pollution) in the i-th area. See Kōgai Kenkō Higai Hōshō Seido no Mondaiten ni Kansuru Sangyōkai no Iken (Opinion of the Industrial Community of the Pollution Health Damage Compensation System) (May 19, 1975).

154. These officials attribute the companies' financial difficulties not simply to the 1973 Act, but also to other unrelated factors. See sec. 3.2.4 for discussion of the present plight of Chisso Company and for a discussion of the deterrent effect of the levy.

155. See 1977 OECD report (n. 116 supra).

156. Twenty-seven Class 2 grievances were initiated by Minamata disease victims. See Quality of the Environment in Japan (Environment Agency, 1977).

157. Article 111 places the Grievance Board under the jurisdiction of the Environment Agency.

158. For example, by 1979 nine Class 2 grievances had been withdrawn and six had been dismissed.

159. A total of 7,039 victims in Kumamoto, Kagoshima, and Niigata Prefectures were reported to be seeking designation at the time of litigation, but only 3,454 cases had been processed. Of these, 1,666 were certified and 877 were still pending. (Mainichi Daily News, Dec. 16, 1976.)

160. Reportedly, Shintarō Ishihara promised that the Environment Agency with the help of doctors from national hospitals and universities would examine one hundred and fifty persons per month, a rate eight times the previous processing speed. Further, in order to deal with borderline cases, Ishihara announced that a team of specialists would be engaged. He stated that the agency would commence administrative guidance of Chisso to assure the flow of compensation payments. Finally, Ishihara noted that the agency might solicit contributions from other corporate sources, and establish a re-serve fund of ¥48 million to assist the prefectural government. (Japan Times, July 2, 1977.)

161. See Quality of the Environment in Japan (n. 156 supra). Although the inadequacy of facilities is not mentioned, the authors learned of the present situation from government officials.

162. Reportedly on the occasion of Sawada's apology, victim protestors demanded that Kumamoto Prefecture not appeal the court's ruling (Japan Times, Dec. 17, 1976; Mainichi Daily News, Dec. 27, 1977). But on December 15, 1978, another group of Minamata victims filed a second suit, seeking $247,000 from the prefectural government for its failure to expedite certification.

163. On March 28, 1979, the Kumamoto District Court recognized the claims of 12 of 14 plaintiffs who had previously been denied certification. The court ruled that Chisso had to pay these victims approximately $750,000. The case is significant because it shows how the courts are clarifying administrative criteria under the act (Int. Environmental Reporter, Apr. 11, 1979).

164. Agency officials at the time reportedly noted: "If we were to ask for police assistance to remove the patients seeking help, that would be tantamount to 'killing' the agency by ourselves. We would like to avoid such an action" (Mainichi Daily News, Mar. 18, 1978). Ultimately, however, several victims were arrested. The sit-in was not the first conflict Environment Agency officials had had with Minamata and other victims. In 1977, 104 organizations of Minamata and other disease victims had called for the resignation of Director General Shintarō Ishihara over Ishihara's televised statement that the revision of the Basic Law in 1970 was an "act instigated by witchcraft." (Mainichi Daily News, Nov. 13, 1977.) The beleaguered Ishihara resigned in November 1977.

165. Daily crude steel output capacity of this plant was reported to be 10,000 tons. (Mainichi Daily News 12, April 19, 1978.)

166. The plaintiffs included seven children and the families of three persons who allegedly had died from air pollution. (Mainichi Daily News, Apr. 21, 1978.)

167. The companies included the Kansai

Electric Power Company, Sumitomo Metal Industries, Nippon Glass Co., and Godo Steel Corporation.

168. The plaintiffs claimed that the concentrations of SO_x, NO_x, and particulates were 1.5–6 times higher than the standards permitted. (Mainichi Daily News, Apr. 22, 1978.) See also the plaintiffs' brief, entitled Osaka Nishi Yodogawa Yūgai Busshitsu Haishitsu Kisei to Seikyū Yōken (Conditions Demanded Regarding Regulation of Osaka Nishi Yodogawa's Harmful Discharged Substances), April 20, 1978. The reader should compare the Osaka litigation with the Yokkaichi case (ch. 3) to examine how the Osaka litigation, in some sense, used the concepts of area designation and victim certification to extend the rationale of the Yokkaichi court.

169. See ch. 7 for the impact of a similar attitude on the implementation of the Law for the Resolution of Pollution-Related Disputes.

170. The act of attacking the compensation system in the courts would undoubtedly be considered even more "inhumane" than industry's present more discrete lobbying efforts.

171. Article 68.

172. Based on interviews of the responsible officials.

173. The situation is somewhat similar to industry's attitude toward tax fraud, although the social consequences of failing to pay the levy may be greater. See the discussion of administrative guidance in ch. 5.

174. Government officials apparently interpret the compact with industry to require the large industrial associations to act as a financial guarantor when individual firms fail to pay the levy. This point, however, is somewhat obscure.

175. Article 46.

176. Some minor services, like providing patients with air purifiers, summer camp outings for children, and small grants to victims for travel, have been made. The government's Minamata Disease Center has recently been established.

177. Designation was used as a negotiating point during the Environment Agency's discussions with steel manufacturers over pollution control in Kawasaki.

178. That neither the government nor industry has undertaken such a study is interesting, and attests to the general view that the system is primarily viewed as a vehicle for compensation, not regulation.

179. This is particularly true in the steel and electric power industries that are among the industries most stringently controlled.

180. Such as the Pollution Control Public Works Cost Allocation Law, various emission standards, or pollution control agreements.

181. Industry officials indicated that firms would seek to install pollution control equipment after the ¥700/m³ point irrespective of market structure. One complicating factor is that some company officials have apparently been concerned about the morality of firms avoiding paying the levy. "Would this not reduce the amount of money available for compensation?" "Is it appropriate for firms to act individually before the society as a whole has addressed how to finance the deficit?" they ask. One should not underestimate the increasing social conscience of some companies. For example, Kōbe Steel has reportedly volunteered to pay 70% of the cost of treating the city's approximately 5,000 air pollution victims. The company's annual allocation may amount to ¥350 million. (Mainichi Daily News, Sept. 3, 1976.)

182. Japan Times 2, June 8, 1978. The International Environmental Reporter 214 (1978) notes that as much as $67,000–$75,000 has been paid to each patient.

183. Daily Yomiuri, June 17, 1978. The firm's compensation payments for the 1978 business year totaled ¥6.3 billion, accounting for 72% of the ¥8.8 billion loss the company reportedly suffered.

184. Japan Times, June 8, 1978.

185. Asahi Evening News, June 3, 1978.

186. See rule 2.4 of the Tokyo Stock Exchange. Delisting of security, however, would not prevent its being sold "over the counter."

187. See Cabinet Resolution, June 20, 1978, Concerning Minamata Disease Countermeasures. Kumamoto's own screening office will also remain open. Int. Environmental Reporter 214 (1978). See also Sadao Togashi, Minamata-Byō Meguru Saikin

no *Jōkyō (Recent Situation Surrounding Minamata Disease)* 673 Jurisuto 22 (1978).

188. The government has assisted Chisso and its affiliates in the past. For example, the Japan Development Bank reportedly lent Chisso ¥2.3 billion in 1975 to assist the company in its many financial troubles. The government also assisted in the reconstruction of the Chisso Petrochemical Corporation, a 100% subsidiary of Chisso, after an explosion at the petrochemical works. One report stated the loan to be in the amount of ¥3,900 million (then about $13 million). (Daily Yomiuri 1, Jan. 26, 1975.)

189. Japan Times, June 8, 1978. At the same time that the government announced its assistance for Chisso, it also announced that it planned to succor Sasebo Heavy Industries (SSK), the nation's eighth largest shipbuilding company. Sasebo, like many other companies in this industry, is in the midst of a shipbuilding slump (as many as 26 firms reportedly collapsed between August 1977 and April 1978) and is threatened with bankruptcy. The government has proposed to assist SSK in making ¥8.3 billion in payments to former workers who voluntarily retired to help the company cut back on personnel costs and in meeting other obligations. The government's assistance to SSK, that employs many workers in Sasebo, is reportedly in exchange for the local Sasebo government's agreement to permit the entry into the port of the controversial nuclear-powered submarine, Mutsu. (Japan Times, June 1978.)

190. Michio Nishihara, *Chisso Kyūsai to Osensha Futan Gensoku (Financial Assistance to Chisso and the "Polluter Pays" Principle)* 673 Jurisuto 17 (1978).

191. Based on interview of Teruo Kawamoto, one of the leaders of the Minamata victims, June, 1978.

191a. Based on discussions with Environment Agency officials.

192. Areas in Nagoya and Higashi Osaka.

193. Based on interviews of industry and government officials. In one noteworthy development the Japanese government persuaded Nippon Kōgyō Company to pay ¥2 million to each patient suffering arsenic poisoning in Sasagatani, Shimane Prefecture, despite the fact that the company

discontinued operations well before the enactment of the compensation system.

194. See, e.g., S. 1480 (1980). See also the hearings on Kepone Contamination, Senate Subcommittee on Agriculture Research and General Legislation of the Committee on Agriculture and Forestry, on Kepone Contamination, held at Hopewell, Virginia, Jan. 22–27, 1976. That we are on the verge of an awakening is clear from the increase in newspaper reports (e.g., *The Chemicals Around Us*, Newsweek, Aug. 21, 1978); warnings such as a special HEW film on asbestos (Honolulu Star Bulletin, Aug. 9, 1978); and the current congressional hearings on PCB poisoning (S. 1531). See also the proposed administrative toxic substance compensation system (H.R. 9616) introduced by Congressman William Brodhead. For a review of these and other cases, see Stephen M. Soble, *A Proposal for the Administrative Compensation of Victims of Toxic Substance Pollution: A Model Act*, 14 Harv. J. of Legislation 683 (1977); Joseph Page and Mary O'Brien, Bitter Wages (1972); and Paul Brodeur, Expendable Americans (1973). An important, more recent controversy concerns the reportedly abnormally high incidence of cancer in the vicinity of areas of high radioactive fallout around Salt Lake City, Utah.

195. The reader may ask why a problem of the magnitude of toxic-substance poisoning has been so long ignored in the United States. The simplest explanation may lie in the perception of the "problem." Many say that the United States has not encountered a disorder on the order of Minamata, and it is true that the apparent absence of bodies belies a sense of crisis. Yet, because of the long latency period, many toxic substances encourage our false feeling of security.

A second influence may be that Americans, unlike the Japanese, do not share a collective experience of victimization from an atomic holocaust. In Japan the memory of this experience undoubtedly influences the society's perception of the problem and perhaps its response (see Robert J. Lifton, Death in Life (1967), a study of survivors of Hiroshima). That victimization due to the atomic bomb is still of concern is indicated by the discussion surrounding the Japanese Supreme Court's ruling that an Osaka-born Korean who suffered from the atomic bomb was also enti- tled to medical treatment under the Law for Atomic Victims (Mainichi Daily News, Mar. 31, 1978). Recently the Committee for the Problem of Human Rights and the Japan Federation of Bar Associations of the Hiroshima Bar Association have also called on the government to establish new laws for the aid of Hiroshima and Nagasaki victims (Daily Yomiuri, July 27, 1977). (An in- teresting parallel in the United States is Representative Norman Y. Mineta's support for a bill aimed at helping Hiroshima and Nagasaki victims now living in the United States.)

A third factor may be the apparent reluc- tance or inability of victims in the United States to use the political process more ef- fectively. Pollution victims, unlike blacks and other minorities, have not been able to form a mass movement that could translate individual grievances into a national inquiry into the maladies of the society as a whole. Indeed, the American civil rights movement has been curiously unsympathetic to the plight of pollution victims, many of whom are also poor, politically powerless, and so- cially disadvantaged. The position of labor unions on pollution control also reflects a bias for jobs over health. Without the aid of these powerful groups, the protests of pollu- tion victims have left successions of stolid administrators unpersuaded.

Fourth, in some respects, the bar, par- ticularly the environmental bar, shares the blame. Generally, environmental lawyers have not rallied to the aid of pollution vic- tims, because their primary attention at least in the early 1970s was devoted to re- forming the administrative process. Simi- larly poverty lawyers have been absorbed in defending their clients' "civil" rights. (Cf. the discussion in ch. 4 of how "environmen- tal" rights are being viewed as a part of civil rights in Japan.) As a result, the pollution injury field has yet to produce its own *Brown v. Board of Education* (349 U.S. 294, 75 S.Ct. 753, 99 L.Ed. 1083 (1955)) that cries to the nation, awake!

Legislative conservatism has also played a part. Somehow pollution prevention is split off from the main body of thought on com- pensation. Although the United States has developed workmen's compensation, and statutory schemes for pollution control,

there is little engineering of compensation within regulatory frameworks.

Finally, and perhaps most basically, American society still has not reached a consensus on the issue of health damage from pollution. For this reason difficult problems of scientific uncertainty appear to hamstring the society's decision to compen- sate. The process is circular. The victim's plight is left unpublished, private and public investment in medical research on pollution remain inadequate, and the administrators lack adequate data to inform their judg- ments. But without an incentive to acquire this data, the increasing evidence on chemi- cally caused health damage is considered by decision makers to be of less significance than it really is.

196. There appear three principal grounds to justify intervention in the United States. The first is essentially a moral argument. Tragedies from toxic substances, unlike natural catastrophes, are made by men. Those who release the chemicals to the en- vironment should bear the cost of rectifica- tion. The second rests on equity. Victims of toxic chemicals seem more entitled to relief than sufferers of natural disasters. Yet the latter receive substantial assistance under present programs like that provided in the Flood Disaster Protection Act of 1973. Similarly, many state workmen's compen- sation statutes cover employees for indus- trial diseases. But the poor man or woman without even the good fortune to hold a job receives nothing. Third, we should compen- sate because, as described, compensation strengthens existing regulatory controls.

Japan's rationale for compensating its vic- tims was first, political (it was a palliative to quell the national crisis of confidence), and, second, economic (industry considered compensation essential to spread the risks of liability). A welfare objective was of only tertiary concern.

197. See Marcia R. Gelpe and A. Dan Tar- lock, *The Uses of Scientific Information in Environmental Decision Making*, 48 South- ern California L.R. 371 (1974) for a general discussion of scientific and legal concepts of proof.

198. From the perspective of efficiency, too much investment in pollution control, of course, is as bad as too little. In either case, there is waste in the sense that resources

might be employed more productively elsewhere.

199. As the budget is depleted, it will become increasingly expensive to aid the additional victims from the pool of atypical cases.

200. For the same reasons given in n. 198 supra.

201. Japan follows the rule of comparative negligence.

202. But see *United States v. Reserve Mining Co.*, 380 F.Supp. 11 (D. Minn. 1974); *Reserve Mining Co. v. U.S.*, 498 F.2d 1073 (8th Cir. 1974); *Reserve Mining Co. v. EPA*, 514 F.2d 492 (8th Cir. 1975).

203. For a recent proposal to this effect, see Julian Gresser, Balancing Industrial Development with Environmental Protection in the Republic of Korea, A Report of Environmental Mission to the World Bank (World Bank, 1977).

204. But note that the incremental benefits from allocating these revenues to alternative uses may be higher.

205. Government, of course, can also explore other measures of adjusting the incidence of cost. One idea, unexplored in the United States, is making voluntary industry contributions to victim relief a charitable contribution. This might also be supplemented through an "incentive tax deduction." Another option would be for government to encourage insurance carriers to underwrite pollution victim insurance. Although industry has thus far been reluctant to assume the risks of liability, a carefully designed government program of reinsurance, as in other areas, might mollify its concerns.

206. P.L. 91-173; P.L. 82-203, now codified at 30 U.S.C. sec. 901 et seq.

207. But excluding coal mine constructors; see *National Independent Coal Operator's Association et al. v. Brennan*, 372 F.Supp. 16 (1974).

208. See 30 U.S.C. 934.

209. Criticisms of Regulations (No. 10 of SSA, 20 CFR Part 410, March 4, 1971, Part II), submitted by the Appalachian Research and Defense Fund, Inc., March 13, 1971, to John Dent.

210. See, e.g., *Turner Elkhorn Mining Co. v. Brennan*, 385 F.Supp. 424 (1974; *Usery v.*

Turner Elkhorn Mining Co., 428 U.S. 1, 96 Sup. Ct. 2882; and *National Independent Coal Operators Assn.* (n. 207 supra) for these and related challenges.

211. Special benefits for disabled coal miners totalled $1,014,925,000 in 1974; $967,782,000 in 1975; $986,271,000 (estimated) in 1976; and $913,897,000 (estimated) in 1977.

212. Based on interviews of officials responsible for overseeing the black lung program.

213. For a catalogue of reforms, see Report of the U.S. National Commission on State Workmen's Compensation Laws (Washington, D.C., GPO 1972).

214. See Mich. Const. of 1963, art. 5, sec. 18, requiring deficit in one year to become a charge against next year's budget, thus effectively prohibiting deficit spending.

215. H.R. 9616, sec. 8.

216. Id., sec. 14(b).

217. Of course a crisis might trigger a similar development. Professor Krier and others have suggested that our entire regulatory approach is predicated on a precipitous reaction to crisis. One such example in the pollution field is the disastrous smog at Donaora, Pennsylvania, in 1948, that sparked some of the first postwar efforts to control air pollution. See James E. Krier and Edmund Ursin, Pollution and Policy (1977).

218. See, e.g., *Cardinal v. University of Rochester*, 187 Misc. 519, 63 N.Y.W.2d 868 (1946).

219. See *Zahn v. International Paper Company*, 414 U.S. 291, 94 S.Ct. 505 (1973), which held that the $10,000 amount in controversy requirement be met by each member of the class in a suit based on diversity jurisdiction. See Hinds, *To Right Mass Wrongs: A Federal Consumer Class Action Act*, 14 Harv. J. Legislation 776 (1976).

220. This was precisely the barrier that the plaintiffs overcame in the Yokkaichi suit. It is not probable that a U.S. court would reach the same result.

221. See *Toxic Substance Control Act Amendments*, Hearings Before Subcommittee on Consumer Protection and Finance of the House Interstate and Foreign Commerce Committee, on H.R. 9616, S. 1531, Serial No. 95-150, March 7, April 26, July 25, 1978.

222. Many of these ideas are being debated

as this book goes to press. See S. 1480 (1980), which provides for liability and compensation to victims of hazardous substances and for the cost of cleanup of inactive hazardous-waste disposal sites. See also *Compensating Victims of Occupational Disease*, 93 Harv. L. Rev. 916 (March 1980).

Chapter 7

1. Kōgai Funsō Shori Hō (Law No. 108, June 1, 1970).

2. The use of mediation and other extrajudicial settlement techniques is a dominant institutional pattern in East Asia generally. See, e.g., Jerome Alan Cohen, *Chinese Mediation on the Eve of Modernization*, 54 California L. Rev. 1201 (1966); F. S. C. Northrop, *The Mediational Approval Theory of Law in American Legal Realism*, 44 Virginia L. Rev. 347 (1958).

3. Richard Beardsley, Village Japan (1959). This excerpt is instructive:

Mediation is the manner in which Japanese farmers have settled their disputes for centuries. Still the approved means of settlement, it derogates from the use of the court system. Also, being completely local in operation, it is a means of reinforcing the community against outsiders and thus contributes to community cohesiveness and self-containedness.

Various more practical considerations sustain and reinforce the predisposition toward local mediation. Historically the farmer's experience with the formal institutions of Tokugawa justice probably did not breed confidence in the courts. There are similar deterrents today. To go to court one must employ a lawyer, and the cost of lawyers is prohibitively high. Lawyers are commonly regarded as interested primarily in maximizing their fees and are generally suspected of prolonging and complicating cases to this end. The financial risks of bringing a civil suit are considerable, particularly if the judgment goes against one. And, anyway, the humble farmer simply feels ill-at-ease and out-of-place in the formal and official atmosphere of a courtroom. All of these are circumstances hardly conducive to the free or effective use by the villager of judicial process as a means of settling his disputes.

Of the considerable number of persons with whom we discussed such matters, many summed up their views by quoting the old adage "chūsai wa toki no ujigami" ("mediation is the gold of the times"). Despite such feelings, however, it was generally held that as a last resort for the solution of their most desperate problems they might with great reluctance and trepidation go to court. When pressed as to the possible nature of such problems, most could conceive of little that could induce them to take so desperate a measure, save the most vital questions affecting the boundaries of agricultural land or irrigation rights.

4. Takeyoshi Kawashima, *Dispute Resolution in Contemporary Japan*, from Arthur T. Von Mehren, editor, Law in Japan 41 (1963). Kawashima describes the livelihood of the jidan-ya as follows:

This emphasis on compromise has produced its own abuses. A special profession, the jidan-ya or makers of compromises, has arisen, particularly in the large cities. Hired by people having difficulty collecting debts, these bill collectors compel payment by intimidation, frequently by violence. This is of course a criminal offense, and prosecution of the jidan-ya is reported from time to time in the newspapers. But their occupation is apparently flourishing. Furthermore, public opinion seems to be favorable or at least neutral concerning this practice; even intimidated debtors thus compelled to pay seem to acquiesce easily and do not indicate strong opposition. This attitude is doubtlessly due to some extent to the delay and expense of litigation, but at the same time the traditional frame of mind regarding extrajudicial means of dispute resolution undoubtedly has had some influence on public opinion toward the jidan-ya. The common use of the term jidan-ya seems to suggest that extrajudicial coercion and compromise are not distinctly differentiated in the minds of the people.

In the same work Kawashima also notes the infrequent resort to arbitration in Japan:

It is characteristic of Japanese culture that arbitration has been a kind of reconcilement. For this very reason, arbitration in the sense of contemporary Western law is alien to Japan. Despite the fact that the Code of Civil Procedure contains provisions for an arbitration procedure, it is seldom used. Clauses specifying that a dispute arising out of a contract shall be settled through arbitration are normally not employed except in agreements with foreign business firms.

5. Walter Ames, Police and Community in Japan (University of California Press, forthcoming). Ames makes the following observation on the role of the police:

People with problems come to police boxes to talk to the police officers, or if the problem is more serious, they go to the police station and talk to policemen in the Crime Prevention Section. The problems vary: the Kurashiki Crime Prevention Section recorded twenty-four cases of formal counselling in 1974 ranging from marital to money management and apartment rental problems. The solutions suggested by the police are non-binding and the police sometimes act as intermediaries between the parties in disputes. Instances of informal counselling with police officers in police boxes are much more frequent, though seldom recorded. For instance, I was in a police box in Kurashiki when a woman brought in a chain letter that threatened death to the receiver should he or she break the chain by not sending out a large number of similar letters to acquaintances. The woman said she would feel better if she gave it to the old police sergeant; he told her to destroy any more that might come and to sprinkle salt around her house for good luck. Sometimes police officers complain that they spend too much time counselling and performing other service functions, and that they would rather be doing "real police work," such as arresting criminals.

6. Dan Fenno Henderson, Conciliation and Japanese Law (2 volumes, 1965).

7. Environment Agency White Paper (1976).

8. For example, in 1960, the monthly average number of articles on pollution was 14; in 1971, 124. In 1960, the percentage of news space devoted to this subject was 0.4%; in 1972, 2.8% (Quality of the Environment in Japan, Environment Agency, 1972).

9. A system of extrajudicial dispute settlement in fact appealed to many different groups. The opposition parties viewed it as a more effective means of dealing with pollution problems; the victims thought it could provide quicker relief than the courts, and industry felt it might undercut the momentum of the nascent citizens' movement.

10. The history of the revision of the Mining Law followed a similar pattern. Initially the government wanted to solve the pollution problem by adopting the principle of no-fault liability, but the government's plan was frustrated by attacks from industry. Mediation was incorporated as a last resort to fend off industrial opposition.

11. Article 21.1.

12. For example, during the period 1958–1968, only 36 applicants for mediation were reported under the Water Quality Conservation Law. Mediation under the Mining Law, however, was more frequent (about 60–70 disputes). One commentator notes at the time that the principal reason these extrajudicial procedures were not employed more often was that most people were unaware of their existence. See Ichiro Katō, et al., *Kenkyūkai: Kōgai no Funso Shori to Higaisha Kyūsai* (Research Group—Pollution Disputes and Victim Relief), 408 Jurisuto 14 (Oct. 15, 1968).

13. See *Kenkyūkai* (n. 12 supra), and also ch. 3.

14. At this point, the reader might consider Northrop's analysis of mediation theory. He argues that mediation makes "the dispute, rather than the (external) legal rule by which similar disputes are to be resolved, the elemental notion in legal science." Under this theory, litigation in terms of positive legal rules or commands from "without" becomes "a minor, special case." "It gives up any attempt to objectify or to externalize the 'good' or 'just' solution that holds prescriptively for any larger group in society than the disputants." He continues, "the holder of the mediational approval theory of law . . . concludes that the point of all law [i.e., settling a particular dispute] . . . is more likely to be achieved efficiently by restricting the disputes to the disputants than by extending it to a whole society in the hope of finding a universal determinate, prescriptive legal sentence of which all, or even a majority, approve. . . ." "Since its method requires that the dispute is to be resolved by the disputants themselves, finding a middle way that both approve, their specification of the solution automatically provides the sanction. . . . By restricting the criterion of justice to the local disputants treating each case as particular and unique, justice is done to the people's approval and disapproval." In

short, "the law escapes the conflict between the political sovereign's command and the approval of minority groups and dissenters" (Northrop, n. 2 supra).

We invite the reader to consider, perhaps after reviewing ch. 2, whether the Dispute Law was not also motivated by this antimajoritarian fear that the courts might someday function as an independent, external, prescriptive source of authority. John O. Haley offers a similar speculation about the motivation behind the conciliation provisions in the 1939 Mining Law, the 1938 Agricultural Land Adjustment Law, the 1939 Personal Status Conciliation Law, and the 1942 Special Wartime Affairs Law. See John O. Haley, *The Myth of the Reluctant Litigant*, 4 J. Japanese Studies 359 (1978).

15. A major part of the debate centered on the National Administrative Organization Law (Kokka Gyōsei Soshiki Hō) (1948, Law No. 120). Article 3 of the Organization Law stipulates that certain organizations recognized by this law would enjoy independent rule-making powers. An example of this kind of organization is the Fair Trade Commission (Kōsei Torihiki Iinkai). Article 8 of the same law notes that article-3 organizations may also possess "affiliated organizations." These provisions have been interpreted to mean that affiliated organizations may not act independently in their own name. This technical debate related to the Dispute Law in that prior to 1972 the "Central Committee" was considered an affiliated (hence non-independent) organization. With the 1972 amendment the committee was given article-3 status. At present, while lodged within the Prime Minister's Office, the Central Committee is essentially an independent administrative body.

16. The Dispute Law was again amended in 1974. The most recent version in part is included in app. 2.

17. See Walter Gelhorn, Ombudsman and Others (1966) for a discussion of the function of the counselor system in other contexts.

18. See art. 49 of the Dispute Law in app. 2.

19. Id.

20. Article 14. These terms are discussed in sec. 1.2.2. The practice of mediation, etc., is very similar at the local and central levels.

21. This limitation has often been questioned since there would seem no reason to exclude this procedure in view of the careful selection of the Review Board's members.

22. Article 20.

23. Article 17(1).

24. Id.

25. Kōgai tō Chōei Iinkai Setchi Hō (1972, Law No. 52).

26. See Article 33. Compare art. 312 of the Civil Procedure Code, the Duty to Present Records (Bunsho Teishitsu Gimu).

27. The Dispute Law also only includes pollution related disputes as defined in art. 2 of the Basic Law.

28. Typical examples in this category are cases of noise pollution involving airplanes or the bullet train.

29. Actually in many cases the distinguishing feature between mediation, conciliation, and arbitration is not the formality of the proceeding, but its legal effect. At times, however, as stated, mediation is less formal in that it can be performed by any number of mediators, whereas the Dispute Law requires three persons in a conciliation proceeding.

30. For a similar description of the role of a "facilitator" in the United States see Donald B. Straus, *Mediating Environment, Energy, and Economic Trade-Offs*, 32 Arbitration J. 96 (1977).

31. The same is true for conciliation, arbitration, and other proceedings.

32. Article 695 states:

(Compromise) A compromise becomes effective when the parties have agreed to terminate a dispute between them by mutual concessions.

Article 696 states:

(Effect) If, in cases where it has, by a compromise, been admitted that one of the parties possesses the right constituting the object of a dispute or that the other party does not possess such right, it has afterwards been established that the former party did not possess the right or that the other party did possess the same, such right shall be treated as having by virtue of the compromise been transferred to the former party or extinguished as the case may be.

33. Under some conditions a court will strike down a mediated agreement as vio-

lative of public policy. For one case in point, see the court's disposition of the mimaikin contract in the Minamata case in ch. 3.

34. Article 27.3.

35. Article 31.

36. Article 33.

37. Article 34. The reader should refer to case 2 (sec. 2.3) for a specific example of the operation of this provision.

38. Conciliated agreements are also governed by art. 695 of the Civil Code. Conciliation under the Dispute Law should also be distinguished from judicial conciliation, for a conciliation settlement conducted under judicial auspices has the identical effect of a court's final judgment. This procedure is discussed by Henderson at great length (n. 6 supra). The effect of a judicial conciliation settlement is also the same as that of an arbitration award.

39. See n. 4 supra.

40. The legal effect of arbitration is governed by art. 800 of the Code of Civil Procedure. *(Effect of award) An award shall have the same effect as a judgment which is final and conclusive between the parties.*

41. Quasi-arbitral determinations are binding (in this sense like an arbitration finding) unless the parties object within a defined time period.

42. Because Japan does not have a jury system, questions of fact are determined by court.

43. Article 42.20.

44. For example, a subcommittee finding on causation would not necessarily create a rebuttable presumption on this issue.

45. For one comparison, see J. Cohen, n. 2 supra.

46. See app. 2.

47. Id.

48. Judicial conciliation (minji chōtei) procedures have also been harmonized to deal with pollution disputes. Article 33.3 of the Law Amending in Part the Civil Conciliation Law (Minji Chōtei Hō Kajishinpan Hō no Ichibuo Kaisei Suru Hōritsu), 1974, Law No. 55, May 24, 1974, states:
Conciliation of Pollution Disputes . . . *(Jurisdiction of Cases of Conciliation of Pollution Disputes, etc.) Cases of conciliation of disputes over pollution or involving harm resulting from an invasion of interests in sunlight or wind, etc., in addition to Article 3 courts, shall be under the jurisdiction of a summary court having jurisdiction over the place of the harm, or the place where the harm may have occurred.*

This provision expanded the traditional basis for venue (i.e., where the wrong was committed or at the head office of the defendant) to facilitate conciliation proceedings for the parties.

49. The text discusses briefly a conciliation proceeding, the mode of dispute settlement currently most favored. In part, this section is an adaptation of an excerpt from Henderson (n. 6 supra) 220–22. Although Henderson's account deals with judicial chōtei, the administrative conciliation proceeding with which we are concerned is sufficiently similar to permit comparison. In practice the key difference between judicial and administrative conciliation is the extent to which administrative conciliators become involved in on-site inspections and the conciliator's statutory authority to contract with experts to perform research. Although this is permitted under judicial conciliation (art. 8, Civil Conciliation Law), it is much less used.

50. Article 11 of Kōgai to Chōsei Iinkai Setchi Hō (Basic Law for the Establishment of a Pollution Settlement Committee) (1972, Law No. 52).

51. Use of court decisions as a basis for a conciliated settlement has proved particularly useful in Minamata-related disputes. See case 4 (sec. 2.5).

52. For an annotated compendium of pollution cases involving conciliation, see Factual Research Report on Compensation for Property Damage from Koñbinātos in 1975 (Shōwa Gojū Nendo Konbināto Kōgai ni Yoru Butteki Higaihoshō ni Kansuru Jittai Chōsa Hōkokusho). An annotated discussion of pollution related complaints is in Kōgai Kujō Shori Jireishū (Compendium of Settled Pollution Complaints) (Pollution Coordination Committee Office, March 1977). See also Naoyuki Kuroda et al., *Kōgai to Chōsei Iinkai ni Okeru Kōgai Funsō Shori (Pollution Disputes Within the Pollution Settlement Committee)*, 836 Hanrei Jiho 13.

53. Cases 1 and 2 and some of the comments thereto are adapted from a report by Kōzō Fujita in 4 Kankyō Hō Kenkyū (Environ-

mental L.J.) 337 (Nov. 1975). Fujita, a judge by profession, served for a while on the Central Committee.

54. This comment is based principally on K. Fujita's (n. 53 supra) analysis.

55. This comment supplements K. Fujita's (n. 53 supra) analysis.

56. Nōyōchi no Dojō no Osen Bōshi tō ni Kansuru Hōritsu (1970, Law No. 139).

57. The reader might compare the payment agreed to under conciliation with the payments established by the court; also with the payments based on direct negotiations, and the settlement reached under the original mimaikin contract.

58. Sections 2.6 and 2.7 are adapted from Hiroshi Takeda's analysis in Kankyō Hō Kenkyū (Environmental L.J.) 270 (no. 7 Mar. 1977).

59. See Pollution Control Public Works Cost Allocation Law.

60. For example, on September 28, 1978, the central, local, and municipal governments and various business groups created a subsidy program to defray the costs of interest payments on operating funds borrowed by fishermen, processors, and distributors suffering losses from mercury and PCB poisoning. Also, on October 1, 1974, the central and local governments established special mutual insurance for red tide damages to fisheries. The program covers mutual insurance premiums. Similarly, on April 1, 1975, the central and local governments, various fishery organizations, and business groups set up an oil spill fishery damage relief fund. Benefits are paid for relief of damages and for clean-up. See Quality of the Environment in Japan (n. 8 supra) 72.

60a. WECPNL is a unit that takes into account the average probable durations of noises as well as their instantaneous intensities. Thus, in this case, its computation involves airport traffic at various times of day, types of aircraft flown at various times, landing and takeoff patterns, etc. For details of the computation see Environmental Policy of Japan 181 (Government of Japan, Nov. 1976).

61. See Environment Agency Directive No. 154, December 20, 1973.

62. According to 1974 statistics, 68% of all cases involved administrative guidance.

63. See Environment Agency White Paper (1975).

64. For example, the Tokyo Metropolitan Government appoints officials in its Pollution Control Bureau to the position of counselors.

65. We cannot, of course, conclude that parties have always been satisfied with their agreements. Also, we do not possess statistics on how many settled disputes have recurred.

66. The Central Committee has begun investigating the causes of victims' unwillingness to use the committee.

67. Victims note that committee members are appointed by the prime minister; Central Committee members reply that the Diet must consent to these appointments. But see ch. 5 for the pro forma nature of Diet concurrence. The reader should refer to ch. 5 for a discussion of the composition of the Central Committee that lends further insight into why many victims are reluctant to have it settle their disputes.

68. To the Central Committee, these victims' actions appear simply "radical" or "political." We suggest that Northrop's analysis (n. 2 supra), however, may provide a deeper insight. The victims want an external prescriptive solution—a rule, not a deal.

69. As noted, there have been only a few Article 34 proceedings.

70. One factor responsible for this may be Cabinet Order 139, promulgated in 1974, that restricts the jurisdiction of the Central Committee to major disputes involving Minamata or itai-itai disease and air and water pollution diseases. Disputes over other injuries are excluded; the committee would need special authorization by Cabinet Order to assert jurisdiction over these disputes.

71. To some extent, conciliation has replaced some of the purposes of the class action. See ch. 3 for a discussion of limitations of "representative suits" under the Code of Civil Procedure.

72. The issue of control over information relates of course to the neutrality of the conciliators. Residents have generally felt that local conciliators' managing of information has been fairer, but this may simply represent their greater confidence in local processes.

73. The reader might consider the theoretical implications of the Dispute Law in light of Ronald Coase's *The Problem of Social Cost*, 3 J. L. and Economics 1 (1960). Coase argues that if bargaining between polluters and victims could occur without costs (i.e., the so-called "transaction" costs of gathering information, costs of facilitating a bargain, costs of implementing it, etc.) such bargaining would lead to an optimal allocation of resources. He illustrates his thesis at one point by an example of a farmer and a cattle rancher who occupy adjacent property. There is a conflict in the use of a resource, land, since the cattle frequently run over and destroy the farmer's crops. Coase argues that it is in the interest of both the farmer and the rancher to bargain.

The parties have several options. If the net profit from the sale of crops is greater than the cost of erecting a fence, or the cost of bribing the rancher to control his cattle, the farmer has an incentive to adopt these latter measures. Conversely, if cattle ranching is very lucrative, the rancher rather than cutting down on his stock might address the farmer's complaint by erecting the fence at his own expense or compensating the farmer in various ways. The parties will explore these and other options and bargain to an "equilibrium solution" of their controversy. An equilibrium point is optimally efficient since neither party can improve its own position through adopting further preventative, compensatory, or avoidance measures without disadvantaging his opponent. Coase demonstrates that even where a legal rule imposes liability on one of the parties, the parties will seek to improve their positions by continuing to bargain.

74. (Implementation Order, art. 17). For example, a court trial involving a $50,000 case might involve filing fees of $240.00. Conciliation of a pollution case involving the same disputed amount would cost about $30.00.

75. For a discussion of the Civil Liberties Commissioners system, see W. Gelhorn, Ombudsmen and Others (1965).

76. The administrative costs of the Environmental Dispute Coordination Commission are not very great. In FY 1976, they were ¥292 million and in FY 1977, ¥300 million (about $1 million). (5 Japan Environmental Summary, no. 6, June 10,

1978). In determining the system's yearly budget, the most complicated issue is the estimation of monies required for investigation purposes (chōsahi) (the appropriation for salary and staff do not generally vary). Generally, the past budget serves as a benchmark for future appropriations. Thus, for 1978, the budget for investigatory purposes was about ¥20 million ($100,000). Should these funds prove insufficient, the Central Committee can obtain short-term supplementary financing relatively easily.

77. In the following discussion, we use "mediatory" to refer to all forms of extrajudicial dispute settlement, not simply mediation.

78. See Robert E. Lutz, *The Laws of Environment Management: A Comparative Study*, 24 American J. Comparative L. 447 (1976). Although Professor Lutz's review refers principally to industrialized societies, extrajudicial dispute settlement techniques seem equally useful for many developing countries confronting environmental problems. The Republic of Korea is an example of one country that is considering modeling its mediation system after Japan's Dispute Law. See Gresser, World Bank Report, n. 129, ch. 1 supra.

79. Interest in applying mediatory procedures to environmental disputes has recently increased in the United States. See *Removing the Rancor from Tough Disputes*, Business Week (Aug. 30, 1976); Gerald W. Cormick, *Mediating Environmental Controversies: Perspectives and First Experience*, 2 Earth L.J. 215 (1976). For an analytical review of the state of the art of mediation in the United States, see Lawrence Susskind and Alan Weinstein, Toward a Theory of Environmental Dispute Resolution—A Working Paper, Department of Urban Studies, MIT (1980).

80. See, e.g., Frank Sander, *Varieties of Dispute Processing*, Pound Conference, 70 F.R.D. 79.

81. 42 U.S.C. sec. 2000(a)3.

82. Based on interviews of service officials in Boston.

83. We do not suggest that environmental disputes are not acrimonious in the United States. As the 1977 Seabrook nuclear power plant case attests, the protests of American environmentalists are also at times violent.

Conflicts between victims of Kepones, PCBs, and other toxic substances and the companies allegedly responsible for these injuries have also been bitter.

84. Formerly Chief Judge of the Sendai Appellate Court. In Japan, judges are usually not lawyers.

85. Including the Former Vice-Minister of Okinawa and high ranking officials in the Prime Minister's Office and the Legal Advisor's Office (Hōseikyoku).

86. The salary of the Chairman of the Central Committee is about $65,000 and investigators make between $40,000 and $45,000.

87. Henderson notes that the Japanese government has expended considerable effort through the courts and administrative offices to educate conciliators; the government sponsors the lectures and meetings of conciliators. Although conciliators receive only nominal fees (travel and hotel expenses, plus a daily stipend), many people are anxious to serve because of the prestige that goes with the position; apparently this is particularly the case in the rural areas, where persons with the greatest prestige in the community frequently serve as conciliators. It has recently been charged that conciliators are too often chosen from certain social strata and that their background colors the operation of chōtei in favor of traditional institutions (Henderson n. 6 supra).

88. Cormick, n. 79 supra.

89. However, Japanese law schools also do not emphasize training in mediatory skills.

90. Cormick, n. 79 supra.

91. At the same time, of course, the reasons for the difficulties of the Central Committee can also be better understood by the fact that they are all high-ranking officials and thus distrusted by many victims.

92. For example, Michigan's Environmental Protection Act (1970), art. 14.528(203) states that the court may "appoint a master or referee, who shall be a disinterested person and technically qualified, to take testimony and make a record and a report of his findings to the court in the action." The Federal Environmental Pesticide Control Act of 1972 (7 U.S.C. sec. 135 et seq. as amended) also employs an administrative court to resolve many disputes that might otherwise go directly to a regular court.

93. See the discussion in ch. 6 of Judge McBridge's creative role in the Kepone case.

94. Some judges may be reluctant to enter too directly into mediation for the fear of tying their hands, or revealing bias. But see ch. 4. Perhaps a good analogy to a non-judicial mediator attached to the courts is the social worker in family law cases.

95. See sec. 9 of the Federal Water Pollution Control Act of 1972, P.L. 92-500, 86 Stat. 816.

96. For example, it was suggested that environmental problems were too complex, technical, and novel for ordinary courts.

97. It was argued that although environmental cases constituted only a small percentage of court dockets, these cases often consumed inordinate time and otherwise overtaxed the system.

98. Environmental court advocates urged that the present system permits inconsistent interpretations of federal statutes such as the National Environmental Policy Act of 1969 (NEPA); and since these statutes embraced the entire federal government's decision-making process, there was an overriding need for consistency of interpretation. A specialized court, it was felt, would meet this need.

99. It was urged that to the extent that a system of environmental courts could devise a rational and consistent interpretation of NEPA and other federal statutes, agency and judicial delays could be reduced. For a complete discussion of these arguments, see Scott C. Whitney, *The Case for Creating a Special Environmental Court System*, 14 William and Mary L. Rev. 473 (spring 1973).

100. For a discussion of the many problems that befell the environmental court proposal, see the note, *Attorney General's Report Rejects Establishing an Environmental Court*, 4 Environmental L. Reporter 10019 (1974).

Chapter 8

1. In this chapter, for example, we do not discuss present U.S.–Japan conflicts relating to pollution control at American bases (n. 79, ch. 4 supra, refers to a discussion of the Yokota base case and related controversies); nor do we analyze some of the problems for

international trade posed by Japan's "higher" environmental standards.

2. This first section is based on Julian Gresser, *Japan's Handling of International Environmental Problems: Contradictions With the Domestic Record*, Proceedings of the 71st Annual Meeting of the American Society of International Law (April 21–23, 1977).

3. The one possible exception has been Japan's opposition to atmospheric nuclear tests. For a discussion of Japan's contribution in the formulation of principle 26 at the Stockholm Conference of 1972 see Louis B. Sohn, *The Stockholm Declaration on the Human Environment*, 14 Harv. Int. L. J. 423 (1973).

4. Paris Convention, February 21, 1974.

5. 93rd Cong. 1st sess., Senate Executive F (November 23, 1972).

6. 93rd Cong., 1st sess., Senate Executive H (March 3, 1973).

7. 92nd Cong., 2d sess., Senate Executive K (December 18, 1971).

8. 93rd Cong., 1st sess., Senate Executive C (December 29, 1972).

9. Note, however, that Japan wished to dump both chemicals in solidified concrete at great depths.

10. See sec. 1.2.2.

11. As of 1979, the Japanese government did *not* impose pollution controls on the fluorocarbon or fluorocarbon aerosol industries. (Of course there are other regulations pertaining to factory sitings, operations, etc., but these are not intended to control, nor do they directly bear on, the environmental effects of fluorocarbon aerosols). The manufacture, processing, or distribution of these substances in commerce was not prohibited; no warnings or special instructions for their use, distribution, or disposal had been issued. Moreover, government officials indicated (in a 1979 interview with one of the authors) that the government did not intend to establish controls for this industry. The environmental issues relating to the manufacture, use, or disposal of fluorocarbon aerosols had scarcely been addressed by the media, environmental or consumer groups, the bar associations, the scientific community, or the opposition parties. That is to say, there had not been any outside stimulus for the government to do anything

about the problem. An interesting international comparison is "Comparative Legal Study of Regulations Banning and Restricting Fluorocarbons," Draft Report of the Natural Resources Defense Council, October 1, 1978.

12. Richard A. Frank, *Environmental Aspects of Deepsea Mining*, 15 Va. Int. L. J. 815 (1976).

13. See sec. 1.2.3.

14. A report by the United States–Japan Panel on Natural Parks (October 1976) notes that:

The observed mortality of sea birds per unit length of gill net, set during experimental fishery cruises in and adjacent to areas where the Japanese salmon fishery was active, was related to the total length of nets set by the Japanese salmon fishery. In the area north of 46°N and west of 175°W, the area known as the "mother ship fishery area," in which 359 catcher boats supported by 11 mother ships set an average of 2,900 miles of net daily during the ca. 65-day fishing season, an annual mortality of 75,000 to 250,000 birds has been estimated. The salmon catch of the mother ship fishery area comprises only one-third of the total Japanese gill net salmon catch. Assuming that salmon catch effort, bird densities, and species composition are similar through the North Pacific, the total estimated mortality of sea birds lies between 214,500 and 715,000 birds.

Moreover, Richard Bakkala of the National Marine Fisheries Service estimates in a paper entitled "Estimated Mortality of Seabirds, Fur Seals, and Porpoise in Japanese Salmon Drift Net Fisheries and Sea Lions in the Eastern Bering Sea Trawl Fishery" that

1. The kill of sea birds from salmon drift net fisheries of Japan in 1970 ranged from 214,000 to 763,000 (all species) depending upon estimation procedure.

2. It is estimated that from 3,150 to 3,750 fur seals are caught and killed annually in the mother ship salmon fishery.

3. An average of about 11,800 Dalls porpoise are killed annually in gill nets of the mother ship salmon fishery. Japanese scientists estimate a kill of about 10,000 animals.

15. See the Convention for the Protection of Migratory Birds and Birds in Danger of Extinction and Their Environment, March 4,

1972. Recently it has been reported that Japan will join the "Ramsar Treaty" for the protection of waterfowl. See Daily Yomiuri 2, Feb. 10, 1978.

16. See, for example, Takaharu Okuda, *Antipollution Movements Get Together to Oppose Japan's Overseas Economic Aggression*, 8 AMPO 10 (1976).

17. These organizations were: Save Palau Organization; Tia Belaund; Natural Resources Defense Council (NRDC); Environmental Defense Fund (EDF); The Sierra Club; Friends of the Earth; Friends of the Earth, International; The Nature Conservancy; National Audubon Society; The World Wildlife Fund-United States Appeal; Pacific Science Association; Micronesia Support Committee; National Wildlife Federation; International Union for the Conservation of Nature (IUCN).

18. Copies of the petition were at the same time submitted to the United Nations Trusteeship Council, the United Nations Environment Programme, and the U.S. president and secretaries of state and interior.

19. See *Temporary Foreign Residents Can Petition*, Japan Times 1, March 12, 1977.

19a. In a recent development, MITI and the Kagoshima prefectural government have agreed to abandon plans to build a CTS in Shibushi Bay because of opposition from environmentalists and local fishermen. See Int. Environmental Reporter 374, Nov. 10, 1978. An Okinawa court, however, has also dismissed a suit by residents against a CTS in Kin Bay (Int. Environmental Reporter, June 13, 1979).

20. The text summarizes the petitioners' arguments in "In the Matter of the Establishment of a Superindustrial Complex in Palau, Micronesia Before the Japan Diet (March 3, 1977)—Petition by the Save Palau Organization, the National Resources Defense Council, and Other Organizations for a Hearing Pursuant to Article 16 of the Japanese Constitution, Article 79 of the Diet Law, and the Petition Law." For a comprehensive discussion of the Palau case see *Palau Deepwater Port*, Hearing Before the Committee on Energy and Natural Resources, U.S. Senate, March 24, 1977, pub. no. 95-24.

21. The petitioners were not able to ascertain whether the Mitre report was prepared at the behest of Nisshō Iwai, Panero, or independently.

22. During this period Bechtel Corporation allegedly also prepared a report for Nisshō Iwai. At the time of the original petition, this report was kept secret.

23. See ch. 3.

24. K. Brower, *To Tempt a Pacific Eden, One Large Oily Apple: Proposed Superport*, 78 Audubon Magazine 56-68 (1976).

25. For some studies on the effects of oil pollution on plants, birds, fish, and marine mammals, see Petroleum in the Marine Environment (National Academy of Sciences, 1975); and Oil Spills in the Marine Environment (Energy Policy Project, Ford Foundation, 1974).

26. See ch. 5.

27. See ch. 5.

28. Suishitsu Odaku Bōshi Hō (1970, Law No. 138).

29. See ch. 5.

30. Kaiyō Osen Bōshi Hō (1970, Law No. 136).

31. See in particular the Yokkaichi decision in ch. 3.

32. See generally, L. B. Sohn, *The Stockholm Declaration on the Human Environment*, 14 Harvard Int. L. J. 3 (1973).

33. See, for example, Department of Transportation, Loop Deep Water Port License Application Final Environmental Impact Statement (1976); Potential On-Shore Effect of Deep Water Terminal-Related Industrial Development, 1975 CEQ Report prepared by Arthur D. Little, Inc.; and Baldwin and Baldwin, On-Shore Impacts of Off-Shore Oil Development (1975).

34. Donald McHenry, Micronesia: Trust Betrayed (1975).

35. In subsequent testimony before the Senate Committee on Energy and Natural Resources, one of the authors proposed the idea of joint hearings between the Diet and the Congress. See Testimony of Julian Gresser, "The Palau Superport from the Perspective of United States-Japan Relations," March 24, 1977.

36. The Petition Office relied on the following legal opinion of the Legal Department of the House of Representatives:
1. Inquiry
April 28, 1949, Sanhoshō, No. 58

From: Okuno Kenichi, Chief, Bureau of Legislation, House of Councilors

To: Director, Research and Opinions, Justice Agency.

The following points have come up requiring an opinion of your office.

(1) Whether art. 16 of the Constitution applies to alien residents in Japan as well as to Japanese citizens. . . .

2. Opinion
April 28, 1949 (Justice Agency Opinions, no. 62)

From: Director, Legal Research Opinions, Kaneko Hajime

To: Chief Legislative Bureau, House of Councilors, Okuno Kenichi
Re: Reply concerning Constitution art. 16 et alia

In response to your inquiry of April 28, 1949, this office is of the following opinion.

Issues: (1) Whether or not art. 16 of the Constitution applies not only to Japanese citizens but also to alien residents in Japan.

Opinion: (1) Article 16 of the Constitution applies to alien residents in Japan [who] submit to Japanese sovereignty, except as specified by a particular provision in a treaty. Although art. 3 of the constitution is entitled "Rights and Duties of Citizens," rights specific to individuals are not necessarily limited to Japanese. It is understood that aliens too may enjoy them, with the exception of those rights limited to Japanese citizens. The clause "any person" in art. 3 of the Constitution is considered to apply to aliens as well; but more concretely, whether each clause applies to aliens or not, must be determined according to the nature of the rights [created by] each article.

Article 16 of the Constitution is the provision concerning petitions. There is an opinion that the content of the right to petition is within the penumbra of the political right to vote. If the right to petition is limited to one's own citizens as under art. 126 of the Weimar Constitution, it [the right to petition] may be viewed in that way. However, since its exercise merely imposes a duty for the nation to receive a petition, it differs in nature from the political right to vote which the citizen as an organ of the nation exercises in public affairs. Further-

more, no matter what its content or purpose, a petition is a mere expression of desire. There is no particular [reason] to prevent aliens from expressing their hopes for the country whose society they live in; considering that the Constitution accords this right to "any person," it is proper to interpret art. 16 as applying also to foreign residents in Japan.

B. Opinion of Satō Tatsuo, former head of the Cabinet Legislative Bureau.

1. "Any person" in art. 16 of the Constitution is any person within the territory to which the force of the Japanese Constitution extends. Alien residents abroad are not included.

2. However, if the Diet Law provides that the Diet can receive the petition of an alien resident abroad, this provision shall not be considered unconstitutional. The reason is that this would be an ex gratia measure.

37. This historical motivation for the Diet policy is obscure. One possible explanation is that there was concern after World War II that Koreans and other nationals of countries surrounding Japan might wish to object to human-rights and other violations, and to frustrate such protests the rule was adopted. Then, too, in the early postwar period few foresaw that Japan would become enmeshed as she now has in an international system of responsibilities and obligations that requires liberalization of procedural access to judicial, administrative, and even legislative tribunals.

38. Emphasis added.

39. See ch. 4 for a discussion of the current uncertainty of the environmental rights doctrine. Even less certain is whether the Migratory Bird Treaty and other international prescriptions create individual rights and interests, and whether plaintiffs had standing to raise them under Japanese law.

40. See Van Loukhuyzen v. Daily News Company, 203 Mich. 570, 170 N.W. (1918) (resident alien has right to petition); also, Graham v. Richardson, 403 U.S. 365, 91 S. Ct. 1848, 29 L.Ed.2d 534 (1971); In re Griffiths, 413 U.S. 717, 93 S. Ct. 2851, 37 L.Ed.2d 910 (1973); Sugarman v. Dougal, 413 U.S. 634, 93 S. Ct. 2842, 37 L.Ed.2d 853 (1973), and cases cited therein. These cases generally disfavor classifications based on alienage. But see Kleindienst v. Mandel,

408 U.S. 753, 92 S. Ct. 2576, 33 L.Ed.2d 683 (1972).

41. We thank Choon Ho Park of the East-West Center, Hawaii, for allowing us to use a preliminary draft of his paper, *A Zonal Approach to Marine Pollution Control: Japan's Position in the Third U.N. Law of the Sea Conference*, upon which this section is based.

41a. Enforcement Measures by Coastal States for the Purpose of Preventing Marine Pollution. (U.N. Doc. A/Ac.138/SC.III/L.49.)

42. For a description of the background in detail, see Katsuhiko Iguchi, *Daisankai Kaiyōhō Kaigi to Kaiyō Osenboshi* (*The Third* [U.N.] *Law of the Sea Conference and the Marine Pollution Prevention Law* [of Japan]), Kaiyō Osen No Gendaiteki Shomodai (Current Problems on Marine Pollution), 7 Environmental L. J. 96–115 (1977).

43. The 1954 Convention, as amended in 1962, would be replaced by the International Convention for the Prevention of Pollution from Ships, 1973, not yet in force as of July 1977; its 1969 and 1971 amendments were incorporated in the 1973 Convention. For further details, see, e.g., 3 Third U.N. Law of the Sea Conference, Official Records (UNCLOS III) 43, 48 (Caracas, 1974).

44. Kaiyō Osen Oyobi Kaijō Saigai no Bōshi ni Kansuru Hōritsu (1970, Law No. 136).

45. Stockholm Conference speech by Dr. Ōishi.

46. K. Iguchi, n. 42 supra, 104–108.

47. K. Iguchi, n. 42 supra, 100–101; in fact, Iguchi groups the positions into four categories; the last two can be grouped into one with a degree of variation.

48. 2 UNCLOS III 370 (Caracas, 1974).

49. 2 UNCLOS III 371.

50. Economist 81, March 12, 1977.

51. Although the OECD report (ch. 5) points out that Japan's efforts in marine pollution control are not equal to its efforts in other areas, the report fails to offer an explanation for this fact.

52. William M. Terry, The Japanese Whaling Industry Prior to 1946 (U.S. Department of the Interior, 1951).

53. Terry reports that 1,312 whales were taken by 1908.

54. The law, however, did not specify minimum lengths nor the conditions on using whale carcasses.

55. Gyogyō Hō (1949, Law No. 267; rev. 1971, Law No. 130).

56. See, e.g., art. 38 of the Ministerial Order Concerning the Permission and Control of Designated Fisheries (Shittei Gyogyō no Kyōka Oyobi Torishimari Nado ni Kansuru Shōrei) (1963, Ministerial Order No. 5). Article 38(2) will probably be revised in 1980 pursuant to the International Whaling Commission recommendations on sperm whales.

57. Id.

58. Id., art. 47.

59. Id., art. 51.

60. Id., arts. 41 and 49.

61. Id., art. 57.

62. Id., arts. 55 and 42.

63. For example, art. 52 authorizes limitations on permits when "it is deemed necessary to take restrictive measures for the purpose of propagation and conservation of aquatic animals and plants, fishery adjustment, or other public interest. . . ." (See also art. 39.) Article 13 of the Fisheries Resource Protection Law (Suisanshigen Hogo Hō) (1951, Law No. 313; as amended in 1952, Law No. 196; 1953, Law No. 213; 1962, Law No. 156; 1963, Law No. 161; 1964, Law No. 168; 1968, Law No. 74) authorizes further limitations on fish species, catch, and time of fishing operations in order to protect the resource, and art. 29 requires scientific surveys on the condition of the stocks.

64. For example, art. 56.2 states: "In case the competent minister grants no license or authorization in accordance with the provision of the preceding paragraph, he shall give the applicant previous notice in writing of the reason therefor, and shall give him or his proxy the opportunity to explain himself at the public hearing and to produce sufficient evidence."

65. It is thus extremely unlikely that a whale conservation group could ever persuade a Japanese court to countermand a decision by the minister under art. 58 on the grounds that the minister had not heeded "conservation limits" in his decision.

66. Michael Benefiel, The Japanese Whaling Industry, 1974–1975 (U.S. Department of Commerce, 1977).

67. Perhaps the best recent article on the subject is by James E. Scarff, *The International Management of Whales, Dolphins, and Porpoises: An Interdisciplinary Assessment*, 6 Ecology L. Quarterly 323 (pt. 1), 571 (p. 2) (1977).

68. Scarff, supra, reports that "When the draft of the resolutions was under deliberation, no scientific discussion was conducted on the current status of the international management exercised over the harvesting of the whale stocks, nor were any reports presented by experts on the status of the whale stocks. That the draft was put to a vote without debate can only be regarded as a political move. No consideration whatsoever was given to the position of the Japanese delegation, which had twice expressed its desire during the deliberations to refer the draft to scientists for discussion."

69. Id.

70. Id.

71. Lawrence A. Friedman, *Legal Aspects of the International Whaling Controversy: Will Jonah Swallow the Whales?* 8 New York University J. Int. L. and Politics 211 (1975).

72. 22 U.S.C. sec. 9, 1978 (Supp. V, 1975).

73. Reported by Friedman, n. 71 supra.

74. Scarff, n. 67 supra.

75. HR 80, 94th Cong., 1st sess. (1975).

76. HJ. Res. 448, 94th Cong., 1st sess. (1975); Senator Warren Magnuson, Chairman of the Joint Commerce Committee, introduced a companion bill, S. J. Res. 81, 94th Cong., 1st sess. (1975), that would have prohibited the import into the United States of any fish or fish product produced by any foreign enterprise engaging in commercial whaling.

77. Yet legislative initiatives continue. Recent examples are various resolutions and proposals urging Japan to adopt comprehensive legislation to protect dolphins. These grew out of reports that Japanese fishermen were massacring dolphins at Iki to protect their catch. See, e.g., the resolution proposed by Senator Lowell Weicker urging Japan to join the international community in protecting sea mammals. The U.S. House of Representatives also was urged to express its "strong concern" over the Iki incident. See. H.R. Res. 1071, March 13, 1978; Res. 538, March 20, 1978; Res. 1101, March 22, 1978; Res. 1102, March 22, 1978; Res. 1109, April 4, 1978; Res. 1129, April 12, 1978.

78. Fishery Conservation and Management Act of 1976, 90 Stat. 331, P.L. 94-265 (1976).

79. Scarff, n. 67 supra, reports that "over the past few years Japan has taken approximately 30% of its sperm whale quota within 200 miles of the U.S. Territories, and in the past month, about 5% of its Bryde's whale catch off the Hawaiian and Midway Islands," 3 Marine Mammal News 4 (Feb. 1977).

80. See, e.g., the Marine Mammal Protection Act of 1972 (16 U.S.C. secs. 1361–62, 1371–84, 1401–07 (Supp. III, 1974)) and the Endangered Species Act of 1973 (16 U.S.C. secs. 1531–43 (1976).

81. Japan Times, June 25, 1977.

82. Japan Times, June 10, 1978. A further development is *Adams v. Vance*, 8 ELR 20160, 570 F.2d 950 (1978) wherein the U.S. Court of Appeals (D.C. Cir.) vacated a lower court order requiring the Secretary of State to object to the International Whaling Commission's ban on Eskimo hunting of bowhead whales. Subsequently, however, an exception was granted by the commission to permit the Eskimos to take a few whales.

83. For a discussion of these accidents see letters submitted by various Japanese environmentalists in connection with the U.S. ERDA's preparation of an environmental impact statement in *Sierra Club v. United States Atomic Energy Commission*, Civil No. 1867-73 (4 ELR 20685) (D.D.C. 1974).

84. *Natural Resources Defense Council v. United States Nuclear Regulatory Commission*, 574 F.2d 633 (1976).

85. See ch. 5.

86. Although more powerful, industry-oriented agencies like MITI must also obtain Cabinet approval, their proposals command greater respect and are rarely rejected.

87. This practice is poetically referred to as *amakudari*, "descent from the heavens."

88. See Japan's Emerging View of International Law, Proceedings of the 71st Annual Meeting of the American Society of International Law, Apr. 21–23, 1977.

89. See generally, L. J. Adams, Theory, Law and Policy of Contemporary Japanese Treaties (1974).

90. See, e.g., the legal opinion of John R. Crook, then of the Legal Adviser's Office to Mr. Christian Herter (Assistant Secretary for Oceans, International Environment and Science) (July 2, 1975). For further discussion of the Migratory Bird Treaty see M. J. Bean, The Evolution of National Wildlife Law (1977) and also *Fund for Animals v. Frizzell*, 6 ELR 20188 (D.C. Cir., Dec. 24, 1975), 402 F. Supp. 35 (D.D.C. 1975), *aff'd.* 530 F.2d 982 (D.C. 1975).

91. There are now many articles and books dealing with this subject; two particularly helpful references are Richard B. Bilder, The Settlement of International Environmental Disputes, University of Wisconsin Sea Grant Report 231 (February 1976), and Jan Schneider, *New Perspectives on International Environmental Law*, 82 Yale L. J. 1659 (1973). For a general reference see L. B. Sohn, *The Shaping of International Law*, 8 Ga. J. Int. and Comp. L. 1 (1978).

91a. Some scholars argue that implementing legislation is not necessary for international custom. The United States also has not considered OECD recommendations to be legally binding. For a general discussion of Japan's obligation under international law to protect the international environment see S. Yamamoto, *Kankyō Songai ni Kansuru Kokka Kokokusai Sekinin* (The International Responsibility of Nations for Environmental Damage), 40 Hōgaku 4; Shoichi Toishikage, *Kokusaiteki Kankyō Hogo to Kokusai Hō* (*International Environmental Protection and International Law*), 681 Jurisuto (1979).

92. *Charter for Economic Rights and Duties of States*, a resolution adopted by the U.N. General Assembly, Jan. 15, 1975.

93. Richard A. Frank, *The Foreign Reach of NEPA*, in International and Trade Aspects of Environmental Law, ALI-ABA Study Course, Jan. 18, 1974. See also Spitzer, *The Extraterritorial Scope of NEPA's Environmental Impact Statement Requirement* 74 Michigan L. Rev. 349 (1975).

94. A programmatic impact statement is a general comprehensive document like that prepared by ERDA in the Sierra Club v. AEC case (n. 83 supra). It is used as a general reference. Agencies generally later prepare more specific environmental analyses.

95. Section 102(2)C of the act permits

exemptions for reasons of national security by reference to 5 U.S.C. sec 552.

96. See generally, N. Robinson, *Environmental Safeguards in Canal Treaty Clarified*, New York L. J. 1, col. 1 (March 28, 1978); also Department of State Draft EIS, New Panama Canal Treaty Between the United States and the Republic of Panama (CEQ No. 71057, Aug. 1977). The process of preparing extraterritorial assessments has been curtailed under Executive Order 12114 (Jan. 4, 1979.) The process of preparing extraterritorial assessments, however, has been substantially simplified and curtailed under Executive Order 12114 (Jan. 4, 1979). In some instances the State Department has been starkly indifferent to environmental impacts abroad. See the Department's failure to assess the impact of Westinghouse Corporation's construction of a nuclear facility in the Phillippines, despite the existence of four reputedly active volcanoes and known hazards of earthquakes and tidal waves.

97. See *The Schooner Exchange v. McFadden*, 11 U.S. 97 (Cranch), 116 (1812); and Declaration of Principles of International Law Concerning Friendly Relations and Cooperation Among States in Accordance with the United Nations Charter (G.A. 2625 XXV, Oct. 24, 1970).

The attitude of many agencies is summarized in *In the Matter of Babcock and Wilcox*, 7 Environmental L. Reporter 30017 (1977).

98. See, e.g., *Sierra Club v. Coleman*, 405 F. Supp. 53, 6 ELR 20051 (D.D.C. 1975) (required Federal Highway Administration to prepare an environmental impact statement on construction of a section of the Pan American Highway through the Darien Gap in Panama and Columbia and to analyze the risk of the spread of aftosa (hoof and mouth disease) to the United States); *Environmental Defense Fund v. AID*, 6 ELR 20121 (D.D.C. 1975). See also: *Forthcoming CEQ Regulations to Determine Whether NEPA Applies to Environmental Impacts Limited to Foreign Countries*, 8 ELR 10111 (1978); *Renewed Controversy Over the International Reach of NEPA*, 7 ELR 10205 (1977); neither the statute nor its legislative history answer adequately the question of NEPA's extraterritorial application.

An executive order leaves the matter even

more uncertain, because the courts will still be free to recognize NEPA's transnational application in the absence of an amendment to the statute. See *Draft Executive Order on Reviewing Environmental Effects of Major Federal Actions Abroad*, 9 Environmental Reporter, Current Developments, 568 (1978).

99. There are several possible bases for controlling the environmental effects of the fluorocarbon aerosol industry in Japan. These are the Law Concerning the Examination and Regulation, etc. of Chemical Substances (1974, Law No. 117); the Air Pollution Control Law (1968, Law No. 97); and the Pharmaceutical Law (1960, Law No. 145). Although each of these laws might be administratively interpreted to cover fluorocarbons and/or aerosols, none of these laws explicitly addresses the chemical properties, uses, and nature of the products of these industries. For this reason, many government officials in Japan believe that either new legislation will be required or that, when necessary, fluorocarbon aerosol "pollution" should be managed via administrative guidance.

100. A related problem concerns the exemptions granted to exports from the United States of hazardous products. See Testimony of Jacob Scheer on the 1977 Reauthorization of the Federal Insecticide, Fungicide, and Rodenticide Act, before the House Committee on Agriculture, March 29, 1977.

101. OECD: Council Recommendations on Principles Concerning Transfrontier Pollution, November 14, 1974; Recommendations of the Council on an Equal Right of Access in Relation to Transfrontier Pollution, May 11, 1976; Recommendations of the Council for the Implementation of a Regime of Equal Right of Access and Non-Discrimination in Relation to Transfrontier Pollution, May 17, 1977.

102. See also Akira Kawamura, *A New Development in Japanese Antitrust Law: Its Application to International Transactions*, 3 L. Asia 167 (1972); Greyson Bryan, *The Tax Implications of Japanese Multinational Corporations*, 8 Int. J. L. and Politics 153 (1975).

103. TIAS (Treaties and International Acts Series, U.S. Department of State) 8172; these cooperative programs include the United States-Japan Conference on Natural Resources Development, Japan-United States Committee on Scientific Cooperation, United States-Japan Cooperative Medical Science Committee, Agreement Between the Governments on Cooperation in the Field of Environmental Protection (1975), Agreements Between the Governments for Cooperation in Civil Uses of Atomic Energy (1968) and Energy Research and Development (1974), Exchange of Notes Between the Governments (1962, 1969, and 1975), Communique of the Joint Japan-United States Cabinet Committee on Trade and Economic Affairs (1969), Memorandum of Understanding Between Department of Housing and Urban Development and Ministry of Construction (1970), and Exchange of Notes between the Governments (1974).

104. It is interesting that some European governments have expressed a similar concern that several U.S. state governments are also granting subsidies.

105. Article 5 states: "The two governments reaffirm that the recommendations of international organizations to which both countries are parties will be taken into account in formulating their respective environmental policies."

Japan's resistance to the "polluter pays" principle in the bilateral negotiations may perplex the reader, since the concept had been an important part of Japanese domestic policy since 1973. The most plausible explanation for this may be that although Japan was willing to adopt the principle domestically, she did not want to invite the misunderstanding that she regarded the principle to be in any way internationally binding. Explicit mention in the agreement might have created this impression. The vagueness of the language in art. 5 and the reference to the "Guiding Principles Concerning International Economic Aspects of Pollution Control" in the Agreed Minutes, rather than in the text, express the differing U.S. and Japanese positions on this issue.

106. A potential conflict undisclosed by the terms of the agreement was the issue of American nuclear weapons. Many Japanese suspected that the United States was concealing nuclear weapons in Japan and wanted the government to insist that the agreement include a provision explicitly forbidding

this. The Japanese government, however, resisted this proposal on the ground that the United States would never agree to it. Shortly after, however, the Spanish government announced that agreement had been reached with the United States on precisely this issue.

107. For a general discussion on progress to date, see Kirk D. Maconaughy, *Sharing Environmental Knowledge with Japan*, 4 EPA J. 24 (1978); Edward Perron, Foreign Development: VI *The Agreement for Cooperation in the Field of Environmental Protection Between the United States and Japan*, 1 Harv. Environmental L. Rev. 574 (1977).

Note that similar agreements have been concluded with the Soviet Union in 1972 and with West Germany in 1974. For an assessment of the U.S.–Soviet agreement, *see* Merritt, *The Soviet–U.S. Environmental Protection Agreement*, 14 Natural Resource J. 275 (1974).

108. For example, in the areas of industrial liquid-waste treatment, solid waste management, stationary source air pollution technology, and photochemical smog. Maconaughy (n. 107 supra) reports that both countries have exchanged research teams. For example, the EPA's Kepone Mitigation Feasibility Task Force visited Japan, where it studied industrial methods of handling toxic sludge.

109. The preamble states:
The Government of the United States of America and the Government of Japan,

Believing that cooperation between two Governments is of mutual advantage in coping with similar problems of environmental protection in each country and is essential in meeting the responsibilities of each Government for the protection and improvement of the global environment, and

Desiring to strengthen further such cooperation and to demonstrate its importance . . .

A separate issue, of course, is whether the agreement compels affirmative action. Although the agreement seems too vague to be judicially enforceable, both governments do recognize some obligation to collaborate toward global environmental improvement, a fact that environmental groups might use to political advantage.

110. Julian Gresser, Balancing Environmental Protection with Industrial Development in the Republic of South Korea, a report of the Environmental Mission to the World Bank (Dec. 22, 1977).

111. ESCAP (The Economic and Social Affairs Commission for Asia and the Pacific) includes Australia, Bangladesh, Cook Islands, Fiji, India, Indonesia, Republic of Nauru, Pakistan, Papua New Guinea, Philippines, Singapore, Sri Lanka, and the Soviet Union, among others. See Jeffrey N. Shane, *Legal Aspects of Environmental Protection in Asia*, prepared for second conference, L. Asia, August 30, 1977; Colin MacAndrew, et al., Development, Economics, and the Environment: The Southeast Asia Experience (1979). See also U.S. Agency for International Development, Environmental and Natural Resource Problems in Developing Countries, March 1979.

112. Shane, *Legal Aspects . . .* 4.

113. ESCAP, *Text of Questionnaire and County Responses in Respect of Environmental Protection Legislation in the ESCAP Region*, pursuant to ESCAP letter of October 4, 1974.

114. Id. 98; also, Government of Papua New Guinea, Environment and Small Industries, a Country Report for the Asia Productivity Organization Symposium on Environment (Tokyo May 26, 1976).

115. OECD, Environmental Policies in Japan (1976); see also ch. 5.

116. See Gresser, n. 110 supra.

117. Jeffrey N. Shane, *Environmental Law and Technical Cooperation: Agenda for Asia and the Pacific*, a paper presented at the ESCAP/UNEP Expert Group Meeting on Environmental Legislation, Bangkok (Dec. 13, 1977).

118. For a critical assessment of the Asia Development Bank's environmental performance, see International Institute for Environment and Development (IIED), Multilateral Aid and the Environment, A Study of the Environmental Procedures and Practices of Nine Development Financing Agencies (1977).

119. Kazuo Sumi, Protection of the Marine Environment in the East Asian Waters (Japan Institution of International Law, 1977); Robert E. Lutz, *National Hegemony and International Suzerainty in the*

Oceans: The Environmental Implications of the Law of the Sea Negotiations, 6 Int. Business Lawyer 1974 (1978). A more recent development is the agreement reached between the Eight Gulf States (Bahrain, Iran, Iraq, Kuwait, Oman, Quatar, Saudi Arabia, and the United Arab Emirates) on Marine Environmental Protection. See 4 Environmental Policy and L. 81 (1978).

120. Mochtar Kusumaatradja and St. Munadjat Danusapurtro, *The Elements of Environmental Policy and Navigation Scheme for Southeast Asia*, Annual Conference of the Law of the Sea Institute, Hawaii (1977).

121. See Geoffrey Kemp, *Threats from the Sea—Sources for Asian Maritime Conflict*, 19 Orbis 1037 (1975), for an analysis of the development of ocean-based military hardware.

122. See Hisashi Kasahara and William Burke, North Pacific Fisheries Management (1973). The social and other impacts, particularly among fishermen, who are often members of the poorest strata of these societies, will be great. See Pamela Baldwin and Malcolm Baldwin, Onshore Planning for Offshore Oil: Lessons from Scotland (1975), for a discussion of the environmental and other impacts of offshore drilling in the North Sea.

Index